IRELAND'S 1916 RISING

Heritage, Culture and Identity

Series Editor: Brian Graham,
School of Environmental Sciences, University of Ulster, UK

Other titles in this series

Ireland's 1916 Rising

Explorations of History-Making, Commemoration & Heritage in Modern Times

MARK McCARTHY

*Lecturer & Programme Chair in Heritage Studies,
Galway-Mayo Institute of Technology, Republic of Ireland*

Routledge
Taylor & Francis Group

LONDON AND NEW YORK

First published 2012 by Ashgate Publishing

2 Park Square, Milton Park, Abingdon, Oxon OX14 4RN
711 Third Avenue, New York, NY 10017, USA

Routledge is an imprint of the Taylor & Francis Group, an informa business

First issued in paperback 2016

British Library Cataloguing in Publication Data
McCarthy, Mark, Dr.
  Ireland's 1916 Rising : explorations of history-making,
  commemoration & heritage in modern times.–(Heritage, culture and identity)
  1. Ireland – History – Easter Rising, 1916 – Anniversaries, etc.
  2. Ireland – History – Easter Rising, 1916 – Influence.
  3. Ireland – History – Easter Rising, 1916 – Historiography. 4. Memorialization – Ireland.
  I. Title II. Series
  941.7'0821–dc23

Library of Congress Cataloging-in-Publication Data
McCarthy, Mark, Dr.
 Ireland's 1916 rising : explorations of history-making, commemoration & heritage in modern times / by Mark McCarthy.
    p. cm. – (Heritage, culture and identity)
 Includes bibliographical references and index.
 ISBN 978-1-4094-3623-2 (hardback) 1. Ireland – History – Easter Rising, 1916. 2. Ireland – History – Easter Rising, 1916 – Anniversaries, etc. 3. Ireland – History – 20th century. I. Title.
 DA962.M457 2012
 941.5082'1–dc23

                                                                                    2012035943

ISBN 978-1-4094-3623-2 (hbk)
ISBN 978-1-138-25335-3 (pbk)

In loving memory of my mother

Be green upon their graves
Oh happy Spring!
For they were young and
eager who are dead.
Of all things that are young
and quivering
With eager life be they
remembered.
They move not here – they are
gone to ... clay.
They shall not die again
for liberty.
Be they remembered of
their land for aye.
Green be their graves
and green their
Memory.

James Stephens, 'Spring 1916'

It is so strange that when someone dies, they literally disappear. Human experience includes all kinds of continuity and discontinuity, closeness and distance. In death, experience reaches the ultimate frontier. The deceased literally falls out of the visible world of form and presence ... The absence of their life, the absence of their voice, face and presence become something that, as Sylvia Plath says, begins to grow beside you like a tree.

John O'Donohue, *Anam Cara: Spiritual Wisdom from the Celtic World*

The struggle of man against power is the struggle of memory against forgetting.

Milan Kundera, *The Book of Laughter and Forgetting*

The reason for the range of work that has been conducted into the questions of what, why ... and how we remember should be apparent ... memory is a key psychological process ... Memory is far more than simply bringing to mind information encountered at some previous time. Whenever the experience of some past event influences someone at a later time, the influence of the previous experience is a reflection of memory for that past event.

Jonathan K. Foster, *Memory: A Very Short Introduction*

Contents

Contents

List of Plates

Acknowledgements

Throughout the course of preparing this book I have accumulated debts of gratitude to many individuals, whose help and backing is acknowledged. My first debt is to all the team at Ashgate who facilitated the publication of this book, especially Caroline Spender, Carolyn Court, Dr. Emma Gallon, and Valerie Rose. The backing and advice of Professor Brian Graham, Editor of the 'Heritage, Culture and Identity' series, is very much appreciated as well.

In working at Galway-Mayo Institute of Technology (GMIT), I have received great encouragement from my colleagues in heritage studies, including the following: Paul Gosling, Dr. Sean Lysaght, Cian Marnell, Margaret O'Riordan, Dr. Brian O'Rourke, Dr. Suzanne O'Shea, Dr. Cilian Roden, Declan Sheridan, John Tunney, and Fiona White. Former graduate students also provided much inspiration. Thanks are especially due to Marta Gergelyova, David Lawlor, Yvonne MacDermott, Michael Quinn, and Dr. John Towler, for the many thought-provoking discussions about Ireland's heritages. Mary Creaven and Martina Linnane provided help in many ways, while Michael Carmody, Dr. Patrick Delassus, Dr. Des Foley, Dr. Noel Harvey, Donal Haughey, Dr. John Lohan, Mary MacCague, Gerard MacMichael, Dr. Seamus McGuinness, Dr. Gavin Murphy, Cáit Noone, Bernard O'Hara, Peadar O'Dowd, and Gerry O'Neill were also supportive and encouraging throughout. I would also like to acknowledge, with thanks, the backing of GMIT's Research Office, which very kindly provided a subvention towards the cost of publishing this book.

For help in locating and reproducing a number of photographs, I wish to thank Glenn Dunne of the Reader Services (Rights and Reproductions) section of the National Library of Ireland (along with Keith Murphy and Bernie Mackoff), Yvonne Oliver of the Image Sales Licensing section of the Imperial War Museum's Photographic Archive, Adam Petitt of Getty Images, Anne-Marie Ryan of Kilmainham Gaol Archives, Irene Stephenson of *The Irish Times*, the staff of Maxwell Photography, and Philip Grant and Peter Howlett of the Press Office of the Department of Foreign Affairs and Trade. Permission to reproduce particular photographic images in this book has been granted courtesy of the National Library of Ireland (Plates 2.1, 2.3, 2.4, 4.1, 4.2, 4.5, 5.1, 5.4, and 5.5), the Trustees of the Imperial War Museum (Plates 2.2, 2.5, 2.6, 2.7, and 3.1), Getty Images (Plate 3.3), Kilmainham Gaol Archives (Plate 3.5), *The Irish Times* (front cover and Plates 6.1 and 7.1), and Maxwell Photography (Plate 7.2).

For particular assistance and for granting permission for the use of quotations from various archival collections, I am grateful to Brother Thomas Connolly and Rosemary King of the Allen Library (Dublin), Comdt. Victor Laing and Lisa Dolan of the Military Archives (Cathal Brugha Barracks), Robert Mitchell of the Army Museum of Western Australia (Freemantle), Helen Langley of the Bodleian Library (Oxford), Mary Lombard and Crónán Ó Doibhlin of the Boole Library (University College Cork), Sandra Powlette of the British Library (London), Debbie Walsh of the Cobh Heritage Centre, Brian McGee of Cork City and County Archives, Dr. Máire Kennedy of the Dublin City Library and Archive, Noelle Dowling of Dublin Diocesan Archives, Anthony Richards of the Imperial War Museum (London), Marie Boran and Kieran Hoare of the James Hardiman Library (NUI Galway), Steve Howell of the J. S. Battye Library of West Australian History (Perth), Anne-Marie Ryan of Kilmainham Gaol Archives and Museum (Dublin), Aideen Ireland and the Director of the National Archives of Ireland (Dublin), Dr. Alastair Massie of the National Army Museum (Chelsea), the Board of the National Library of Ireland (Dublin), the Trustees of the National Library of Scotland (Edinburgh), Rosalind Leake of the National Maritime Museum (Greenwich), Tal Nadan of the New York Public Library, Mari Takayanagi of the Parliamentary Archives (London), Brian Crowley of the Pearse Museum (Dublin), Father Albert McDonnell of the Pontifical Irish College (Rome), Don C. Skemer of Princeton University Library, the Deputy Keeper of the Records at the Public Record Office of Northern Ireland (Belfast), Petra Schnabel of the Royal Irish Academy (Dublin), Ellen O'Flaherty of Trinity College Dublin, and Seamus Helferty of University College Dublin Archives.

I also wish to record my gratitude to the Duke of Bedford and the Trustees of the Bedford Estates for authorisation to quote from the Arnold White Papers, the Bonham Carter Trustees for approval to cite material from the MS. Asquith collection, Anthony Beater for consent to reproduce an extract from his father's diary, Mrs. M. S. Nathan for sanction to quote from her uncle's diary, Douglas Sealy for allowing the use of an extract from Douglas Hyde's diary, John Mullen for permitting access to the Archives of the Fianna Fáil Party, and Eileen Kelly for facilitating usage of the Archives of the Fine Gael Party. I am also grateful to Tom Gillespie of *The Connaught Telegraph*, for granting permission to reproduce the text of Martin Neary's poem, 'The Nineteen-Sixteen Men'. For further assistance with my research endeavours, I am indebted to the staff of the GMIT Library (Dublin Road campus), the Local Studies Room in Cork City Central Library, the Library of Congress (Washington, DC), the National Archives (Kew), and the United States National Archives and Records Administration (College Park).

Additional thanks are due to audiences at Belfast, Castlebar, Derry, Galway, and Québec, who offered constructive criticism on conference and seminar presentations that I delivered on the subject matter of 1916, when this book was in the very early stages of preparation. My thanks also go to Professor Steven Ellis, for inviting me to participate in a stimulating European Social Fund Exploratory Workshop in Bristol, on the theme of 'Region, Memory and Agency in Eastern and Western Europe'. As my research neared completion, extremely valuable help was obtained from a number of people. For cartographic assistance, I am much indebted to Dr. Siubhán Comer, who drew the map of Ireland at relatively short notice. Additionally, I am grateful to Terence Fitzgerald in Washington, DC, who read the manuscript in its entirety and proved to be abidingly perceptive in his feedback. Furthermore, I wish to acknowledge the generosity and skill of the following, who read through specific chapters or extracts: Kieran Hoare, Rory MacGowan, Barry McMillan, Cian Marnell, Dr. Diarmuid Scully, and John Tunney. All views expressed herein are my own and any errors, omissions or misinterpretations are entirely my own responsibility.

As both a mentor and friend, Professor William J. Smyth has offered magnificent support throughout my academic life. So too have Dr. Maura Cronin, Professor Kazuhito Kawashima, the late Professor Breandán MacAodha, Dr. John McDonagh, Professor Patrick O'Flanagan, and Dr. Simon Potter. On a personal level, I am greatly indebted to my father for moral support and encouragement, and to my brothers, relatives and friends for their kind hospitality, camaraderie and forbearance during the prolonged gestation of this book. This book is dedicated to the memory of my mother, for inspiring me throughout all my years of study and work.

Map of Ireland

Chapter 1

Introduction

On a bright spring day in Dublin, on 9 April 2006, the Taoiseach of the Republic of Ireland, Bertie Ahern, delivered a speech entitled 'Remembrance, Reconciliation, Renewal' to a select gathering of people, consisting largely of media personnel, politicians and public servants. With the country reaping the benefits of the Peace Process, the 54-year-old leader of the 29th Dáil Éireann called for 'a great national conversation on what it means to be Irish' and said that the state needed to properly remember its own people from the past, so that 'we, her citizens, can more clearly understand the present and better plan for our shared future'. 'We do this', he added, 'to build upon the enduring legacy bequeathed to us by the living generations of Irish men and Irish women who in nine decades of struggle, independence and achievement built a stable and a democratic Republic'.[1] The Taoiseach's speech was delivered at the opening of an exhibition, entitled 'Understanding 1916: The Easter Rising', at the National Museum Collins Barracks, to mark the 90th anniversary of the 1916 Rising. This nostalgic event was followed a week later by the revival of an impressive military parade along Dublin's O'Connell Street on Easter Sunday by Óglaigh na hÉireann, the Irish Army. Up to 120,000 spectators turned out to watch the parade, which had not taken place since 1971 as a consequence of the worsening state of the Troubles in Northern Ireland. Any fears that the Taoiseach and his advisors may have had about resurrecting the ghosts of 1916 proved unfounded in 2006. Encouraged by the much-improved political situation in Northern Ireland, goodwill was abundant and the legacy of the Rising was recast in a more positive light for a new generation.

The events of Easter 2006 contrasted sharply with the situation that prevailed during the dark years of the Troubles, when the story of 1916 was approached with great caution by successive governments in the Republic and regularly panned by revisionist historians. However, with the 90th anniversary military parade winning overwhelming public approval, the Rising was elevated to a position of greater prominence on the commemorationist calendar. Once the parade finished, the Irish people looked towards the upcoming centenary with optimism. It was not long, however, before the exuberance that oozed

---

[1]   *The Irish Times*, 10 April 2006.

from the materialistic spoils of the Republic's Celtic Tiger economy became a thing of the past. Confidence quickly evaporated from late 2008 onwards, when Ireland experienced a bust of titanic proportions, caused by a lethal cocktail of 'domestic fiscal and banking crises and global financial and economic downturns'.[2] The government's decision to guarantee nearly all of the liabilities of six financial institutions on 30 September 2008 served as a defining moment in Irish economic history. All of a sudden, as Conor McCabe has written, 'the losses of the banks were now the losses of the people' and the 'sins of the father had been laid upon the children'.[3] These banking difficulties culminated in a sovereign-debt crisis in November 2010, when the Irish state was frozen out of international bond markets and forced to accept a rescue package of €85 billion from an external troika – comprising the European Union, the European Central Bank and the International Monetary Fund.

News of the bailout was greeted with universal dismay throughout the Republic. There was, as Tommy Graham has observed, 'much hand-wringing and bewailing of the loss of the sovereignty struggled for by previous generations'.[4] In a Dáil debate on 18 November 2010, memories of the rebels of 1916 were invoked by Fine Gael's Michael Noonan, who lamented the loss of the sovereignty won by 'the patriot dead' who 'fought and died' for independence. The next day, the *Irish Examiner* printed a mock 'Proclamation of Dependence' on its cover page. Employing the same design as the iconic 1916 Proclamation, the satirical piece by Shaun Connolly summoned the people of Ireland to its 'financial sovereign funeral' and bemoaned how the 'outsourcing' of freedom had caused the 'destruction of the ... people's hope'.[5] The ghosts of 1916 were very much in evidence again at a major national demonstration in Dublin on 27 November 2010, organised by the Irish Congress of Trade Unions. In freezing winter conditions, an actress read excerpts from the 1916 Proclamation from a stage in front of the General Post Office (GPO), while journalist and writer Fintan O'Toole told over 50,000 protestors on O'Connell Street that the

2   E. O'Leary, 'Reflecting on the "Celtic Tiger": Before, During and After', *Irish Economic and Social History*, Vol. 38 (2011), p. 74.

3   C. McCabe, *Sins of the Father: Tracing the Decisions that Shaped the Irish Economy* (The History Press Ireland, Dublin, 2011), pp. 10, 173.

4   T. Graham, 'Counting Down to 2016', *History Ireland*, Vol. 19, No. 1 (2011), p. 3.

5   *Dáil Debates*, Vol. 722, 18 November 2010; *Irish Examiner*, 19 November 2010. In an article written for the *Financial Times*, 25 November 2010, the Minister for Finance, Brian Lenihan, was particularly forthright about the solemnity of the situation, acknowledging that there was 'no denying the reputational damage Ireland has endured'.

country was facing a monumental crisis in democracy: 'We are not subjects, we are citizens, and we want our Republic back.'[6]

The significance of this cataclysmic episode of Irish history, coming as it did less than five and a half years before Easter 2016, was not lost on one of Europe's leading historians. 'The Republic', wrote Norman Davies, 'found itself in intensive care; a land of smiles became a land of woe, and its image as a brave pioneer evaporated'.[7] However, as a national conversation began to take place on the way that politics operated, calls for a renewal of the Republic became a recurrent feature of Irish socio-political discourse in the years leading up to the Rising's centenary. The idea of an Irish rebirth featured strongly in Michael D. Higgins's book, *Renewing the Republic*, which called for the building of 'an entirely different kind of society' by 2016 and the establishment of 'a floor of citizenship below which people would not be allowed to fall'.[8] Soon after the book's publication, Higgins was elected the ninth President of Ireland. In his inaugural speech at Dublin Castle on 11 November 2011, he displayed a strong sense of patriotism by recalling how one of the 1916 leaders, James Connolly, 'believed that Ireland was ... a country still to be fully imagined and invented – and that the future was exhilarating precisely in the sense that it was not fully knowable'. In dwelling further on the links between the past, present and future, Higgins emphasised the important role of creativity in society, drew attention to 'the ethics and politics of memory' and called for the construction of a *bona fide* Republic that would be 'inclusive ... in its fullest sense'. He also spoke about Ireland's 'rich heritage' and one of the fundamental challenges that public history would pose for citizens in the near future: 'A decade of commemorations lies ahead [from 2012–2022] – a decade that will require us to honestly explore and reflect on key episodes in our modern history as a nation [from 1912–1922].'[9]

In looking ahead to the Rising's 100th anniversary in 2016, it is clear at this point that many Irish citizens seem increasingly appreciative of the sacrifices made long ago by those who fought and died for freedom. Besides acting as a test of the political standing of those invested with the power to determine the Republic's future destiny, the centenary will offer many opportunities to look back upon the accomplishments (and shortcomings) of independence

---

[6] *The Sunday Independent*, 28 November 2010; *The Irish Times*, 29 November 2010; S. Kay, *Celtic Revival? The Rise, Fall and Renewal of Global Ireland* (Rowman & Littlefield Publishers, Lanham, 2011), p. 92.

[7] N. Davies, *Vanished Kingdoms: The History of Half-Forgotten Europe* (Allen Lane, London, 2011), p. 677.

[8] M. D. Higgins, *Renewing the Republic* (Liberties Press, Dublin, 2011), pp. 12–13.

[9] *The Irish Times*, 12 November 2011; http://www.president.ie/speeches/ (accessed on 6 August 2012).

and to ponder the shifting meanings of Irishness. Accordingly, the format of the upcoming commemorative events will likely be a subject of great curiosity not only for the Irish living and working at home, but for the 70 million-strong Diaspora scattered abroad. It is for reasons such as these that the time seems ripe to ask: why, how and in what ways has memory of the Rising persisted over the decades in the consciousness of the Irish? In pursuing answers to these questions, which are not only of historical concern, but of contemporary political and cultural importance, this book furnishes an in-depth account of the powerful influence that 1916 has exerted upon the emergence of modern Ireland, as reflected through chronicles of history-making, commemoration and heritage in the twentieth and early twenty-first centuries.

The Resilience of the Story of 1916

Alan Titley once remarked that the 1916 Rising gave Ireland 'a creation story as good as any' it was 'ever likely to get'.[10] Although statements of this kind have aroused much inquisitiveness (and heated debate) over the years, the observation certainly deserves further scrutiny at this moment in time. For nearly a century now, the story of 1916 has displayed considerable staying power as part of the mythology of the Irish nation state. As a direct result of its position as a momentous turning point in modern Irish history, memory of the Rising has crossed and recrossed, in myriad ways, with negotiations concerning the place of the past in the present. Nationalists and republicans have long regarded 1916 as the key revolutionary moment that set in train a series of events that culminated in the waging of the War of Independence from 1919–1921. This in turn heralded the final political separation of the majority of the island of Ireland from Great Britain, through the creation of the 26-county Irish Free State in 1922. Inevitably, many of those holding political positions in Ireland in the decades after the granting of independence played a powerful role in ensuring that the events of Easter Week 1916 were destined to be committed to perpetual memory.

Although the bloodshed inherent in the 1916 Rising has been stigmatised by many critics over the years (especially by those who cast dark shadows over its legacy by pointing to parallels with violence by republican paramilitaries during the Troubles), there has certainly been no shortage of enthusiasts at the other end of the spectrum, who have customarily portrayed the deeds of the rebels in a

---

[10] A. Titley, 'The Brass Tacks of the Situation', in D. Bolger (ed.), *Letters from the New Island: 16 on 16. Irish Writers on the Easter Rising* (Raven Arts Press, Dublin, 1988), p. 27.

heroic fashion. During the 2000s, for example, a range of well-known writers of a nationalist/republican persuasion continued to trumpet the significance of the Rising. Tim Pat Coogan, for example, unflinchingly asserted that 1916 still held 'all the sacrificial significance of High Mass' for Irish republicans, adding that its meaning 'is as vitally important today as it was then'. The Fianna Fáil politician, Martin Mansergh, proudly remarked that the event retained an 'iconic status at the pinnacle of the republican tradition'. Likewise, Eoin Neeson, a former Director of the Government Information Bureau, suggested that the Rising was still 'the pivotal and formative event of modern Irish history', given both its role 'in ... national and sovereign development' and its wide-ranging rationale, which extended 'beyond any basic military or political purpose'.[11]

These writers were not alone in acknowledging the enduring legacy of 1916 at the outset of the twenty-first century. Despite the odd detractor, there was no denying the growing historiographical acceptance that a fundamental reassessment of the Rising's legacy was needed. Within the field of Irish historical studies, a pronounced move by historians towards a post-revisionist paradigm (buoyed on by the success of the Good Friday Agreement in 1998), coupled with the release of new records to researchers (including the court martial records in Kew and the Bureau of Military History collection in Dublin), resulted in a resurgence of interest during the 2000s in matters related to the history of the Rising. Diarmuid Ferriter's *The Transformation of Ireland 1900–2000*, published in 2004, was significant in broadening the parameters of Irish historiography. His book, which garnered very positive reviews, sought to move discourse beyond the controversies generated by the revisionist debates on the writing of Irish history. The 'sniping', he felt, had become too 'fractious, heated and narrowly focused'. There had also been a risk, he added, 'of revisionists making the same mistakes they were criticising – establishing a framework of historical interpretation based on history as a morality tale of wrong and right'.

---

[11]   T. P. Coogan, *1916: The Easter Rising* (Cassell & Co., London, 2001), p. 7; M. Mansergh, 'The Easter Proclamation of 1916 and the Democratic Programme', in M. Jones (ed.), *The Republic: Essays from RTÉ Radio's The Thomas Davis Lecture Series* (Mercier Press, Cork, 2005), p. 58; E. Neeson, *Myths from Easter 1916* (Aubane Historical Society, Cork, 2007), pp. 71–72. For a more critical view of what the 1916 Rising did (and did not) achieve, see R. F. Foster, *Luck & the Irish: A Brief History of Change, 1970–2000* (Allen Lane, London, 2007), p. 144. Whilst recognising that it was the Free State's 'founding event', he also notes that it 'started a process that put paid to any possibility of an autonomous Ireland that might include the North'. Also see P. Hart, 'What Did the Easter Rising Really Change?', in T. E. Hachey (ed.), *Turning Points in Twentieth-Century Irish History* (Irish Academic Press, Dublin, 2011), pp. 7, 20. Although conceding that 1916 'was a "turning point" in Irish history', he argues that 'a united and utterly sovereign republic' was far away from the rebels' reach, as the unionists 'were not going to accept inclusion in a Catholic-majority state without a fight'.

However, he did concede that 'the writing of the history of 1916 ... will remain controversial', given the enduring influence of political agendas on the shaping of historical mindsets.[12]

In terms of its long-term and wider global significance, the story of 1916 was also elevated during the 2000s to a much higher platform within the canon of modern international history. In his book, *Empire: How Britain Made the Modern World*, Niall Ferguson drew attention to the fact that 'the principal threats to the stability of the [British] Empire appeared to come from within rather than from without', and that the Irish Rising had been 'the first tremor' to shake its very foundations.[13] Ferguson was by no means alone in recognising the repercussions that the Rising had for the subsequent course of British imperial history. As Piers Brendon acknowledged towards the end of the 2000s in his narrative history, *The Decline and Fall of the British Empire 1781–1997*, the Rising was far more than the 'mere riot' that the authorities had portrayed it at the time. It 'inspired independence movements' throughout Britannia's colonies (especially in India and Egypt) and had the outcome 'of blasting the widest breach in the ramparts of the British Empire since Yorktown'.[14] A similar conclusion was reached by Clair Wills, who articulated the view that 1916 'was intended to be world-historical from the start' and that 'its legacy was felt in anti-imperial movements throughout the first half of the twentieth century'. As a major episode 'in revolutionary history', she pointed out that 1916 'has been hailed as the world's first anti-colonial revolt, a spur for anti-colonial movements throughout the world'.[15]

The 2000s were also significant in other regards. The 'old-style "drum and trumpet" approach' to Irish military history, as Ian Beckett has noted, no longer reigned supreme in historiographical approaches to revoluntionary history. While military history at 'its worst' was once mainly preoccupied 'with battle narrative and great commanders', this was ultimately superceded by a paradigm that sought to integrate military narratives within 'wider political,

12   D. Ferriter, *The Transformation of Modern Ireland 1900–2000* (Profile Books Ltd., London, 2004), pp. 19, 23. In an interview with Belinda McKeon, published in *The Irish Times*, 2 October 2004, Ferriter also insisted that revisionism did not place 'enough emphasis on Britain's failure to understand, or to know, Ireland'. Further details on post-revisionism in Irish historiography can be found in K. Whelan, 'Come all you Staunch Revisionists: Towards a Post-Revisionist Agenda for Irish History', *Irish Reporter* (1991), pp. 23–26.

13   N. Ferguson, *Empire: How Britain Made the Modern World* (Allen Lane, London, 2003), pp. 323, 328.

14   P. Brendon, *The Decline and Fall of the British Empire 1781–1997* (Alfred A. Knopf, New York, 2008), pp. 259, 295, 309–310.

15   C. Wills, *Dublin 1916: The Siege of the GPO* (Profile Books Ltd., London, 2009), pp. 4–5.

socio-economic and cultural' contexts.[16] Consequently, as W. H. Kautt has observed, the remit of military history was broadened beyond the realms of 'battle history, or worse, on mere battle chronology', to a wider study of 'war, its causes and conduct, and its effect on societies'.[17] Recent years have also witnessed a significant intensification of research activity within a number of embryonic fields of interdisciplinary enquiry – including memory, heritage and commemorationist studies. In light of these trends and developments, the time now seems fitting to contemplate how scholarly analysis in the field of Irish historical studies can benefit from a fresh and innovative synthesis of the key legacies of 1916. As will be seen in the following pages, the mixing of a rich vein of interdisciplinary perspectives can facilitate a writing style that seeks to be both creative and wide-ranging in its appraisal of the Rising's special place within popular metanarratives (or 'big stories') concerning history-making, commemoration and heritage.

Writing Memory and Heritage

Before elaborating further upon the particulars of this book's main aims, methods, contents, and chapter structure, it is necessary in the first instance to untangle some of the complexities surrounding the meanings of memory and heritage, and to then pinpoint some of the key historiographical developments that have characterised the advancement of Irish commemorationist enquiries. In doing so, it might be possible to arrive at a better-rounded conceptualisation of the quintessential nature of Irish heritage, culture and identity. A fundamental task in this regard is to illuminate how various histories and memories of 1916 have been constructed and reshaped by a range of writers. The story of 1916 has been the subject of much scrutiny and debate over the years, with analysis of the events of Easter Week preoccupying the minds and labours of many commentators, especially around the time of significant anniversaries. Until relatively recently, however, scrutiny of issues surrounding various commemorations and heritages, including the Rising's impact upon the forging of modern memory and cultural identity, had only gained a perilous foothold in the canon of Irish historical studies. Why was this the case?

One of the reasons for the lack of extended treatments of the remembrance of the Rising was due to the fact that studies of memory, heritage and

---

16   I. F. W. Beckett, 'Review Article: War, Identity and Memory in Ireland', *Irish Economic and Social History*, Vol. 36 (2009), p. 64.

17   W. H. Kautt, 'Studying the Irish Revolution as Military History: Ambushes and Armour', *The Irish Sword: The Journal of the Military History Society of Ireland*, Vol. 27 (2010), p. 253.

commemoration were still more or less in their infancy in Ireland at the dawning of the twenty-first century. In recent years, however, a number of seminal studies have made important advances in knowledge and understanding of the legacies of 1916. These works, which will be identified later on, were much encouraged by the stimulus that was provided by advances in international scholarship from the 1980s onwards – most especially by the far-reaching influence of David Lowenthal's authoritative study, *The Past is a Foreign Country*. Writing in 1985, Lowenthal demonstrated how knowledge 'of the past' was essential to people's 'well-being', as they 'remember things, read or hear stories or chronicles, and live among relics from previous times'. 'All present awareness', he argued, 'is grounded on past perceptions and acts', with tradition underlying 'every instant of perception and creation' and permeating 'not only artefacts and culture', but the 'very cells' of people's 'bodies'. Furthermore, he added that 'the past remains integral to us all, individually and collectively', and that places needed to be conceded to the ancients, 'not simply back ... in a separate and foreign country', but rather 'assimilated in ourselves, and resurrected into an ever-changing present'.[18]

Memory studies were also boosted by the multi-volume *Les Lieux de Mémoire*, which was first published in 1992 under the direction of Pierre Nora and subsequently republished in English as *Realms of Memory*. This 'polyphonic' tome, which owed its origins to a seminar at the Ecole des Hautes Etudes en Sciences Sociales in Paris, was innovative for the ways in which its contributors examined 'national feeling not in the traditional thematic or chronological manner', but by scrutinising 'the principal *lieux* ... in which collective memory was rooted, in order to create a vast topology of French symbolism'.[19] In the spring of 1999, international scholarship on memory received another stimulus through the foundation at Tel Aviv University of *History and Memory*, a twice-yearly journal focusing on the construction of historical consciousness and collective memory. Reflecting on the growth of memory studies from the late twentieth century onwards, Charles Withers pronounced in 2004 that memory 'is, as it were, everywhere', in several fields of scholarly endeavour concerned with

---

[18]   D. Lowenthal, *The Past is a Foreign Country* (Cambridge University Press, Cambridge, 1985), pp. 185, 412.

[19]   P. Nora, 'From *Lieux de Mémoire* to *Realms of Memory*', in L. D. Kritzman (ed.), *Realms of Memory: The Construction of the French Past. Volume I: Conflicts and Divisions* (Columbia University Press, New York, 1996), pp. xv, xxiii. Of influence too was Simon Schama's work on the endurance of inherited landscape memories. See S. Schama, *Landscape and Memory* (Fontana Press, London, 1996).

issues of memorialisation.[20] Half a decade later, in her insightful introduction to *Memory: An Anthology*, A. S. Byatt attempted to take stock of the varied workings of memory throughout the course of human existence – including its functioning as an archive and binding strand for people's heritages and identities. Memory, she noted, invokes questions about an important conundrum – namely the challenge of how to illuminate both 'the ways people have thought about memory' and the 'things they have remembered'. 'To remember', she added, 'is to have two selves, one in the memory, one thinking about the memory, but the two are not precisely distinct, and separating them can be dizzying'.[21]

In many respects, the idea of heritage overlaps and interlocks with many of the themes common to studies of memory. So what exactly then is the meaning of the term 'heritage' and how have its definitions evolved? In *Webster's Third New International Dictionary*, published in 1976, it is defined as 'something transmitted by or acquired from a predecessor'. The explanation furnished for its etymology is that it stems from the French word *heriter* (which means 'to inherit'), which in turn derives from the Late Latin word *hereditare*, which ultimately comes from the Latin word *heres* (which means 'heir').[22] Since then, however, many writers have pointed to the emergence of heritage as a form of popular/public history – one that illuminates how the past finds representation in the present, in a multiplicity of ways. According to John Carman and Marie Louise Stig Sørensen, it was during the 1980s that trends such as postcolonialism, poststructuralism and postmodernity impacted upon the emergence of heritage studies as an academic discipline in its own right. As a consequence, the concept of heritage 'shifted from being a taken-for-granted field of meanings and practices to becoming an area calling out for investigation and analysis aiming to understand how heritage becomes constituted, what it is and does, and how different groups engage with it'.[23] Following its foundation in 1994, the *International Journal of Heritage Studies* began to play a key role in the discipline's evolution, by becoming the leading outlet for the worldwide dissemination of peer-reviewed research by academics and practitioners sharing a mutual interest in heritage matters.

---

[20] C. Withers, 'Memory and the History of Geographical Knowledge: The Commemoration of Mungo Park, African Explorer', *Journal of Historical Geography*, Vol. 30, Issue 2 (2004), p. 318.

[21] A. S. Byatt, 'Introduction', in H. Harvey Wood and A. S. Byatt (eds), *Memory: An Anthology* (Vintage Books, London, 2009), pp. xii–xiii, xx.

[22] P. B. Gove (ed.), *Webster's Third New International Dictionary of the English Language Unabridged* (G. and C. Merriam Company, Springfield, 1976), p. 1059.

[23] J. Carman and M. L. S. Sørensen, 'Heritage Studies: An Outline', in J. Carman and M. L. S. Sørensen (eds), *Heritage Studies: Methods and Approaches* (Routledge, London, 2009), p. 17.

In the Republic of Ireland, the government's passing of the Heritage Act in 1995 ushered in a more formal approach to matters concerning heritage management, conservation and promotion. Under Section Five of the Act, the quasi-autonomous and apolitical Heritage Council (An Chomhairle Oidhreachta) was given statutory functions 'to propose policies and priorities for the identification, protection, preservation and enhancement of the national heritage'. This was defined in a very tangible fashion, as 'including monuments, archaeological objects, heritage objects, architectural heritage, flora, fauna, wildlife habitats, landscapes, seascapes, wrecks, geology, heritage gardens and parks and inland waterways'.[24] In addition to the introduction of important legislation, the 1990s also witnessed the rapid growth of Ireland's heritage industry. Although the early development of the industry, according to Ruth McManus, tapped into 'an immeasurable wealth', the enterprise was certainly 'about more than making money' from tourists. There was also a growing realisation that accomplishment in heritage management also depended upon 'conservation ... complemented by education, with enjoyable, responsible representation', which could 'preserve ... heritage for future generations'.[25]

From a pedagogic perspective, the appointment in 1995 of poet and philosopher John O'Donohue to a lectureship in Humanities at Galway Regional Technical College (renamed Galway-Mayo Institute of Technology, or GMIT, in 1998) is worthy of mention. It was during his two years there, as Lelia Doolan has noted, that O'Donohue 'designed and began teaching' on a 'wide-ranging' programme in heritage studies (at National Certificate level) and also published an international bestseller, *Anam Cara* (Irish for: *Soul Friend*).[26] In the years that followed, the heritage studies presence at GMIT expanded

---

[24] Government of Ireland, *The Heritage Act, 1995* (The Stationery Office, Dublin, 1995), pp. 6–7; M. O'Hanrahan, 'The Heritage Council', *Group for the Study of Irish Historic Settlement Newsletter*, No. 6 (1996), p. 19.

[25] R. McManus, 'Heritage and Tourism in Ireland: An Unholy Alliance?', *Irish Geography*, Vol. 30, No. 2 (1997), p. 98. Further discussion of the embryonic years of the Irish heritage industry can be found in K. Whelan, 'The Power of Place', *The Irish Review*, No. 12 (1992), pp. 13–20. For a treatment of conservation matters in heritage studies, see P. Howard, *Heritage: Management, Interpretation, Identity* (Continuum, London, 2003), p. 1, who writes that heritage 'is all pervasive, and concerns everyone', and 'is taken to include everything that people want to save ... including material culture and nature'.

[26] L. Doolan, 'Foreword', in J. O'Donohue, *Echoes of Memory*, New Edition (Transworld Ireland, London, 2009), p. 6. Further commentary on the essence of O'Donohue's scholarship can be found in *The Irish Times*, 14 November 2009, in which Michael D. Higgins outlines how the 'vindication of memory' was a persistent theme in 'all his works ... sometimes as sensory recall of sensations lost, of beauty that was ephemeral ... [or] as a sacred repository of ... the promise of possibility'.

considerably. In 2008, it became the only Irish higher education institution to offer a full range of awards in the discipline, at BA, BA (Honours), MA and PhD levels. It is to this suite of programmes, characterised by their distinctive academic/applied mix, that this book owes much of its original stimulus to. Externally, inspiration and enlightenment has come from the actions of the United Nations Educational, Scientific and Cultural Organisation (UNESCO). At the 32nd session of its General Conference in Paris in 2003, it adopted the *Convention for the Safeguarding of the Intangible Cultural Heritage*. The notion of 'intangible cultural heritage', as used in the convention, refers to 'the practices, representations, expressions, knowledge, skills ... that communities, groups and, in some cases, individuals recognise as part of their cultural heritage'.[27] Scholarly direction for this book has also come from a burgeoning international literature on heritage studies, especially from a number of seminal English-language texts that were first published in the late 2000s. In a similar fashion to UNESCO, these produced a much more liberal set of definitions of heritage than previously, with writers routinely stressing its present-centred nature and drawing attention to its variety of higher cultural meanings (for example, its relationship to forms of identity like nationalism, republicanism and patriotism).

In *Pluralising Pasts*, Gregory Ashworth, Brian Graham and John Tunbridge argued that rather than directly engaging with the past, 'heritage is present-centred and is created, shaped and managed by, and in response to, the demands of the present'. 'As such', they added, 'it is open to constant revision and change'.[28] In *The Ashgate Companion to Heritage and Identity*, David Harvey aired a similar viewpoint by drawing attention to heritage 'as a present-centred phenomenon' and pointing to the fact that people in all ages 'used retrospective memories as resources of the past to convey a fabricated sense of destiny for the future'.[29] Likewise, in *The Heritage Reader*, Rodney Harrison and others observed how 'heritage ... draws on the power of the past to produce the present and shape the future'. The rapid rise of a nostalgic 'attachment to the past' in the Westernised world, they noted, was 'fostering the notion of heritage as a shared and collective thing that binds society (or perhaps more accurately, parts of society) together'. The concern with the past, they added, had also manifested itself by encouraging persons 'to keep mementos', to become obsessed with matters to

---

27   UNESCO, cited in G. Corsane, 'Issues in Heritage, Museums and Galleries: A Brief Introduction', in G. Corsane (ed.), *Heritage, Museums and Galleries: An Introductory Reader* (Routledge, London, 2005), p. 6.

28   G. J. Ashworth, B. Graham and J. E. Tunbridge, *Pluralising Pasts: Heritage, Identity and Place in Multicultural Societies* (Pluto Press, London, 2007), p. 3.

29   D. Harvey, 'The History of Heritage', in B. Graham and P. Howard (eds), *The Ashgate Companion to Heritage and Identity* (Ashgate, Farnham, 2008), p. 22.

do with preservation and to become more resistant to 'change at a larger scale'.[30] Trends and developments in the heritage sector did not go unnoticed by leading historians. In *Why History Matters*, John Tosh touched upon some of the aforementioned themes, explaining how the fundamental business of heritage had focused more and more on 'the conservation, transmission and enjoyment of visible survivals from the past'. It had also expressed itself, he added, very visibly at two levels – namely local appreciation of the past (as frequently articulated through amateur pursuit) and state-centred policy on the past (as expressed through the agendas of government institutions and agencies).[31] As will be seen in the following chapters, this combination of local and national spheres has been of crucial significance in the forging of many Irish heritage experiences, not least the storylines associated with 1916.

Historiographies of Commemoration

In *The Idea of History*, Richard Collingwood once referred to history as 'a certain kind of organised or inferential knowledge'.[32] Thus there can be no engagement with history, as Guy Beiner has written, 'without historiography' (that is, the systematic study of the writing of history). 'Historical understanding', he adds, 'is founded on an accumulative chain of self-referential interpretative reconstructions produced by skilled practitioners'.[33] So what then can be said about the historiography of Irish commemorationist studies in recent times? Influenced by the aforementioned trends and developments in memory and heritage studies, it is impossible to escape the fact that from the late 1990s onwards, the remit of Irish historical understanding was broadened considerably by the publication of various edited collections featuring significant input by historians – including Laurence McBride's *Images, Icons and the Irish Nationalist*

[30]   R. Harrison, G. Fairclough, J. H. Jameson Jnr., and J. Schofield, 'Introduction: Heritage, Memory and Modernity', in G. Fairclough, R. Harrison, J. H. Jameson Jnr., and J. Schofield (eds), *The Heritage Reader* (Routledge, London, 2008), pp. 1–2.

[31]   J. Tosh, *Why History Matters* (Palgrave Macmillan, Basingstoke, 2008), p. 10.

[32]   R. G. Collingwood, *The Idea of History* (Oxford University Press, Oxford, 1970), p. 252.

[33]   G. Beiner, *Remembering the Year of the French: Irish Folk History and Social Memory* (The University of Wisconsin Press, Madison, 2007), pp. 8–9. Further commentary on the essence of historiography can be found in J. Tosh with S. Lang, *The Pursuit of History*, 4th Edition (Pearson Education Ltd., Harlow, 2006), p. 8, who define it as 'the values and assumptions expressed in the writing of history itself'. Also see M. Donnelly and C. Norton, *Doing History* (Routledge, London, 2011), p. 16, who note that historiography involves 'thinking about the particular ways in which ... relevant specialist historians have conceptualised, researched and written about the subject at hand'.

Imagination, Laurence Geary's *Rebellion and Remembrance in Modern Ireland*, Ian MacBride's *History and Memory in Modern Ireland*, and Edmund Bort's *Commemorating Ireland: History, Politics, Culture*.[34] These books uncovered many historical incidences of individual and collective remembering, and illuminated some of the cultural, political and economic processes inherent in the ways that households, communities, regions, and nation states practiced commemoration in times gone by.

Considering the significance of the event, it is worth mentioning at this juncture that the amount of commemorationist studies of the 1916 Rising by historians and scholars from other disciplines have not been plentiful to date. For many years, the occasion of significant anniversaries witnessed the publication of new histories of 1916. However, very little analytical commentary appeared in print on the actual nature of the commemorations themselves. The Golden Jubilee of the Rising in 1966 inspired a number of historians to add their own views to the debates on 1916, but none of the edited collections that appeared by the end of the 1960s (from scholars like Father F. X. Martin, Owen Dudley Edwards, Fergus Pyle, and Kevin B. Nowlan), offered any meaningful commentary on the significance of the events that had been staged for the special anniversary.[35] The one publication that could truly be classified as commemorationist in nature was *Cuimhneachán 1916–1966: A Record of Ireland's Commemoration of the 1916 Rising*. This souvenir-type publication, which was produced by the Department of External Affairs, offered a thorough official record of the events that had been staged throughout the state for the Golden Jubilee.[36] What it lacked, however, was any critical analysis of what had actually taken place. Rather unsurprisingly, commemorationist perspectives were also in short supply following the low-key 75th anniversary in 1991. As the Troubles waged in Northern Ireland, the topic of 1916 was purposefully avoided by most historians. Consequently, it was left to others to debate the anniversary's meanings, especially those from the literary and artistic world.

---

[34] See L. W. McBride (ed.), *Images, Icons and the Irish Nationalist Imagination* (Four Courts Press, Dublin, 1999); L. M. Geary (ed.), *Rebellion and Remembrance in Modern Ireland* (Four Courts Press, Dublin, 2001); I. McBride (ed.), *History and Memory in Modern Ireland* (Cambridge University Press, Cambridge, 2001); and E. Bort (ed.), *Commemorating Ireland: History, Politics, Culture* (Irish Academic Press, Dublin, 2004).

[35] See F. X. Martin (ed.), *Leaders and Men of the Easter Rising: Dublin 1916* (Methuen & Co. Ltd., London, 1967); O. Dudley Edwards and F. Pyle (eds), *1916: The Easter Rising* (MacGibbon & Kee, London, 1968) and K. B. Nowlan (ed.), *The Making of 1916: Studies in the History of the Rising* (The Stationery Office, Dublin, 1969).

[36] Department of External Affairs, *Cuimhneachán 1916–1966: A Record of Ireland's Commemoration of the 1916 Rising* (An Roinn Gnóthaí, Dublin, 1966).

Significant in this regard was Mairín Ní Dhonnchadha and Theo Dorgan's edited collection, *Revising the Rising*.[37] Published by Field Day, it was a book small in size yet rich in debate.

While the occasion of the 90th anniversary in 2006 generated new publications on the history of the Rising, it also witnessed the welcome appearance of a few works that sought to illustrate how knowledge and understanding of the event could be advanced by a commemorationist perspective on matters of remembrance. Some of these works were published in advance of the anniversary. Towards the end of 2005, for example, a discussion of aspects of the memory theme was offered by James Moran's *Staging the Easter Rising: 1916 as Theatre*, which dealt mainly with the specialised subject of how dramatists represented the Rising in their plays.[38] A short epilogue to Charles Townshend's *Easter 1916: The Irish Rebellion*, entitled 'The Rebellion in History', also added to the debates by offering some insightful reflections on the interplay between historiography and commemoration.[39] In the same year, Róisín Higgins contributed a short chapter on the Golden Jubilee commemoration of 1966 to an edited collection of essays, entitled *Ireland: Space, Text, Time*.[40]

Early in 2006, to mark the commencement of the year of the 90th anniversary, the History Department of University College Cork organised a major academic conference on 1916. This opened with a much-publicised keynote speech by the President, Mary McAleese. Her words sparked off a heated debate on the significance of 1916 in the letters page of *The Irish Times* and other outlets. The proceedings of the conference, which addressed some matters to do with commemoration, were published as a voluminous collection of essays edited by historians Dermot Keogh and Gabriel Doherty, entitled *1916: The Long Revolution*.[41] The 90th anniversary was also marked by the release of a special commemorative issue of the popular bimonthly magazine, *History Ireland*, featuring a number of short articles on both the history and remembrance of 1916. In his editorial, Tommy Graham welcomed the government's decision to reinstate the military parade on Easter Sunday 2006 and called for 'more debate, more discussion, and not less, about the legacy of

---

[37]   M. Ní Dhonnchadha and T. Dorgan (eds), *Revising the Rising* (Field Day, Derry, 1991).

[38]   J. Moran, *Staging the Easter Rising: 1916 as Theatre* (Cork University Press, Cork, 2005).

[39]   C. Townshend, *Easter 1916: The Irish Rebellion* (Allen Lane, London, 2005), pp. 344–59.

[40]   R. Higgins, '"The Constant Reality Running through Our Lives": Commemorating Easter 1916', in L. Harte, Y. Whelan and P. Crotty (eds), *Ireland: Space, Text, Time* (The Liffey Press, Dublin, 2005), pp. 45–56.

[41]   G. Doherty and D. Keogh (eds), *1916: The Long Revolution* (Mercier Press, Cork, 2007).

1916'.[42] The Defence Forces also marked the occasion by publishing a double-sized issue of its magazine, *An Cosantóir*, edited by Seargent William Braine. This commenced with a message from the Minister for Defence, Willie O'Dea, who reflected upon the state's 'roots in Easter Week 1916' and remarked that the 90th anniversary was also about honouring and remembering 'what it inspired – our independence and freedom'.[43]

After the 90th anniversary, a growing number of scholars of 1916 turned their attention to matters of commemoration. Towards the end of 2007, a North-South research collaboration between the History Departments of University College Dublin and The Queen's University of Belfast reached its fruition with the publication of Mary Daly and Margaret O'Callaghan's edited volume on the Golden Jubilee year, *1916 in 1966: Commemorating the Easter Rising*.[44] This was followed in 2009 by Roisín Higgins and Regina Uí Chollatáin's edited collection, *The Life and After-Life of P. H. Pearse*.[45] In reflecting on the 1916 leader's contemporary reputation, the editors noted that 'the passing of time has seen Pearse both revered and reviled', to the extent that 'he has become one of the most contested figures in Irish history' with a 'reputation ... determined by the politics of the present rather than the past'.[46] In the same year, the theme of commemoration also occupied a full chapter of Clair Wills's *Dublin 1916: The Siege of the GPO*, which documented the transformations that the rebel headquarters had undergone since the Rising took place.[47] The concluding chapter of Ferghal McGarry's *The Rising. Ireland: Easter 1916*, published in 2010, also touched upon memory issues. In a brief appraisal of the long-term aftermath of the Rising, he ascertained that its 'most controversial aspect

---

[42] T. Graham, 'From the Editor', *History Ireland*, Vol. 14, No. 2 (2006), p. 4. Also see P. Bew, '"Why Did Jimmie Die?" A Critique of Official 1916 Commemorations', *History Ireland*, Vol. 14, No. 2 (2006), pp. 37–39; R. Higgins, C. Holohan, and C. O'Donnell, '1966 and All That: The 50th Anniversary Commemorations', *History Ireland*, Vol. 14, No. 2 (2006), pp. 31–36 and C. Townshend, 'Making Sense of Easter 1916', *History Ireland*, Vol. 14, No. 2 (2006), pp. 40–45.

[43] W. O'Dea, 'Message from the Minister for Defence', *An Cosantóir: The Defence Forces Magazine*, Vol. 66, No. 3 (2006), p. 4.

[44] M. Daly and M. O'Callaghan (eds), *1916 in 1966: Commemorating the Easter Rising* (Royal Irish Academy, Dublin, 2007).

[45] R. Higgins and R. Uí Chollatáin (eds), *The Life and After-Life of P. H. Pearse* (Irish Academic Press, Dublin, 2009).

[46] R. Higgins and R. Uí Chollatáin, 'Introduction', in Higgins and Uí Chollatáin (eds), *The Life and After-Life of P. H. Pearse*, p. xvii.

[47] Wills, *Dublin 1916*, pp. 133–71.

... was its legacy', especially 'its place in the fragmented political discourse of independent Ireland'.[48]

Aims, Methods and Contents of this Book

From a historiographical perspective, the aforementioned publications have been of undoubted significance in so far as they have enabled studies of memory, heritage and commemoration to gain an important foothold in scholarship on the 1916 Rising's legacies. Nonetheless, whilst important foundations have been laid, many significant gaps still remain. Consequently, a *central aim* of this book is to break new ground by offering a wide-ranging exploration of the making and remembrance of the story of 1916 in modern times. Drawing together many complex threads – especially the interlocking dimensions of history-making, commemoration and heritage – several examples and case studies are used to reveal the Rising's undeniable influence upon modern Ireland's evolution, both instantaneous and long-term. In addition to a chapter that furnishes an in-depth history of the tumultuous events of Easter 1916, this work mainly concentrates on illuminating the evolving relationship between the Irish past and present. In doing so, the last five chapters unearth the far-reaching political impacts and deep-seated cultural legacies of the actions taken by the rebels, as evidenced by the most pivotal episodes in the Rising's commemoration and the myriad varieties of heritage associated with its memory. The narrative is, naturally enough, the product of its own unique series of inclusions and exclusions. Moreover, the book also contains a small amount of conjecture on topics where the jury is still out – more so in the case of the treatment that is presented for more contemporary occurrences.

In terms of *methodology*, an attempt is made to engage with new perspectives, directions, topics, and ideas concerning the story of 1916. In order to facilitate this, a wide range of source materials are utilised, including rare manuscripts, letters, diaries, pamphlets, eye witness and participant accounts, speeches, newspaper reportage, parliamentary debates, published histories, memoirs, biographies, poems, songs, novels, photographs, film and television, opinion polls, census returns, museum exhibitions, and fieldwork. Beyond the realm of historical reconstruction, attention is paid to how acts of commemoration have been fashioned by political ideologies and disputes, how a heritage-politics rhetoric has intertwined with constructs of nationhood, how the legacy of the

---

48 F. McGarry, *The Rising. Ireland: Easter 1916* (Oxford University Press, Oxford, 2010), pp. 286–87.

rebel past has impinged upon cultural and economic activity in areas ranging from museums to tourism, and how the persistence of memory has manifested itself in the form of heritage objects. This book is also cognisant of the notion of 'commemorative trajectory' – which relates to Withers's argument that 'memory in the guise of its representation changes over time, over space and between different peoples in different ways'.[49] The chapters that follow also take heed of the importance of the idea of 'historical trajectory'. As Rodney Harrison and others have noted in an overview of the nature and practice of cultural heritage management throughout Westernised societies, the heritage experience 'is not a "given" but rather the particular product of a particular historical trajectory in particular social and cultural contexts'.[50]

In terms of *content*, this book contains a series of carefully chosen examples and case studies to explore fundamental aspects of the Rising's history, commemoration and heritage. The narrative commences with an historical account of the causes, course and consequences of the 1916 Rising. Thereafter, the focus switches to scrutinising assorted acts of commemoration and varied manifestations of heritage. In addition to examining Sinn Féin's celebration of the Rising's first anniversary in 1917 and the official tributes that were instigated by the Cumann na nGaedheal government from 1924 onwards, much attention is paid in this book to a succession of noteworthy anniversary commemorations (many of which involved considerable input from Fianna Fáil-led governments). These include: the 19th and 20th anniversaries in 1935 and 1936 (which were staged at a time in which the Free State faced internal pressures from a resurgent Irish Republican Army, or IRA), the Silver Jubilee commemoration during the so-called Emergency in 1941 (when the state faced serious external threats, as a result of World War II), the momentous Golden Jubilee in 1966 (which was celebrated in the Republic with much official gusto), the toned-down commemorations in the 1970s and 1980s (including 1972, when the Easter Sunday military parade was suspended due to the Troubles), the subdued 75th anniversary in 1991 (when the Troubles were still raging), the more upbeat 85th anniversary in 2001 (when the story of 1916 resurfaced on television screens), the revitalised 90th anniversary in 2006 (when the military parade resumed, following the success of the Peace Process), and the sombre 94th and 95th anniversaries in 2010 and 2011 (the latter of which was commemorated by a newly-elected Fine Gael-Labour government, at a time of unprecedented economic turmoil). Plans and preparations for the century of the Rising in 2016 are also put under the spotlight.

[49] C. Withers, 'Memory and the History of Geographical Knowledge', p. 318.
[50] Harrison, Fairclough, Jameson Jnr., and Schofield, 'Heritage, Memory and Modernity', p. 1.

Besides the cross-sectional (or snapshot) focus on some of the key anniversary commemorations in the past, this book also strives to connect each of the major case studies together by incorporating a temporal perspective that concentrates on change over time and makes reference to political developments in every decade since the Rising took place. A clear pattern to the approach taken to 1916 by successive Irish governments is revealed as the book progresses, with the Rising's key commemorative milestones being strongly accentuated when the state faced internal pressures or external threats, celebrated on occasions of comparative peace and harmony, or downplayed during eras marred by prolonged paramilitary violence. Attention is also paid to the plethora of sentimental rituals associated with the nature and practice of commemoration. As will be seen, anniversaries of 1916 have proven to be emotion-laden affairs – typically involving the recital of prayers at church services, the militaristic routines associated with the pomp and ceremony of parading soldiers, the lowering of the National Flag (the Tricolour), the reciting of the 509 words of the Proclamation, moments of silence following wreath-laying, the raising of the Tricolour following the 'Last Post', the firing of 21 gun salutes, and the singing of the National Anthem (or 'Soldier's Song').

An in-depth exploration of the Rising's heritage is also offered. Over time and space, this has found perpetual expression in hundreds of historical and literary texts, a plethora of newspaper articles and supplements, and copious amounts of political speeches and debates. Reverence for the Rising has also been perpetuated in public by means of monuments and plaques, exhibitions at the National Museum and the Pearse Museum, visitor attractions such as Kilmainham Gaol and the Pearse cottage at Rosmuc, bus and walking tours of Dublin city centre, and more recently, through the waging of heritage conservation campaigns (such as 'Save Number 16 Moore Street'). The Rising's heritage has also manifested itself in landscape archaeologies (for example, graveyards, bullet holes in buildings, trench foundations, and ammunition buried in soil), place imagery (for example, wall murals) and different kinds of 'histotainment' (for example, concerts, dance shows, dramatic plays, festivals, films, musicals, pageants, and reenactments). Other expressions of its heritage have materialised in the following forms: art exhibitions, auctions of rebel memorabilia, reunion dinners, sound recordings, oral customs, private pilgrimages, songs, poems, television documentaries, radio programmes, websites, conferences, summer schools, educational syllabi, lesson packs, essay competitions, and awards of scholarships and honorary degrees.

An assortment of heritage/souvenir objects have also perpetuated reminiscences of the 1916 Rising amongst the populace – including items such as manuscripts, illuminated scrolls, uniforms, guns, flags, drums, coins, cups, stamps, postcards, greeting cards, photographs, maps, jewellery, mourning badges, Easter lilies, stickers, cartoons, calendars, medals, paintings, posters, sculpture, textiles, printed T-shirts, CDs, and DVDs. The story of 1916 has also found articulation in the names given to streets (such as Pearse Street and Cathal Brugha Street in Dublin), bridges (like the Michael O'Hanrahan Bridge in New Ross), public parks (like O'Farrell Park in Dublin's docklands and Páirc an Piarsaig in Tralee), Army barracks (like Galway's Dún Uí Mhaoilíosa), housing estates and terraces (for instance, Connolly Terrace in Galway), pubs (like the Padraig Pearse on Dublin's Pearse Street), apartment tower blocks (for example, Dublin's seven Ballymun Towers), gyms (such as the Markievicz Leisure Centre off Dublin's Tara Street), and 15 railway stations dotted throughout the Republic. Its memory has also been immortalised in the naming of pipe bands, educational institutions (like Mellows Agricultural College), historical societies, republican halls, branches of political parties (especially Fianna Fáil), and Gaelic Athletic Association teams and spaces (for example, Liam Mellows GAA Club, Casement Park and Croke Park's Hill 16).

Arrangement of Chapters

The chapters of this book are arranged in an easy-to-follow chronological structure. They present a narrative, descriptive and analytical history of the making and remembrance of the story of 1916, with each going far and wide in time and space in order to address a miscellany of thematic issues.[51] Chapter 2 presents an updated history of the Rising, by taking account of various new histories and by utilising previously-neglected source materials. The analysis is situated within the context of the key political issues that dominated Ireland in the mid-1910s – namely the Home Rule question (and unionist opposition), the implications of Irish involvement in the Great War (or World War I) and the emergence of militant nationalism as a force to be reckoned with. Despite stringent press censorship and the best attempts of British intelligence officers to curtail sedition against the Crown, a small number of militants within the Irish Volunteers, led by Patrick Pearse, acted on the dictum that 'England's difficulty' was 'Ireland's opportunity' and staged a six-day Rising in Dublin – to the surprise of many. Support was also forthcoming from the Irish Citizen Army (led by socialist James Connolly) and the Irish Republican Brotherhood

---

[51] The use of all three modes, as Peter Beck has noted, is well established in presenting the past in writing. See P. J. Beck, *Presenting History: Past & Present* (Palgrave Macmillan, Basingstoke, 2012), p. 24, in which he states that 'description and narrative help to recreate the past', while 'analysis is important for the purposes of exposition, explanation, discussion, interpretation and argument'.

(led by militants such as Thomas Clarke and Seán MacDiarmada). As it proceeds, this chapter provides an in-depth account of the principal events of the Rising in Dublin city – from the occupation of the GPO on Easter Monday 24 April 1916 to the eventual surrender by Pearse the following Saturday. A geographical sketch/local history of the Rising throughout the provinces is also furnished. In addition to scrutinising the crucial role played by the leaders of the Rising, reference is made to the lesser-known part played by the rank-and-file rebels in the Irish Volunteers and Irish Citizen Army. As the rich content of the aforementioned Bureau of Military History collection demonstrates, the testimonies and memories of those who acted as foot-soldiers offers a revealing insight into the strategies and ideologies that characterised the fighting in Easter Week. This chapter also draws upon sources that illuminate the experiences of some of those in the Crown forces who suppressed the Rising. Like so many paths to nationhood, 1916 proved to be a bloody and messy affair, and resulted in casualties on both sides of the fighting. Tragically, the Rising also led to many civilian deaths and injuries, while significant damage was caused to buildings in Dublin city centre.

Chapter 3 assesses both the immediate and long-term implications of the 1916 Rising. Initial reaction on the streets of Dublin and in the press proved to be extremely hostile to the rebels. However, the subsequent response from the British Army, commanded by General John Maxwell, was severe. Having been found guilty in their court martial proceedings, Pearse and the chief leaders of the rebels – some of whom equated their struggle and beliefs to a 'Holy War' – were executed by firing squad. Ultimately, the executions proved to be a blunder on the part of the British authorities, while the arrest and imprisonment of thousands of suspects also turned the tide of Irish public opinion in favour of the rebels. By the time the first anniversary of Rising was commemorated in 1917, support for a more radicalised Sinn Féin party had grown considerably. In the General Election of the following year, it won over 70% of Irish seats at Westminster (but subsequently refused to take them). From 1919–1921, a War of Independence (or Anglo-Irish War) was waged throughout Ireland. Following the signing of a Treaty, the 26-county Free State came into existence in 1922, headed by Cumann na nGaedheal. Despite much acrimony over the Civil War of 1922–1923, the first official commemoration of the Rising by the Free State was held in Dublin city in 1924, while the GPO was finally reopened to the public in 1929. In 1932, Fianna Fáil came to power, with veteran rebel Eamon de Valera as President of the Executive Council. In 1935, he gained some tactical ground on a resurgent IRA by staging a grandiose celebration of the Rising's 19th anniversary, which included the unveiling of a bronze statue of Cúchulainn inside the GPO. For the 20th anniversary in the following year, small parades of

Army units were held on Easter Sunday in Cork, Galway, Limerick, Athlone, Letterkenny, Carlow, and Dundalk. After World War II broke out in 1939, Éire/Ireland adopted a policy of military neutrality. The threat of invasion, however, prompted the government to stage a major parade of 25,000 military personnel past the GPO for the Rising's 25th anniversary in 1941. During the 1950s, notable acts of commemoration included the following: the naming of a GAA stadium in Belfast after Roger Casement, the unveiling of a statue of Liam Mellows in Galway, the unveiling of a bust of Countess Markievicz in St. Stephen's Green, and the completion of a fitting memorial at the Arbour Hill burial plot. The late 1950s also saw the drafting of an action plan to restore Kilmainham Gaol – the site where the leaders of the Rising were executed.

The Rising's memory and the ideals enshrined in the Proclamation, as the in-depth case study furnished in Chapter 4 illustrates, were cultivated to unprecedented levels during the Golden Jubilee celebrations that were staged in 1966 in several locations across the Republic (and elsewhere). Dublin city was the centre of the action, hosting a major military parade on Easter Sunday that was watched on O'Connell Street by the sole surviving leader of the 1916 Rising – President Eamon de Valera – who took part later that day in an emotional wreath-laying ceremony at Kilmainham Gaol. A range of new cultural landscapes were also created for the 50th anniversary, including the Garden of Remembrance at Parnell Square, which was opened by de Valera on Easter Monday. Other events held at Eastertime in the capital included an exhibition at the National Museum of Ireland and a commemorative heritage pageant at Croke Park, entitled *Aiséirí-réim na Cásca* ('Resurrection, the Easter Pageant'). RTÉ broadcast a range of commemorative programmes on radio and television (including a much-hyped four-hour television drama called *Insurrection*), while memory of the Rising was also perpetuated by the release for sale of several objects of commemorative memorabilia – including stamps, coins, Sword of Light badges, newspaper supplements, and history books. This chapter also offers a geographical account of the commemorations that were held throughout the Republic's regions (at locations such as Cork, Limerick, Galway, Athlone, and Westport), the Border counties, Northern Ireland, England, and the United States of America.

Soon after the celebrations of Easter 1966, the course of modern Irish history took a turn for the worse. Chapter 5 investigates the repercussions of the outbreak of the Northern Ireland Troubles in the late 1960s for subsequent commemorations of the Rising. On 11 April 1971, the Republic commemorated the Rising's 55th anniversary with a military parade of around 1,800 military personnel past the GPO. It was not until 35 years later that the Army again paraded through O'Connell Street on Easter Sunday. As the Provisional IRA began appropriating the 1916 mythology in order to justify its own prolonged armed struggle, the powers that be in the Republic deemed that 1916 needed to be commemorated in a much more sensitive fashion than previously. At the same time, the Rising's place in history began to be questioned by revisionist historians, who expressed unease about the legacy of Easter Week. Despite the objections of those who held disparaging views about Pearse's association with bloodshed, the centenary of the 1916 leader's birth was marked by a range of events in 1979 – including the opening of the Pearse Museum at St. Enda's Park. By the time of the Rising's 75th anniversary in 1991, however, the government decided to run a much scaled-down commemoration. This involved a short ceremony outside the GPO on Easter Sunday, presided over by the President, Mary Robinson and the Taoiseach, Charles Haughey. However, this was supplemented by a range of unofficial events organised by members of the creative arts community, under the banner of 'Reclaim the Spirit of 1916'. The chapter ends by outlining how the cessation of paramilitary violence in Northern Ireland, followed by the brokering of the Good Friday Agreement in 1998, heralded the dawn of a new era of respect and understanding between different political traditions across the island of Ireland. As the story of Irish nationalists who fought in the Great War (for Home Rule) received greater acknowledgment from the government of the Republic, Bertie Ahern (who became Taoiseach in 1997) also set about rehabilitating the legacy of the Rising. Rather gradually, sentiment towards the Rising improved. Its 85th anniversary in 2001 was marked by the screening on RTÉ and BBC of *Rebel Heart*, a major television drama series based on the historic events of Easter Week.

The occasion of the 90th anniversary in 2006 witnessed the defining moment of the Rising's remarkable comeback to the stage of Irish historical commemoration. Chapter 6 overviews various high-profile events that were held to mark this anniversary, including the above-mentioned academic conference at University College Cork, the exhibition at the National Museum and the parade by the Defence Forces on Easter Sunday. The anniversary was also marked in the capital by the National Library of Ireland, which eagerly embraced the emergent trend of remembrance by means of social media, by launching a special 1916 website. This carried Polish and Chinese translations of the Proclamation, in recognition of the Republic's new multiculturalism. Throughout the state, members of different political parties and interested onlookers also came out in force to mark the anniversary at a range of church services, parades and wreath-laying ceremonies. Filmmakers and publishers also played a substantial role in the commemorationist activities, by releasing a range of old films about the independence struggle on DVD and issuing reprints of old classics, while auction houses also profited considerably from the sale of rare manuscripts and memorabilia. Historians also embraced the occasion wholeheartedly by producing an ample amount of new reading material for consumption by an eager public. The chapter finishes by considering the wider social, cultural and political significance of what transpired in 2006.

In Chapter 7, the book draws to a conclusion by bringing the narrative up to the early 2010s and by synopsising some of the key heritage stories that made headlines in Ireland in the lead-up to the 100th anniversary of the Rising. Government preparations to mark the centenary were set in motion in 2006, and from that time onwards, more and more proposals were mooted by a range of interested parties. In trying to offer some preliminary observations about the pathway towards 2016, this final chapter makes a number of determinations about the grand position of history, commemoration and heritage in Irish politics and society. The implications of the passage of time are considered for the functioning of memory, heritage conservation/restoration, battlefield archaeology, and tourism. Attention is also drawn to the prominence that the government increasingly accorded to pluralist pasts in both peace and reconciliation initiatives and in the continued renegotiation of North-South and Anglo-Irish relations. The book ends by highlighting the prospects and possibilities that unfolded for heritage, culture and identity as 2016 drew closer, with Ireland seeking to come to terms with its greatest economic crisis since the granting of independence. Following on from an examination of the predominantly solemn commemoration of the Rising's 95th anniversary on Easter Sunday 2011, the book ends by looking at the therapeutic goodwill that sprung a few weeks afterwards from the state visit to the Republic of Ireland by Britain's Queen Elizabeth II – who movingly observed a minute's silence at the Garden of Remembrance for those who fought for Irish freedom.

Chapter 2

Making Irish History:
The Easter Rising, 1916

No understanding of the place of the 1916 Rising in the making of Irish history would be complete without probing the complex reasons behind the outbreak of six days of intense fighting that shook Dublin city to its very core, resulted in skirmishes around the country and had immense long-term consequences for the history of Ireland and its subsequent relations with Britain. To reach some degree of understanding about 1916 obviously requires an appreciation and grasp of the multifaceted nature of Irish (and European) history and geopolitics at the beginning of the twentieth century. At the time, 'Irish-Ireland', according to Kevin Nowlan, was a complex web of interests which were from time to time in conflict with one another, 'but they all helped to give a new and exciting quality to Irish political and intellectual life'.[1] Although nation states, as Niall Ferguson has demonstrated, 'were a comparative novelty in European history', the Irish (like the Poles) 'saw nationhood as an alternative to subjugation by unsympathetic empires'.[2] The rapid growth since the late nineteenth century in the phenomenon of Irish cultural nationalism – as exemplified by the activities of the GAA, the Gaelic League and the literary revival pioneered by poet William Butler Yeats and others – was in many respects a manifestation of a growing resistance to the process of colonialism. What many nationalists opposed was 'Anglicisation', which went hand-in-hand with the domination of Ireland by the might and power of the expanding British Empire.[3]

---

[1]   K. B. Nowlan, 'Introduction', in K. B. Nowlan (ed.), *The Making of 1916: Studies in the History of the Rising* (The Stationery Office, Dublin, 1969), p. xi.

[2]   N. Ferguson, *The War of the World: History's Age of Hatred* (Allen Lane, London, 2006), pp. 74–75.

[3]   For a definition of the process of colonialism, see J. Ruane, 'Colonialism and the Interpretation of Irish Historical Development', in P. Gulliver and M. Silverman (eds), *Approaching the Past: Historical Anthropology through Irish Case Studies* (Columbia University Press, New York, 1992), p. 295, who defines it as 'the intrusion into and conquest of an inhabited territory by the representatives ... of an external power; the displacement of the native inhabitants ... from resources and positions of power; the subsequent exercise of economic, political, and cultural control over the territory and native population'.

Colonisation gathered considerable pace with the New English Protestant plantations of the sixteenth and early seventeenth centuries (especially after the Munster Plantation in 1586 and the Ulster Plantation in 1609), the Cromwellian Wars of 1649–1653, the Williamite victory at the Battle of the Boyne in 1690, and the imposition of assorted penal laws against Catholics in the years that followed. An array of new fortifications, mapping schemes, territorial arrangements, hyphenated identities, and commercial ventures were ushered in by the expanding British Empire, but as William J. Smyth has demonstrated, the early modern conquest and settlement of Ireland also brought 'stresses and indignities' for the Gaelic Irish (and Old English). These included 'the break-up and fragmentation of the old power structures' and 'the pain that followed the rupturing of a society's psychic moorings, the undermining of a people's sense of place and identity'.[4] The 'taking of Irish resources into British hands', according to Gerry Kearns, was formalised through 'a succession of new orderings of property and territory' and the colonial downgrading of the natives to 'something much less than political subjects'. Nowhere else in Europe, he adds, did 'a minority oppress a majority in such a fashion'. The resultant 'asymmetries of power' proved to be a recipe for intermittent political violence.[5] Growing resistance to British hegemony culminated with the outbreak of the 1798 Rebellion in the counties of Wexford, Wicklow, Meath, Carlow, Kildare, Antrim, Down, and Mayo. This was waged by the pikemen of the United Irishmen – including Protestant gentry, Presbyterian radicals, rural Catholic Defenders, and the urban Catholic middle class – who drew inspiration from the American War of Independence and the French Revolution. The 1798 Rebellion, however, ended in failure with the loss of over 30,000 lives. It also signalled the end of parliamentary activity in Dublin following the creation of the United Kingdom of Great Britain and Ireland under the Act of Union, which came into effect in 1801.

Thereafter, Irish laws were made at Westminster in London. As Charles Townshend has illustrated, their everyday implementation was conducted by a Dublin-based Executive. Its titular head was the so-called Viceroy (or the Lord Lieutenant and Governor-General of Ireland), whose function was 'partly decorative' and held by a nobleman. The actual day-to-day work of government during the nineteenth and early twentieth centuries, however, was done by a team overseen by the Chief Secretary and Under Secretary, the former of whom spoke on issues pertaining to Ireland in the House of Commons. As the Chief

---

[4]   W. J. Smyth, *Map-Making, Landscapes and Memory: A Geography of Colonial and Early Modern Ireland* c. *1530–1750* (Cork University Press, Cork, 2006), pp. 456–57, 468.

[5]   G. Kearns, 'Bare Life, Political Violence and the Territorial Structure of Britain and Ireland', in D. Gregory and A. Pred (eds), *Violent Geographies: Fear, Terror and Political Violence* (Routledge, New York, 2007), pp. 9, 12–13, 17, 30.

Secretary's political status had grown considerably by the early 1900s, he was more likely to be a member of the British Cabinet than the Viceroy. While both the Viceroy and the Chief Secretary were political appointees who resided at stylish lodges in Phoenix Park, much of the real administrative power resided in Dublin Castle, run by the Under Secretary, who was a civil servant. Through the agency of the armed 10,000-strong Royal Irish Constabulary, which was distributed in barracks located throughout the entire island, and its intelligence unit, the Crime Special Branch, Dublin Castle symbolised 'the overwhelming power of Britain'. In the capital itself, policing duties were in the hands of 'a British-style unarmed force', the 1,500-strong Dublin Metropolitan Police, which also ran a detective unit known as the G Division. However, Britain's 'giant', according to Townshend, 'had feet of clay', and problems in the overall coordination of intelligence were compounded by indifference 'at the level of social psychology and ideology'.[6]

In addition to the 1798 Rebellion, further episodes of armed resistance to British rule were waged by Robert Emmet in 1803, the Young Irelanders in 1848 and the Fenians in 1867. But nowhere was the resilience of Irish defiance of British hegemony more dramatically seen than in the momentous events that occurred at the end of April 1916, when militant republican forces staged an armed Rising over six days, mainly in Dublin city centre. Set within the broader international context, the direction of Irish history in the lead-up to the Rising was significantly distorted by the events of the Great War, with Britain at conflict with Germany. Although the war, according to Thomas Bartlett, 'was, initially, a popular cause in Ireland' and 'boosted the Irish economy', jingoism was far less evident in Ireland than it was in Britain.[7] Rather gradually, as F. S. L. Lyons has written, a feeling arose 'that the war, while undeniably good for business, was not Ireland's affair', especially when the future 'seemed so obscure'.[8] In the run-up to Easter 1916, a waning of Irish enthusiasm was accompanied by the resurfacing in certain locations of the old Fenian adage that 'England's difficulty' was 'Ireland's opportunity'. 'Republican separatists', as Townshend has noted, 'believed that they had both a duty and a right to strike on behalf of the Irish people, because Ireland was at a critical point in its history'. Fearing that their 'sense of nationality' was under threat from the encroachment of British power

---

[6] C. Townshend, *Easter 1916: The Irish Rebellion* (Allen Lane, London, 2005), pp. 24–26.

[7] T. Bartlett, *Ireland: A History* (Cambridge University Press, Cambridge, 2010), pp. 379, 382. Those holding sceptical views about the war were able to make their views known in a range of newspapers. An example can be seen from an article that appeared in *Sinn Féin*, 22 August 1914, which declared that the war on the continent was 'none of our seeking' and urged readers not to 'stand idle' in a time of 'rampant' jingoism.

[8] F. S. L. Lyons, *Ireland since the Famine* (Fontana Press, London, 1986), p. 361.

and principles (or 'Anglicisation'), Patrick Pearse and his fellow Irish-Irelanders felt that the time was ripe 'to take up the torch of revolt' at a time in which the British state had been weakened by its involvement in the war on the continent.[9]

Although many Irish people at the time believed that British power 'was virtually invincible', Eoghan O'Neill has noted that one of the main objectives of the Rising 'was the intention to prove that foreign domination was neither invulnerable nor inevitable'.[10] So what then were the characteristics of the predominant reasoning behind the challenge to British hegemony? Whilst the grievances of many of the 1916 rebels were economic (especially those who came from the ranks of working class Dubliners), it has been argued by Declan Kiberd that the aggravations 'of *all* the fighters were cultural', as they sought 'a land in which Gaelic traditions would be fully honoured'. 'What they rejected', he believes, 'was not England but the British imperial system, which denied expressive freedom to its colonial subjects'.[11] However, after the surrender of the rebels, the 1916 Rising was initially condemned by much of the population in Ireland – who felt that it was not only undemocratic, but that it had represented the views of a minority of militants who were misguided in their actions. This was especially the case with those who saw the rebels' collusion with the Germans as a stab in the back for the thousands of nationalists who had signed up with the British Army during the Great War, in the hope that the long saga concerning Home Rule would finally be resolved. As Michael Laffan has noted, 'there was widespread bewilderment and even dismay' amongst many of the inhabitants of Dublin in particular, especially from those outraged at its negative implications for the prospect of obtaining Home Rule after the war on the continent was over.[12]

The death of innocent civilians and the widespread destruction to property in Dublin city also added to the public outcry throughout Ireland. However, the execution of the main leaders of the Rising by the British authorities dramatically changed the tide of Irish public opinion into widespread retrospective support for the actions of the rebels, thus leading to a romanticisation of the doctrine of blood sacrifice and a huge upsurge in public support for the separatist ideals enshrined in the Proclamation that Pearse had read outside the GPO.

---

[9]   C. Townshend, 'Making Sense of Easter 1916', *History Ireland*, Vol. 14, No. 2 (2006), pp. 40–41.

[10]   E. O'Neill, 'The Battle of Dublin 1916: A Military Evaluation of Easter Week', *An Cosantóir: The Irish Defence Journal*, Vol. 26, No. 5 (1966), p. 215.

[11]   D. Kiberd, '1916: The Idea and the Action', in K. Devine (ed.), *Modern Irish Writers and the Wars* (Colin Smythe Ltd., Gerrard's Cross, 1999), pp. 20–21.

[12]   M. Laffan, *The Resurrection of Ireland: The Sinn Féin Party 1916–1923* (Cambridge University Press, Cambridge, 2005), p. 47.

Even though it failed in its immediate mission of establishing an Irish Republic of 32 counties, the 1916 Rising was unquestionably the spark which ignited the War of Independence of 1919–1921. This in turn culminated in the granting of independence from British rule to most of the island of Ireland after the formation of the 26-county Free State in 1922 – an event that impinged at a variety of levels upon the subsequent break-up of the British Empire's hegemony over one quarter of the world's territory.

The Home Rule Crisis

In the years that that immediately preceded the Rising, the Irish political climate changed profoundly, due in no small part to the militarisation of nationalist and unionist politics and the concurrent Home Rule crisis. Irish Home Rule, as Roy Jenkins has noted, 'was the dominating domestic political issue of the period'.[13] That it remained the central topic in British political discussion until the outbreak of the Great War in 1914, is attested to by the fact that it was one of the most recurrent topics of debate in *Punch*, the weekly magazine of political analysis and satire. Analysis by Joseph Finnan has shown that the magazine, which had a circulation of 100,000 and was one of Britain's 'national institutions', published a total of 77 full-page cartoons on Irish themes between January 1910 and July 1914. These accounted for 13.1% of all of the large cartoons that it featured between January 1910 and July 1914.[14] So what then accounted for the fascination with Irish politics? The answer can be found back in the 1870s, when the Irish Parliamentary Party began to escalate its campaign for the granting of Home Rule – that is, some degree of autonomy for Ireland from Britain. For an Empire that had territories scattered all over the world, a demand like this so close to home was a serious matter indeed.

Irish Home Rule was the subject of a lot of political debate in Westminster during the late nineteenth and early twentieth centuries. Despite previous setbacks, political efforts to grant some form of self-government to the island of Ireland appeared to be achieving tangible results with the introduction by the Prime Minister, Herbert Asquith, of the Third Home Rule Bill at the House of Commons on 11 April 1912. The Bill's main provision, as Kieran Rankin has noted, 'was a modest devolutionary measure for an Irish parliament that would still be subordinate to Westminster'.[15] The disabling of the House of

---

[13]   R. Jenkins, *Churchill* (Pan Books, London, 2002), p. 233.

[14]   J. P. Finnan, '*Punch's* Portrayal of Redmond, Carson and the Irish Question, 1910–18', *Irish Historical Studies*, Vol. 33, No. 132 (2003), pp. 424–25.

[15]   K. Rankin, 'The Search for "Statutory Ulster"', *History Ireland*, Vol. 17, No. 3 (2009), p. 28.

Lords veto under the Parliament Act the year beforehand, combined with the Liberal government's parliamentary dependency on Irish MPs, meant that the possibility of Home Rule being granted looked better than previous efforts in 1886 and 1893. However, this set alarm bells ringing back in the north of Ireland, where widespread opposition came from the Ulster Unionists, who believed that the establishment of a Dublin parliament would inevitably lead to 'Rome Rule'.

As Richard Toye has noted, unionists opposed 'Rome Rule' not only because they worried about 'subjugation by the Catholic majority in Ireland as a whole', but also because they believed that 'it might damage the prosperity of Ulster', which was Ireland's most prosperous and most industrialised province (largely as a result of the vibrancy of the linen and shipbuilding industries).[16] For all the economic arguments, however, it is clear, as Andrew Gailey has shown, that the burning issue to the majority of unionists 'was a struggle for mastery between two communities'. The threat of 'Rome Rule' aroused fears 'that a Catholic-dominated Ireland would endanger not simply the Protestant faith but also the Protestant way of life and the *predominance* of their values'. In such an emotive atmosphere, evangelicalism and fundamentalism surfaced amongst the Ulster Protestants. The Trinity College Dublin-educated barrister, Sir Edward Carson, who spoke with a 'rich' Dublin accent, 'was presented as "the saviour of his tribe": a political Second Coming in the manner of another outsider – William of Orange – in 1690'.[17]

Ultimately, the Home Rule crisis, according to Paul Bew, manifested itself through an increased 'intensity of communal feeling' throughout Ireland, with sharp divisions emerging between unionists and nationalists over both 'religious-ideological' and 'material-economic' issues.[18] The signing of the Solemn League and Covenant by 250,000 unionists at Belfast's City Hall on 28 September 1912, according to Gabriel Doherty, 'signalled a determination at all costs to maintain the Union intact'. There were several other reasons, he notes, as to why the campaign of Ulster resistance to Home Rule during the years 1912–1914 turned out to be 'increasingly bellicose in tone' and noticeably 'more bitter than those of the previous generation'. The Liberal party had been clipped of its unionist wing and so a recurrence of the internal splits of 1886 was not to be expected, while a repeat of the defeat of the Second Home Rule Bill of 1893 seemed unlikely as the veto of the House of Lords had been removed.

---

[16]   R. Toye, *Lloyd George & Churchill: Rivals for Greatness* (Pan Books, London, 2008), p. 113.

[17]   A. Gailey, 'King Carson: An Essay on the Invention of Leadership', *Irish Historical Studies*, Vol. 30, No. 117 (1996), pp. 67–68, 73–74.

[18]   P. Bew, *Ideology and the Irish Question: Ulster Unionism and Irish Nationalism 1912–1916* (Clarendon Press, Oxford, 2002), pp. 27, 34.

From a geographical and demographic perspective, unionism had also become more intense in the province of Ulster, especially in Counties Antrim, Down and Derry – which all recorded Protestant majorities in the 1911 Census.[19] Fears of Home Rule also led to the formation of the Ulster Volunteer Force by Carson and James Craig on 31 January 1913, whose numbers swelled to nearly 100,000 by the year's end. This organisation resolved to use 'all means which may be found necessary', including physical force, 'to defeat the present conspiracy' to implement the Home Rule Bill.[20] In spite of being plagued by a range of teething problems, including truancy, equipment deficiencies and uncertainty about the sort of engagements that it was being prepared for, it still had the capacity 'to impress journalists and even undercover RIC officers with spectacular public demonstrations'.[21]

At times, unionist resolve to resist Home Rule was also evident outside of Ulster, as demonstrated by an event at St. Stephen's Green in Dublin, where 39 bodies representing the Irish Unionist Alliance attended a meeting of the unionists of the three southern provinces on 28 November 1913. The meeting of unionist representatives was addressed by the leader of the Conservatives, Andrew Bonar Law, who reaffirmed his party's opposition 'as completely as ever to the whole idea of any separation of the United Kingdom'. The southern unionists also organised a major political demonstration at 8.00 pm that evening in the capital's Theatre Royal and the adjoining Winter Gardens. According to a report published in *Notes from Ireland*, Law (along with Carson), 'received a most enthusiastic and deafening welcome, the entire audience rising and cheering, and waving small Union Jacks, with which all the seats were provided'.[22] Unionist resolve was again demonstrated with the Larne gun-running on 24–25 April 1914, when the Ulster Volunteer Force imported 24,600 rifles and three million rounds of ammunition from Germany. When viewed within the context of a country already policed by thousands of armed members of the Royal Irish Constabulary and garrisoned by the British Army, Brendan O'Shea and Gerry White have noted that 'the emergence of the Ulster Volunteer Force was the first step on a perilous road to the militarisation of Irish society'.[23]

---

[19]   G. Doherty, 'Modern Ireland', in S. Duffy (ed.), *Atlas of Irish History*, 2nd Edition (Gill and Macmillan, Dublin, 2000), p. 110.

[20]   Cited in R. Dudley Edwards with B. Hourican, *An Atlas of Irish History*, 3rd Edition (Routledge, London, 2005), p. 54.

[21]   T. Bowman, 'The Ulster Volunteer Force and the Formation of the 36th (Ulster) Division', *Irish Historical Studies*, Vol. 32, No. 128 (2001), pp. 498–500.

[22]   *Notes from Ireland*, 1 December 1913.

[23]   B. O'Shea and G. White, 'The Road to Rebellion', *An Cosantóir: The Defence Forces Magazine*, Vol. 66, No. 3 (2006), p. 6.

Developments in Ulster were viewed with suspicion by Bulmer Hobson and other members of the Irish Republican Brotherhood's Supreme Council, an oath-bound organisation formed in 1858 and dedicated to the goal of creating an independent Irish Republic by the use of physical force. In seeking out a respectable figure to head up a new nationalist volunteer force, Hobson approached the University College Dublin academic, Eoin MacNeill, who subsequently agreed to act as leader. Soon afterwards, the forces of Irish nationalism were roused when a provisional committee of Irish Volunteers was established at a meeting at Wynn's Hotel at Lower Abbey Street in Dublin on 11 November 1913. The committee's manifesto was promulgated at a mass meeting held at the Rotunda Rink on the night of 25 November, at which the Irish Volunteers officially came into existence, with 3,000 men enlisting immediately. Its manifesto, reported *Irish Freedom* (a republican newspaper that first appeared in 1911), carried an appeal 'to secure and maintain the rights and liberties common to all the people of Ireland'. The duties of members, it added, 'will be defensive and protective, and they will not contemplate either aggression or domination'.[24] Membership numbers swelled in the months that followed, especially after news spread about the failure of the security forces to quell the events at Larne.

Another crucial dynamic that fed into Irish nationalism in the years leading up to the Rising was the increasing radicalisation that emanated from discontentment with the state of social and economic affairs. Many families in the slums of Dublin and other cities, for example, suffered from extreme poverty and high mortality rates. 'The non-industrial base of Dublin', as Roy Foster has noted, 'was one of the main reasons for the precarious and extremely impoverished condition of it proletariat', while many of those who lived in the 'warrens of indescribably squalid tenements ... bereft of water or sanitation', depended upon casual labour for work.[25] Dissatisfaction with working and living conditions, unprincipled employers and the pitiable state of workers' rights eventually gave rise to socialist and labour movements. Dublin's 1913 Lockout, a protest rally led by Jim Larkin (who had founded the Irish Transport and General Workers' Union in 1908), served as a graphic reminder of the tensions in the capital between unskilled casual workers and well-heeled moguls such as William Martin Murphy. Many of Dublin's disgruntled workmen joined up with the Irish Citizen Army, which was established in early November 1913 by James Connolly, three weeks prior to the formation of the Irish Volunteers. Connolly,

---

24   *Irish Freedom*, January 1914.
25   R. F. Foster, *Modern Ireland 1600–1972* (Allen Lane, London, 1988), pp. 436–37.

as Proinsias MacAonghusa has noted, was 'the most notable spokesman of the Irish working class' and 'lived for socialism'.[26]

In Desmond Ryan's view, Connolly was the labour movement's 'greatest brain', who was equally 'at home in a library or on a barricade' and who 'wanted nothing more in heaven and on earth than a social and political revolution for Ireland'.[27] In founding the Irish Citizen Army amidst the Lockout, Connolly was especially determined to protect the interests of organised labour by defending workers during scuffles with the Dublin Metropolitan Police. As one of Ireland's most illustrious writers and thinkers, Connolly had been instrumental in establishing the Irish Socialist Republican Party in 1896 and its mouthpiece, *The Workers' Republic*. One of his most significant books was *Labour in Irish History*, which was first published in 1910. According to Richard Killeen, it reinterpreted Ireland's past 'in Marxist terms, as a series of internal class struggles in which the economic interests of the weakest were serially betrayed by the strong' and also sought, 'to rescue an imagined communitarian Irish tradition' from a 'distant Celtic past'.[28]

In contrast to the forces of conservative nationalism and the Catholic middle-class orientation of the Irish Volunteers, Connolly's labour movement was more radical in spirit and also counted Countess Markievicz amongst its members. Its leadership before 1916, as Townshend has noted, was 'energetic, vocal and unambiguously revolutionary – whether in syndicalist or the Marxist mode', while Markievicz's 'colourful if quirky presence' gave the movement a 'plausibly feminist' touch.[29] Time and again, however, tensions surfaced between members of the Irish Citizen Army and the Irish Volunteers. As Ann Matthews has noted, 'the relationship between them was often, to say the least, fraught'. During the foundation of the Irish Volunteers at the Rotunda, for example, Victorian 'social mores' abounded and dictated that women could not join, while 'very serious class issues' also surfaced when members of the Irish Citizens Army entered the meeting and interrupted a speech by Laurence Kettle, on the grounds that his father, a farmer, had been involved in a labour dispute with his

---

[26] P. MacAonghusa (ed.), *What Connolly Said: James Connolly's Writings* (New Island Books, Dublin, 1995), pp. 10, 14.

[27] D. Ryan, *The Rising: The Complete Story of Easter Week* (Golden Eagle Books Ltd., Dublin, 1949), p. 11.

[28] R. Killeen, *A Short History of the Irish Revolution 1912 to 1927* (Gill and Macmillan, Dublin, 2007), p. 30.

[29] C. Townshend, 'Historiography: Telling the Irish Revolution', in J. Augusteijn (ed.), *The Irish Revolution, 1913–1923* (Palgrave, Basingstoke, 2002), p. 5.

workforce.[30] On the other hand, many of the Irish Volunteers had reservations too about the socialists. In a family memoir that was originally written in Irish, Frank Henderson, who hailed from Fairview in Dublin and rose to the rank of Commandant of the Second Battalion, recalled a period during the 1910s that he had spent 'prowling around a while in their midst' and hearing them 'spouting long extracts' derived from socialist texts. Finally, one of their members informed him that he could not be both a Catholic and a socialist 'at the same time'. From that time onwards, he recalled, 'I had nothing more to do with them', although he did hold dear to the view that the influence of capitalists over workers 'would have to be shaken'.[31]

The state of Irish politics was further altered in the years leading up to the Rising by the foundation of Na Fianna Éireann (the Irish National Boy Scouts) by Constance Markievicz and Bulmer Hobson in 1909. This pseudo-military youth organisation, which had a strong sense of discipline, made an important military and educational input into the movement for Irish independence.[32] Another important nationalist group was the female auxiliary of the Irish Volunteers, Cumann na mBan (the Irish Women's Council). Its inaugural public meeting was also held at Wynn's Hotel at South Brunswick Street in Dublin on 2 April 1914. As Cal McCarthy has noted, it went on 'to become one of the most uncompromisingly nationalist (and later republican) groups' and remained firm in its 'resolve to assist in securing Irish self-rule'.[33] Then in June 1914, following protracted negotiations, the Irish Parliamentary Party's John Redmond offered his support to the Irish Volunteers, while in the following month, Erskine Childers managed to import 900 Mauser rifles and 25,000 rounds of ammunition for the organisation at Howth. These were brought from Germany on his yacht, the Asgard, and subsequently distributed amongst followers throughout Ireland. Thereafter, the Irish Volunteers became increasingly widespread and numbers swelled to over 180,000. As *Éire-Ireland* reported, the movement made 'rapid headway in all parts of Ireland' in its quest 'to establish a national defence force'.[34]

---

[30] A. Matthews, 'Vanguard of the Revolution? The Irish Citizen Army, 1916', in R. O'Donnell (ed.), *The Impact of the 1916 Rising: Among the Nations* (Irish Academic Press, Dublin, 2007), p. 26.

[31] M. Hopkinson (ed.), *Frank Henderson's Easter Rising: Recollections of a Dublin Volunteer* (Cork University Press, Cork, 1998), p. 27.

[32] M. Hay, 'The Foundation and Development of Na Fianna Éireann, 1909–16', *Irish Historical Studies*, Vol. 36, No. 141 (2008), pp. 53, 71.

[33] C. McCarthy, *Cumann na mBan and the Irish Revolution* (The Collins Press, Cork, 2007), p. 16.

[34] *Éire-Ireland*, 26 October 1914.

In the lead-up to the outbreak of the Great War on 1 August 1914, both the unofficial armies of the Ulster Volunteer Force and the Irish Volunteers exercised in public in military formations bearing arms, with many Volunteers wearing their own uniforms. The Third Home Rule Bill was passed through the Westminster parliament in the summer of 1914 under Asquith's Liberal government, in spite of vehement opposition from outraged unionists, who had the backing of Bonar Law, the leader of the Conservatives. The Bill was signed into law by King George V on 18 September and placed on the statute books, while an accompanying Act suspended its functioning for one year, or until the end of the war, when it would be reviewed with a view to securing the general consent of Ireland and the United Kingdom. Two days later, Redmond, who was widely expected to be the first Prime Minister of the new Home Rule parliament, called upon members of the Irish Volunteers to enlist in the British Army at a speech in Woodenbridge, County Wicklow. He pledged his support for the British war effort and called upon members of the Irish Volunteers to defend Ireland's shores against potential aggression from Germany. The time had come for them, he said, to account for themselves as men 'not only in Ireland itself, but wherever the firing line extends, in defence of right, of freedom and religion in this war'.[35]

The organisation split following the Woodenbridge event, with the majority of members siding with Redmond's new organisation, the National Volunteers. On the other hand, about 13,000 of the 180,000 retained the Irish Volunteers title (and its Irish translation, Óglaigh na hÉireann) and refused to support the British war effort. The first convention of the downsized Irish Volunteers was held at the Abbey Theatre in Dublin on 25 October 1914, where Eoin MacNeill was elected Chairman and was received with large cheers by an enthusiastic crowd that included Pearse. A declaration of policy was passed unanimously to 'unite the people of Ireland on the basis of Irish nationality', to abolish 'the system of governing Ireland through Dublin Castle' and to resist any attempt at conscripting Irishmen into military service with the British Army.[36] Writing in *Éire-Ireland* (a daily bulletin costing one halfpenny and edited by Arthur Griffith) the next day, Pearse reaffirmed the objective of securing and guarding 'the rights and liberties of Ireland'. He also tried to put his own spin on matters by declaring that 'individual Volunteers would be untrue to their Volunteer pledge if they were to enlist for foreign service' because 'such an enlistment would, in fact, constitute desertion from the Irish army at a moment when the Irish army needs every man in Ireland'. He concluded by noting that any attempt

---

[35]   Redmond, cited in O'Shea and White, 'The Road to Rebellion', p. 8.
[36]   *Éire-Ireland*, 26 October 1914.

to end 'that traditional national attitude' of resisting 'duties or responsibilities within the British Empire' would subsequently lead to the abandonment of 'our national claims'.[37]

According to British intelligence reports, support for MacNeill's Irish Volunteers mainly came from militant members who were 'already connected with other disloyal and revolutionary societies'.[38] As the grouping did not possess any established recruiting offices, it had to rely on periodicals such as *The Irish Volunteer* to publicise its aims, while it also attracted recruits by taking to the streets with pipe bands on holidays such as Saint Patrick's Day. Once attested, members were required to pay a weekly subscription which enabled local units to rent offices and halls for both organisational and training purposes.[39] Much support for the downsized Irish Volunteers emanated from members of the Irish Republican Brotherhood, several of whom went on to assume prominent positions within its leadership.

The Great War: 'England's Difficulty … Ireland's Opportunity'

At a wider level, it is vital at this early juncture to seek to arrive at a better understanding of the 1916 Rising by situating it within the wider context of the Great War of 1914–1918. The conflict, as Niall Ferguson has suggested in his classic work, *The Pity of War*, was 'all horror and misery' and 'at once piteous, in the poet's sense, and "a pity"' – owing to the dreadfulness of the conditions soldiers had to put up with in the trenches and the huge death toll, which surpassed nine million overall.[40] After the heir to the Austrian-Hungarian Crown, Archduke Franz Ferdinand and his wife Sophie, were assassinated on 28 June 1914 during their state visit to Sarajevo by the Bosnian Serb nationalist, Gavrilo Princip, a series of events were set in motion which resulted in a colossal loss of human life and unprecedented levels of barbarity across the European continent. The war directly affected people in every part of Ireland and also impacted profoundly upon the lives of the Irish Diaspora living in Britain and other English-speaking countries such as the USA, Canada, Australia, and New Zealand. The outbreak of the conflict, after Germany declared war on Russia and France on 1 and 3 August 1914 respectively (with Britain declaring war on

---

[37]   Ibid.

[38]   Intelligence Notes, 1916, Chief Secretary's Office: Judicial Division, National Archives, Kew (hereafter NA, Kew), CO 903/19/2.

[39]   G. White and B. O'Shea, *Irish Volunteer Soldier 1913–23* (Osprey Publishing Ltd., Oxford, 2003), p. 10.

[40]   N. Ferguson, *The Pity of War* (Penguin Books, London, 1999), pp. xlii, 340, 343, 462.

Germany on 4 August, following its invasion of neutral Belgium), cast a huge shadow on the affairs of the world. As soon as the fighting commenced, people all around Europe braced themselves for the difficult times that lay ahead. 'The setback ... will be terrible', wrote a London antiques dealer on 12 September 1914, 'and goodness knows where and when it will end'.[41] Most people, however, were convinced that the fighting would be finished by Christmas.

Whilst 'England's Difficulty' came to be seen as 'Ireland's Opportunity' by militant nationalists, Home Rulers regarded the Great War as a chance to further the objectives of Irish constitutional nationalism. This in turn added a new dimension to the intense rivalries that existed at the time between the forces of unionism and nationalism. As matters transpired, thousands of men heeded Redmond's call and sided with the British war effort, in the belief that it would be over in a matter of months and that Home Rule would then be granted. However, as the months turned into years, Irish people of all religious affiliations increasingly resigned themselves to the fact that the conflict would run for much longer than originally anticipated. As the number of Irish casualties increased with the passage of time, the publication in newspapers of 'Rolls of Honour' listing those killed in action no doubt added to people's sorrows and woes. Adding to tensions at the end of 1914 were the regular sightings of German submarines off the Irish coast.[42] The mission of these submarines was to target unarmed merchant ships without warning, so as to disrupt Britain's economy, which had become heavily dependent on external food supplies since the late nineteenth century.[43] As the war progressed and efforts to secure victory over the Germans intensified, many locations throughout Ireland became profoundly militarised. As Philip Orr has noted, roads, railways and ports 'were busy with troop movements', newspapers were full of wartime stories, recruitment marches entered several towns and villages, and munitions factories hummed 'with wartime work'.[44]

On 7 May 1915, the brutality of war was demonstrated when a Cunard transatlantic liner, the Lusitania, was torpedoed around 10 miles off the Old Head of Kinsale. Of the 1,959 people on board, a total of 1,198 died (including many British and American citizens), while many of the civilians who survived

---

[41]   A. W. Bahr to John Quinn, 12 September 1914, New York Public Library, Manuscripts and Archives Division (hereafter NYPLMAD), John Quinn Papers.

[42]   See, for example, the evidence presented in Journal of Walter S. Burt, Midshipman, HMS Prince George, 9 December 1914, National Maritime Museum, JOD/201, who spotted a submarine near Queenstown.

[43]   C. Pointing, *World History: A New Perspective* (Pimlico, London, 2001), p. 353.

[44]   P. Orr, '"Across the Hawthorn Hedge the Noise of Bugles"', *History Ireland*, Vol. 17, No. 1 (2009), pp. 37–38.

were shown 'kindness ... one could never repay' when they were brought to
Queenstown and Kinsale in County Cork by rescue vessels.[45] This catastrophe
put extra pressure on the USA to enter the war, and also copper-fastened
anti-German sentiment throughout Britain and Ireland. In a letter that he
sent to a well-connected relation named Tadie at the end of May, 19-year-old
Irishman Toby Moore (who was studying in London and living in Kensington),
highlighted the peer pressure that was being exerted upon him to fight against
the Germans. In asking her to get him 'a commission in the Irish Brigade', he
noted that three-quarters of his peers 'have already gone' to serve in the Army,
adding that it was unpleasant 'to be asked why one hasn't joined, by everybody
one meets'.[46] Back in Ireland itself, the Lusitania tragedy served to deter many
people from travelling across the Atlantic by liner. This can be seen from the
case of individuals such as Margaret O'Donovan, an American citizen who had
moved to Ireland in 1913 in order to care for her elderly mother. A few years
afterwards, in an affidavit sworn at the American Consul in Queenstown in
order to secure a new passport (and to overcome presumption of expatriation),
O'Donovan explained that she had stayed on in Ireland for longer than
anticipated, because of her 'fear of submarines'. She also noted that some of her
friends had died in the Lusitania tragedy.[47] A similar story surrounded another
American citizen, Patrick O'Brien, who had moved to Ireland in 1913 in order
to recuperate from injuries that he had suffered in a car accident in Boston. In an
affidavit also sworn at the American Consul, O'Brien stated that he had stayed
put in Dungarvan in County Waterford due to 'fear of submarines and other
accidents which might occur on the sea'.[48]

Outrage against Germany's wartime actions remained high throughout
the years 1914–1918.[49] In the months following the sinking of the Lusitania,
thousands more Irishmen joined the British Army following an intensive bout of

---

[45] Winifred Hull to Mrs. Swanton, 25 May 1916, Cobh Heritage Centre.

[46] Toby Moore to Tadie, 30 May 1915, National Library of Ireland (hereafter NLI),
MS. 10566/4, Colonel Maurice Moore Papers.

[47] Department Passport Application, Affidavit Sworn at Queenstown by Margaret
O'Donovan of Boston, Mass., 17 June 1916, United States National Archives and Records
Administration, College Park (hereafter USNARACP), RG 84: 350: 21/1/3, Records of the
Foreign Service Posts of the Department of State, US Consular Records for Cork.

[48] Department Passport Application, Affidavit Sworn at Queenstown by Patrick O'Brien
of Boston, Mass., 21 August 1916, USNARACP, RG 84: 350: 21/1/3, Records of the Foreign
Service Posts of the Department of State, US Consular Records for Cork.

[49] See, for example, Elizabeth Dowden to Dr. Clara Barrus, 6 April 1918, NYPLMAD,
Clara Barrus Papers. Writing to a friend in New York almost three years after the sinking of the
Lusitania, Dowden forthrightly expressed her hope 'that the forces of evil – the Devil and his
Germans will not gain final victory over the Right'. As a lady who had 'immediate relations and

recruiting, during which colour posters laden with iconic images of the sinking of the liner encouraged Ireland's sons to avenge the incident. Recruitment bands also stirred up a sense of adventure and duty amongst elements of the populace. Despite major political divisions between unionists and nationalists, thousands of Irishmen of all political persuasions ended up enlisting in the fight against the Germans, thus swelling the ranks of regiments such as the Royal Inniskilling Fusiliers, the Royal Irish Rifles, the Royal Munster Fusiliers, the Connaught Rangers, the Royal Irish Regiment, the Royal Irish Fusiliers, the Leinster Regiment, and the Royal Dublin Fusiliers. Irish enlistment in the British Army during the course of the Great War was encouraged due to a number of reasons. For a long time, a deep-seated military custom had existed in Ireland, with the British Army offering an opportune conduit for young men who had an interest in soldiering as a career. For many sons of the landed gentry, whose families had a long tradition of serving in the military, it was a rallying to do one's duty for king, flag and Empire at a time of need. Such reasons were especially behind the decision of thousands of unionists, especially from Ulster, to enlist.

At a wider level, there was also a very strong belief held amongst many members of the British Army that they were fighting for the greater good of humankind. For example, the Scotsman, Major Arthur C. Murray, wrote from the Western Front that he was 'fighting for elementary principles on which are based not only the comity of nations but the whole social fabric of civilised mankind'.[50] Throughout the island of Ireland, catchment areas were spatially defined for local regiments, while advertisements aimed at recruiting Irishmen for the cause were frequently placed by the Department of Recruiting in pro-unionist newspapers such as *The Irish Times*, promising that those who enlisted would be 'fed, clothed and boarded', and proclaiming that all Irishmen 'should answer the Call – farmers' sons, merchants, men in shops and offices'.[51] To hammer the message home in the heady days of recruitment, long lists of those volunteering were published in house magazines, journals and trade papers.[52] For some individuals, signing up offered the chance to obtain a smart

friends in the fighting', she also let it be known that she was suffering 'a time of terrible anxiety' due to the severe risks they were encountering in battle.

[50]   Major Arthur C. Murray to Mr. MacDonald, 10 January 1917, National Library of Scotland (hereafter NLS), MS. 8805, Correspondence of Major Arthur C. Murray (Afterwards Viscount Elibank).

[51]   Scrapbook of Newscuttings, Royal Irish Academy, Dublin, MS. 23.P.27; *The Irish Times*, 30 October 1915.

[52]   P. Gough, 'Corporations and Commemoration: First World War Remembrance, Lloyds TSB and the National Memorial Arboretum' *International Journal of Heritage Studies*, Vol. 10, No. 5 (2004), p. 438.

uniform and to travel abroad for what may have initially seemed like a great adventure overseas. Upon reaching the frontlines, of course, the grim reality of staring death straight in the face soon became apparent.[53] In County Galway, people were motivated to enlist having heard about stories circulating in the press that Irish farms and bank accounts would fall into German hands if rumours of an invasion of Ireland came true. Economic factors also came into play, and many recruits were attracted by the chance of simply obtaining a wage. The dependants' Separation Allowance encouraged many men from the hard-up working class to enlist. In cities like Cork, where one in every seven families had a member fighting, songs like 'Salonika' demonstrated the cultural impact of the Great War. The song, according to Donal Ó Drisceoil, was 'written from the perspective of a "separa", a woman in receipt of the allowance'.[54] Collectively, they were known as Separation Allowance Women.

Nationalist Ireland, which provided the majority of the recruits, was particularly taken by the rallying call to defend France and small countries such as Belgium and Serbia against the forces of the Kaiser. While recruits from farms, villages, towns, and cities throughout Ireland came from every political and religious persuasion, the majority were working class Catholic nationalists. As Daniel Grace has shown, the town of Nenagh in County Tipperary supplied around 1,500 recruits, mainly to the Royal Irish Regiment. The woes of Belgium were reported upon extensively in the *Nenagh News* and collections for Belgian refugees were conducted throughout the locality. Furthermore, recruiting meetings were normally attended by a soldier from the Front, who would speak emotionally about the horrors inflicted by the Germans on the men and women of Belgium (scenes of which were depicted through magic lantern shows). One such meeting at Nenagh in November 1915 was attended by Captain Deane of the Duke of Cornwall's Light Infantry, who horrified his audience by telling them that he knew from experience that the German troops 'outrage the women, kill the children, put the innocent up against the wall and have them shot'.[55]

While those residing in Ireland during the years 1914–1918 escaped the mass destruction witnessed on the continent, many acts of fearlessness and bravery featured in the overall Irish contribution to the Great War. While serving in

---

[53]   What is particularly eye-catching, according to Bartlett, *Ireland: A History*, p. 383, are the 'numbers who continued to volunteer for military service long after the lengthy casualty lists and the ghastly realities of trench warfare had revealed that such an undertaking was exceptionally hazardous'.

[54]   D. Ó Drisceoil, 'Conflict and War, 1914–1923', in J. S. Crowley, R. J. N. Devoy, D. Linehan, and T. P. O'Flanagan (eds), *Atlas of Cork City* (Cork University Press, Cork, 2005), p. 258.

[55]   D. Grace, 'Soldiers from Nenagh and District in World War 1', *Tipperary Historical Journal*, No. 12 (1999), p. 45.

France with the Royal Munster Fusiliers, J. S. Meehan experienced the horrors and challenges of war – vermin in the trenches, fear, tiredness, dehydration, and a leg injury which led to the threat of his foot being amputated as gangrene set in. In April 1915, Meehan got lost in action in No Man's Land near Neuve Chapelle, which he recalled as follows:

> I was lost between the lines and lost all sense of direction as to the enemy lines and our own. Any way which I might travel between the lines in order to reach our own lines might, and again might not, be the proper way. The night fogs and morning mists obscured all sense of direction. I lay down in a shell hole and slept from sheer exhaustion. I had no rations or water with me. For two days and three nights I lay stupefied between the lines not knowing where lay our lines or where the enemy lines. All sense of direction had vanished. Fear gripped me and thirst tortured me. I licked everything that was damp, my tongue swelled in my mouth, and eventually I could not close my eyes ... I was rescued by a search party on the third night and brought back to my unit. Medical treatment was given [to] me and I was fit for duty in a couple of weeks.[56]

Advances in technology meant that troops in France had the added burden of coping with machinery that was capable of inflicting large scale devastation on opposing sides. 'This is a scientific war', wrote Lance-Corporal George Ramage of the 1st Battalion Gordon Highlanders, who added that troops were constantly preoccupied with the challenges posed by 'inoculation, aeroplane, chlorine, deadly accurate artillery, sandbags, machine guns, grenades, mining, periscopes, telephones, [and] wireless motors'.[57]

Several vivid accounts have survived of the unflinching courage and endurance of Irishmen involved in the war effort, or the so-called 'Great Adventure', as Captain Beater of the Royal Dublin Fusiliers once called it.[58] For many Irish soldiers, however, the journey from Ireland to the continent ended in disaster. Not long before the outbreak of the 1916 Rising, the death toll on

---

[56] War Services of J. S. Meehan, *1914–18*, National Army Museum, MS. 6307/86/3, Documents Relating to the Royal Munster Fusiliers. For a definition of No Man's Land, see M. Brown, *The Imperial War Museum Book of the First World War* (Pan Books, London, 2002), pp. 57, 59, who notes that it was 'the terrain between the front lines of opposing armies' and 'was almost always a place of horror and desolation'.

[57] War Diary of Lance-Corporal George Ramage, 1st Battalion Gordon Highlanders, 8th Brigade, 3rd Division, on the Western Front, 25 April 1915, NLS, MS. 944.

[58] War Diary of Captain O. L. Beater, 9th (Service) Battalion Royal Dublin Fusiliers (48th Brigade, 16th Division), 6 October 1915, Imperial War Museum (hereafter IWM), MS. 86/65/1, The Papers of Captain O. L. Beater.

the continent continued to mount by the day. Although the huge loss of life and carnage in the Western Front was widely castigated as time went on – by distinguished pacifists, by those writing for seditious publications in Ireland and by many conscientious objectors in Britain itself (including trade unionists and feminists) – there was still no shortage of newspapers that contained glowing reports about the Irish contribution to the war effort. One of these was *The Eye-Opener*. On 1 April 1916, almost 20 months after Britain first declared war on Germany, this newspaper went to great lengths to highlight the deeds of Irishmen 'on the battlefields the world over'. In doing so, it also pointed to the thousands of 'homes in desolation' across Ireland as 'permanent proof' of the fact that they 'were never cowards or slackers when fighting had to be done'.[59]

The Path to the Rising

Notwithstanding the resolve of Ulster Protestants to resist at all costs the severing of the connection to the Crown, the sequence of events that culminated in the 1916 Rising must be viewed against the fact that disillusionment with the nature of British power in Ireland was particularly pronounced amongst certain Catholic nationalists. Despite existing for over 115 years since the Act of Union, Townshend has contended that the British state 'never fully incorporated its Catholic Irish subjects', that its 'attitudes to Ireland were an odd mixture of bafflement, arrogance and ignorance' and that its rule was characterised 'by a deep-seated, pervasive prejudice'.[60] In the lead-up to the Rising, divisions between the British government and disillusioned elements of the Irish population were further exacerbated by a growing doubt over the issue of Home Rule for all of Ireland, uncertainty over whether or not conscription would be introduced and dissatisfaction with the mounting death toll on the continent. Frustration with the establishment was very much in evidence in republican publications such as *The Spark*. On 7 February 1915, for example, it reported that a Belgian refugee, 'hot from the seat of war', had 'declared how appalled he was on beholding the poverty and distress amongst the Donegal peasantry'. It suggested to its readers that it was therefore best for Ireland to concentrate instead on 'the work of reviving our own Gaelic culture'. Furthermore, it went so far as to recommend that Irish people should consent to letting 'the "civilised nations" kill and maim each other to their heart's content – we cannot alter the issue of the war by a hair's breadth'. 'The work which we owe to our own people

---

59   *The Eye-Opener*, 1 April 1916.
60   Townshend, *Easter 1916*, p. 26.

and country', it declared, 'is at our side; let us turn to it'.[61] The inaugural issue of *Fianna*, published in February 1915, hinted that patriotic sacrifices in the name of Irish freedom might be necessary. It called upon the youthful officers of Na Fianna Éireann to 'stand by your country and help her fight for her liberties'. 'You have a cause to fight for', it continued, urging members to 'be prepared to die for your cause, and keep before you always "Ireland first"'.[62]

Amidst a climate of mounting censorship at the beginning of 1915, it was reported in *Scissors and Paste* (a bi-weekly newspaper based at Middle Abbey Street in Dublin), that newspapers in Germany such as the *Lakalanzeiger* 'never tired of predicting that there will be an Irish rebellion either tomorrow or the day after'.[63] By the summer of 1915, as Joe Lee has noted, Redmond was 'losing his grip on nationalist Ireland as the war turned into carnage' and Home Rule 'receded into an uncertain future'.[64] It was around this time that Pearse's words and actions hinted more and more at the impending revolutionary turmoil. As Russell Rees has pointed out, Pearse's writings 'displayed a powerful Messianic strain', mixing religious testimony with prophecy about an impending Rising by depicting 'Ireland's struggle for redemption in terms of Christ's sacrifice at Calvary'.[65] *Irish Opinion* later characterised Pearse as 'a creature of infinite diversity', and noted how spoke 'an acquired Gaelic with such mastery that the racy older native-speakers' could be heard rejoicing 'over the rich new combinations he would suddenly fling out in a speech as his passion caught fire from an idea'.[66] A qualified barrister, Pearse served as editor of *An Claidheamh Soluis* (the newspaper of the Gaelic League) from 1903–1909. In 1910, he founded St. Enda's in Rathfarnham, an Irish language boarding school for boys. The school's mission, as stated in one of its prospectuses, was to train 'useful citizens for a free Ireland', and to kindle 'in their breasts something of the old spiritual and heroic enthusiasm of the Gael'.[67] In a letter to Seán T. O'Kelly,

---

[61]   *The Spark*, 7 February 1915.

[62]   *Fianna*, February 1915.

[63]   *Scissors and Paste*, 2 January 1915.

[64]   J. J. Lee, *Ireland 1912–1985: Politics and Society* (Cambridge University Press, Cambridge, 1989), p. 23.

[65]   R. Rees, *Ireland 1905–25: Volume 1, Text & Historiography* (Colourpoint Books, Newtownards, 1998), p. 199.

[66]   *Irish Opinion*, 1 July 1916.

[67]   St. Enda's College Prospectus, Allen Library (hereafter AL), Box 187, P. H. Pearse Letters, Newscuttings, etc.

Pearse let it be known that he took great personal pride in 'the great work for Irish education' which the school was attempting.[68]

To understand Pearse's militaristic mindset in the time leading up to Easter Week, and the respect that he gained from others in the long term, one must look at what inspired the politics of this complex character who eventually styled himself as 'Commandant General of the Army of the Irish Republic' and 'President of the Provisional Government'.[69] For some people like E. E. Speight, Pearse was simply the 'shining one'.[70] Others, such as Seán MacGiollarnáth (who also edited *An Claidheamh Soluis*), recalled that Pearse was a nationalist 'in the fullest sense' and 'was committed to a policy of fight'. Anybody 'who understands his character', he noted, 'will easily understand how he became leader of an insurrection'.[71] Much of Pearse's thinking, as Joost Augusteijn has noted, can 'be understood in the context of his time', as his ideas 'were primarily a personal interpretation of currents in Irish and international thought'. His opinions about using physical force for political ends, for example, partly reflected Western society's 'growing militarisation'.[72] Ideologically, Pearse's goal of independence for Ireland was strongly rooted in the discourse of Theobald Wolfe Tone, the father of Irish Republican thought in the late eighteenth century and a leading advocate for the rejection of an outside sovereign. One of his heroes was Cúchulainn, the youthful warrior of Irish sagas who figured so prominently in the plays of Yeats. Pearse also drew much inspiration from previous generations who had relied solely upon military means – in 1641, 1690, 1798, 1803, 1848, and 1867. He saw himself as belonging to a heroic generation of Gaels that had been brought up in an era marked by a renaissance in Irish culture, sport, theatre, and literature.[73]

---

[68]   P. H. Pearse to Seán T. O'Kelly, 24 August 1912, AL, Box 187, Folder 3, P. H. Pearse Letters, Newscuttings, etc.

[69]   General John Grenfell Maxwell to Lord Basil Blackwood, 11 May 1916, Bodleian Library, Oxford University (hereafter BLOU), MS. Asquith 43/28–33.

[70]   Poem by E. E. Speight, Entitled 'A Rann I Wrought' (in Memory of Patrick Pearse), Trinity College Dublin (hereafter TCD), MS. 4640/2381/3.

[71]   S. MacGiollarnáth, 'Patrick H. Pearse: A Sketch of His Life', *Journal of the Galway Archaeological and Historical Society*, Vol. 57 (2005), p. 150. This article was initially published as a pamphlet in 1916, under MacGiollarnáth's *nom-de-plume*, Coilin.

[72]   J. Augusteijn, *Patrick Pearse: The Making of a Revolutionary* (Palgrave Macmillan, Basingstoke, 2010), pp. 342–43.

[73]   Department of External Affairs, *Cuimhneachán 1916–1966: A Record of Ireland's Commemoration of the 1916 Rising* (An Roinn Gnóthaí, Dublin, 1966), p. 16. The introduction in 1908 of Irish history into the school curriculum also proved to be a significant cultural milestone. Thereafter, as noted by M. Bourke, *The Story of Irish Museums 1790–2000: Culture, Identity and*

This link was clearly evident from the powerful oration that Pearse famously delivered at Glasnevin Cemetery on 1 August 1915, at the funeral of the veteran Fenian, Jeremiah O'Donovan Rossa (who had died in the USA the previous June). At O'Donovan Rossa's graveside, Pearse (Plate 2.1) gave his endorsement of an Ireland 'not free merely, but Gaelic aswell; not Gaelic merely, but free as well'.[74] The funeral oration, which was drafted in his summer cottage in Rosmuc (located in the heart of the Connemara Gaeltacht), had a profound impact upon the making of Irish history. The Right Reverend Monsignor Michael Curran (who served at the time as Secretary to the Archbishop of Dublin, William J. Walsh, and was subsequently appointed as Rector of the Irish College in Rome), remarked at a later stage that the O'Donovan Rossa funeral represented 'the date that publicly revealed that a new political era had begun', one that served as 'the prelude to ... 1916'.[75] In an analysis of the inspiration that Pearse derived from the west, Pat Sheeran has argued that his retreats to Connemara were akin to an 'inward stroke', with the Gaeltacht representing a 'fountainhead of renewal' and the Irish language playing a 'vital role ... in halting the progress of cultural imperialism'.[76] In Seán Farrell Moran's view, the oration represented 'the zenith' of Pearse's 'political speaking career' and proved to be 'an immediate national sensation'. From thereon, Pearse became, in the public's mind, 'the official spokesman' for nationalists who advocated the tradition of physical force.[77]

Contemporary newspaper accounts highlight the immediate impact of the oration. Six days after it was delivered, *The Irish Volunteer* reported that those gathered around O'Donovan Rossa's grave had 'pledged to Ireland their love and to English rule in Ireland their hate'.[78] On 15 August 1915, *The Spark* was particularly forthright in its criticism of the Redmondites, labelling their catchcry of 'Trust England' as the most shambolic 'of all the great betrayals ever heard of on this earth'.[79] In the months that followed, this attitude became more commonplace in different parts of Ireland.

Education (Cork University Press, Cork, 2011), p. 328, teachers fostered 'an awareness in their pupils of "Irishness", as distinct from "Englishness"'.

[74] P. H. Pearse's Oration at the Graveside of O'Donovan Rossa, Glasnevin Cemetery, August 1915, Pearse Museum; Commemoration Card for Thomas J. Clarke, P. H. Pearse and Thomas MacDonagh, with Quotations from Each, TCD, MS. 4649/4825.

[75] Right Reverend Monsignor Michael Curran, 17 June 1952, National Archives of Ireland (hereafter NAI), Bureau of Military History, Witness Statement No. 687.

[76] P. F. Sheeran, 'The Absence of Galway City from the Literature of the Revival', in D. Ó Cearbhaill (ed.), *Galway: Town & Gown 1484–1984* (Gill and Macmillan, Dublin, 1984), p. 241.

[77] S. Farrell Moran, *Patrick Pearse and the Politics of Redemption: The Mind of the Easter Rising, 1916* (The Catholic University of America Press, Washington, 1997), pp. 146–47.

[78] *The Irish Volunteer*, 7 August 1915.

[79] *The Spark*, 15 August 1915.

Plate 2.1   Patrick Pearse (to the right of centre, in uniform with a white lapel badge and cap in his hand) at the funeral of Jeremiah O'Donovan Rossa, Glasnevin Cemetery, 1 August 1915

Source: National Library of Ireland, Keogh Brothers Ltd. Collection, KE 234.

At the outset of 1916, the Irish Volunteers continued drilling in Ireland, much to the dismay of the intelligence agents working for the Royal Irish Constabulary's Crime Special Branch and the detectives in the Dublin Metropolitan Police's G Division. Under the direction of Major Price, Dublin Castle's intelligence service also operated a system of postal censorship in Ireland, which entailed the restriction and confiscation of seditious mail that was being sent by radical nationalists. As matters transpired, all those who were destined to lead the Rising, with the noteworthy exception of Pearse, had his or her mail censored at some stage in the first 21 months of the Great War. However, Price only had a small team of assistants, and pleas to the War Office in London for more resources often fell on deaf ears. In the end, some of the priceless intelligence that was so earnestly garnered by the Irish postal censors was never fully exploited.[80]

---

80   B. Novick, 'Postal Censorship in Ireland, 1914–16', *Irish Historical Studies*, Vol. 31, No. 123 (1999), pp. 343, 346, 351, 356.

It is clear, therefore, that besides the failure to curtail the smuggling of arms and ammunition, there were other major shortcomings in British policy towards Ireland in the years prior to the Rising. One of the most serious of these, as W. Alison Phillips has demonstrated, was the failure to curtail the influx of 'an astonishing mass of seditious literature'. Rather ironically, the escalation of anti-British sentiment was facilitated to a certain extent by the censorship regime's own procedures, insofar as it 'took notice only of matter which was judged to be of military importance'. Consequently, articles ridiculing the British or boldly praising the Germans' 'gallantry and humanity', curiously went on to gain 'an enhanced authority as having been "passed by censor"'. The censor was also fighting a losing battle in other respects. Despite the suppression of seditious newspapers such as *Scissors and Paste* in March 1915, it was not long before others stepped into the fold throughout the rest of the year and into the early months of 1916 (including, for example, Connolly's *Workers' Republic* and Arthur Griffith's *Nationality*, which first appeared on 26 June 1915). Even though most of these newspapers did not enjoy high circulation figures and were printed on bad quality paper, the fact that they were 'passed from hand to hand' meant that 'their cumulative effect upon the ... masses was necessarily great'. Increasingly, therefore, it was 'becoming clear to all those who could read the somewhat obvious signs of the times that matters were coming to a crisis'.[81]

Opposition to the British Army's recruitment campaign was evident in the months leading up to the outbreak of the Rising. At the same time, the column inches of publications like *The Gael* reaffirmed the arguments of Pearse. On 29 January 1916, one of its contributors emphasised the need to adhere to 'unchanging principles of Irish patriotism' which stressed the need for protecting the Irish language, along with 'literature, political faith and the social feelings of the Gael'. 'No nation', added the contributor, 'has a history, a genius, or virtues or failings like our own, and it is with careful consideration of these facts that the future Ireland must be developed'.[82] The expression of sentiments like this made things increasingly difficult for the authorities. One graphic indication of the hostility that some British troops were encountering in Ireland during the early months of 1916 can be seen from a letter written by Captain Leslie Horridge to his mother in Earl's Court, London. While based in the Curragh, County Kildare, Horridge observed on 12 February that it was

---

[81] W. Alison Phillips, *The Revolution in Ireland 1906–1923*, 2nd Edition (Longmans, Green & Co. Ltd., London, 1926), pp. 88–90, 95.

[82] *The Gael*, 29 January 1916.

'quite easy to see that the Irish are not very friendly ... and it is quite likely that something will be thrown at you in the villages if you are in public'.[83]

On 26 February, a contributor to *The Gael* expressed her distaste for the recruitment posters that decorated the walls of the Dublin quays by declaring: 'England may "get" the soldiers of Ireland in way she does not anticipate.' 'She may find them', she added, 'lined up not to fight for her, but against her, and she may find them rather good at driving her from the land, as her German cousins are rather good at driving her from the sea'.[84] *An Claideamh Soluis* conveyed equally-strong views about the possibility of revolutionary action, declaring in no uncertain terms on 11 March that the Gaelic League programme of 'the Irishing of Ireland' warranted radical measures in order to achieve 'the setting free of those forces now circumscribed or misdirected by the necessities imposed by foreign rule, for the regeneration of an Irish nation on Irish lines'.[85] Five days after this conspicuous call to arms, the intelligence services gathered some more worrying information. An agent known as 'Chalk' reported that the young men of the Irish Volunteers were 'very anxious to start "business" at once, and ... they are being backed up strongly by Connolly and the Citizen Army, and things look as if they are coming to a crisis, as each man has been served out with a package of lint and surgical dressing' along with 'a tin of food similar to that issued to soldiers'.[86] In response, some British soldiers were put on a rota at various bases, 'to be ready to turn out at a moment's notice'.[87]

Even though the authorities remained alert to the dangers of Irish sedition, there was no respite in the antagonism faced by the British Army and the Redmondites. A graphic example can be seen from John J. Scollan's lament to 'Ireland's Fighting Men', published in *The Hibernian* on 25 March:

> in these decadent days of spurious Catholicity and Nationality ... the young men of Ireland are asked by political tricksters to give up their homes, their wives, their children, their sweethearts, parents, brothers, sisters, all that is worth living for – aye, their very lives! What fools some of us Irish are! Half the Irishmen who were murdered at Gallipoli, Flanders or France would have been quite sufficient to have

---

[83]   Captain J. L. Horridge to his Mother, 12 February 1916, IWM, PP/MCR/235, The Papers of Captain J. L. Horridge.

[84]   *The Gael*, 26 February 1916.

[85]   *An Claideamh Soluis*, 11 March 1916.

[86]   'Chalk' to the Detective Department, 16 March 1916, NA, Kew, CO 904/23/3.

[87]   Memories of Service in Ireland with the 6th Reserve Regiment of Cavalry in 1916, IWM, MS. 75/92/1, The Papers of A. C. Hannat.

driven the enemy bag and baggage from this country years ago – and they would not have died in vain![88]

Despite the mounting tensions, British intelligence officers such as 'Granite' were still of the opinion at the end of March 1916 that 'there is at present no fear of any Rising by the Volunteers – standing alone they are not prepared for any encounter with the forces of the Crown, and the majority of them are practically untrained'.[89]

Behind the scenes and in public, however, emotions and tensions were coming to the boil. On 1 April, in an opinion piece in *Honesty: An Outspoken Scrap of Paper*, entitled 'Among Fleas and Soreheads', Editor Gilbert Galbraith poured scorn on both the British Army's recruitment campaign and the tax burden that the war effort had imposed upon the Irish population. He also denounced certain segments of Ireland's 'reptile' press, charging that it was 'subsidised by the British government under the guise of payments for recruiting advertisements'. England's war, Galbraith alleged, was fuelled by a desire to destroy 'Germany's supremacy as a trade rival' and Irish lives were being wasted by virtue of an 'enormous blood tax ... of further fodder for German shot and shell'. 'Ireland's surest hope', he argued, 'lies in the Irish Volunteer Force ... that has stood, and is standing, between Ireland and conscription'. He further proclaimed: 'Every man and woman knows that the Irish Volunteers are no toy soldiers and that when the occasion arises they will prove their worth to the satisfaction of the statesman who once referred to Ireland as "a country of whipped curs."'[90]

On 8 April, *Honesty* reported that a raid (sanctioned by a military warrant) had taken place eight days previously on the headquarters of the Gaelic Press at Upper Liffey Street. In its assessment of the situation, the paper proceeded to ridicule the British intelligence service and warned of reprisals. 'Dublin Castle', it noted, 'may maintain an extensive spy system of self-fancied efficiency ... backed by ... bullets, bayonets ... and batons ..., but the forces standing for truth, justice and liberty in Ireland today ... were never in a better position than the present hour ... to return "like for like".'[91] In spite of continuing attempts by the authorities to restrain seditious discourse by curtailing the freedom of the press in the weeks before Easter Monday, many more commentaries of disaffection continued to appear in print, openly ridiculing British attempts at curtailing disloyalty to the Crown. Connolly's *The Workers' Republic* reported how

---

[88]   *The Hibernian*, 25 March 1916.
[89]   'Granite' to the Detective Department, 27 March 1916, NA, Kew, CO 904/23/3.
[90]   *Honesty: An Outspoken Scrap of Paper*, 1 April 1916.
[91]   Ibid., 8 April 1916.

'men and women' had been moved 'to tears of joy and thanksgiving' on 16 April, after the Irish Citizen Army 'hoisted and unfurled the Green Flag of Ireland, emblazoned with the Harp without the Crown' on the top of the headquarters of the Irish Transport Workers' Union at Liberty Hall. The flag, it noted, represented 'the sacred emblem of Ireland's unconquered soul', and people were prepared to sacrifice their lives, if necessary, in order to keep it flying.[92] As the month of April progressed, the atmosphere in the city became increasingly strained as shopkeepers were warned of selling nationalist newspapers that were likely to cause disaffection, contrary to the Defence of the Realm regulations. On 22 April, however, *Honesty* lambasted the measure as overly draconian, calling it an 'insidious effort' aimed at curtailing the newsagents' 'liberty of action'.[93] John J. Scollan, writing in *The Hibernian* on the same day, openly criticised the Catholic Justice Kenny and charged him with relying too heavily on police reports about 'goings-on of a seditious character' and with inciting servants of the Crown 'to commit breaches of the peace'.[94]

Despite the attempted suppression of a number of nationalist newspapers by the military authorities, the findings of a Royal Commission on the Rebellion in Ireland was to conclude at a later stage (after examining 29 witnesses during five meetings in London and four in Dublin) that the years leading up to the Rising had proved to be a time during which 'lawlessness was allowed to grow up unchecked'. This, it added, was largely as a result of a policy in which Ireland had 'been administered on the principal that it was safer and more expedient to leave law in abeyance if collision with any faction of the Irish people could thereby be avoided'. 'Such a policy', it contended, was 'the negation of that cardinal rule ... which demands that the enforcement of law and the preservation of order should always be independent of political expediency'. Besides shortcomings in the government's 'reluctance ... to repress by prosecution written and spoken seditious utterances', the Royal Commission also blamed the outbreak of the Rising on 'the importation of large quantities of arms into Ireland' and a subsequent disinclination 'to suppress the drilling and manoeuvring of armed forces ... who were openly declaring their hostility to Your Majesty's Government and their readiness to welcome and assist Your Majesty's enemies'.[95]

---

92   *The Workers' Republic*, 22 April 1916.
93   *Honesty: An Outspoken Scrap of Paper*, 22 April 1916.
94   *The Hibernian*, 22 April 1916.
95   Royal Commission on the Rebellion in Ireland, *Report of the Commission: Presented to Both Houses of Parliament by Command of His Majesty* (His Majesty's Stationery Office, London, 1916), pp. 3, 12–13.

The Rising in Dublin

On Easter Monday 24 April 1916, matters eventually came to a head with
the outbreak of the Rising in Dublin city, which lasted for six days. The first
day witnessed weather that 'was unseasonably warm' and many people took
advantage of the Bank Holiday to venture to the horse races and festivity at
Fairyhouse.[96] In O'Connell Street itself, the scene was one of relative peace and
calm, as 'tramcars rattled along on their well-worn slender metal strips', horses
'pulled a variety of carts', women in shawls 'ambled here and there with a sense
of purpose', 'flat-capped men ... engaged in aimless corner-boy conversation', and
bowler-hatted men walked leisurely 'with long-skirted ladies and discussed more
respectable things'.[97] That morning, unaware of what was about to happen in the
city centre, the British garrison in the capital totalled more than 2,400 officers
and men, deployed at Marlborough Barracks, Royal Barracks, Richmond
Barracks, and Portobello Barracks.[98] As Pearse departed from the grounds of
St. Enda's with his brother, Ruth Dudley Edwards has observed that he 'had
paid his usual attention to costume' in what proved to be 'his last dramatic
performance'. In contrast to many of his fellow rebels who 'were carelessly
turned-out', Pearse carried with him 'a repeating pistol, ammunition pouch and
canteen, and wore his smart green Volunteer uniform with a matching slouch hat
and a sword'.[99] The core members of the newly-formed provisional government
then met up at Liberty Hall in the city centre, from where they soon executed
their plan of attack.

The streets of the capital were for the most part deserted when Pearse,
along with Connolly, was driven from Liberty Hall to the GPO in an open
touring car driven by The O'Rahilly.[100] Approximately 1,000 men of the Irish
Volunteers and the Irish Citizen Army, under the cover of a customary parade,
seized key strategic points across the city – including the GPO (commanded by
Patrick Pearse) on O'Connell Street (then known as Sackville Street), Jacob's
Factory (commanded by Thomas MacDonagh), the Four Courts (commanded
by Edward Daly), the South Dublin Workhouse (commanded by Eamonn
Ceannt), and Boland's Mill (commanded by Eamon de Valera). Many of the
buildings occupied by the rebels were elevated, thus proving particularly useful

---

[96]   M. McNally, *Easter Rising 1916: Birth of the Irish Republic* (Osprey Publishing Ltd.,
Oxford, 2007), p. 37.

[97]   T. Reilly, *Joe Stanley: Printer to the Rising* (Brandon, Dingle, 2005), p. 9.

[98]   McNally, *Easter Rising 1916*, p. 37.

[99]   R. Dudley Edwards, *Patrick Pearse: The Triumph of Failure* (Irish Academic Press, Dublin,
2006), p. 275.

[100]   A. Ryan, *Witnesses: Inside the Easter Rising* (Liberties Press, Dublin, 2005), p. 153.

for snipers. However, they failed to take other key sites such as the Magazine Fort in Phoenix Park, Dublin Castle and Trinity College Dublin.[101] In the latter institution, for example, immediate steps were taken to secure the grounds as soon as word was heard that a Rising was taking place. The Lincoln Place Gate was locked and sentries were posted there, while the doorway leading to College Park and the side door leading to Westland Row were also shut.[102]

One of the few large open-air public spaces that the rebels did manage to take was St. Stephen's Green. Under the command of Michael Mallin they dug trenches and constructed barricades, but they retreated to the Royal College of Surgeons the following day after coming under machine gun fire from the tops of the Shelbourne Hotel and the United Service Club. Many years later, when writing *The Easter Rebellion*, historian Max Clifford had the advantage of being able to interview Romance languages lecturer, Professor Liam Ó Briain, who along with Harry Nicholls, had been one of the rebels involved in digging the trenches at St. Stephen's Green on Easter Monday. Commandant Michael Mallin, recalled Ó Briain, ordered Nicholls to dig a trench at the corner of the green that faced Cuffe Street. Ó Briain himself was handed a pick by Lieutenant Bob de Coeur and instructed to dig. He recollected that it was the first time in his life that he had done 'a decent day's work', but it irritated him 'to see Citizen Army fellows, most of whom were accustomed to using their hands and muscles in their various trades', more keen to do sentry duty than dig for Ireland. While embroiled in digging, Ó Briain recalled that he was distracted by a 'beautiful' woman passing by, with a 'deep, Protestant look about her'. Lieutenant de Coeur, however, interrupted the moment by shouting: 'Get back to your digging, man – that trench needs to be wider. Sure yer giving yourself no arse room!'[103]

Throughout Easter Week, the main plan of the rebels, as Shane Hegarty and Fintan O'Toole have observed, was 'to establish strongholds in key areas of the city which would allow them to control or cut off communications with the rest of the country'. Due to shortcomings in their planning and reduced numbers,

---

[101]   The College of Surgeons and St. Stephen's Green: Description of the Fighting, Pamphlet, Undated, Dublin City Library and Archive (hereafter DCLA), Birth of the Republic, BOR 25/10; P. Cottrell, *The Anglo-Irish War: The Troubles of 1913–1922* (Osprey Publishing Ltd., Oxford, 2006), pp. 33–35.

[102]   Report by Corporal Mein, Dublin University Officer Training Corps on the Sinn Féin Rebellion, 24 April 1916 and the Defence of Trinity College, TCD, MS. 4873.

[103]   M. Caulfield, *The Easter Rebellion*, New Edition (Gill and Macmillan, Dublin, 1995), p. 96. Further archaeological details about the dugouts that were made in St. Stephen's Green on Easter Monday can be found in M. L. Hamilton Norway, *The Sinn Féin Rebellion As I Saw It* (Smith, Elder & Co., London, 1916), p. 11, who notes that 'just inside the railings among the shrubberies, the rebels had dug deep pits or holes, and in every hole were three men'.

however, 'their success was patchy'.[104] This was especially so on the outskirts of Dublin city, where little could be done to prevent the incursion of British troops or the arrival of a gun ship up the River Liffey. Owing to the confusion and loss of support that resulted from Eoin MacNeill's countermanding order against mobilisation on Easter Sunday, G. A. Hayes-McCoy has speculated that the leaders of the Rising 'may originally have intended a more aggressive action', but that one 'can never be quite certain what was in their minds'. An even greater mystery, he adds, is how the seven leaders 'could have come to count ... on the ultimate support of men whom they had either kept in the dark as to their intentions or had deliberately misled'.[105]

Pearse was convinced that he was doing the right thing for the greater good of Ireland in the long term. As the Commander in Chief of rebel forces, he based his Easter Monday actions on the age-old dictum that 'England's difficulty' was 'Ireland's opportunity'. In the heat of the field of battle, however, it soon became clear that the rebels faced impossible odds against the military might of the British Army (whose numbers in the capital rose to over 16,000 by the end of the week of fighting) and that military defeat was inevitable. Connolly, according to C. Desmond Greaves, was well aware of the impossibility of succeeding as a result of the countermanding order. Above all, however, he wanted to proceed so as to ensure that the Rising did not 'die whimpering amid arrests, ridicule and recrimination'.[106] He was by no means alone when it came to the decision to strike a blow for independence, despite the low chances of defeating the British. As Brian Barton has argued, the seven leaders of the Rising 'were all suffused with a boundless energy, borne of a passionate desire to achieve the goal of Irish independence, and were unshakable in their conviction that it could only be achieved by force'. Accomplishing this objective, he adds, 'had come to give meaning to their lives'.[107]

During the very early stages of the Rising, resistance from the authorities was less than forthcoming, but it was not long before the situation changed. After the rebel forces first marched into Dublin's main thoroughfare, they easily took control of the GPO shortly before noon, without a shot being

---

[104]　S. Hegarty and F. O'Toole, *The Irish Times Book of the 1916 Rising* (Gill and Macmillan, Dublin, 2006), p. 42.

[105]　G. A. Hayes-McCoy, 'A Military History of the 1916 Rising', in K. B. Nowlan (ed.), *The Making of 1916: Studies in the History of the Rising* (The Stationery Office, Dublin, 1969), pp. 256–57.

[106]　C. Desmond Greaves, *The Life and Times of James Connolly*, New Edition (Lawrence and Wishart Ltd., London, 1986), p. 410.

[107]　B. Barton, *From Behind a Closed Door: Secret Court Martial Records of the 1916 Easter Rising* (The Blackstaff Press, Belfast, 2002), p. 3.

fired.[108] From the entrance of the GPO headquarters, and to the bemusement of astonished onlookers, Pearse read the Proclamation just after noon on behalf of the provisional government of the Irish Republic to the people of Ireland. In drawing deeply upon 'the history of Irish resistance to British rule', Townshend has noted that Pearse drew upon 'an impressive genealogy of revolt', yet it only stretched back to the sixteenth century, in contrast to the '"eight centuries" of resistance' invoked by other nationalists.[109] Copies of the Proclamation, which were displayed at prominent locations throughout the streets of Dublin, outlined Ireland's history as the rebels saw it – a nation's struggle for independence from British rule and the rejection of constitutional means in favour of armed revolt. The Proclamation, which was addressed to 'Irishmen and Irishwomen', declared the rights of the Irish people to be sovereign, affirmed a free Irish Republic where all individuals could be free to accomplish their potential, and aspired to the formation of a native government elected on the democratic principles of self-determination and government by consensus: 'In the name of God and of the dead generations from which she receives her old tradition of nationhood, Ireland, through us, summons her children to her flag and strikes her freedom.' 'We declare', it added, 'the right of the people of Ireland to the ownership of Ireland, and to the unfettered control of Irish destinies, to be sovereign and indefeasible'.[110] The rebels visualised a new democratic system completely independent from Britain, one which, according to the Proclamation, would guarantee 'religious and civil liberty, equal rights and equal opportunities to all its citizens'. In addition to Pearse, who was the author of the Proclamation, the other six signatories were James Connolly, Thomas MacDonagh, Seán MacDiarmada, Thomas Clarke, Eamonn Ceannt, and Joseph Plunkett.

The poet, Desmond Fitzgerald, who was part of the GPO garrison, remarked that Pearse 'looked rather graver than usual' on Easter Monday. He sensed that even though Pearse had displayed some degree of ecstasy, 'there was also a heavy

---

[108] The decision to occupy the GPO rather than Dublin Castle should not be underestimated, as the latter was fortified. Also worthy of consideration, as Duncan Campbell-Smith has shown, is the fact that the GPO stood as a significant emblem of British power. See D. Campbell-Smith, *Masters of the Post: The Authorised History of the Royal Mail* (Allen Lane, London, 2011), p. 259, in which he points out that the GPO's 'grandeur made it a symbol of British rule in Ireland second only to Dublin Castle itself'. The fact that the rebels did not take immediate control of the Telegraph Office in the upstairs of the building suggests that the GPO was deliberately 'targeted more for its symbolic value than for its practical importance'.

[109] Townshend, *Easter 1916*, p. 1.

[110] For the full text of the Proclamation, see Department of External Affairs, *Cuimhneachán 1916–1966*, p. 12.

sense of responsibility' for what was going on.[111] Pearse, of course, realised that his actions would not win immediate approval from the public. One onlooker, who listened as he read the words of the Proclamation, later stated that he had felt sorry for him, as the response from the gathered crowd had been quite nerve-jangling. Nonetheless, as Max Caulfield has demonstrated, a small number of 'ragged cheers hung in the air' after Pearse had finished his reading. However, there 'were no wild hurrahs ... no scenes reminiscent of the excitement which had gripped the French mob before they stormed the Bastille'. The crowd that listened to Pearse simply 'shrugged their shoulders, or sniggered a little, and then glanced round to see if the police were coming'. Then, as the crowd thickened and confusion took hold on O'Connell Street early in the afternoon, a number of poor 'shawlie' women (clad in black shawls) from tenement dwellings began to verbally abuse the rebels ensconced inside the GPO, 'who answered them in kind'.[112] It was not long before news of the occupation of the GPO spread around the city centre and panic set in. As he arrived into the city centre from the south suburbs around noon, Eamonn Bulfin of the Rathfarnham Company Irish Volunteers noted that the tram on which he had travelled did not proceed any further than the junction of Dame Street and George's Street. 'There was terrible excitement', he noted, 'and a great deal of rushing and scurrying about ... The tram driver and conductor simply abandoned ship and fled'.[113]

With a few exceptions, many of the names of the leaders of the 1916 Rising were not widely known to the general public until their names appeared in the newspapers after their executions. Their militant actions also came as a surprise to many contemporary journalists. Pearse, according to *The Leader*, 'was one of the last men you would have thought of as a soldier ... He was a poet and a dreamer, and a man who tackled big projects with a light heart'. He was also 'earnest to solemnity', it continued, 'and adopted rather the chanting style of eloquence'.[114] According to his fellow rebel, Desmond Ryan, Pearse 'had the power of the enkindling word'. Rather crucially, he also possessed the ability to 'persuade, convince, inspire', especially when it came to influencing the rank-and-file members of the Irish Volunteers.[115] Although harsh criticisms of Pearse's bloodlusting tendencies were made by numerous offended contemporaries, such views were vehemently challenged by his supporters, including his mother, Margaret, who took pride in the fact that 'the war for Ireland's freedom was

---

[111] D. Fitzgerald, *Desmond's Rising: Memoirs 1913 to Easter 1916* (Liberties Press, Dublin, 2006), p. 133.

[112] Caulfield, *The Easter Rebellion*, pp. 73–74.

[113] Eamonn Bulfin, Undated, NAI, Bureau of Military History, Witness Statement No. 497.

[114] *The Leader*, 20 May 1916.

[115] Ryan, *The Rising*, p. 11.

the most holy of wars'. 'My good woman', she once told a critic of the Irish Volunteers, 'Don't argue with me about ambushes. Why you will find ambushes in the Bible'.[116] However, Pearse himself was by no means immune from criticism by his fellow rebels, including one of the Rising's leaders. For example, Edward Daly, who was eventually executed for his part in the Rising, said at his trial that when he had heard of plans for the fighting, he held a meeting with his officers and formed the opinion 'that the whole thing was foolish but that being orders we had no option but to obey'.[117]

Unsurprisingly, given the secrecy that had surrounded its planning and preparation, the timing of the Rising left the vast majority of Dublin city's inhabitants rather dumbfounded. James Stephens, who kept a daily journal during the fighting, vividly captured the sense of shock felt by Dubliners as their city was shook to its very core on Easter Monday: 'This has taken everyone by surprise ... today, our peaceful city is no longer peaceful; guns are sounding, or rolling and crackling from different directions, and, although rarely, the rattle of machine guns can be heard also.' 'Around me', he added, 'as I walked the rumour of war and death was in the air'.[118] The Gaelic League's former President, Douglas Hyde, was also taken aback by the events that transpired around him, whilst on a bicycle trip through the city centre on Easter Monday. As he turned out of Dawson Street, he heard what he believed was 'the tyre of a motor burst loudly in front of the Shelbourne Hotel, and then another burst and then another'. What sounded like a 'great mortality among tyres' was in fact the Rising in full swing. How extraordinary, he felt, 'that a couple of thousand of men armed with rifles only should venture to try and hold up an Empire'.[119] 'There has been nothing in Dublin like this in our generation', wrote a bewildered Alfred Fannin the next day.[120] One lady, Nelly O'Brien, thought that the initial turmoil in the city may have been 'in the nature of a demonstration against conscription as it had been announced that the Volunteers would resist disarmament'.[121] For Officer A. A. Luce of the Royal Irish Rifles, who had returned to Dublin on

[116] Typescript Obituary Notice of Mrs. Margaret Pearse, NLI, MS. 21092, Pearse Papers.

[117] Proceedings of the Trial of Edward Daly, 4 May 1916, NA, Kew, WO 71/344.

[118] J. Stephens, *The Insurrection in Dublin* (Colin Smythe Ltd., Gerrard's Cross, 2000), pp. 1, 14. This contemporary account was first published in October 1916.

[119] Diary of Easter Week Dictated by Douglas Hyde and Typed by Nelly O'Brien, TCD, MS. 10343/7.

[120] A. Warwick-Haller and S. Warwick-Haller (eds), *Letters from Dublin, Easter 1916: Alfred Fannin's Diary of the Rising* (Irish Academic Press, Dublin, 1995), p. 19.

[121] E. L. (Nelly) O'Brien's Account of Her Experiences during Easter Week, TCD, MS. 10343/1.

leave from the front in France, 'news of the Rising ... was a bolt from the blue ... to law-abiding citizens'.[122]

As the fighting escalated, various claims and counter-claims were made by each of the sides involved. Orders were promptly issued by the British authorities on Easter Monday, warning 'all citizens of the danger of unnecessarily frequenting the streets or public places' where His Majesty's forces were engaged in the suppression of unrest.[123] The first and only issue of the republican *Irish War News*, issued from the GPO the next day, reported on the 'heavy and continuous fighting' in the first 24 hours of the insurrection. Designed to raise the morale of the insurgents, it contained a propagandist assertion from Pearse that Dublin's population 'are plainly with the Republic, and the officers and men are everywhere cheered as they march through the streets'.[124] By contrast, Tuesday's *The Irish Times* was vehemently opposed to the Rising, and lamented how it had 'never been published in stranger circumstances than those which obtain today'. 'At this critical moment', it wrote, an effort 'has been made to overthrow the constitutional government of Ireland'. Expressing its wish for a speedy end to what it called 'this desperate episode in Irish history', it praised the courage and calmness of 'indifferent spectators' caught up in the area 'of the fiercest fighting', and outlined its hope for 'the speedy triumph of the forces of law and order'. Whilst conceding that there had been 'singing and shouting' amongst some young pedestrians in the city centre, 'apparently much excited by the events happening around them', it took solace in the fact that these 'perambulating groups of irresponsibles' had eventually dispersed.[125]

On Wednesday, *The Times* of London registered its disapproval of events in Dublin thus: 'Today we see the Sinn Féin conspiracy seizing control for the moment of a great part of the Irish capital in league with our enemies. Such are the fruits of truckling to sedition and making light of contempt for the law.'[126] Outside of the city centre, the seriousness of the rebels' intentions were illustrated by the fact that they managed to seize two police barracks and a Post Office in Swords and its surrounds on the same day, along with a quantity of arms and ammunition. They later attacked a railway station and

---

[122] Recollections of Easter Monday 1916, by A. A. Luce at the time Lieut. 12th Royal Irish Rifles, Later Captain, 14 October 1965, TCD, MS. 4874/2.

[123] Proclamation by Wimbourne Seeking Law-Abiding Citizens to Abstain from any Acts or Conduct which might Interfere with the Action of the Executive Government, 24 April 1916, NA, Kew, CO 904/23/2b.

[124] *Irish War News*, 25 April 1916.

[125] *The Irish Times*, 25 April 1916.

[126] *The Times*, 26 April 1916.

blew up a bridge at Rodgerstown, 'injuring the incoming line to Dublin'.[127] By Thursday morning, the main positions taken up by the insurgents back in the city centre were still intact. *The Irish Times* of that day reaffirmed its standing as a bastion of respectability, by impressing upon its readers the need for wise and loyal Dubliners to rigidly observe the regulations imposed by Martial Law. Acknowledging the inconvenience imposed upon 'many respectable families' by the 'novel problem' of having to stay in their houses between the hours of 7.30 pm and 5.30 am, it suggested that they should make productive use of their free time by either cultivating 'a habit of easy conversation', putting their gardens 'into a state of decency that will hold promise of beauty', doing 'some useful mending or painting about the house', or engaging in 'the art of reading'. Rather incredulously, *The Irish Times* also asked whether any better situation could be afforded for reading the works of Shakespeare, 'than the coincidence of enforced domesticity with the poet's tercentenary'.[128]

Back in the area of the fighting in the city centre, Pearse and his fellow rebels were faced with the substantial problem of advancing British troops and their increasingly vigorous shelling of insurgent positions as the week progressed. At times, back-up to the Volunteers was provided by the youthful members of Na Fianna Éireann, with 'some serving as commanders and fighters' and 'others engaged in carrying dispatches, scouting and reconnoitring'. A total of seven of their members died during the fighting, while Eoin MacNeill's son, Niall, was later court martialled by the organisation for not taking up arms (a charge from which he was eventually cleared). Bulmer Hobson, the co-founder of Na Fianna Éireann, also decided not to fight, having felt that German aid was not strong enough. Instead, he had favoured operations involving guerrilla warfare.[129] Given the difficult circumstances within which they were operating, many attempts were made by the leaders of the Rising to keep up morale amongst the rank-and-file rebels. Efforts were also made to enlist the public's support. On Thursday, for example, Pearse issued a bulletin calling upon Dublin civilians who believed 'in Ireland's right to be free' to help his cause 'by building barricades in the streets to impede the advance of the British troops'.[130]

---

[127] *The Irish Independent*, 5 May 1916.

[128] *The Irish Times*, 27 April 1916. For a commentary on the newspaper's stance, see D. Kiberd, *Inventing Ireland: The Literature of the Modern Nation* (Vintage Books, London, 2006), p. 268, who explains that its high editorial advice during the 'moment of crisis' was aimed at highlighting 'the culture which their soldiers were fighting to defend'.

[129] Hay, 'Na Fianna Éireann', pp. 53, 69.

[130] Photostat Copy of a Draft of a War Bulletin by P. H. Pearse, Commandant General of the Army of the Irish Republic, 27 April 1916, NLI, MS. 8499.

On Friday, in his final dispatch from the GPO, Pearse said that for four days, throughout intervals in the fighting, his men had sung freedom songs such as 'The Memory of the Dead', 'God Save Ireland' and the 'Soldier's Song'. Such rousing songs, according to Georges D. Zimmermann, 'no doubt ... contributed to create the spirit of martial enthusiasm which sustained the Volunteers'.[131] On the same day, however, *The Cork Examiner* ominously warned that even though 'the trouble is nearing its end', the repercussions were 'fraught with more dangerous possibilities', including 'further bloodshed evil in itself, but disastrous to the peace of Ireland perhaps for the ensuing decade'.[132] A communiqué issued by the authorities on Saturday 29 April confirmed that the end was very near. Although it wrongly stated that Connolly 'is reported killed', it confirmed that troops were 'gradually overcoming resistance' and that reinforcements were 'constantly arriving'.[133] With the carnage mounting, Pearse finally surrendered to the British Army before the day's end, following almost a week of 'grave loss of life and destruction of property'.[134] A report filed by a special correspondent of *The Times* the next morning noted how 'Dublin last night and in the early hours of this morning was as peaceful as a graveyard', so much so 'that one could hear the barking of a dog and the lowing of cattle ... in a field near the North Wall'.[135] By this time, however, communications between the various rebel positions around the city of Dublin were cut off from each other, so it took a while longer before all of the leaders were convinced of the authenticity of Pearse's order to surrender. Boland's Mill, for example, was not evacuated until early on the morning of Monday 1 May.

Communications between Ireland and the rest of the world were also severely curtailed, thus resulting in a degree of confusion as to what exactly had transpired. *The Gaelic American* of 29 April, a New York-based paper devoted to the cause of Irish independence, contained a brief report on the first few days of the Rising. 'Ireland is fighting gallantly for her independence', it declared, whilst pointing out to its readers that it had gone to press three days beforehand and was operating in a climate of 'garbled and censored cable despatches which have been coming, by a roundabout route from London'.[136] German newspapers

---

[131]   G. D. Zimmermann, *Songs of Irish Rebellion: Irish Political Street Ballads and Rebel Songs, 1780–1900* (Four Courts Press, Dublin, 2002), p. 68.

[132]   *The Cork Examiner*, 28 April 1916.

[133]   Communique, 29 April 1916, Dublin Diocesan Archives (hereafter DDA), Monsignor Michael Curran Papers (Political).

[134]   *Weekly Irish Times, Sinn Féin Rebellion Handbook, Easter 1916* (Fred Hanna Ltd., Dublin, 1917), p. 7.

[135]   *The Times*, 2 May 1916.

[136]   *The Gaelic American*, 29 April 1916.

unsurprisingly took an interest in the fighting on the streets of Dublin. A short article in the *Frankfurter Zeitung* of 26 April passed comment that the Rising could not be regarded as being of the utmost importance 'for the course of the war [on the continent]', but it did concede that the turbulence was 'certainly no child's play'.[137] As the Rising drew to a close, the *Frankfurter Zeitung* of 29 April reported 'that there is in Germany the most lively sympathy with this ill-treated, plundered, downtrodden and despairing people'. 'If the Irish succeed in winning independence', it stated that 'Germany assuredly will not grudge it them, and our best wishes go with the arms of the Irish fighters for freedom'.[138] Likewise, an issue of the *Cologne Gazette* published towards the end of April, conveyed its admiration of the Irish. It declared that Sir Roger Casement's actions on Good Friday (namely his attempt to land arms at Banna Strand in County Kerry, from a small boat into which he had climbed from a German submarine, the U-19) had made him 'a better Irishman than Redmond and the other politicians who are trying to keep their down-trodden country under the knout of Ireland's worst enemy'.[139]

Back in Ireland, *The Irish Times*, which had to suspend publication for a number of days towards the end of the Rising, appeared in print again as a special four-page edition. It was snapped up by eager crowds who read that 'the back of the insurrection' had been broken by a 'cordon of troops which was flung round the city' and which 'narrowed its relentless circle until further resistance became impossible'. Various rebels, it added, 'came in dejectedly under the white flag'.[140] Many of the surrendering Volunteer units marched in formation when laying down their weapons, while some rather despondent-looking rebel prisoners were escorted away on their own by armed guards of five soldiers (Plate 2.2). In declaring an unconditional surrender, Pearse later revealed that his objective had been 'to prevent the further slaughter of the civil population' and also to spare the lives of his followers who were surrounded, many without food.[141]

---

[137] In London, the details of this report were reproduced three days later in *The Times*, 29 April, 1916.

[138] Cited in ibid., 3 May 1916.

[139] Cited in ibid.

[140] *The Irish Times*, 28 and 29 April, and 1 May 1916.

[141] Photostat of Patrick Pearse's Last Statement/Message before his Execution, 2 May 1916, NAI, Department of the Taoiseach, 99/1/467, Leaders of Rising of Easter Week, 1916.

Plate 2.2   Rounding up the rebels – a prisoner being conveyed to
              Dublin Castle, April 1916

Source: Imperial War Museum, HU 55529.

A special correspondent for *The Times*, who viewed 489 prisoners embarking at the North Quay Wall for Holyhead on the night of Sunday 30 April, passed comment on how 'utterly exhausted' they were: 'Heavy eyes, haggard cheeks, drooping shoulders, and a general limpness of body told their tale of sleepless nights and harassed days.'[142]

The Rising in the Provinces

That the 1916 Rising occurred in Dublin was unsurprising. As Owen McGee has remarked, it 'was not a national insurrection', and its occurrence in the capital reflected the fact that 'it was the only town or city in the country where republicans' level of influence remained consistent in preceding decades'. The Rising itself, he adds, was far more than 'a purely IRB affair' aimed at the old principle of calling on the people to propagate 'a great democratic republic amongst themselves'. Instead, he notes that the public felt that the rebels had died for the ideals of the Irish-Irelanders' generation and Pearse's envisaged resurgence of a Gaelic

---

[142] *The Times*, 3 May 1916.

civilisation.[143] Outside of Dublin city and county, the rest of Ireland witnessed very little action by the Irish Volunteers. Many contemporaries, however, were acutely aware of just how close the entire country had come to experiencing a more widespread Rising. In a letter to John Hagan (the Vice Rector of the Pontifical Irish College, Rome) on 8 May 1916, Monsignor Curran remarked that without MacNeill's countermanding order, 'we would have had a score of Dublins over the country' and 'all would have thrown in their lot as the young fellows in Dublin did'. Under such a scenario, he speculated that the entire island would have been 'inundated with [British] soldiers'.[144]

Besides the confusion caused by the publication of the countermanding order in *The Sunday Independent*, the containment of the Rising can also be attributed to the failure of Casement's mission to import arms and distribute them around the country. Plans for a provincial Rising were also hampered by an array of misunderstandings – caused by the confidentiality surrounding the military plans, disparities between the highest-ranking rebels inside and outside the capital, and vagueness in various communications after the fighting had started. Much uncertainty, as Desmond Ryan has noted, prevailed outside of the capital, as did 'broken plans, divided counsels ... and inaction'.[145] In some cases, according to Mick O'Farrell, 'instructions cancelling one set of orders were received well before the original orders themselves were delivered'. Large-scale confusion abounded 'with various on-again, off-again messages being delivered to rebel leaders'.[146] Furthermore, as Ferghal McGarry has pointed out, 'more deep-rooted differences' between the organisers of the Rising and the provincial leaders concerning the feasibility of fighting 'also played an important part in the collapse of the Rising in rural Ireland'.[147]

While the Rising in the provinces, according to Townshend, 'may not have amounted to an emergency', it was enough to give reason for Martial Law to be extended across the whole of the island of Ireland by the authorities, while it also meant that the containment of rebel activity 'would eventually reach far beyond Dublin'.[148] In places where orders did get through and the Irish Volunteers rose (namely Counties Meath, Wexford and Galway), one of the

---

[143] O. McGee, *The IRB: The Irish Republican Brotherhood from the Land League to Sinn Féin* (Four Courts Press, Dublin, 2005), pp. 355–57.

[144] Michael J. Curran to John Hagan, 8 May 1916, Pontifical Irish College Archives, Rome (hereafter PICAR), The Papers of John Hagan, HAG 1/1916/62.

[145] Ryan, *The Rising*, p. 228.

[146] M. O'Farrell, *50 Things You Didn't Know About 1916* (Mercier Press, Cork, 2009), p. 45.

[147] F. McGarry, *The Rising. Ireland: Easter 1916* (Oxford University Press, Oxford, 2010), p. 210.

[148] Townshend, *Easter 1916*, p. 242.

main tactics adopted was to attack barracks of the Royal Irish Constabulary. This police force, according to W. J. Lowe, represented 'the manifestation of British presence and authority that Irish people encountered most commonly, probably on a daily basis'. While officer positions in the force were predominantly held by Protestants, the religious composition of its rank-and-file bore closer resemblance to 'that of the Irish population at large', containing a mixture of Catholics and Protestants.[149] Consequently, it had a good standing in the community. However, in addition to maintaining law and order all over Ireland's four provinces, the Royal Irish Constabulary also 'played a particularly conspicuous role in the accumulation and synthesis of political intelligence'.[150]

In light of both its policing and intelligence-gathering functions, the Royal Irish Constabulary was a prime target for rebel attacks during Easter Week. In the end, however, only a few police barracks were attacked by the rebels. In a number of instances, the rebels scored some minor victories. In County Meath, for example, a little barracks was captured in Kilmoon. According to *The Irish Independent*, the incident 'was very grave'. It reported that the episode resulted in the deaths of two sergeants and four constables, and the wounding of 18 constables. Furthermore, a chauffeur named Keep had to have his leg amputated after being shot by an explosive bullet, but he died soon afterwards.[151] On Friday 28 April, a barracks commanded by a Sergeant Toomey was also attacked and captured a short distance outside the village of Ashbourne by a group of rebels led by Thomas Ashe. A total of 27 police officers lost their lives in the fighting, along with two rebels and two civilians. One of the eye witnesses to the fighting, postal worker John Austin, noted that Ashe and his followers 'were very excited after their victory and were cheering, as men would after a football match'. That evening, he noted that the road outside the barracks 'was a terrible sight with blood and bandages strewn on it'.[152]

In County Wexford, minor incidents associated with the Rising occurred in Ferns, Gorey and Enniscorthy. Pearse's announcement that Wexford had risen proved 'a resonant one' and its heartening effect upon the garrison in the GPO

---

[149] W. J. Lowe, 'Irish Constabulary Officers, 1837–1922: Profile of a Professional Elite', *Irish Economic and Social History*, Vol. 32 (2005), p. 19.

[150] K. Fedorowich, 'The Problems of Disbandment: The Royal Irish Constabulary and Imperial Migration, 1919–29', *Irish Historical Studies*, Vol. 30, No. 117 (1996), p. 89.

[151] *The Irish Independent*, 26, 27, 28 and 29 April, and 1, 2, 3 and 4 May 1916.

[152] John Austin, 23 November 1953, NAI, Bureau of Military History, Witness Statement No. 904. In the longer-term, Townshend, *Easter 1916*, p. 215, notes that the military significance of the fighting at Ashbourne was that the guerrilla tactics utilised proved to be a harbinger 'of the methods to be adopted in a later and very different republican insurgency', namely the War of Independence.

'was visceral'. As many as 600 Volunteers gathered in Enniscorthy on Tuesday 25 April and inexorably, 'as if commemorating Wexford's epic 1798 history', Vinegar Hill was taken over for a while. While shots were exchanged with the Royal Irish Constabulary, a direct attack on a nearby barracks was abandoned.[153] The rebels eventually managed to seize the principal streets of Enniscorthy, along with the Anthenaeum in Castle Street, the railway station and a castle. They formally surrendered the following Monday morning when news reached them of the surrender in Dublin and Colonel French arrived in the town with 2,000 troops.[154] Initially, when news of Pearse's surrender reached them, the Enniscorthy rebels refused to give in. Consequently, the local British Army commander authorised two Irish Volunteers to travel to Arbour Hill prison in Dublin, where Pearse ordered them to surrender.[155] One of the Volunteers who ended up meeting Pearse in his cell was Séamus Doyle, who later recalled that the leader had 'seemed physically exhausted but spiritually exultant'. Writing materials were then handed to Pearse, who gave the order for rebels in the districts 'to lay down arms or disband'.[156] Pressure exerted from local clergy also persuaded the Volunteers in Wexford to put an end to their actions.

In the west of Ireland, Pearse's original plan relied upon the receipt of around 3,000 rifles, followed by a tactical manoeuvre to hold the line of the River Shannon. In the absence of the rifles, however, the strategy disintegrated and as Townshend has noted, the 'vague directive seemed even more impractical than it might originally have been'.[157] As matters transpired, rebel activity throughout County Galway was coordinated by Lancashire-born Liam Mellows (or Ó Maoilíosa in Irish). Ann Healy has described him as 'a larger than life sort of character', while C. Desmond Greaves has noted that his 'youth and ... infectious enthusiasm' proved to be vital in enabling him to secure the support

---

[153] Townshend, *Easter 1916*, pp. 240–41.

[154] *The Irish Independent*, 6 May 1916.

[155] M. Foy and B. Barton, *The Easter Rising* (Sutton Publishing Ltd., Stroud, 1999), p. 327.

[156] Doyle, cited in P. F. MacLochlainn (ed.), *Last Words: Letters and Statements of the Leaders Executed After the Rising at Easter 1916* (Office of Public Works, Dublin, 2005), p. 17.

[157] Townshend, *Easter 1916*, p. 228. Before the countermanding order was received in the west, the Volunteers in Galway seemed acutely aware that the odds were stacked against them. See F. X. Martin (ed.), 'Select Documents XX. Eoin MacNeill on the 1916 Rising', *Irish Historical Studies*, Vol. 12, No. 47 (1961), p. 250, in which the following memorandum by MacNeill is reproduced: 'On [Easter] Sunday morning, a messenger from Galway came to my house [in Dublin] and told me that the arrangements planned for Galway (without my knowledge) were without any chance of military success. This message was sent before my message calling off the Rising had reached Galway.'

of many locals, especially those who had grown up in small farms.[158] Despite not being a native of the area, Mellows had been very active in republican politics in the west since the spring of 1915, working as the chief organiser for the Irish Volunteers in County Galway. A deeply religious person, he was admired by the local Catholic clergy and well-respected in the community. Mellows was also known to be a strict disciplinarian, a reputation derived from his earlier work with Na Fianna Éireann. Prior to the outbreak of the Rising, he managed to escape from Reading Gaol and was smuggled back from exile in Britain, with the help of Connolly's second daughter, Nora. During Easter Week, Mellows managed to rally his troops in the county and carry out offensive operations in Clarinbridge, Oranmore and Athenry, before retreating to Moyode Castle and finally disbanding at Limepark.[159]

The Irish Independent, in describing the Rising in County Galway, reported that there had been a number of 'thrilling encounters' between the rebels and Crown forces.[160] The village of Killeeneen, located in the east of the county, served as the starting point of the Rising in the west. Having mobilised his men, who were dressed in an assortment of attire (including forage hats and Sam Browne belts), Mellows led a party of Volunteers to the Royal Irish Constabulary barracks at Clarenbridge on the morning of Tuesday 25 April. Although they took two policemen who had been on patrol as prisoners, they failed to take the barracks.[161] Throughout the incident, spiritual assistance was provided by a 27-year-old Catholic priest named Father Harry Feeney, who acted as Chaplain to the rebels.[162] Another incident on Tuesday involved a group of 106 rebels, led by Joe Howley, who attacked the Royal Irish Constabulary barracks at Oranmore. After an unsuccessful attempt to take the barracks, they joined up with Mellows's men on the road between Oranmore and Clarenbridge. The combined force of 200 rebels then renewed the attack on Oranmore, during which they cut sections of the Galway-Oranmore railway line, damaged a

---

[158] A. Healy, *Athenry: A Brief History and Guide* (The Connacht Tribune Ltd., Galway, 1989), p. 33; C. Desmond Greaves, *Liam Mellows and the Irish Revolution* (An Ghlór Gafa, Belfast, 2004), p. 75.

[159] N. Ó Gadhra, *Civil War in Connaught 1922–1923* (Mercier Press, Cork, 1999), p. 52; Ú. Newell, 'The Rising of the Moon in Galway', *Journal of the Galway Archaeological and Historical Society*, Vol. 58 (2006), pp. 121–23; L. Collins, *16 Lives: James Connolly* (The O'Brien Press, Dublin, 2012), p. 342.

[160] *The Irish Independent*, 8 May 1916.

[161] V. Whitmarsh, *Shadows on Glass. Galway 1895–1960: A Pictorial Record* (Privately Published, Galway, 2003), p. 318; Newell, 'The Rising of the Moon', p. 125.

[162] Secret Memo Written by Neville Chamberlain re. Rev. H. Feeney, BLOU, MS. Asquith 43/121; P. Ó Laoi, *History of Castlegar Parish* (The Connacht Tribune Ltd., Galway, Undated), pp. 147–48.

bridge and sabotaged telephone lines.[163] Unluckily for the rebels, however, the police soon appeared on the scene, accompanied by a party of 10 soldiers from the Connaught Rangers. Furthermore, a British war ship named the HMS Gloucester was stationed in Galway Bay, and as the Press Association's Galway correspondent bluntly reported, the rebels 'were checked by shells from a destroyer'.[164] After coming under heavy bombardment, they soon withdrew in the direction of Athenry.

In the town centre of Galway itself, the authorities quickly secured the main public buildings, having got word of what was going on in the capital. When he turned up for work at the GPO on Eglington Street early on Tuesday morning, postal worker Thomas Courtney (who was the Intelligence Officer with the Castlegar Company of the Irish Volunteers) noted that the door was opened by a policeman. Upon entering, he witnessed three armed policemen in the hallway, while others were 'scattered about the office'. 'I knew', he added, 'from the number of empty stout bottles I saw everywhere that the police must have been there all night'. After leaving the building, Courtney proceeded to Eyre Square, where he 'saw that armed police were standing at the corner and others in doorways'. Besides the strong police presence in Galway town, another factor that aided its defence was the propinquity of the Connaught Rangers depot in Renmore, located a short distance to the east of the town. The menacing presence of the HMS Gloucester also thwarted a rebel attack on the town, thus sparing its buildings from the scenes of utter devastation that the capital suffered as a result of shelling from a gunboat in the River Liffey. When he enquired about the HMS Gloucester after it had begun firing in the direction of Oranmore, Courtney was warned of its capabilities by a British sailor: 'They are not big guns, they are only 4 inch. The real big ones will shake the town.'[165]

The first (and only) casualty of the Rising in Galway was Patrick Whelan, a 38-year-old policeman who was stationed at Eglington Street Barracks. Whelan was killed by a gunshot at Carnmore crossroads early in the morning of Wednesday 26 April, during a skirmish between the Royal Irish Constabulary (assisted by soldiers commanded by Colonel Bodkin) and around 60 rebels attached to the Castlegar and Claregalway Volunteers (commanded by Captain Brian Molloy).[166] On the same day, a large group of around 500 Irish Volunteers

163 Newell, 'The Rising of the Moon', pp. 125–26.

164 *The Freeman's Journal*, 26, 27, 28 and 29 April, and 1, 2, 3, 4 and 5 May 1916.

165 Thomas Courtney, Ex Castlegar Company Irish Volunteers, Undated, NAI, Bureau of Military History, Witness Statement No. 447.

166 Ó Laoi, *History of Castlegar Parish*, p. 139; Newell, 'The Rising of the Moon', p. 126; M. A. Vaughan, 'Finding Constable Whelan: An Incident from 1916', *Ossary, Laois and Leinster*, Vol. 4, (2010), pp. 231–36. On 27 April 1916, Whelan was laid to rest in Section G, Row 3 of

gathered at the Department of Agriculture and Technical Instruction's experimental (or model) farm close to Athenry, but they were poorly armed with just 25 rifles, 60 revolvers, 60 pikes, and 300 shotguns. Most of them consisted of 'young Catholic men from small farm, labouring and artisan backgrounds'. Because their position was exposed and open to attack, they decided to retreat southwards to the deserted Moyode Castle (the former home of the Persse family, located five miles south-east of Athenry). This was easily taken as it was only protected by a caretaker.[167] In the end, the rebels ended up staying at the castle for two nights, with food being supplied by some locals. To keep up morale, Fr. Feeney heard confessions and also delivered general absolution to those who were given the task of going on scouting missions to gain intelligence.[168] Furthermore, Thomas Davis's song, 'A Nation Once Again', was sung next to the camp fire at Moyode, 'in an atmosphere tense with optimism and excitement'.[169]

Such sanguinity, however, was short lived. It was reported by *The Freeman's Journal* that 'things were quiet' in Galway town by Friday 28 April, after 'a large number of soldiers had been landed from a war vessel'.[170] In addition to the British reinforcements in the town, news reached the rebels that soldiers were also gathering in Loughrea, Ballinasloe and Athlone. They therefore retreated to an empty big house at Limepark near Peterswell, where they took the decision to disband. Most of the rebels returned to their homes by Saturday 29 April, while the main leaders opted to go on the run rather than hold out in their positions, thus sparing the county from scenes involving 'an honourable blood sacrifice'.[171] Once matters quietened down, Galway County Council passed a resolution at its meeting on 3 May, expressing 'condemnation of the recent disturbances of social order brought about by irresponsible persons whereby great damage has been done to the material prosperity and prospects of Ireland'.[172]

Mellows, who was a wanted man, went on the run and crossed into County Clare with Frank Hynes and Alfie Monaghan. The trio then made their way

---

Bohermore Cemetery in Galway. At a later stage, a limestone Celtic cross was erected over his burial place by 'officers and men of the RIC and many sympathetic friends'.

[167] F. Campbell, 'The Easter Rising in Galway', *History Ireland*, Vol. 14, No. 2 (2006), pp. 23–24.

[168] Newell, 'The Rising of the Moon', p. 127.

[169] C. Desmond Greaves, *The Easter Rising in Song and Ballad* (Kahn and Averill, London, 1980), p. 65.

[170] *The Freeman's Journal*, 26, 27, 28 and 29 April, and 1, 2, 3, 4 and 5 May 1916.

[171] Newell, 'The Rising of the Moon', pp. 127–28. The decision to disband was not taken lightly. See M. Neilan, 'The Rising in Galway', *The Capuchin Annual 1966*, No. 33 (1966), p. 326, who notes that 'it was with the greatest reluctance' that the Volunteers 'began to break up' and that a lot of them actually 'wept openly'.

[172] *The Freeman's Journal*, 8 May 1916.

through the Derrybrien Mountains until they met Michael Moloney, an IRB man, who put them up in a cattle shelter on his lands in the Balloughtra Mountains until the end of the autumn.[173] Eventually, Mellows was moved to the home of Father Crowe, a Catholic priest at Ruan. From there he disguised himself as a nun and fled to Cork. According to Annie Fanning, he was accompanied by 'Fr. Crowe, Bluebell Powell, as a novice ... [and] another girl from Galway ... dressed ... in nun's clothing'.[174] Having managed to evade the authorities in Ireland, Mellows finally escaped to New York. Tom Kenny of Craughwell, the most prominent member of the IRB in County Galway, fled to Boston (where he renewed his camaraderie with Jim Larkin). During the Rising, he had unsuccessfully attempted to get Mellows to follow a more radical course by proposing strikes against bourgeois Home Rulers within the locality of Moyode and urging further agitation through the seizure of land and cattle.[175] After the Rising ended, he too availed of refuge in Mellows's hideout in the Balloughtra Mountains.

In Mayo, intervention by a priest in Castlebar managed to stifle a smaller mobilisation by the Volunteers and plans for surprise attacks on various barracks of the Royal Irish Constabulary throughout the county during the middle of Easter Week.[176] However, even after Pearse's surrender, defiance was reported at Westport on 30 April, where at least 100 Irish Volunteers demonstrated, carrying 17 rook rifles, 4 shotguns, and one Lee-Enfield rifle. In a field beyond the town around 50 of this party were instructed in field movements and hand grenade exercises by their leader, Joseph MacBride. The entire rally 'was most defiant', according to the police, who expected to have their barracks attacked by them. On the previous night, the police took down notices which had announced the demonstration and had called on all Irishmen to join the Irish Volunteers. Having been informed that a much larger demonstration was scheduled to take place on 2 May, the police arrested MacBride and other prominent Volunteers

---

[173] Seán O'Keeffe, 29 September 1955, NAI, Bureau of Military History, Witness Statement No. 1,261.

[174] Annie Fanning, 29 January 1949, NAI, Bureau of Military History, Witness Statement No. 187.

[175] Campbell, 'Easter Rising in Galway', pp. 24–25; F. Campbell, '"Reign of Terror at Craughwell": Tom Kenny and the McGoldrick Murder of 1909', *History Ireland*, Vol. 18, No. 1 (2010), p. 29. Kenny's role in the Rising has also been analysed in T. Kenny, *Galway: Politics and Society, 1910–23* (Four Courts Press, Dublin, 2011), p. 18, who describes him as 'the face of the more radical body politic in Galway', who 'wanted economic freedom through land redistribution and was prepared to fight to get it'.

[176] Townshend, *Easter 1916*, pp. 230–31.

on that date.[177] In the counties of Longford and Roscommon, extensive searches of the countryside for rebels were conducted in early May by a flying column of the Sherwood Foresters, who were guided by the Royal Irish Constabulary and patrols of cavalry of the King Edward's Horse. Many villages in the area were surrounded and searches were conducted in suspect houses.[178]

Despite the existence of a vigorous republican tradition in Belfast, the mobilisation of the Irish Volunteers in Ulster was 'not unexpectedly ... short-lived', according to Townshend, owing to the difficulties of operating within a hostile unionist environment and the rejection on Good Friday by the Tyrone Volunteers (apparently on the grounds that it was unfeasible) of a military plan to march with the Belfast Volunteers to Galway and join up there with Mellows's men.[179] As originally envisaged, the various groups of Irish Volunteers from Ulster were to have first assembled at Belcoo in County Fermanagh. From there, they were to have set off on attacks against police stations and army barracks. Upon joining up with Mellows and his men, they were then supposed to have held the line of the Shannon against advancing British troops.[180] Like much of the rest of the country, however, inadequate communications with Dublin about issues of strategy proved crucial in Ulster's failure to rise in armed revolt, as did the absence of German support.

On the day before Pearse's surrender, Ulsterman Patrick McCartan, who was a member of the IRB's Supreme Council, was clear in his verdict of where the fault lay. In a letter that he wrote from his hideout in a barn in the mountains of County Tyrone to a friend based in Philadelphia in the USA, he was critical of the republican leaders in the capital city for not providing for a 'system of communication', for leaving things until 'the last minute' and for not thinking through the 'details'. Rather exasperatingly, he added: 'They may say we are cowards but, God knows, it is harder for me at least, and I know it is the same with the others, to be here in hiding than in the field with a rifle ... The feeling that we are doing nothing ... is ... hell.'[181] A member of the Belfast Volunteers, Joseph Connolly, also noted that it had been impossible to mobilise by the time that news of the fighting in Dublin reached the city of Belfast. From Easter Monday onwards, he noted that the men in Belfast had 'many conflicting

---

[177] Summary of a Police Report of Irish Volunteer Demonstration at Westport, 30 April 1916, BLOU, MS. Asquith 43/99.

[178] War Diary of A. E. Slack, 2/6th Battalion Sherwood Foresters (59th Division), May 1916, IWM, MS. P.278, The Papers of A. E. Slack.

[179] Townshend, *Easter 1916*, p. 225–27.

[180] F. X. Martin (ed.), 'Easter 1916: An Inside Report on Ulster', *Clogher Record*, Vol. 12, No. 2 (1986), pp. 192, 194–95.

[181] Ibid., pp. 199–200, 205.

discussions about the rights and wrongs of MacNeill's order, the Rising itself and somewhat bitter recriminations about the divided and conflicting lines of the two schools of thought and direction at Headquarters'.[182]

Down in the south of the country, all remained quiet in the city of Cork during the course of Easter Week. Tensions had been high there in the weeks leading up to Easter Monday. Tadg McCarthy, who was attached to the Donoughmore Company Irish Volunteers, noted that there had been 'several clashes' in the city between the Volunteers and the Separation Allowance Women on the evening of St. Patrick's Day.[183] Not long before the Rising, there were 46 companies of Irish Volunteers spread across County Cork, comprising around 1,500 men. In the initial plans for the Rising, the Cork Brigade was to have joined with units from Limerick and Kerry along the Cork-Kerry border, in order to obtain their allocation of weapons.[184] The Cork city unit of the Irish Volunteers, which was commanded by Tomás MacCurtain (and his able deputy, Terence MacSwiney), was considered 'a reasonably strong brigade', having been put through an active training programme in the early months of 1916. A total of 221 of the city battalion did mobilise on Easter Sunday, whereupon they marched to Capwell Station and took a train to Crookstown in West Cork, in an operation that was apparently linked to Casement's attempt to land arms in nearby Kerry. On Easter Monday, however, great confusion reigned at the Volunteer Hall headquarters on Sheares Street in the city. When it was decided that no action was to be taken, many of the Cork Volunteers went back to work in the days afterwards – some rather disgruntled.[185]

Ultimately, the issuing of MacNeill's countermanding order kept the peace in Cork, as did pressure exerted by the local clergy. Furthermore, great uncertainty and confusion had arisen over inconsistent dispatches and orders from Dublin. As had been the case in other areas throughout the provinces, Casement's capture proved a major setback for the Cork city Volunteers, as they had come to believe 'that ample supplies of arms, ammunition ... and light artillery would be made available from Germany'. Thus whilst the Dublin Volunteers, 'as they showed, had sufficient arms and ammunition to maintain a fight for a week ... the Cork Volunteers had scarcely enough to last five minutes'.[186] Other factors

---

[182] Joseph Connolly, Ex Belfast Irish Volunteers, 9 April 1948, NAI, Bureau of Military History, Witness Statement No. 124.

[183] Tadg McCarthy, 29 June 1954, NAI, Bureau of Military History, Witness Statement No. 965.

[184] Ó Drisceoil, 'Conflict and War', p. 259.

[185] Townshend, *Easter 1916*, pp. 234–36.

[186] Seán Murphy, Thomas Barry, Patrick Canton, and James Wickham, 27 March 1957, NAI, Bureau of Military History, Witness Statement No. 1598.

also came into play. The topography of Cork city and its surrounds added to the decision of the poorly-armed Cork Volunteers not to rise, as the low-lying city centre (originally built on a marshy estuary of the River Lee – hence the Irish name 'Corcach', meaning 'Marsh') would have proven an easy target in the event that they seized public buildings. Had they done so, the British would probably have responded by opting for shelling. As Desmond Ryan later explained, 'the guns of the British command' dominated the low-lying city from the hilly ground above.[187] There was a certain sense of *déjà vu* about all of this, because during the siege of Cork by the Williamite forces back in 1690, Cork city's physical geography had proved enormously detrimental to the Jacobite cause. In the end, not a single drop of blood was shed in Cork city in Easter 1916. The city also escaped the large-scale physical destruction that the centre of Dublin experienced.

For the most part, the authorities in Cork city and county encountered very little turmoil during Easter Week. Many companies of Volunteers throughout the county had followed instructions to parade at various locations on Easter Sunday, but in the absence of new supplies of arms, little else could be done when the fighting in Dublin started the next day. The military authorities, in any case, were quick to mobilise throughout County Cork. In an account of his service with the 3rd Battalion Connaught Rangers in Ireland, Captain C. A. Brett later recalled how news of the Rising had led to 'manifest fears that it would spread'. His 2,800-strong battalion at Kinsale, which had been training soldiers to replace casualties in the fighting battalions on the continent, produced a flying column of about 400 men which marched to Crosshaven. From there they were ferried by two tugs to Cobh (then known as Queenstown), from where they then marched to the Fota Island estate on the outskirts of Cork city – a location in which they 'were obviously well placed, had there been any trouble in Cork city'. However, no trouble arose and the battalion moved on to Dungarvan, County Waterford, where they were joined by some Royal Engineers and Royal Artillery 'with a field gun and limber drawn by a steam traction engine'.[188]

Another soldier involved in contingency action in Cork city and county was Captain Edward Frederick Chapman of the 14th Battalion Fusiliers. On Easter Monday, he was sitting in a smoke room in Cork city, 'wishing it would stop raining', when he heard news of the Rising in Dublin. He expected trouble, but as nothing happened he was sent two days later to the town of Fermoy, located in the north of the county. While there he was placed on a bridge as a guard, which made him feel like a London policeman, 'stopping and examining every

---

[187]   Ryan, *The Rising*, p. 230.

[188]   Transcript Account of Service in Ireland with the 3rd Battalion Connaught Rangers from 1914–1916, IWM, MS. 76/134/1, The Papers of Captain C. A. Brett.

cart or motor that passed, to see if it contained arms or ammunition'. From the bridge he heard 'all sorts of rumours' that rebels were marching on Fermoy from Mitchelstown, 'and others similar'. Remaining alert, Chapman recalled how his platoon 'were quite expecting to have a scrap, especially when we heard a shot fired by the sentry'. As it turned out, however, the sentry 'had fired a shot at a shadow that looked like a man'.[189] Nearby in Mallow, the military took precautions by heavily guarding the rail station and by erecting 'shelters for guns on the railway viaduct over the Blackwater', which also controlled the Dublin to Killarney road. On the road itself, barbed wire obstacles were put in place so as to prevent the passage of motor traffic, while 'special permits were required by any motorist using roads in the vicinity'.[190]

In the west of County Cork, nothing of consequence happened during the course of Easter Week, despite the mooting of some audacious proposals. Commandant Séamus McCarthy of the Eyries Company Irish Volunteers recalled: 'I do not remember that we got any orders from Cork or elsewhere during Easter Week. No arms were surrendered by the Eyries Corps, and no arrests were made in the parish.'[191] Others had different ideas. On the night of Wednesday 26 April, for example, a communication was sent from Tom Hales in Bandon to the Dunmanway Company of Irish Volunteers, proposing that they join up with the Bandon and Macroom Companies to attack the elevated British Army positions that surrounded the city of Cork. The plan, which would have involved an attack from the north on the British positions, was ultimately rejected two days later on the grounds that 'it was then too late to take any action'. Even though the Dunmanway Company had been 'kept ... on the alert and ready for action' throughout Easter Week, all that it was able to do in the end was to 'safeguard' its diminutive supply of arms – which included one revolver, two pistols, five rifles, and 28 shot guns.[192]

On 29 April, *The Cork Examiner* reported an announcement from the military authorities 'that the situation generally in the South of Ireland command is good', and that 'adequate precautions have been taken to deal with any disturbances that may arise'.[193] In the end, the only shots fired in resistance in County Cork were by the Kent brothers – Thomas, Richard, David,

---

[189] Captain Edward Frederick Chapman to his Sister, Hilda, 30 April 1916, IWM, MS. 92/3/1, The Papers of Captain Edward Frederick Chapman.

[190] *The Irish Independent*, 6 May 1916.

[191] Séamus McCarthy, 17 July 1948, NAI, Bureau of Military History, Witness Statement No. 137.

[192] Con Ahern, 13 October 1947, NAI, Bureau of Military History, Witness Statement No. 59.

[193] *The Cork Examiner*, 29 April 1916.

Edmund, and William – of Bawnard House in Castlelyons, four miles from Fermoy. During the course of Easter Week the five brothers did not sleep in their home at night, as Thomas was the Commandant of the Galtee Brigade Irish Volunteers. However, when they returned on the night of 1 May, they awoke the following morning to find the house surrounded by the Royal Irish Constabulary. After refusing to surrender, the police, according to P. J. Power, 'fired a volley after which a fierce conflict ensued'.[194] The Kents, who were armed with three shotguns and a rifle, resisted arrest for three hours, during which they mortally wounded Head Constable Rowe and injured a number of police officers. After the shootout ended, David was badly wounded while Bawnard House itself 'was completely wrecked'. Its inside 'was "tattooed" with the marks of the rifle bullets', but miraculously, its 'altar and statues in the beautiful oratory alone escaped destruction'.[195] In attempting to defy arrest after the shootout, Richard was shot and later died in Fermoy Military Hospital.

In the weeks after Easter Week, police raids commenced throughout the southern counties, thus leading to an atmosphere of tension between the authorities and those with connections to the Irish Volunteers. Anxiety was felt by some of the British soldiers who went to Limerick on 2 May. Richard Pedler, a Londoner serving with the Queen's Westminster Rifles, noted that while marching through its streets under the cover of darkness, 'we found we were walking over old tin cans, broken bottles and such like articles and we all thought that we were in for a lively time'. For the most part, however, he noted that despite being 'much stared at', he and his fellow officers 'were treated quite decently during our stay there'.[196] By 3 May, Major General F. C. Shaw reported that the situation in the south was 'quiet', with 'steady progress ... being made towards the restoration of normal conditions'. For Ulster, he reported that 'the situation ... is normal'.[197] Arrangements were soon made in many locations for the Volunteers to give up their arms. This happened in the Town Hall in Limerick on 5 May, in the presence of the Mayor and Town Clerk, Lieutenant Colonel Gordon Clark of the Queen's Westminster Rifles and various personnel from the Leinster Regiment. As each Volunteer handed in his weapons, their details were taken down by the Town Clerk and handed over to the military. One local, Mr S. O'Mara, expressed his gratitude for the Volunteers' actions by

---

[194] P. J. Power, 'The Kents and their Fight for Freedom', in B. Ó Conchubhair (ed.), *Rebel Cork's Fighting Story 1916–21: Told by the Men Who Made It* (Mercier Press, Cork, 2009), p. 109.

[195] Ibid., p. 110.

[196] R. A. Pedler to his Brother, Percy, 14 May 1916, IWM, The Papers of Richard Albert Pedler.

[197] Memorandum by Major General F. C. Shaw, May 1916, BLOU, MS. Asquith 42/69.

sending them a cheque 'to help to reimburse them', out of his belief that 'many of these rifles were bought out of their hard-earned savings'.[198]

Rage against the Rebels

As he reached the Irish coastline towards the end of Easter Week, General John Grenfell Maxwell, who had been appointed Commander of the British Army in Ireland, observed an almost apocalyptic-like scene from his vessel. In a letter that he sent to his wife, Louise, on 28 April 1916, he recalled that from his vantage point out on the sea, it had 'looked as if the entire city of Dublin was in flames'. Although 'it was not quite so bad as that' by the time that he reached the North Wall, he still observed that 'a great deal of the part north of the Liffey was burning'. 'Bullets were flying about', he added, while the hazardous fire of machine guns was 'breaking out every other minute'.[199] In the days following Pearse's surrender, there was a high level of public revulsion in Dublin at the immense loss life, the destruction of property (which was estimated at £2.5 million) and the perceived pro-German stance of the rebels. An iconic view of Dublin from Nelson's Pillar after the Rising (Plate 2.3), which later featured on postcards, vividly portrayed the destruction caused to the GPO and surrounding buildings by the intense shelling and gunfire. Much of the inside of the building was badly damaged, including the telegraph rooms on the top floor. The exterior of the building also remained in very bad shape after its evacuation, while the scene all along the southern end of O'Connell Street, according to *The Irish Times*, was 'one of utter desolation', with the skeletons of many buildings still standing 'with gaping, blank interiors'.[200] So great was the 'mass of ruins' along the principal thoroughfare, wrote William L. O'Farrell, that Dublin resembled some sort of 'Ypres on the Liffey'.[201] The view to the north-east from O'Connell Bridge after the Rising was also one of sheer devastation, with flames emanating from buildings along the quays (Plate 2.4). If ever there was an image which conveyed the fact that the Rising was over and done with, it was an iconic image of a contented group of Army officers pictured with the captured rebel flag (Plate 2.5).

---

[198] *The Irish Independent*, 9 May 1916.

[199] General Maxwell to His Wife, Louise Selina Bonynge Maxwell, 28 April 1916, Princeton University Library, Department of Rare Books and Special Collections (hereafter PULDRBSC), Sir John Grenfell Maxwell Papers, Box 6, Folder 9.

[200] *The Irish Times*, 28 and 29 April, and 1 May 1916.

[201] William L. O'Farrell to John Hagan, 2 June 1916, PICAR, The Papers of John Hagan, HAG 1/1916/69.

Plate 2.3   A view of Dublin from Nelson's Pillar in the days following the
1916 Rising

Source: National Library of Ireland, Valentine Collection, R 27448.

Plate 2.4   A view from O'Connell Bridge of buildings on fire along the
              northern Liffey quays after the Rising
Source: National Library of Ireland, Valentine Collection, R 27448.

Plate 2.5   A group of Army officers pictured in Dublin city with the
              captured rebel flag, May 1916
Source: Imperial War Museum, Q 82351.

Recalling his experiences in the GPO, J. J. McElligott noted that when news arrived of the order to surrender, 'there was general surprise but no protest. We were all black and grimy, unshaven and hungry'. As he and the other rebels walked out into O'Connell Street, he saw 'brisk fires ... burning over a large area'. En route to Richmond Barracks, McElligott witnessed 'quite a number of people along the South Quays mostly jeering and booing'.[202] Liam O'Flaherty of the Dublin Brigade Irish Volunteers also experienced hostility from onlookers as he and his fellow prisoners were marched from Richmond Barracks to the North Wall, via the quays. 'The crowds who lined the route', he noted, 'did not seem very fond of us (the exceptions were our own immediate friends)'.[203] It took some time before public indignation towards the rebels dissipated. Within both the civilian population and the print media in general, many people expressed outrage at the carnage that had occurred on the streets of Dublin as a result of the Rising. In addition to the deaths of rebels and British forces, tensions were considerably heightened by the fact that many innocent civilians had been among the casualties or had suffered injuries.

The business community was taken completely by surprise by the actions of the rebels. A manager of a window cleaning business in Dublin, for example, was traumatised after having had a revolver pointed at his mouth by one of the rebels at Nelson's Pillar. He later recalled how his blood had boiled following the incident, and how he had luckily managed to escape 'at least 50 or 60 shots' fired from the GPO as he made his way southwards down O'Connell Street.[204] Further anguish was reported by a jeweller from Grafton Street, who complained that his premises had been attacked 'with stones' and then 'looted in wholesale fashion'.[205] In addition to incidences of pillaging, the fighting on the streets of Dublin caused considerable damage not only to numerous commercial buildings, but also to the city's infrastructure in general. This impacted considerably upon the livelihoods of all sectors of the community – including businesspeople, shoppers, commuters, and residents. Consequently, one of the immediate tasks of the authorities after Pearse's surrender was to cope with the humanitarian crisis that took hold as a result of the destitution caused by the Rising.

While a Local Relief Committee was quickly provided with funding for charitable aid, the Board of Guardians of the North and South Dublin Unions

---

[202] Typescript Copy of Dr. J. J. McElligott's Experiences in the General Post Office, Dublin in 1916, 3 August 1962, University College Dublin Archives (hereafter UCDA), MS. LA1/G/125, Eoin MacNeill Papers.

[203] Liam O'Flaherty, Undated, NAI, Bureau of Military History, Witness Statement No. 248.

[204] Transcript by the Manager of a Window Cleaning Business in Dublin, Describing the Events in the City from 24–27 April 1916, IWM, MS. P.462, The Papers of W. Jones.

[205] *The Times*, 1 May 1916.

also weighed in on the relief effort. In early May, they applied to the local authorities in Dublin, asking them to put in force Section 13 of the Local Government Act of 1898, so as to enable them 'to obtain overdrafts for the purpose of relief, or to administer outdoor relief to able-bodied persons'. At a special meeting of Dublin Corporation in the City Hall on 10 May, it was unanimously agreed by the 38 aldermen and councillors present to apply to the Local Government Board to make an Order authorising the two Board of Guardians 'to administer relief' in accordance with the Local Government Act. The corporation then appointed Councillor Andrew Beattie to become an additional member of the Board of Guardians of the South Dublin Union for a fixed term, while Councillor John Ryan was appointed as an additional member of the Board of Guardians of the North Dublin Union. At Dublin Corporation's next monthly meeting, held on 5 June, the following motion was put to members and carried: 'That the deepest sympathy ... be extended to the relatives of the citizens who lost their lives during the recent rebellion.'[206] Outside of the capital city, Dublin County Council also acted swiftly to alleviate distress in the areas under its jurisdiction.

Details of the Rising's impact upon the civilian population can be found in a number of eyewitness accounts. One witness to the Rising was an unarmed soldier named James Glen, who was 'on leave in Dublin' at the time. After walking past the GPO on Easter Monday with 'another young artillery officer', he witnessed a sad scene. 'After we passed', he recalled, 'some rifle shots were fired in our direction, a child beside me was hit and was immediately carried off into a side street by someone in the crowd.'[207] Another bystander, Samuel Bolingbrooke Reede, recalled the shock and bewilderment amongst ordinary citizens who had been going about their daily business on Easter Monday:

> In the afternoon I went in sight of Sackville Street & could see the looting of several shops in the distance. Here I met a friend coming up from town who advised me to go no further, by this time people were standing in crowds and all sorts and descriptions of people chatting as if they knew each other all their lives and telling all sorts of stories and rumours they had heard ... Everything went off fairly quiet but one could hear volleys of rifle firing here and there all night and snipers were busy. Several civilians and soldiers off duty or home on leave were shot.[208]

---

[206] Municipal Council of the City of Dublin, *Minutes of the Municipal Council of the City of Dublin, from the 1st January to 31st December 1916* (Sealy, Bryers & Walker, Dublin, 1918), pp. 251–56.

[207] James A. Glen's Account of Trinity College Dublin in Easter Week 1916, 1 December 1967, TCD, MS. 4456.

[208] Samuel Bolingbrooke Reede's Recollection of the Easter Rising, TCD, MS. 10848.

Fatal injuries to civilians were also attested to in a report filed by a member of the advertising staff of *The Times*, who was in Dublin on Easter Monday and who witnessed the opening scenes of the Rising as they unfolded. Within the vicinity of St. Stephen's Green, he noted that 'drivers of vehicles were summoned to a halt' and that they were subsequently 'seized and overturned'. In one case, he alleged that 'a driver who refused to leave his van was shot by the rebels without the slightest hesitation'.[209]

The loss of civilian life was also alluded to in Mary Lousia Hamilton Norway's *The Sinn Féin Rebellion as I Saw It*, which furnished an 'hour by hour' account of Easter Week through letters that were initially 'written during a period of extraordinary strain for family perusal only'. Her husband at the time, Arthur, was Secretary for the Post Office in Ireland, and she herself was based in the Royal Hibernian Hotel on Dawson Street. This premises was filled with 'excitement and consternation' on Easter Monday. 'Every moment', she wrote, 'people were coming in with tales of civilians being shot in the streets', while owners of 'houses commanding wide thoroughfares and prominent positions ... were told [by the rebels] with a revolver at their heads that the house was required by the Irish Republic for strategic purposes'. She also gave a graphic account of 'a terrible case' involving a civilian who died in hospital the day afterwards, having been found lying on a garden seat in St. Stephen's Green, 'with all his lower jaw blown away and bleeding profusely'.[210] Members of the clergy were also mortally wounded on the streets during the Rising. Nelly O'Brien recalled how 'a priest who had gone over to help a wounded soldier in Dame Street got a bullet through his forehead'. She complained about 'how reckless' the rebels were at the beginning, and 'how antiquated' their 'notions of modern warfare were'.[211] On the other hand, members of the British Army were also singled out for reckless behaviour during the fighting. Towards the end of the Rising, for example, a soldier on sentry duty at Patrick's Close near Kevin's Street Barracks had to be arrested by an officer of the Dublin Metropolitan Police, who felt that he was a 'danger to civilians'. The police officer in question, Patrick Bermingham, recalled how he had witnessed the soldier 'very much intoxicated firing indiscriminately in the air and at windows and doors'. To make matters worse, Bermingham added that the volatile soldier 'would lay down his rifle and light his cigarette and then carry on firing as before'.[212]

[209] *The Times*, 29 April 1916.

[210] Hamilton Norway, *The Sinn Féin Rebellion As I Saw It*, pp. v, 6–7, 21–22.

[211] E. L. (Nelly) O'Brien's Account of Her Experiences during Easter Week, TCD, MS. 10343/1.

[212] Patrick J. Bermingham, 26 June 1952, NAI, Bureau of Military History, Witness Statement No. 697.

As the Rising drew to a conclusion, *The Times* was well aware of the gravity of what had transpired on the streets of Dublin and elsewhere throughout Ireland. Despite only scant news emanating from Ireland, its issue of 29 April used its influential position to demand dramatic action from British politicians and military leaders to quell the unrest in Ireland and to prevent disaffection spreading to other quarters of the populace. 'Nothing can be more cruel or more unwise in insurrections', it declared, 'than half-hearted measures of suppression'. So as to prevent 'fresh dupes' joining the ranks of the rebels and deter 'men hovering on the brink of treason', it demanded 'for the sake of all loyal classes in Ireland ... complete, strong, and drastic measures ... against the insurgents without hesitation and without delay'.[213] As the staging of the Rising had come as a complete surprise to the mainstream Irish national and provincial press, its initial reactions were also extremely negative. 'It all appears as if one had had a nightmare', announced the pro-Home Rule paper, *The Leader* (whose office on Lower Abbey Street was ruined by the fighting). The 'mad escapade', it complained, had originated 'from nowhere ... without warning'.[214]

While acknowledging the bravery that the rebels had demonstrated in fighting for their cause, *The Irish Times* of 1 May was particularly scathing in its verdict on the fighting:

> So ends the criminal advenure [*sic*] of the men who declared that they 'striking in full confidence of victory', and told their dupes that they would be 'supported by gallant allies in Europe' ... The Dublin Insurrection of 1916 will pass into history with the equally unsuccessful insurrections of the past ... The story of last week in Dublin is a record of crime, horror, and destruction, shot with many gleams of the highest valour and devotion. We do not deny a certain desperate courage to many of the wretched men who today are in their graves or awaiting the sentence of their country's laws. The real valour, however, and the real sacrifices were offered on the altar of Ireland's safety and honour. The first tribute must be paid to the gallant soldiers who were poured into Dublin, including at least two battalions of famous Irsh [*sic*] regiments ... but the cost of success has been terrible. Innocent civilians have been murdered in cold blood. The casualities [*sic*] among the troops have been heavy...The destruction of property has been wanton and enormous ... several of the most important business establishments in the city has [*sic*] vanished in flame.[215]

---

213 *The Times*, 29 April 1916.
214 *The Leader*, 29 April, and 6 and 13 May 1916.
215 *The Irish Times*, 28 and 29 April, and 1 May 1916.

The destruction of property in Dublin was also a source of major ire to *The Times* in London, which bemoaned: 'The people of Dublin are now able to gaze upon the smoking ruins to which the wicked folly of the Sinn Féiners has reduced one of the chief commercial quarters of their city.'[216]

Disruption to gas supplies during the fighting interrupted publication of *The Irish Independent*, which had a readership of over 100,000. When it eventually appeared on 4 May, it too expressed its outrage at the Rising, which it disapprovingly referred to as 'this terrible episode of Irish history'. It also lamented that one of 'the most pathetic spectacles' witnessed on Dublin city's streets had been the sight of 'the solitary hearses unaccompanied by mourners proceeding towards the cemeteries'. It also grumbled about the fact that the 'shops on one of the finest thoroughfares in Europe', namely the lower part of O'Connell Street, were 'a shapeless mass of ruins', while most of Henry Street, Middle Abbey Street, Earl Street, Eden Quay, and Prince's Street were 'in a similar condition'. It was also appalled by the indiscriminate activities of mobs of looters, who used the cover of darkness to raid 'many boot, provision, jewellery, and tobacco shops' on O'Connell Street, despite attempts by the insurgents to curtail them by firing blank shots at intervals over their heads. Whole suites of furniture were also robbed from shops, while even costly pianos were seen being 'rolled along the roadway in Henry Street'.[217] The *Irish Independent* also took particular exception to one 'disgraceful incident' witnessed by two of its staff outside Taaffe's outfitting establishment, during which one 'looter went inside and threw out the goods, one of those outside being heard to ask for No. 14 collars'. It also expressed its dismay at what happened in suburban areas such as Ringsend and Ballsbridge, which witnessed 'continual interchanges of shots between the combatants, and many casualties'. This, it noted, was a recipe for 'wild scenes' of looting of 'flour, meal, rice, whiskey, brandy, wines, soaps, tobacco, and miscellaneous articles of merchandise' from 'the great shipping companies' stores'. The paper also remarked that it was also 'a sad spectacle to see the grief-stricken groups in the suburbs who had lost all in the insurrection, and were made absolutely penniless'.[218]

Due to a fire in its Prince's Street premises, which destroyed all of its machinery, publication of *The Freeman's Journal* was suspended after Easter Monday. The next issue, which did not appear until 5 May, also conveyed a sense of shock at the 'stunning horror' of the 'mad enterprise' that had transpired on the streets of Dublin. It took particular exception to the 'reckless and barren

---

[216] *The Times*, 2 May 1916.
[217] *The Irish Independent*, 26, 27, 28 and 29 April, and 1, 2, 3 and 4 May 1916.
[218] Ibid.

waste of life', blaming it on the 'vain promises and delusive lies' of 'men without authority, representative character, or practical sanity'. Moreover, it complained that O'Connell Street 'has been reduced to a smoking reproduction of the ruin wrought in Ypres by the mercilessness of the Hun'.[219] Despite the devastation inflicted on many buildings, traders in the city centre did their best to resume business in the week following Pearse's surrender. The *Irish Independent* of 5 May reported that 'business ... is gradually returning to normal ... and quiet has been established'. Food distribution was proceeding freely, shops were being reopened and the streets were 'being rapidly cleared of the debris by gangs of labourers' who were also demolishing 'toppling buildings'. The city was also full of sightseers, who packed into the vicinity of O'Connell Street.[220] Many difficulties remained, however, and the next day's issue noted how Dublin was 'still completely isolated as regards general telephonic and telegraphic communications'. Additionally, postal deliveries were 'a week behind time' and deliveries of railway parcels were 'disorganised and belated'.[221] Over the next week, efforts to clear the street debris continued apace and *The Irish Independent* reported on 13 May that the lower end of O'Connell Street 'now presents a spectacle similar to a huge building site'.[222]

Upon hearing the news of what happened on the streets of Dublin, many of Ireland's local newspapers followed the line of the national papers by registering their disapproval of the rebels' actions, although some were initially misinformed when pointing out where culpability lay. *The Cork Examiner* published a number of news items about the Rising on a daily basis, including reports on the arrest of Casement in Kerry and various incidents in Tipperary. As the fighting in Dublin was drawing to a close, the leading article of its issue on 28 April expressed concern that the fighting had 'succeeded in spreading consternation all over the country and has cut off all communication from outside with the Irish capital, which so far as food and coal are concerned must be reduced to the extremity of a beleaguered city'.[223] After hearing the news that the leaders of the Rising had surrendered, *The Cork Examiner* of 1 May reported that the ending of 'this mad enterprise' had been greeted 'with relief by every genuine lover of Ireland'.[224] Then, as more vivid accounts of who was responsible for the events in Dublin filtered through, it adopted an increasingly sombre mood on 4 May, commenting that 'the mad orgy of disorder that raged in that city since Easter'

[219]   *The Freeman's Journal*, 26, 27, 28 and 29 April, and 1, 2, 3, 4 and 5 May 1916.
[220]   *The Irish Independent*, 5 May 1916.
[221]   Ibid., 6 May 1916.
[222]   Ibid., 13 May 1916.
[223]   *The Cork Examiner*, 28 April 1916.
[224]   Ibid., 1 May 1916.

had exceeded all expectations. 'Our capital', it continued, 'resembles the pictures we see of some of the Belgian cities ... The humiliating and torturing truth remains that all this devastation was caused by men professing to be Irishmen and acting in the name of Ireland'. It also proclaimed that the signatories to the Proclamation 'have no representative weight whatsoever ... Their actions must be condemned by every right-thinking man'.[225]

Another Cork newspaper, *The Cork Weekly News*, was especially forthright in its abhorrence of the Rising. Its leading article on 6 May, entitled 'The Crushed Irish Rebellion', was exceptionally sarcastic. 'Once again', it proclaimed, 'the folly and wickedness of ... an armed rebellion in Ireland has been demonstrated'. The rebels, it suggested, had suffered a 'rude awakening', as they had 'overlooked the fact that we are no longer living in times when croppy-pikes and blunderbusses were deemed very effective lethal weapons'.[226] The same negative tone was apparent from the individual testimonies of various people around the country. Michael Finucane, for example, later recalled that the people of Kerry were generally unaware at the time of any potential Rising, because the events in Dublin had been kept secret. The general mood of the people in Kerry at the time, he noted, 'was that they didn't want any Rising, they were quite happy with the way life was, and there was no mention of the word "republicanism"'.[227] On 10 May, Gerard Fitzgibbon, an eyewitness to the devastation in Dublin, wrote about his disgust at how 'the finest street in the city is a ... heap of ruins, and for two or three blocks on either side everything is burnt out ... Many bodies have been buried in the ruins'.[228]

The backlash against the Rising was also particularly strong in parts of England, including areas with large concentrations of Irish. Many branches of the United Irish League criticised the actions of the rebels, including the Liverpool and District Committee, which unanimously passed a resolution condemning 'the insane action of a small section of irresponsible Irishmen in Ireland'. It went on to denounce the Rising as 'a treacherous outrage upon Ireland and Ireland's cause, and also against that true liberty in defence of which in this war against German barbarism thousands of Irishmen have died, and ... thousands of Irishmen are fighting at present'.[229] Some English people who visited Ireland to specifically investigate the damage caused by the Rising were especially shocked by what they had encountered in Dublin city centre.

---

[225] Ibid., 4 May 1916.

[226] *The Cork Weekly News*, 6 May 1916.

[227] J. O'Hea O'Keeffe and M. O'Keeffe (eds), *Recollections of 1916 and its Aftermath: Echoes from History* (Privately Published, Kerry, 2005), p. 22.

[228] Gerard Fitzgibbon to William Hume Blake, 10 May 1916, TCD, MS. 11107.

[229] The text of this resolution was published in *The Freeman's Journal*, 6 May 1916.

From his base in the Shelbourne Hotel, Cecil Harmsworth spent days investigating the damage and reported that the 'damage done in Dublin during the recent rebellion' had proven to be 'much greater' than he had been led to believe, and had surpassed 'the whole effect of all the Zeppelin raids that have taken place in England'.[230]

Anger at the actions of the rebels was also apparent amongst troops fighting with the British Army on the continent. This can be seen from a letter written by one artillery officer serving in France, who lamented on 1 May that he did not 'for a moment believe that ... we should have been treated to this humiliating affair in Ireland'.[231] Back in Ireland on the same day, rumours were spreading like wildfire that conscription would be introduced by the government. This caused great anxiety amongst the population, including those who considered themselves as foreign nationals. One of these was Robert O'Neill – a 23-year-old immigrant, who moved from the USA to Ireland in 1901 and who was living in Campile in County Wexford at the time of the Rising. In a letter that was dispatched on 1 May to his Toronto-based father (who counted several American congressmen and senators as his friends), Ohio-born O'Neill complained that a local police sergeant 'claims me to be a British Subject and says I'm under the Compulsion Act, which has now been applied to Ireland'. O'Neill then outlined his fears that conscription would be in force in Ireland 'a fortnight [*sic*] from now' and proceeded to implore his well-connected father for urgent assistance in applying for a passport at the American Consul at Queenstown, which would allow him to travel back to the country of his birth.[232]

The Executions: Reactions and Political Implications

In the aftermath of the surrender, Pearse was executed by a British Army firing squad in Kilmainham Gaol's Stonebreakers' Yard on 3 May 1916. In addition, 14 other rebels were court martialled and executed at Kilmainham by the same means between 3–12 May. Both blindfolds and white cards were used during the executions, with the latter 'pinned over the condemned men's hearts as aiming marks for the soldiers'. Some, like Thomas Clarke, opted not to wear a blindfold. On the morning of the first three executions, the same firing squad

---

[230] Cecil Harmsworth to Alfred Northcliffe, 23 May 1916, Parliamentary Archives, Houses of Parliament, London (hereafter PAHPL), The Lloyd George Papers, LG/D/14/1/8.

[231] Officer Westrope to His Parents, 1 May 1916, IWM, MS. 93/25/1, The Papers of H. L. Franklin.

[232] Robert O'Neill to F. O'Neill, 1 May 1916, USNARACP, RG 84: 350: 21/1/3, Records of the Foreign Service Posts of the Department of State, US Consular Records for Cork.

was used. However, a different firing squad was used 'for each man shot' on all of the following mornings.[233] Afterwards, the bodies of the executed men were buried in Arbour Hill and covered over in quicklime. As punishment for his role in the affray at Bawnard House, Thomas Kent was sentenced to death on 4 May. Days afterwards, on 9 May, he was executed by a firing squad at Cork city's Victoria Barracks, where his remains were buried. This made him 'the only non-Dublin Volunteer' who was executed by the British authorities in the aftermath of the Rising.[234]

Not long after this sequence of dramatic events had unfolded, a tidal change occurred in Irish public opinion about the actions of those who had fought and died in the 1916 Rising, which in turn was to have profound implications for the subsequent course of Irish history. Besides Pearse, those executed at Kilmainham by order of the British government had included the other six signatories of the Proclamation, including a badly wounded James Connolly, who was injured in his arm and leg. In reply to a question put to him as he was carried to his execution, propped up in a chair, Connolly said: 'I will say a prayer for all brave men who do their duty according to their lights.'[235] Outrage at Connolly's execution was graphically evident from a banner that was suspended across Liberty Hall, declaring: 'James Connolly Murdered' (Plate 2.6). Anger was further fuelled by the execution of Joseph Plunkett (who was severely ill), William Pearse (largely for being Patrick Pearse's brother) and Major John MacBride (who had formerly attracted British antagonism during the Boer War of 1899–1902). Among the executed rebel leaders were several intellectuals who would possibly have played an influential political role in the years following 1916, had their lives been spared. Not long after their deaths, Patrick O'Connor lamented: 'it is pitiable to think of men of such splendid ability and genius being cut off in the fullness of their powers.'[236]

Tempers also frayed as a result of the harsh treatment that the authorities had dished out to certain elements of the civilian population. While the statutory powers of the Defence of the Realm Act were normally used in dealings with civilians, Roy Foster has noted that 'reprisals against the innocent and defenceless mounted ominously', as for example when householders in Dublin's North King Street were 'murdered by soldiers' during the course of the Rising.

---

[233] Michael T. Soughley, 25 January 1949, NAI, Bureau of Military History, Witness Statement No. 189.

[234] Townshend, *Easter 1916*, p. 238.

[235] 'James Connolly's Last Words', A Talk by Ina Connolly Heron, Undated, DCLA, Birth of the Republic, BOR OSF 02/05.

[236] Patrick O'Connor to Frank W. Poulter, 20 June 1916, NLI, MS. 4615.

Plate 2.6   A republican flag flying over Liberty Hall, Dublin (and a banner
               suspended across it declaring: 'James Connolly Murdered')
Source: Imperial War Museum, Q 82392.

Anti-British sentiment also increased following the controversial execution of
Francis Sheehy-Skeffington – the well known writer who was shot by a firing
squad at Portobello Barracks under the orders of a British Army Captain,
J. C. Bowen-Colthurst, without the formality of a trial. In Roy Foster's opinion,
Sheehy-Skeffington had possessed a 'popular and saintly' aura, and his 'murder'
represented one of the most high-profile of 'a succession of appalling incidents'
that took place under Martial Law.[237] Back in London, the implications of the
Sheehy-Skeffington incident were soon realised by the British government. Not
long afterwards, Asquith wrote a secret memorandum for the Cabinet, in which
he determined that the occurrence was 'so far as I can judge ... the worst blot on
the proceedings of the military'.[238]

    To make matters worse for the government, Sheehy-Skeffington was well-
known in Ireland as a leading pacifist. Before his demise, he had built up a
formidable reputation as one of the key figures in the anti-recruiting campaign.

---

[237] Foster, *Modern Ireland*, p. 484.

[238] The Irish Rebellion, Cabinet Print on Ireland, Part 1: The Actual Situation, 19 May 1916,
PAHPL, The Bonar Law Papers, BL/63/C5.

Besides his distaste for the Great War, Sheehy-Skeffington was also a critic of the militarism in the Irish Volunteers. As an official enquiry into his death revealed, he was known by eminent trade unionists 'to have been a strenuous opponent to anything like an armed Rising' and was 'consistently opposed to military action on all hands'.[239] The *New Ireland* journal, for whom Sheehy-Skeffington wrote, offered its 'humble tribute of respect' following his death. Although under a military censorship, it did point out that 'in the English press alone tributes to him have appeared since his murder became publicly known'.[240] *The Daily Mail* was particularly concerned about the execution of Sheehy-Skeffington, and demanded that 'every word of the evidence at the inquiry into it must be given in public, and in the presence of newspaper reporters'.[241] *Truth* described the Sheehy-Skeffington affair as 'incomprehensible' and remarked that it 'did not excite less horror in this country [England] than in Ireland'.[242] Maurice Wilkins, writing in *The Irish Citizen*, made the point of highlighting Sheehy-Skeffington's personal links with a raft of progressive organisations, and conveyed his optimism that the tragic circumstances and painfulness of his death would be alleviated somewhat in time by the fact that 'the cause of anti-militarism is infinitely strengthened by his death'.[243]

It was the executed leaders of the Rising who were destined to command most sympathy from the Irish public in the long-run. Once the first few executions had occurred, many English newspapers verged on the side of caution. In a leading article entitled 'The Irish Executions', for example, the London-based *Daily News and Leader* of 6 May expressed its hope that 'we have now heard the last of the death penalties in these Dublin courts-martial, not merely on the ground of humanity ... but as a matter of policy. Anything resembling a reign of terror would be a grievous mistake'.[244] In addition to denunciations of the executions by Irish MPs in the House of Commons, numerous appeals were also sent from concerned Irish citizens to the Prime Minister. On 9 May, for example, a petition was signed by Cork's leading citizens (including the Lord Mayor of Cork, the Assistant Bishop of Cork, the City High Sheriff, and key members of the Cork City Executive of the United Irish League) and then sent by telegram to Asquith. Copies were also sent to John Redmond and to the Lord Lieutenant. It protested 'most strongly against any further shootings ...

---

[239] Thomas R. Johnson and David R. Campbell to Arthur Henderson, MP, 5 May 1916, NA, Kew, WO 35/67.

[240] *New Ireland: An Irish Weekly Review*, 24 June 1916.

[241] Cited in *The Irish Independent*, 12 May 1916.

[242] Cited in *Irish Opinion*, 24 June 1916.

[243] *The Irish Citizen*, 15 July 1916.

[244] Cited in *The Leader*, 29 April, and 6 and 13 May 1916.

and against indiscriminate arrests', as they were 'having a most injurious effect on the feelings of the Irish people'. The petition also warned that the persistence of such policies 'may be extremely prejudicial to the peace and future harmony of Ireland, and seriously imperil the future friendly relations between Ireland and England'.[245] On the same day, during an animated question time in the House of Commons, Mr. Holt, the Liberal MP for Hexham, conveyed similar sentiments when he spoke of 'the grave concern with which many people in England look upon these military executions in Ireland'. Asquith himself conceded that 'no one is more anxious ... than the government that there should be no undue severity in the execution of the law'.[246]

The following day, Earl Loreburn delivered a speech in the House of Lords about the disorder in Ireland. He expressed hope that 'the circumstances will incline the government towards clemency, for no sure foundation was ever laid in the blood of the scaffold'.[247] The English media also called for a softening of the government's response to the security situation. On 12 May, for example, a special correspondent wrote a far-sighted piece in the *Manchester Guardian*, in which he made an urgent plea for immediate restraint, sensing the long-term consequences of the executions:

> It seems to be urgent that the British nation and Empire and the King should be able to realise, before we go too far, a situation in which things are being done within a space of days, even hours, may be determining a long sequence of history in a vitally right or a vitally wrong direction ... The things which the Generals, acting with the best intentions, may do within the next two or three days might very well precipitate the whole course of Irish history for a generation in a wrong direction, and might most dangerously influence the position of the Empire in the war.[248]

The Cork Examiner of 12 May also called upon the government to bring an immediate halt to the executions, by suggesting that 'a policy of clemency ... is also a policy of wisdom'. It also expressed the 'hope that out of the tragedy already enacted Ireland's aspirations will spring afresh and that a brighter future will help to obliterate a sad and bitter memory'.[249] These pronouncements did not fall on deaf ears, for the last of the executions (with the exception of Casement) took place on that day.

---

[245]   The full text of this petition was published in *The Cork Examiner*, 11 May 1916. Due to illness, the Bishop of Cork, Most Rev. Dr. O'Callaghan, was unable to sign it.

[246]   *The Times*, 10 May 1916.

[247]   Ibid., 11 May 1916.

[248]   Cited in *The Irish Independent*, 12 May 1916.

[249]   *The Cork Examiner*, 12 May 1916.

The knock-on effects of the Rising quickly manifested themselves in the political scene, where major changes were swiftly announced in the line up of the British government's administration of Ireland. From the outset of the fighting in 1916, the English media looked for scapegoats. Without hesitation, it leading mouthpieces placed the blame for the Rising firmly upon the powers that be in Dublin, especially the Chief Secretary, Augustine Birrell (who had held the position since 1908), the Under Secretary, Sir Mathew Nathan (who had previously served as Governor of Hong Kong) and the Viceroy, Ivor Churchill Guest (who was also known as Lord Wimborne). *The Times* of 27 April published a vitriolic leading piece entitled 'The Government and Ireland'. It criticised Birrell and his administration for constantly disregarding 'ample warnings of coming trouble' and allowing 'themselves to be caught napping in the face of strongest evidence that a seditious Rising was being prepared under their eyes'. It also asked whether there was 'any reason why politicians who conspicuously fail should be allowed to remain in office?'[250] Pressure on the administration continued unabated from *The Times* on 1 May, which slammed the 'irrepressible ... humour' that Birrell had apparently conveyed to a party of journalists, when making 'airy references' to the Rising.[251]

The resignation of Birrell and his replacement by Lewis Harcourt was first announced in the House of Commons on 4 May. On the same day, various high-profile members of the House of Lords vented their fury at the administration in Ireland, charging it with gross incompetence. Writing in private to John St. Loe Strachey, the editor and proprietor of *The Spectator*, Lord Midleton, who was leader of the southern unionists, singled out the Chief Secretary for particularly harsh criticism: 'If Birrell had been a German spy, which he is not, he could not have done more against his country.'[252] On the same day, General Maxwell, in a letter to his wife, painted a more sympathetic portrait of Birrell as a person. However, he still had critical words to say about the Chief Secretary and his administration: 'I may say that he is not as bad as is thought but like so many politicians he does not put into effect what he preaches.' He also contended that the administration's failure to curtail breaches of the law in the years leading up to the Rising had ultimately caused the 'loss of life' that occurred during Easter Week. 'These Sinn Féiners', he lamented, 'have been allowed to arm ... and train, parade openly ... in uniform and with arms practice street fighting.'[253] Although

---

[250] *The Times*, 27 April 1916.

[251] Ibid., 1 May 1916.

[252] Lord Midleton to John St. Loe Strachey, 4 May 1916, PAHPL, Papers of John St. Loe Strachey, STR/21/1/20a.

[253] General Maxwell to His Wife, 4 May 1916, PULDRBSC, Sir John Grenfell Maxwell Papers, Box 6, Folder 9.

The Times of 8 May gave Birrell some credit for the National University of Ireland Act of 1908, it had little sympathy for his failure to quell political unrest in Ireland, which it attributed to 'his inability to govern a restless and mercurial people with a strong hand'. Furthermore, it took him to task for increased absences from Ireland and for having spent far too much time in Westminster. Rather acerbically, it quipped: 'Like the puppet in the old-fashioned barometer, his disappearance was always a sign of stormy weather in Ireland.'[254]

Later on in 1916, the findings of a Royal Commission on the Rebellion in Ireland unsurprisingly placed the finger of blame on both Birrell and Nathan. It determined that 'the Chief Secretary as the administrative head of Your Majesty's Government in Ireland is primarily responsible for the situation that was allowed to arise and the outbreak that occurred'. Furthermore, it found that the Under Secretary 'did not sufficiently impress upon the Chief Secretary during the latter's prolonged absences from Dublin the necessity for more active measures to remedy the situation in Ireland which on December 18th [1915] ... in a letter to the Chief Secretary he described as "most serious and menacing".'[255] While the nature of Birrell's position demanded that he typically had to spend up to half the year in London, Townshend has offered a more compassionate assessment of his plight, noting that 'he liked Ireland (and professed to like the Irish)', while in Ireland itself, he 'was widely admired – certainly by nationalists – as a humane, intelligent and sympathetic minister, on whom the label "ruler" sat less comfortably than it had on some of his predecessors.'[256] Nathan, by contrast, proved to be an unpopular Under Secretary, having been responsible for implementing a series of harsh budgetary cutbacks. According to Seán Cronin, he was 'a cautious, quiet, methodical man', a disposition suited to the work of a 'career official'. One of his main tasks at Dublin Castle, he notes, had involved the assessment of intelligence reports that had been sent by letter, telephone and telegram. It had also been his duty to liaise with the Irish Office at Old Queen Street in London and to report on the state of affairs in Ireland.[257]

In the end, Nathan lost his position after the Rising and was replaced by Sir Robert Chalmers, who had been Governor and Commissioner-in-Chief of Ceylon in 1913.[258] Lord Wimborne also resigned from his ceremonial position as the Viceroy after the Rising. A little more than a week after the executions,

[254] *The Times*, 8 May 1916.
[255] Royal Commission on the Rebellion in Ireland, *Report of the Commission*, p. 13.
[256] Townshend, *Easter 1916*, p. 24.
[257] S. Cronin, *Our Own Red Blood: The Story of the 1916 Rising* (Irish Freedom Press, Dublin, 2006), p. 38.
[258] *The Times*, 5 May 1916; *The Freeman's Journal*, 6 May 1916; *The Irish Independent*, 6 May 1916.

Asquith himself reached the conclusion that the Irish Viceroyalty had 'become a costly and futile anachronism'. In a secret memorandum dating to 21 May, he went so far as to recommend to his Cabinet that the position should be abolished altogether. Its uselessness, he argued, had 'never been more clearly demonstrated than during the last few months'. The office, he further complained, had 'been reduced to a cipher, to the rôle of a *vice-roi fainéant*, clothed with apparent power and responsibility, but obliged in practice to content himself day by day with the crumbs that may fall from the Under-Secretary's table'. 'It is not, under actual conditions', he added, 'an office that any self-respecting man of parts and ability can be asked to undertake'.[259]

In addition to hostility from the print media, elements from within the forces of Irish constitutional nationalism also blamed the British government for not doing enough to prevent the Rising. Some anger was expressed about the arming of the Ulster Volunteers some three years previously – a situation which had also led to the proliferation of arms outside of Ulster. *The Freeman's Journal*, which was a vocal supporter of the Irish Party, pointed the finger of blame in a number of directions. The leading article of its issue of 8 May consisted of a lengthy exposition of 'The True Causes of the Sinn Féin Insurrection'. It attributed a large part of the blame for 'Dublin's week of horrors' on Carson and the arming of the Ulster Volunteer Force, and flatly accused the government of 'cowering' before threats such as 'armed defiance of the law' by the Carsonites. *The Freeman's Journal* also became embroiled in a bitter war of words with other Irish newspapers over who else was responsible for the outbreak of the Rising. It denounced, for example, the way in which Redmond and the forces of constitutional nationalism had been weakened by certain organs of the press – including *The Irish Independent* – which it accused of scaremongering over the thorny issues of conscription, overtaxation and land taxes. Such 'distrust and despair', it charged, 'was worked to the very last hair among their younger dupes by the insurgents'. Despite earlier stern denials from *The Irish Independent*, *The Freeman's Journal* remained adamant that were it not for the 'propaganda' of 'the *Independent* and what has come to be known as the Mosquito Press', the insurgents would not have been as successful in recruiting so many Volunteers to their ranks.[260]

Not surprisingly, the events of the Rising and the subsequent executions resulted in a series of sharp and frank exchanges between politicians, government officials, the military, and the media. While conceding in a speech delivered to the House of Commons at Westminster on 3 May that the Rising

---

[259] The Irish Rebellion, Cabinet Print on Ireland, Part 2: The Future and Part 3: Transitional, 21 May 1916, PAHPL, The Bonar Law Papers, BL/63/C8.

[260] *The Freeman's Journal*, 8 May 1916.

was a source of personal 'misery and ... heartbreak', John Redmond made an emotional appeal for leniency for the majority of the rebels. He pleaded with the government 'not to show undue hardship or severity to the great mass of those who are implicated', and asked for 'only such action as will leave the least bitterness in the minds of the Irish people, both in Ireland and elsewhere throughout the world'.[261] Redmond, however, came up against stiff resistance from the media in London. After the first three executions, *The Times* of 4 May applauded the government for acting, 'for once, with a salutary swiftness; the prompter the punishment of the leaders, the lesser the need for dealing harshly with their dupes'.[262] At the same time, those with unionist sympathies were eager to commend the military authorities for suppressing the Rising. In a letter sent to Maxwell on 5 May, Walter Long of the Local Government Board (who served as Chief Secretary for nearly nine months in 1905) claimed that Ireland had been 'saved ... from an awful disaster'. He also thanked Maxwell by expressing 'gratitude for the splendid work you have done', adding that it was 'no exaggeration to say ... that your arrival marked a complete change in the scene'.[263] The hardline reaction was also commended by Mr. H. Samuel. In a speech at the National Liberal Club in the English capital, he described the Rising as a 'feather-headed rebellion', characterised by the cold-blooded and deliberate murder of policemen and civilians, and went on to condone the stern punishment that the government had dispensed.[264]

Some in the British administration, however, felt uneasy about the treatment that was dished out to the leaders of the Rising. Before his resignation, for example, Lord Wimborne registered his opposition to the execution of 'comparatively unknown insurgents' and expressed his dismay about the proceedings in a letter to Maxwell on 8 May. In it, he stated that the executions were 'capable of producing disastrous consequences'.[265] Privately, even Asquith himself was fearful of the 'grave danger of ... bitter resentment' arising from the executions, and wrote a memorandum in early May for the purpose of seeking a 'reassuring statement ... without delay'.[266] Maxwell, however, seemed somewhat taken aback by the Prime Minister's anxiety. In a letter to his wife on 9 May, he observed that the government was 'getting very cold feet' over the punishment of

---

[261] *The Times*, 4 May 1916.

[262] Ibid.

[263] Walter Long to General Maxwell, 5 May 1916, PULDRBSC, Sir John Grenfell Maxwell Papers, Box 4, Folder 12.

[264] *The Irish Independent*, 11 May 1916.

[265] Ivor Churchill Guest, 2nd Baron and Afterwards 1st Viscount Wimborne, Viceroy of Ireland to General Maxwell, 8 May 1916, British Library (hereafter BL), ADD. MS. 58372/R.

[266] Asquith Memorandum, 10 Downing Street, 4 or 8 May 1916, BLOU, MS. Asquith 43/9.

the rebel leaders and becoming 'afraid'. Additionally, he registered his frustration at the pressure being put on him at 'every moment' to adopt a softer stance, by not awarding 'death sentences'. He then sought to justify the executions that had taken place by then, arguing that 'some must suffer for their crimes'.[267] In a telegram to Asquith the next day, Maxwell was equally forthright in his defence of the executions. He explained that in view of the gravity of the Rising 'and its connection with German intrigue and propaganda, and in view of the great loss of life and destruction of property resulting therefrom', it was found 'imperative to inflict the most severe sentences on the known organisers of this detestable Rising and of those Commanders who took an active part in the actual fighting which occurred'. He also expressed hope that the executions 'will be sufficient to act as a deterrent to intrigues and bring ... home to them [the rebels] that the murder of His Majesty's subjects and other acts calculated to imperil the safety of the realm will not be tolerated'.[268]

Inevitably, those found guilty 'of hostile associations ... against His Majesty' came under scrutiny from the British authorities.[269] In the weeks and months that followed the Rising, the British resorted to using a range of severe measures to alleviate their security concerns. On the day of Pearse's surrender, for example, the Lord Lieutenant once again proclaimed Martial Law so as to counteract attempts 'to subvert the supremacy of the Crown in Ireland'.[270] Despite the 'ominous sound' of Martial Law and the severity of the regulations, the Dublin-based correspondent for *The Times* noted that the inconvenience of having to be indoors by 8.30 pm was 'counterbalanced by the peace and safety' that came with it.[271] Another measure aimed at tightening the security situation was introduced on 11 May, when Maxwell issued a new order. This stipulated that under no circumstances was a parade, procession, political meeting, football or hurling match, or athletics meeting to be held anywhere in Ireland 'without the written authority, previously obtained, of the Lord County Inspector of Royal Irish Constabulary, or, in Dublin City, of the Chief Commissioner of the Dublin Metropolitan Police'.[272]

---

[267] General Maxwell to His Wife, 9 May 1916, PULDRBSC, Sir John Grenfell Maxwell Papers, Box 6, Folder 9.

[268] General Maxwell to Herbert Asquith, 10 May 1916, BLOU, MS. Asquith 43/17–18.

[269] Joesph McGill's Internment Order, Kilmainham Gaol Museum.

[270] Proclamation of Martial Law throughout Ireland by the Lord Lieutenant, 29 April 1916, NA, Kew, CO 904/23/2b.

[271] *The Times*, 10 May 1916.

[272] Public Notice from General Maxwell Relating to Political Meetings, Parades or Processions, 11 May 1916, NA, Kew, CO 904/23/3.

Pressure continued to be exerted from many quarters for the Liberal government to maintain a firm response to the Irish situation, and a series of letters sent to the Editor of *The Times* went so far as to advocate that the captured rebels should be conscripted and sent to the fight in the trenches on the continent. Such action, suggested one letter-writer on 9 May, would enable the rebels 'find out for themselves what manner of man the German is'.[273] Another letter on the same day, signed by H.S.C, suggested more dramatic measures by urging the continuance of Martial Law. The letter also recommended 'that by means of a general conscription every Irishman of military age may learn the inestimable benefit of self-discipline and submission to authority'. This notion was also supported in a letter from the Archbishop of Armagh, who pleaded: 'Surely the bitter cry from the trenches for more and more Irish soldiers will not be disregarded by the Government, especially in view of recent terrible happenings in Dublin and elsewhere throughout the country.'[274] Thus for many of those who managed to evade arrest and deportation following Pearse's surrender, the days and months following the Rising were very testing. Some members of the Irish Volunteers managed to escape from Dublin city centre through pure luck, by taking a chance at obtaining a safe passage through the barricades that had been erected by the British Army. As he made his escape through one cordon after another, Thomas Smart of the Dublin Brigade noted how he at times felt 'like a rat in a trap'. Finally, after a number of close encounters with the authorities, he noted that his life 'was practically that of a hermit' for around three months.[275]

A leading article in *The Times* of 10 May conceded that the application of conscription to Ireland through the extension of the Military Service Bill 'would be more trouble than it is worth', and suggested that a more appropriate course of action would be 'a thorough round-up of the disloyal organisations throughout the country'.[276] This was indeed what transpired, as conscription was never introduced into Ireland. In total, 90 people were sentenced to death by court martial but in only slightly more than a dozen cases were the penalties implemented. The sentences in the other cases, including those of Eamon de Valera (who went on to serve as President of the Executive of the Irish Free State in 1932) and Countess Markievicz, were later commuted to various terms of penal servitude.[277] De Valera's life was only spared by virtue of the fact that he held an American passport. The son of a Spanish exile, Juan Vivian de Valera and an

---

[273] *The Times*, 9 May 1916.

[274] Ibid.

[275] Thomas Smart, 25 May 1949, NAI, Bureau of Military History, Witness Statement No. 255.

[276] *The Times*, 10 May 1916.

[277] Intelligence Notes, 1916, NA, Kew, CO 903/19/2.

Irish emigrant, Catherine Coll, he was born on 14 October 1882 in the Nursery and Child's Hospital, located in the eastside of Manhattan, New York. After his father's death, his uncle brought him to Ireland to live with his grandmother at the age of three.[278] On the other hand, Markievicz was exempted from execution because of 'the gender specific values which impacted upon the politics of the era'.[279] However, thousands of Irish Volunteers were rounded up and were sent to jails across Britain. This course of action was staunchly supported by Maxwell, who was of the opinion that 'it would be most prejudicial to the safety of the Realm to allow those who were so recently engaged in open rebellion to be at large while the war lasts'.[280]

The recollections of some of the rebels taken into custody attest to the fact that they were treated in different ways following their arrest. At Richmond Barracks in Dublin, it was left to the men of Royal Dublin Fusiliers to guard republican prisoners captured during the Rising and to search their visiting relations and friends (Plate 2.7). Writing from there on 12 May, Bernard O'Rourke noted how he was 'fairly comfortable', despite having 'plenty of company', with 35 men sharing his room.[281] Asquith himself visited the prisoners confined in Richmond, where he 'enquired carefully into their prison treatment ... and invited complaints', but was purportedly told by 'one and all ... that they had none'.[282] Although tempers did fray as a result of the shipment of large amounts of prisoners overseas, some stories emerged of the Volunteers being treated in a humane way. *The Cork Weekly News* of 13 May, for example, carried a detailed report of the transfer of rebel prisoners to a ship at North Quay Wall in Dublin for an unknown destination across the Irish Sea. Its reporter noted that when all of the captives were below decks, their fatigued captors 'had to carry aboard heavy boxes of rations for the prisoners', who 'were given exactly the same fare as the escort' – including tinned beef, good army bread and hot cocoa.[283]

---

[278] J. M. Silinonte, 'Vivian de Valera: The Search Continues', *Irish Roots*, Issue No. 49, No. 1 (2004), pp. 17–19.

[279] C. McCarthy, *Cumann na mBan and the Irish Revolution* (The Collins Press, Cork, 2007), p. 73.

[280] General Maxwell to the Attorney General, 9 May 1916, BL, ADD. MS. 62461, Viscount George Cave Papers, Vol. 7.

[281] Bernard O'Rourke to Clare, 12 May 1916, UCDA, MS. P.117/2, Bernard O'Rourke Papers.

[282] The Irish Rebellion, Cabinet Print on Ireland, Part 1: The Actual Situation, 19 May 1916, PAHPL, The Bonar Law Papers, BL/63/C5.

[283] *The Cork Weekly News*, 13 May 1916.

Plate 2.7　　Men of the Royal Dublin Fusiliers search friends and relations
　　　　　　of republican prisoners captured during the Rising, Richmond
　　　　　　Barracks, Dublin, May 1916
Source: Imperial War Museum, HU 73487.

Although they sometimes encountered a hostile reaction from various
members of the public when they arrived in England and Wales, some of the
rebel prisoners spoke well of their captors (some of whom were Irish). Michael
Newell of the Castlegar Company Irish Volunteers, for example, recalled being
helped during a stop at the train station in Nottingham, whilst en route from
Ireland to prison in Frongoch:

> We were assembled on the station platform under a very strong escort of soldiers with
> rifles and fixed bayonets. We were surrounded by a very hostile crowd of both men and
> women, who jeered us, called us nasty names; they also spat at us. One of the soldiers
> dropped his rifle to the trail position and struck three of the hostile crowd, knocking
> them out. He then shouted, 'Up Carraroe, Up Connemara'. He was John Keane, a
> native of Carraroe. I heard afterwards that he was tried for this assault and sentenced

to two years' imprisonment, but instead was sent with a draft to the Dardenalles. He was not heard of again.[284]

Eoin MacNeill likewise spoke well of his Irish-born captors. After being escorted with 10 others to Dartmoor by the Royal Dublin Fusiliers, he recalled at the end of May that they were 'good decent men, a pleasure to meet'. 'We are all in good form ... in excellent spirits', he wrote, as he recalled how one of the officers guarding him came from Tipperary and was the brother of 'a distinguished science student'.[285]

Others, by contrast, painted a rather different picture of their time in captivity. Michael Hynes of the Kinvara Company Irish Volunteers, who spent a week in Richmond in early May, complained that provisions had been quite scarce. 'We got no food', he protested, 'except a few hard biscuits each day', while cocoa was served due to the absence of tea. For the most part, the prison seemed to be completely overwhelmed and caught unawares by the large numbers of surprise detainees. As there was 'no beds or bed clothes', many of the men ended up sleeping on the floors in the clothes that they were wearing.[286] Michael Lynch from Dublin, who was imprisoned in Knutsford Barracks with other rebels, also complained of being treated harshly by the British authorities. In a letter to a Catholic priest on 1 June, he protested at being 'put to a cruel test' during the 'early stages of our confinement'. 'All the ingenious devices of that civilisation which has been the hypocritical cant of England', he protested, 'were employed in an endeavour to break our spirit'. 'Solitary confinement, jeers, insults, ill treatment ... insufficient food ... [and a lack of] bed clothing', he added, 'were employed with almost fiendish ingenuity'. What sustained them all, however, 'was fervent prayer' and recitations of the Rosary, so much so that their captors were 'obviously impressed'.[287]

Back in Ireland, the weeks and months following the Rising saw no let-up in efforts by the authorities to distinguish between the loyal and seditious elements of the populace. Even though he came to have his suspicions against some of its junior clergy, Maxwell wrote to Archbishop Walsh on 8 May, in order to convey his gratitude for what he perceived as the 'practically universal' services provided by priests of the Catholic Church 'during the recent disturbances in Dublin'.

---

[284] Michael Newell, Undated, NAI, Bureau of Military History, Witness Statement No. 342.

[285] Eoin MacNeill to His Wife, Taddie, 31 May 1916, UCDA, MS. LA1/G/132, Eoin MacNeill Papers.

[286] Michael Hynes, 26 May 1955, NAI, Bureau of Military History, Witness Statement No. 1,173.

[287] Michael Lynch to Father Nevin, 1 June 1916, PICAR, The Papers of John Hagan, HAG 1/1916/68.

Furthermore, he invited Walsh 'to bring to notice individual cases of special gallantry or devotion'.[288] By contrast, the names of 48 people were added to a list of Post Office staff suspected of disloyalty. But when placed within the context of 17,000 workers, this was a miniscule figure and represented a little more than a quarter of 1% of the total workforce. Conversely, the actions of those who had sought to disrupt the plans of the rebels during the Rising, and who helped to restore public services after it ended, did not go unnoticed (or unrewarded) by the authorities. Many staff members belonging to the Post Office were among those who were singled out for special praise and had their service records amended accordingly. In a letter sent to Sir Evelyn Murray in London on 21 June, Arthur Hamilton Norway was strongly grateful for the efforts made by his Dublin-based staff to keep communications lines open. The Secretary for the Post Office alluded to the personal dangers that those on duty during the Rising had faced and made the suggestion that they should be given extra time off as a mark of gratitude for their allegiance.[289]

Efforts were also made to recognise the services rendered by the Royal Irish Constabulary and the Dublin Metropolitan Police in suppressing the Rising. One notable scheme in this regard was the Irish Police and Constabulary Recognition Fund, which was overseen by the Earl of Meath as Honorary Treasurer. In a letter seeking subscriptions to a Bank of Ireland account in early October 1916, its Honorary Secretary, V. C. Le Fanu, commended the police for saving 'not only loyal Irishmen and women from untold sufferings and terrors, but indirectly the whole Empire from very serious trouble, and possibly from disaster'. All funds raised, he continued, would go to 'rewarding those who distinguished themselves, and by giving generous compensation to the widows and dependent relatives of those who were killed or seriously injured' in Easter Week. Among those who contributed to the scheme were the Special Constables of London, who sent 'a cheque for £100, collected voluntarily'.[290] Members of various British Army regiments were also given due recognition for their part in suppressing the Rising. The Nottingham and Derbyshire regiments, for example, received a special message of thanks from Maxwell for their part in the battle for Mount Street Bridge. A range of honours were also handed out to their soldiers, including the Order of St. Michael and St. George, the Distinguished Service Order, the Distinguished Conduct Medal, and the Military Cross.

---

[288] General Maxwell to Archbishop William J. Walsh, 8 May 1916, DDA, Archbishop William J. Walsh Papers (Laity), 385/7.

[289] S. Ferguson, *'Self Respect and a Little Extra Leave': GPO Staff in 1916* (An Post, Dublin, 2005), pp. 51–53, 55.

[290] V. C. Le Fanu to Archbishop Walsh, 5 October 1916, DDA, Archbishop William J. Walsh Papers (Laity), 385/6.

Many other soldiers, according to Paul O'Brien, 'were mentioned in despatches, their deeds of gallantry being temporarily recorded but later forgotten as the casualties from the Great War mounted'.[291]

Towards the end of 1916, the authorities in Trinity College Dublin rewarded members of the Dublin University Officer Training Corps for defending the college buildings from the rebels. Under the chairmanship of Lewis Beatty of Millar and Beatty Ltd. on Grafton Street, an Officer Training Corps (OTC) citizens' commemorative fund was established, and on 24 November the Citizens' Committee presented the OTC with two large cups – one of which was personally presented to Cadet Robert Tweedy. Furthermore, it also presented 132 small duplicates (supplied by West & Sons at £1. 4s. 6d. each) of these to members of the OTC who had been actively involved in protecting the college buildings.[292] An illuminated scroll commemorating the presentation of the two silver cups thanked the OTC and other inmates of Trinity College, noting how the grounds 'were effectively protected from attacks of the rebels and loss of life and destruction of property'.[293] Captain Ernest Henry Alton (who went on to serve as Provost of Trinity College from 1942–1952), was rewarded the Military Cross by the British authorities for defending the college, which he accepted 'not so much as a personal honour, but as a tribute to the OTC and its auxiliaries'.[294] Outside of the hallowed halls of unionist bastions such as Trinity College, however, the public's attitude towards the defeated rebels had by then changed rather dramatically across much of the island of Ireland. This startling metamorphosis had its roots in the weeks that immediately followed the suppression of the Rising. Its implications for the course of modern Ireland's emergence were extraordinarily profound, especially for the subsequent nature of history-making, commemoration and heritage.

---

[291] P. O'Brien, *Blood on the Streets: 1916 and the Battle for Mount Street Bridge* (Mercier Press, Cork, 2008), pp. 91–93.

[292] Dublin University Officer Training Corps, Silver Cup Presented to Cadet Robert N. Tweedy for the Defence of TCD, Sinn Féin Rebellion, Easter 1916, TCD, MS. 10937.

[293] Dublin University Officer Training Corps, Illuminated Scroll Commemorating the Presentation to the DUOTC of Two Silver Cups by Dublin Businessmen in Recognition of the Protection Afforded to Streets Near College by the OTC during the 1916 Rising, 5 August 1916, TCD, MS. 5949.

[294] Recollections of Easter Monday 1916, by A. A. Luce at the time Lieut. 12th Royal Irish Rifles, Later Captain, 14 October 1965, TCD, MS. 4874/2.

Chapter 3

Memory-Making and Cultural Politics, 1916–1965

According to official figures, the intense fighting during the course of the 1916 Rising resulted in up to 452 deaths. Among the casualties were 60 or 62 rebels, a maximum of 132 Crown forces (comprising three Dublin Metropolitan Police constables, 13 members of the Royal Irish Constabulary, 17 Army officers, and a maximum of 99 other ranks) and 256 or 258 civilians. A total of 368 Crown forces were wounded, while the combined numbers of rebels and civilians injured reached 2,217.[1] Notwithstanding the blame that was placed upon them for a significant proportion of the loss of life that had occurred, the days that followed the ending of the Rising proved to be a time when some of its leaders momentarily cast aside their own personal woes, by focusing their attention on how the events of Easter Week would be viewed and remembered for generations to come. In a letter that he wrote to his mother from his cell in Arbour Hill on 1 May 1916 (but which did not come to the attention of historians until 1965), Patrick Pearse maintained that the 'deeds' carried out by the Irish Volunteers during the Rising had been 'the most splendid in Ireland's history'. With an eye to the future, the rebel leader was all too aware of his impending fate in his court martial proceedings: 'People will say hard things of us now, but we shall be remembered by posterity and blessed by unborn generations.'[2]

Despite the initial chorus of public disapproval of the death and destruction wreaked upon Dublin city centre and elsewhere by the fighting, Pearse's longer-term vision of how he and his fellow rebels would be remembered in Ireland was not far off the mark. To its utter dismay, the British government and its most fervent supporters quickly realised the profoundly negative implications of the 15 executions carried out from 3–12 May 1916. The day after the last of the executions, a plainly alarmed Lord Midleton wrote that the government

---

[1] These figures have been derived from C. Townshend, *Easter 1916: The Irish Rebellion* (Allen Lane, London, 2005), pp. 270, 393, who relies on figures furnished in a report that General Maxwell sent to the Secretary of State for War on 26 May 1916, along with modified figures that were subsequently produced by military intelligence.

[2] Pearse, cited in P. F. MacLochlainn (ed.), *Last Words: Letters and Statements of the Leaders Executed After the Rising at Easter 1916* (Office of Public Works, Dublin, 2005), p. 19.

had 'let loose forces of which they are not aware, and Ireland has gone back many years in loyalty in consequence'.[3] A confidential memorandum, drafted by Neville Chamberlain on 15 May and then circulated to members of the Cabinet by the Prime Minister, Herbert Asquith, worryingly reported 'that while public opinion generally throughout Ulster remains opposed to the rebellion, National sympathy is inclining towards the rebels, and particularly in favour of leniency towards the rank and file'. Chamberlain then ominously concluded by noting: 'in some counties ... a significant sign has appeared in a sudden unfriendliness or even hostility towards the police, while ... in many places arms have not been given up by disaffected persons, or ... those handed in have been old and practically useless'.[4] Communications from government workers based in Dublin also drew attention to the repercussions of the executions. As T. P. Gill of the Department of Agriculture and Technical Instruction noted in a memorandum that he sent to Asquith on 20 May: 'The aftermath of the rebellion, arousing old passions and suspicions, has made widespread the belief that faith is not being kept with Redmond over Home Rule.'[5]

The impact of the executions continued to be the subject of much commentary in the weeks that followed. On 23 May, from his sumptuous base in the Shelbourne Hotel, Cecil Harmsworth added credence to a belief 'that popular sentiment has swung around' in the rebels' favour. 'Among the populace', he added, 'there seems to be a curious feeling of pride in the devastation created by a few determined rebels', a sentiment that was further heightened by reports that the casualties amongst the soldiers had exceeded those of the rebel forces.[6] A month later, General Maxwell was equally downbeat when assessing the extent of Ireland's rapid transformation. 'The rebellion', he noted, 'has taken place and been suppressed ... The leaders have been removed ... From one cause or another a revulsion of feeling set in – one of sympathy for the rebels ... the executed leaders have become martyrs and the rank and file "patriots"'. 'There is little sympathy', he lamented, 'for the civilians, police, or soldiers who were murdered or killed in the rebellion'.[7] In stark contrast, the degree to which the 1916 leaders

[3]   Lord Midleton to John St. Loe Strachey, 13 May 1916, Parliamentary Archives, Houses of Parliament (hereafter PAHPL), Papers of John St. Loe Strachey, STR/21/1/20d.

[4]   The Irish Rebellion, Cabinet Print on Public Attitude and Opinion in Ireland as to the Recent Outbreak, 15 May 1916, PAHPL, The Bonar Law Papers, BL/63/C3.

[5]   T. P. Gill to Herbert Asquith, 20 May 1916, PAHPL, The Lloyd George Papers, LG/D/14/1/3.

[6]   Cecil Harmsworth to Alfred Northcliffe, 23 May 1916, PAHPL, The Lloyd George Papers, LG/D/14/1/8.

[7]   The Irish Rebellion, Cabinet Print, Report on the State of Ireland Since the Rebellion, 24 June 1916, PAHPL, The Bonar Law Papers, BL/63/C61.

were revered after their executions can be seen from an emotive poem written by M. J. McManus and printed in *Irish Opinion* on 1 July. The executed rebels, he wrote, had 'brought to this material age ... a meteor-flash of glory', while their passing had left 'a darkened land' and a 'haunting sense of desolation'.[8]

Rising from the Ashes: 1916 as a 'Holy War'

It was not long before the Rising's implications became the subject of intense scrutiny from writers. In *A History of the Irish Rebellion of 1916*, which was first published in 1916, W. B. Wells and N. Marlowe outlined what they claimed were various causes which 'conspired to defeat the prospect that the Rebellion would rapidly become an unhappy incident of the past in Irish history'. They argued that the execution of the leaders in May 1916, coupled with the failure by some to appreciate the gravity of the Rising as nothing more than a mere 'street riot on an extensive scale', gave rise to circumstances that turned a large volume of Irish public opinion 'into a channel of emotional sympathy with the rebels and of strong hostility to the British connexion'.[9] Ultimately, as Roy Jenkins has noted, Ireland was plunged into 'a febrile state' after the Rising.[10] Or as Richard Toye has observed, while the rebels were 'defeated ... without difficulty' by the British military authorities, they 'mishandled the aftermath' with a 'clumsy' reaction that radicalised nationalist attitudes. By the time that Asquith called a halt to the executions, the damage was already done as the production of martyrs had provoked 'a widespread sympathy for the extremists in Ireland that had previously been lacking'.[11] Whilst the executed leaders' sacrifices were ultimately responsible for the persistence of memories of the story of 1916, it must also be stressed that the heavy treatment dished out to the rebels who lived on after Easter Week was also responsible for garnering much sympathy throughout Ireland for the cause of physical force nationalism.

In the aftermath of the Rising, many commentators expressed the view that too many arrests had been made in the rounding-up process that took place throughout Ireland. *The Irish Independent* of 13 May carried an opinion piece entitled 'After the Rebellion', which conveyed the view that the rank-and-file Volunteers who 'took part in the rebellion occupy a position entirely different

---

[8]   *Irish Opinion*, 1 July 1916.

[9]   W. B. Wells and N. Marlowe, *A History of the Irish Rebellion of 1916* (Maunsel and Company Ltd., Dublin, 1916), p. 203.

[10]   R. Jenkins, *Churchill* (Pan Books, London, 2002), p. 310.

[11]   R. Toye, *Lloyd George & Churchill: Rivals for Greatness* (Pan Books, London, 2008), pp. 164, 220.

from the leaders'. It called for them to 'be treated as leniently as possible' and also expressed its opposition to the arrest of men 'who were known to be sympathisers with ... the Irish Volunteers, but who were not connected with the rebellion either as instigators or participants'.[12] Two days later, the predicament facing the British government was highlighted in a confidential memorandum drafted for the Cabinet by the Home Office. By then, 1,800 people had been deported from Ireland. Although recommending that it would be 'inadvisable to set these men free' for the time being, for fear of 'fresh disturbances', it did concede 'that the present state of things cannot continue indefinitely'.[13] Over a week later, Asquith's wife, Margot, invited the Minister of Munitions, Lloyd George, to dinner at 1.45 pm. In seeking some reassurance about attempts to quell growing unrest in Ireland, she urged him to 'please' herself and her husband by doing 'a big thing' in settling Ireland. 'Anyone with wit ... [and] sense of humour', she added, 'must enjoy Ireland ...[,] trying as the Irish are'. She also stated that the entire British Empire would 'be grateful' if he could 'settle it'.[14]

One of the major strategies for tackling the security situation in Ireland was to detain those who had been arrested for over seven months in prisons such as Frongoch Camp in Wales. However, the detentions only added to the sense of public anger against the clampdown on rebel sympathisers. The overall number of people arrested (but not tried by court martial) on a county basis throughout the island of Ireland was 3,343 – over treble the number who participated in the Rising. Orders of internment were made by the Secretary of State in 1,841 cases. Of the 3,343 arrests, the Dublin Metropolitan Police District accounted for 1,696, while the rest of County Dublin accounted for 135. For the rest of Ireland outside of Dublin, a total of 1,512 arrests were made, with the majority of these being made in the province of Connaught – where there were 664 arrests (with 530 of these in County Galway alone). Further south in the province of Munster 308 arrests were made (with County Cork accounting for 140), while in the north of the country only 83 arrests were made in the province of Ulster (with Belfast city accounting for 26). Of the 1,841 people interned, 1,272 were subsequently discharged by Christmas 1916 on the recommendation of an Advisory Committee after their cases had been investigated.[15] Many of the arrests of radical nationalists after the Rising, as Joost Augusteijn has highlighted, were

---

[12]   *The Irish Independent*, 13 May 1916.

[13]   The Irish Rebellion, Cabinet Print on Irish Rebels Interned in England, 15 May 1916, PAHPL, The Bonar Law Papers, BL/63/C4.

[14]   Margot Asquith to Lloyd George, 23 May 1916, PAHPL, The Lloyd George Papers, LG/D/14/1/7.

[15]   Intelligence Notes, 1916, Chief Secretary's Office: Judicial Division, National Archives, Kew (hereafter NA, Kew), CO 903/19/2.

'indiscriminate', and were destined to be amongst the reasons for the return 'of the wish to use physical force after the failure of the Easter Rising'.[16]

Throughout the weeks and months that followed the Rising, passions and emotions ran high. *An Claideamh Soluis* expressed its irritability on 3 May at how 'Martial Law and certain doings ordered by the government have so disturbed the public mind', and immediately called for prisoner releases so as to 'allay public anxiety as to the future'.[17] *The Irish Independent* of 15 May pointed out that the combination of Martial Law and the large amounts of arrests had done nothing to win over Irish hearts and minds. The ongoing 'reign of reprisals and the punishment by penal servitude of hundreds of youngsters who were only dupes', it added, would lead to 'deplorable' results. It also predicted that 'a feeling of revulsion would set in, and sympathy would arise in favour of the prisoners who are now gone'.[18] Cardinal Michael Logue, the Catholic Primate of All Ireland, also expressed deep unease with the extreme response of the government. In a speech to the Maynooth Union on 23 June, he lambasted the authorities for carrying out arbitrary arrests and mass deportations, saying that they 'should have let the matter die out like a bad dream'.[19] Various local authorities also expressed unease with the draconian response to the Rising. The minutes of a quarterly meeting of Dublin Corporation, held on 3 July, recorded the following:

> His Lordship [the Lord Mayor, Councillor James M. Gallagher] ... referred to the fact that hundreds of citizens, men, women, and boys, had been arrested upon suspicion in connection with the insurrection, and were confined in detention camps; and he stated that only the promptest action on the part of the Government, in restoring to their homes people against whom there was no serious evidence, could allay the public resentment which had thus been occasioned.[20]

There was, however, no speedy resolution of the difficulties created by the deportations. Thus in many senses, as Declan Kiberd has surmised, the rebels managed to obtain a type of 'retrospective status as people's heroes' following

---

16   J. Augusteijn, 'Accounting for the Emergence of Violent Activism Among Irish Revolutionaries, 1916–21', *Irish Historical Studies*, Vol. 35, No. 139 (2007), p. 330.

17   *An Claideamh Soluis*, 3 May 1916.

18   *The Irish Independent*, 15 May 1916.

19   Logue, cited in J. Privilege, *Michael Logue and the Catholic Church in Ireland, 1879–1925* (Manchester University Press, Manchester, 2011), p. 114.

20   Municipal Council of the City of Dublin, *Minutes of the Municipal Council of the City of Dublin, from the 1st January to 31st December 1916* (Sealy, Bryers & Walker, Dublin, 1918), pp. 299, 301.

the harsh response of the British authorities.[21] Furthermore, as Conor Cruise O'Brien has noted, 'the general disillusionment that accompanied the enormous casualties of the Great War' led to a discrediting of Home Rule politicians 'who had supported the British war-effort', especially in Catholic areas.[22]

Another significant factor that fed into the rebels' popularity was the way in which leaders such as Pearse, MacDonagh and Plunkett had mixed politics with religion. These men, as Paul Bew has noted, 'were imbued with an overriding principle – that of sacrificial patriotism'.[23] Thomas MacDonagh, for example, noted in his last letter prior to his execution that he was 'ready to die, and thank God that I am ready to die in so holy a cause. My country will reward my deed readily ... God approves of our deed ... It is a great and glorious thing to die for Ireland'.[24] According to David Thornley, the 'concept of the cleansing effect of bloodshed' had enabled Pearse to develop 'a vision of the overthrow of injustice by the sacrificial death of virtue'. His nationalism, he adds, 'had a strongly religious, even Messianic quality', while his loss of life 'elevated him into the most sacred realms of national mythology'.[25] Many of those who met Pearse in the run-up to the Rising had been struck by the passion of his religious commitment. On one occasion, for example, Pearse visited Basin Lane Convent and met some of the nuns who were stationed there. One of these was Sister Francesca (Thomas MacDonagh' sister, Mary), who later recalled: 'I had no idea about the Rising and I asked Pearse why he would not join a monastery.'[26] But Pearse saw a more belligerent role for himself and had strongly promulgated the virtues of a heroic blood sacrifice in the lead-up to the Rising. He appeared convinced that a Christ-like gesture would ultimately lead to the redemption of the nation. In his famous oration at the graveside of O'Donovan Rossa in 1915, for example, Pearse had stated: 'I hold it a Christian thing ... to hate evil, to hate untruth, to hate oppression; and, hating them, to strive to overthrow them.' 'Life springs from death', he continued, 'and from the graves of patriot men and

---

[21]   D. Kiberd, *Inventing Ireland: The Literature of the Modern Irish Nation* (Vintage Books, London, 2006), p. 193.

[22]   C. Cruise O'Brien, *Memoir: My Life and Times* (Poolbeg Press Ltd, Dublin, 1998), p. 83.

[23]   P. Bew, *Ireland: The Politics of Enmity 1789–2006* (Oxford University Press, Oxford, 2007), p. 375.

[24]   Extracts from the Last Letter of Thomas MacDonagh, Kilmainham, Midnight, 2 May 1916, Trinity College Dublin (hereafter TCD), MS. 4631/359.

[25]   D. Thornley, 'Patrick Pearse', *Studies: An Irish Quarterly Review*, Vol. 55, No. 217 (1966), pp. 10, 18.

[26]   Sister Francesca (Mary McDonagh), Undated, National Archives of Ireland (hereafter NAI), Bureau of Military History, Witness Statement No. 717.

women spring living nations'.[27] In the same year, Pearse's play, *The Singer*, proved particularly revealing about his beliefs. It featured a messianic hero, MacDara, who welcomed death and who also stirred up the people with rebellious songs.

The notion of sacrifice, so that Ireland could be reborn, was a repetitive element of the leader's revolutionary manifesto. 'Bloodshed is a cleansing and sanctifying thing', he once wrote, adding that it would necessitate 'the blood of the sons of Ireland to redeem Ireland'. In writing about Robert Emmet's death in 1803, Pearse played strongly upon the notion of religious martyrdom, noting that the patriot had delivered 'a sacrifice Christ-like in its perfection ... dying that his people might live, even as Christ died'.[28] Furthermore, in his last message before his execution, Pearse recalled how he, as a 10-year-old child, 'went down on my bare knees by my bedside one night and promised God that I should devote my life to an effort to free my country'.[29] *An Claideamh Soluis* conveyed similar sentiments during the time of the execution of the 1916 leaders, proclaiming on 10 May 1916 that 'God never deserts a people who hunger for their destiny and who are willing to work and wait'.[30] To contemporary observers such as T. P. O'Connor from London, it was clear that the Rising had 'taken on a semi-religious aspect'. This phenomenon, he noted, was exacerbated in the weeks after the executions by the holding of overcrowded Masses in Dublin's Catholic churches, which typically ended 'in a political demonstration after the service'.[31]

In a letter written 'without reserve' and sent to Lloyd George on 3 June, John Dublin from the Protestant Archbishop of Dublin's palace, warned that future stability in Ireland would partly depend 'upon the line taken by the Roman Catholic priesthood ... [who] represent, as things are, the strongest moral force in the south of Ireland'. Whilst noting that 'the tradition of the Roman Church in their countries is to support law and authority', he cautioned that there was 'abundant evidence that in too many cases the younger Roman Catholic priests favoured the mad revolt of Easter week'.[32] Dublin's fears were by no means

---

[27]   P. H. Pearse's Oration at the Graveside of O'Donovan Rossa, Glasnevin Cemetery, August 1915, Pearse Museum.

[28]   Pearse, cited in G. D. Zimmerman, *Songs of Irish Rebellion: Irish Political Street Ballads and Rebel Songs, 1780–1900* (Four Courts Press, Dublin, 2002), p. 71.

[29]   Photostat of Patrick Pearse's Last Statement/Message before his Execution, 2 May 1916, NAI, Department of the Taoiseach, 99/1/467, Leaders of Rising of Easter Week, 1916.

[30]   *An Claideamh Soluis*, 10 May 1916.

[31]   T. P. O'Connor to Lloyd George, June 1916, PAHPL, The Lloyd George Papers, LG/D/14/2/35.

[32]   John Dublin to Lloyd George, 3 June 1916, PAHPL, The Lloyd George Papers, LG/D/14/2/7.

unfounded, notwithstanding efforts by elements of the Catholic hierarchy to alleviate the concerns of the authorities. A particularly graphic indication of the way in which Pearse was revered by the lower ranks of the Catholic clergy, can be seen from a letter written on 22 June by Reverend James Campbell and forwarded to his parents. Campbell, who was a witness to Pearse's execution the previous month, wrote his dispatch shortly after a meeting that he subsequently had with Pearse's mother. She had spoken to him of how glorious her sons' deaths for Ireland were. In a letter tinged with emotion, Campbell conveyed his admiration of Patrick and William Pearse, describing them as

> saintly souls who worked for the Holy Cause – and such a Cause, that breeds such
> 'Immortal Souls' can never die ... She [Pearse's mother] ... spoke to me so simply of
> [the executions] ... 'I tried hard not to cry ... and said I give you both to our God, the
> God of Ireland and isn't it strange Father I then said a little prayer over him [William]
> to the Blessed Virgin offering up my two sons as she had offered her's' ... I loved Mrs.
> Pearse the moment I laid eyes on her ... We must not lose all hope – a cause that can
> produce such adherents and attract such souls is a holy noble one, it is a sacred thing
> to live and die for.[33]

Thus for many within Ireland, the events of the Rising gradually took on a quasi-religious significance, and the quest for freedom came to resemble something akin to a 'Holy War'.

Stories also emerged about the way in which various members of the Irish Volunteers had sought comfort from their religion as the Rising proceeded. When he met Pearse inside the GPO on Easter Monday, Monsignor Michael Curran enquired whether there was anything he could possibly do. 'No', said Pearse, 'but some of the boys would like to go to Confession and I would be delighted if you would send over word to the [Pro-] Cathedral'.[34] The same scene was repeated in several of the buildings occupied by the rebels. In his recollections of the occupation of the Marrowbone Lane Distillery by the 4th Battalion Dublin Brigade, Séamus Kenny recalled that 'some of the fellows began to cry when they heard shots, because they were a long time from Confession'. As a result, two priests were recruited from Mount Argus 'to come and hear their Confessions'.[35] There was no let-up in the rebels' religious

---

[33]   Rev. James Campbell to his Parents, 22 June 1916, Boole Library, University College Cork (hereafter BLUCC), MS. U.205.

[34]   Right Reverend Monsignor Michael Curran, 17 June 1952, NAI, Bureau of Military History, Witness Statement No. 687.

[35]   Séamus Kenny, 27 October 1948, NAI, Bureau of Military History, Witness Statement No. 158.

commitment after the Rising had ended. Those imprisoned in Frongoch, for example, saw prayer as an essential ingredient for keeping up morale. In the end, a Conference of St. Vincent de Paul Society was established in the camp, so as to see to the spiritual needs of the Irish inmates. One of its representatives was Henry Dixon, who wrote to Archbishop William J. Walsh on 12 July, in order to highlight the dearth of religious facilities on offer. He complained that impediments such as stringent War Office rules had 'interposed against spiritual administrations', thus leading to 'the breaking of the habit of frequent Confession, and the reception of Holy Communion'. As a remedy, he sought the appointment of a full-time chaplain at the camp.[36]

The importance accorded to religion was also evident from the poetic writings of certain individuals, who graphically encapsulated the memory of the executed leaders. Séamus O'Sullivan's poem in honour of the executed Seán MacDiarmada, written in 1916, lamented that the British 'have slain you ... never more the eyes will greet ... as your stick goes tapping down the heavenly pavement'.[37] Another poem written in 1916 by O'Sullivan, spoke of a 'lordier requiem' for those who had died in the Rising.[38] Religious symbolism was also evident in a poem written by Lionel Johnson, who asked the Mother of God, 'whom our fathers called the Queen of Ireland', to 'save Ireland'. Johnson also asked the Lord to grant Ireland 'a great war for the liberties of our people'.[39] Several ballads were also produced, with many of them containing the line 'Who fears to speak of Easter Week?' at the beginning. Initially, these were either copied by hand or typewritten, and then passed around. By the late 1920s, ballads celebrating the rebels of 1916 formed part of 'the standard repertoire of ... patriot singers'.[40]

The Rising and the Wider World

Outside of Ireland, the suppression of the Rising and the subsequent executions and arrests led to much comment throughout Europe and the USA. Critically, the fighting in Dublin also impacted upon the livelihoods of the thousands of Irishmen who had enlisted in the British Army from 1914 onwards. Up until

---

[36] Henry Dixon to Archbishop William J. Walsh, 12 July 1916, Dublin Diocesan Archives (hereafter DDA), Archbishop William J. Walsh Papers (Laity), 385/6.

[37] Poem by Séamus O'Sullivan, Untitled, TCD, MS. 4631/370a.

[38] Poem by Séamus O'Sullivan, Untitled, TCD, MS. 4631/398a.

[39] Memorial Card for Thomas J. Clarke, John Daly and John Edward Daly (died 1916) with a Litany and a Poem by Lionel Johnson, TCD, MS. 4649/4735.

[40] Zimmermann, *Songs of Irish Rebellion*, pp. 72–73.

1916, argued Monsignor Curran, 'nothing could convince most continentals [in Europe] but that Ireland was a kind of province of England'. 'The Rising', he added, 'showed the difference. It could not be explained away'.[41] The actions of the Irish Volunteers garnered much attention from the Catholic Church in Europe. As a result of regular communications from Ireland, clerics in the Pontifical Irish College in Rome were very much aware of the significance of what had happened in Ireland. In a diary entry dated 26 May 1916, Curran observed the following:

> Irish ecclesiastics returning from Rome described the effects of the news of the Irish Rising. It produced a profound impression there. For the first time in many quarters, the difference of the two countries [Ireland and England] was realised and in a way no other method could effect.[42]

Within the Vatican itself, Pope Benedict XV was well aware of the gravity of what had happened on the streets of Dublin. Before the outbreak of the Rising, he was visited by Count Plunkett, who allegedly briefed him on Pearse's intentions and asked him for his Blessing for the Irish Volunteers. The Pope, according to Curran, demonstrated 'great perturbation' and 'profound anxiety', and asked whether their objectives could not be achieved by peaceful means. He then counselled Plunkett to see Archbishop Walsh back in Dublin. When he returned to Ireland, Plunkett failed in his attempt to have an audience with the Archbishop, after showing up at his residence at around 12.00 noon on Easter Monday. The meeting never took place, according to Curran, as Walsh 'was ill in bed' with eczema.[43]

Not surprisingly, it was in the USA that the large Irish Diaspora became most passionate and vocal about the events that had transpired on the streets of Dublin. From the outset, this did not go unnoticed in London, where *The Times* of 28 April worryingly observed that the 'armed conspirators in Dublin ... are strongly supported by the Irish secret societies in America, who are doubtless reaping a golden harvest from German sources'.[44] Throughout many

---

   [41]   Right Reverend Monsignor Michael Curran, 17 June 1952, NAI, Bureau of Military History, Witness Statement No. 687.

   [42]   Ibid.

   [43]   Ibid. For an insight into the seriousness of the Archbishop's health problems, see T. J. Morrissey, *William J. Walsh, Archbishop of Dublin, 1841–1921: No Uncertain Voice* (Four Courts Press, Dublin, 2000), p. 279, who notes that 'he was physically almost entirely cut off from outside contact during nearly all of 1916'. The eczema, he adds, was of a 'disfiguring and weakening nature' and proved to be 'a disheartening and depressing experience' for Walsh.

   [44]   *The Times*, 28 April 1916.

parts of the USA with high concentrations of Irish, there was a rapid escalation of anti-British sentiments once the executions had commenced. On 6 May, the New York-based paper, *The Gaelic American*, lampooned 'the mouthings of the Anglomaniac press' in the city and asserted that 'Ireland's claim to nationhood has been established before the world'. It also declared that 'the American people have been touched by the splendid gallantry of the Irish Volunteers', and pointed out that an enthusiastic show of support for the objectives of the Rising was evident during a mass meeting held at the Cohan Theatre, located at Broadway and 43rd Street. Attendees, it noted, were 'deeply moved by the news from Ireland' and 'enthusiasm was of the hottest and most fervid kind' until a speech being delivered by one of several Catholic priests present, Father Power, was interrupted by an Englishwoman seated near the stage. Thereafter, it noted that 'the poor, trembling, fat creature was promptly escorted to the door', while the priest went on to make 'probably the best speech of his life'.[45]

For the most part, as Jay Dolan has explained, reaction to the Rising amongst the Irish community in the USA 'was similar to that in Ireland – shock and sadness at such a foolhardy attempt to overthrow the world's most powerful government', followed by a swift change in public attitudes following the executions.[46] Such shifting sentiments were reflected in the headline of *The Gaelic American* of 13 May 1916, which read: 'Military Massacre in Dublin Under the Rule of an English Courtmartial'. The shooting dead of the rebel leaders by firing squad, it noted, 'added to the glorious roll of Irish martyrs ... names which will be cherished for all time by the Irish race ... Irishmen the world over, aroused to bitter anger, will exact heavy retribution'.[47] The anger generated by the executions was clear for all to see only a few days later, when 25,000 people turned up to a meeting at Carnegie Hall. Only 5,000 of them were able to gain admission to the hall, where they listened to popular Irish airs such as 'God Save Ireland' and speeches by eminent individuals such as Justice Edward Gavegan of the US Supreme Court. In the 'high state of tension' that gripped the venue, 'cheer after cheer and shout after shout could be heard for blocks away', amidst a mounting 'wave of anger and indignation'.[48]

Organisations in the USA such as the Friends of Irish Freedom, which had been established in New York's Hotel Astor on 4 March to promote Irish interests in America, welcomed news of the Rising with open arms. As Michael Doorley has noted, the outbreak of the fighting in Dublin presented it with 'a golden opportunity' to spread its message to thousands of its fellow Americans. After

---

[45]   *The Gaelic American*, 6 May 1916.

[46]   J. P. Dolan, *The Irish Americans: A History* (Bloomsbury Press, New York, 2008), p. 201.

[47]   *The Gaelic American*, 13 May 1916.

[48]   Ibid., 20 May 1916.

the executions, it organised mass meetings in locations with high concentrations of Irish-Americans, condemning the actions of the British authorities and calling for the prolongation of American neutrality in the Great War.[49] On 22 May, up to 5,000 people gathered at two halls in Providence, Rhode Island, for meetings organised by the organisation's St. Enda's Branch. Addresses were made to both meetings by Senator Albert B. West and former Congressman Joseph F. O'Connell, whose words were greeted with enthusiasm and cheers. According to *The Gaelic American*, the Chairman of the meeting held at the Opera House 'declared that as long as there is any pure Irish blood anywhere on earth, Ireland will never be satisfied with being a part of the British domain'.[50]

Following the executions, the Friends of Irish Freedom also organised speaking tours for '"exiles" from Ireland', including Liam Mellows, Hanna Sheehy-Skeffington (the wife of Francis) and Nora Connolly (the wife of James).[51] Sheehy-Skeffington toured the United States in December 1916, speaking to Irish communities in towns such as Butte in Montana, while in the following month she presented Woodrow Wilson with a petition that highlighted the issue of Ireland's 'claims as a small nation governed without consent'.[52] The lobbying for American support persisted for some time afterwards. Almost a year after the Rising broke out, for example, the James Connolly branch of the Friends of Irish Freedom remained very active in perpetuating the memory of the executed leader of the Irish Citizen Army, by hosting dances and a number of other activities for Irish-Americans.[53] In the years that followed, the transformation which came over Ireland and its far-flung Diaspora after the Rising proved to be a matter of great interest to the American author and journalist, Charles Edward Russell, who wrote colourfully about the impact of the executions. At one stage, in a draft of an article on Ireland's rebellious past, he claimed that Pearse's death had amounted to an 'intolerable perversion of justice'. But 'the story of Ireland', he avowed, was 'without a parallel', given 'what it has meant as a torch to light the nations struggling up to freedom'.[54]

---

[49]   M. Doorley, 'The Friends of Irish Freedom: A Case Study in Irish-American Nationalism, 1916–21', *History Ireland*, Vol. 16, No. 2 (2008), p. 25.

[50]   *The Gaelic American*, 27 May 1916.

[51]   Doorley, 'The Friends of Irish Freedom', p. 25.

[52]   Cited in M. Campbell, 'Emigrant Responses to War and Revolution, 1914–21: Irish Opinion in the United States and Australia', *Irish Historical Studies*, Vol. 32, No. 125 (2000), p. 83.

[53]   Dance Programme, Euchre and Dance of the James Connolly Branch, Friends of Irish Freedom, 19 April 1917, Allen Library (hereafter AL), Box 184/Casement Box 2, Folder 6, Monteith/Casement Papers.

[54]   Article on Ireland, Undated, Library of Congress, Manuscript Division (hereafter LCMD), Charles Edward Russell Papers, Container 37, Writings File.

Back in Ireland itself, the needs of the imprisoned rebels were by no means overlooked in the weeks following the executions. At a meeting in the Dolphin Hotel on 18 May, a decision was taken to establish The Political Prisoners' Families Aid Association, with the objective of engaging in fundraising for the 'relief of distress' among those imprisoned and for 'the making of suitable provision' to support their dependants.[55] In the same month, the Irish National Aid Association and the Volunteer Dependents Fund were also founded to raise funds for the families and dependents of the rebels. Both of these organisations were subsequently amalgamated into a single entity known as the Irish National Aid and Volunteer Dependents' Fund, following a meeting convened by John Archdeacon Murphy, a representative of the Irish Relief Fund Committee of America. In addition to its Irish activities, the new organisation, with Reverend Richard Bowden of the Pro-Cathedral as Chairman, called for the establishment of branches 'in every place abroad where Irish exiles gather'.[56]

Nationalist newspapers did their utmost to flag the significance of the support that was forthcoming for the rebels on a transnational basis. The inaugural issue of *The Irish Nation*, which appeared on 24 June, was particularly vocal in appealing to 'every reading man and woman in Ireland' for support. In language of almost biblical proportions, it alluded to 'volcanic lava streams of political passion, long latent' coming 'once more into play with the same destructiveness and ferocity' for 'the common good of all Irishmen'.[57] J. Clerc Sheridan, writing for *New Ireland* on the same day, spoke of the 'great surprises' that 'have overtaken the Irish people', and noted that Ireland 'finds itself in bewilderment produced by abrupt change'.[58] The sense of transformation was also apparent in many other part of the wider world, including the antipodean British colonies. Three days later in New Zealand, for example, the Irish-born and Rome-educated Catholic priest, James Kelly, focused on British culpability, telling a friend that 'Maxwellian monstrosities have made all over the world heroes of those who were only fools'.[59] Although initially sceptical about the actions of the Irish Volunteers, Kelly went on to put forward a redemptive elucidation of the Rising by using his writings in the weekly newspaper, the

---

[55]   Notes from a Meeting in Dolphin Hotel re. Political Prisoners, 18 May 1916, DDA, Monsignor Michael Curran Papers (Political).

[56]   Report by the Irish National Aid and Volunteer Dependents' Fund, October 1916, DDA, Archbishop William J. Walsh Papers (Laity), 385/5.

[57]   *The Irish Nation*, 24 June 1916.

[58]   *New Ireland: An Irish Weekly Review*, 24 June 1916.

[59]   Kelly, cited in R. Sweetman, '"Waving the Green Flag" in the Southern Hemisphere: The Kellys and the Irish College, Rome', in D. Keogh and A. McDonnell (eds), *The Irish College, Rome and its World* (Four Courts Press, Dublin, 2008), p. 220.

New Zealand Tablet, to express outrage at the denial of Irish self-determination. As Rory Sweetman has pointed out, Kelly succeeded in transforming the paper into 'a vigorous apologist for the new Irish nationalist orthodoxy'.[60]

By no means, however, did such views meet with universal acceptance. Indeed the state of affairs in Ireland remained very delicate throughout the summer of 1916, with nationalists and unionists expressing contradictory views about the Rising. In England, supporters of Carson lobbied Bonar Law, the Secretary of State for the Colonies, to help the cause of unionism. Charles Dunkley from Earl's Barton in Northampton, a member of the Unionist Party, wrote to Law on 25 July, urging that it was time 'to seize this unique opportunity for a settlement' in Ireland that would be satisfactory to the unionists of Ulster. 'I am', he wrote, 'convinced that nothing now but generous treatment will heal the wounds that have been inflicted in the past'. Noting that he had previously broken rank with the Liberals on the Home Rule issue during William Ewart Gladstone's time, he stressed that 'the situation is so different now' and insisted that Carson should be treated kindly.[61]

In Ireland and certain parts of the wider world, one of the main talking points of the summer of 1916 was the fate of Sir Roger Casement, who had attempted to land arms three days before the outbreak of the Rising. In light of the fact that he had worked with the British Consular service up until 1913, news of his links with Germany in 1916 provoked very heated reaction from a number of newspapers. The incident, according to *The Times* of 25 April, earned Casement 'remarkable notoriety'.[62] Four days later, *The Cork Weekly News* conveyed to its readers its disgust at Casement's 'mischievous' activities and his German links. It also rejoiced in the fact that they had 'at last been brought to a full-stop'. The charge of treason, it quipped, was 'a nasty charge to have to answer, and, a particularly odious one in the case of person who is a retired member of the British Consular Service'.[63] Unsurprisingly, given his high-profile status (which derived from his track-record in humanitarian issues in the Congo and the Amazon, where he campaigned against the oppression of the indigenous people), many appeals on Casement's behalf were made to clergymen and politicians alike.

In a heart-rending letter from Buswell's Hotel on 5 July, Agnes O'Farrelly and Maurice Moore informed Archbishop Walsh that they were seeking his assistance to petition the authorities 'to show mercy' for Casement, as they as

---

[60]   Ibid., p. 221.
[61]   Charles Dunkley to Bonar Law, 25 July 1916, PAHPL, The Bonar Law Papers, BL/53/4/6.
[62]   *The Times*, 25 April 1916.
[63]   *The Cork Weekly News*, 29 April 1916.

they believed 'that any further shedding of blood' would damage Anglo-Irish relations and exacerbate 'the irritation which unfortunately exists'.[64] On the same day, C. P. Scott wrote to Lloyd George, asking whether it was 'politically possible' to save Casement from death. He went on to recommend that clemency would 'be in the highest degree politically expedient'. 'I write to you', he wrote, 'because you have a freer mind than anybody else' and 'because you have been chosen by the Cabinet as pacifier of Ireland ... [and] ought to be listened to on such a matter'.[65] Another Cabinet member to be lobbied on the Casement case was the Home Secretary. William Butler Yeats wrote to him on 14 July, warning him that the execution of Casement would have an 'evil ... effect' in Ireland and beyond:

> young people, on whom perhaps the intellectual life of Ireland depend, are less likely to be restrained by fear than excited by sympathy. There is such a thing as the vertigo of self sacrifice. The evil has been done, it cannot be undone, but it needs not to be aggravated weeks afterwards with every circumstance of deliberation. I am convinced that the execution would have an evil effect in America. I had a letter a couple of weeks ago from a keen unpolitical observer in New York describing the execution of the fifteen Irish leaders as greater a shock to the American opinion than the sinking of the Lusitania.[66]

When discussing Casement's situation a couple of months beforehand, the Cabinet had given consideration to the probable result that his execution would have on public opinion in the USA. A confidential Foreign Office memorandum, dated 13 May, was circulated to its members by E. Grey. This was compiled by 'E.P.', who had met with two 'prominent' Irish-Americans based in New York on 28 April, namely the art dealer, John Quinn and Bourke Cochrane. 'Their view', he reported, 'was that Casement's execution would 'rouse sympathy for the rebellion amongst the floating mass of Irish-American malcontents' and 'lend dignity to an absurd adventure which ought to have been smothered in ridicule'. Based on views obtained from 'various leading politicians at Washington' on

---

[64] Agnes O'Farrelly and Maurice Moore to Archbishop Walsh, 5 July 1916, DDA, Archbishop William J. Walsh Papers (Laity), 385/7.

[65] C. P. Scott to Lloyd George, 5 July 1916, PAHPL, The Lloyd George Papers, LG/D/18/15/14.

[66] Appeal by W. B. Yeats for Clemency in the Trial of Roger Casement, 14 July 1916, NA, Kew, HO 144/1636/311643/45.

27 April, 'E.P.' predicted that Irish-Americans would 'be unable to understand his execution except as a piece of [British] vindictiveness'.[67]

Whilst opinion in England was eventually split on what sort of punishment Casement should receive, demands for clemency did not succeed. Casement was finally convicted at a trial in London on 30 July, where he passionately defended his actions and attributed them to 'loyalty for Ireland'. He also spoke of how they were based 'on a ruthless sincerity that forced me to attempt in time and season to carry out in action what I said in words'.[68] After being sentenced to death by hanging, many of his supporters again pleaded for clemency, citing his reputation for 'unflinching judgement in matters of tyranny over the oppressed'.[69] Despite protestations, however, Casement's fate was sealed and he was eventually hanged in Pentonville Prison, London on 3 August. On the night prior to his execution, he wrote from his cell that his dominating thought 'was to keep Ireland out of war. England has no claim on us, Law or Morality or Right'. Ireland, he continued, 'shall not sell her soul for any mess of Empire'.[70] His hanging came as a surprise to many, who felt that his life could have been spared at the eleventh hour. In its monthly issue, issued nearly a fortnight after Casement's death, one writer for *The Irishman* fervently lambasted the British daily press for indulging in a frame of mind, which it alleged, had recited 'distorted details concerning the private character of a man, in order to prejudice public feeling against him'.[71] Following his execution, Casement's body was interred in England, where it lay for the next 49 years.

According to Michael Laffan, 'nationalist impressions of British vindictiveness seemed to be confirmed' following the hanging.[72] Ultimately, the heavy-handed approach to dealing with the leaders of the Rising in May and August 1916, was destined to prove a disaster for the British government in the court of public opinion, not only in Ireland, but in other parts of the world too. Casement's execution, according to Dr. Edward Nolan, Secretary of

---

[67]   The Irish Rebellion, Cabinet Print on the American Point of View Respecting Casement, 13 May 1916, PAHPL, The Bonar Law Papers, BL/63/C2.

[68]   *Irish Opinion*, 8 July 1916.

[69]   Anon., *The Sinn Féin Leaders of 1916. With Fourteen Illustrations and Complete Lists of Deportees, Casualties, Etc.* (Cahill & Co., Dublin, 1917), p. 3.

[70]   Father James McCarroll, Passages Taken from the Manuscript Written by Roger Casement in the Condemned Cell at Pentonville Prison, Pamphlet Produced for Private Circulation among the Friends of Roger Casement, 1950, AL, Box 184/Casement Box 2, Folder 1, Monteith/Casement Papers.

[71]   *The Irishman*, 15 August 1916.

[72]   M. Laffan, *The Resurrection of Ireland: The Sinn Féin Party 1916–1923* (Cambridge University Press, Cambridge, 2005), p. 56.

the Academy of Natural Sciences in Philadelphia, 'was another British blunder'. In a letter that he wrote to John Quinn on 23 August, Nolan stated that were it 'not that her defeat would involve France and Belgium I would rejoice in the chastening of England by Germany for the next century or two'.[73] The controversy surrounding Casement's death was still proving to be a very topical issue in American politics almost two months after the hanging. Those at the very top of the political spectrum were much aware of both the magnitude and implications of what had transpired at Pentonville, as it had become a febrile issue in the Presidential election campaign.

On 29 September, for example, Joseph Tumulty, the secretary to President Woodrow Wilson, wrote a letter to Robert Lansing, the Secretary of State, enquiring whether there was 'anything new' regarding the fallout from the administration's failure to gain clemency for Casement. 'The Casement matter', he cautioned, had 'played a prominent part in the recent New Jersey primaries and resulted in bringing many voters to Martine's standard'.[74] On the same day, Michael Francis Doyle, an exasperated Democratic supporter based in the Irish-American stronghold of Philadelphia, wrote a rather panicky letter to Tumulty. Highlighting the 'unfortunate situation' which he felt had arisen from the wider political fallout from the Rising, he complained that the Republicans were making a campaign question 'of the delay in sending the Senate Resolutions' that could have saved Casement's life, had they been presented to the British in time. Doyle also expressed his dissatisfaction 'that the great bulk of Catholics and those of Irish descent' had been 'opposing the President', due to the mishandling of the Casement affair.[75]

Back in Europe, the consequences of the actions taken by the rebels of Easter Week became very evident in the sphere of Anglo-Irish relations. Matters also became quite complicated for the thousands of Irish nationalists who had enlisted in the British Army in order to fight against the Germans. As Clair Wills has explained, by erupting during the middle of the Great War, the 1916 Rising 'put the issue of national and imperial allegiance sharply into focus'.[76] One very notable impact of the changed political circumstances that soon followed in the wake of the Rising was the atmosphere of hostility that emerged in many circles towards the alternative pathway towards Home Rule that Redmond had

---

[73]   Dr. Edward Nolan to John Quinn, 23 August 1916, New York Public Library, Manuscripts and Archives Division, John Quinn Papers.

[74]   Joseph Tumulty to Robert Lansing, 29 September 1916, LCMD, Robert Lansing Papers, Container 21.

[75]   Michael Francis Doyle to Joseph Tumulty, 29 September 1916, LCMD, Robert Lansing Papers, Container 21.

[76]   C. Wills, *Dublin 1916: The Siege of the GPO* (Profile Books Ltd., London, 2009), p. 4.

advocated through his support for the war against the Germans. On 23 June, for example, a leaflet was widely circulated among around 5,000 people (largely of Sinn Féin sympathies) attending a meeting in Cork city. Addressed to the 'Citizens of Cork', it angrily denounced the press censorship and Martial Law that was in operation, and lambasted the Redmondites for proving false 'to their trust'. The leaflet went on to praise the actions of those who had participated in the Rising, while simultaneously proclaiming that 'Ireland was made the poorer by the loss of thousands of her brave sons' when nationality was 'prostituted' on the 'battlefields of Flanders'.[77]

As the slaughter on the continent continued unabated, many Irish people's disaffection with the British government heightened, particularly after the executions in Dublin. But elements of the Irish public also proceeded to become more and more estranged from thousands of their fellow countrymen serving with the British Army on the continent. To add to their woes, a number of Irish-born troops on the Western Front were also subjected to a variety of opportunistic mind games over Britain's swift suppression of the Rising. Early in May, for example, it was reported in *The Times* that Captain Willie Redmond, MP, who was with the 16th Irish Division at the front, had witnessed the Germans in the opposite trenches erecting a taunting notice while the Rising in Dublin was in full swing: 'Irishmen, in Ireland revolution. English guns firing on your wives and children. English Military Bill refused. Sir Roger Casement is persecuted. Throw your arms away. We give you a hearty welcome.'[78] The Royal Munster Fusiliers were also subjected to ridicule from various placards that were erected above the German trenches at the beginning of May. One of these carried the statement: 'English guns are firing at your wifes [sic] and children.' *The Cork Examiner* eventually published a photograph of the placard in question, expressing satisfaction that its damaged condition was due to the fact that it had been 'riddled by the volley of bullets with which the Munsters greeted its appearance'.[79]

Following the suppression of the Rising, further antagonism of Irish troops occurred when propaganda leaflets were dropped from a German aircraft into British lines in France, ridiculing England's offer of Home Rule:

> The Easter Week Rebellion in Ireland was one of the most important events in the world's war. It showed the world the hollow pretence of England, and saved Ireland

[77] Copy of a Leaflet Entitled 'Citizens of Cork! Do You Repudiate the Men who Died for Ireland in Dublin Lately?', 23 June 1916, British Library, ADD. MS. 62461, Viscount George Cave Papers, Vol. 7.
[78] *The Times*, 6 May 1916.
[79] *The Cork Examiner*, 19 May 1916.

from the ignominy of surrender by her corrupt representatives at Westminster. It proved to the world that the Irish had not lost their spirit of nationality – that they had not become the willing participants in the crimes of England; that the Irish would not, at the behest of Redmond, Dillon, Devlin, and other place-hunting politicians, surrender the ideals of the Irish race for a mess of pottage or a miserable pittance called Home Rule; that the Irish were no slave-minded West Britons, ready to participate in England's dark doings in the hope of sharing in her spoils; that the Irish did not wish their land to be an integral part of the British Empire but a separate and independent nation.[80]

After the Rising, Irish troops were also faced with the pressure of being hassled by their own commanders. As John Morrissey has observed, 'they were ... victims of alienation and propaganda within the British Army', particularly by certain sections of the military high command. This mistrust 'added to the extent of Othering of Irish troops arguably pre-existing in the British military'.[81]

In spite of what happened in Dublin, there was certainly no slackening off by the Irish fighting against the Germans on the Western Front. Tipperary-born Martin O'Meara, for example, was awarded the Victoria Cross for the part that he played as a stretcher-bearer with the 16th Battalion Australian Infantry Force during the Battle of Pozieres in France during the summer of 1916. The citation for O'Meara commended him 'for great gallantry and devotion to duty in rescuing wounded men under intense shell fire, and for voluntarily carrying ammunition and bombs to a portion of the trenches being heavily bombarded'.[82] For the Irish regiments of the British Army, political ideology was often the very last thing on their minds when it came to the harsh struggle for survival and victory against the so-called 'gallant allies' of the Irish Volunteers. Survival took precedence over everything else. C. A. Brett of the 6th Battalion Connaught Rangers, who was sent to France in September 1916 at the tender age of 19, wrote vividly about the constant danger of being killed in action. He recalled how there was a saying in France, namely that 'no bullet (or shell) will hit you unless it has your name on it'. The converse was also applicable, namely: 'If a shell (or bullet) has your name on it, you can't avoid it.' All one could do, he

---

[80]   Pamphlet Entitled 'Who is the Real Foe of the Irish?', with Accompanying Leaflet Dropped from a German Aircraft into the English Lines in France, AL, Box 184/Casement Box 2, Folder 1, Monteith/Casement Papers.

[81]   J. Morrissey, 'A Lost Heritage: The Connaught Rangers and Multivocal Irishness', in M. McCarthy (ed.), *Ireland's Heritages: Critical Perspectives on Memory and Identity* (Ashgate, Aldershot, 2005), pp. 78–79.

[82]   Copy of Recommendations Leading to the Victoria Cross Award, 16 August 1916, Army Museum of Western Australia (hereafter AMWA), Box 33, PD261, Pte. Martin O'Meara Papers.

added, was 'pull yourself together and do your duty, however frightened you may be'.[83] For the rest of the war, there was no let-up in the dangers faced by those stationed in the trenches. As one Scottish soldier graphically noted, death and danger were 'taken very much as a matter of course' by all troops on active service on the Western Front, who were faced with the nightmare prospect of living 'day in day out for months, amidst screaming shells and whining, zipping bullets'.[84]

In the long-run, the Great War went on to assume a prominent place in the forging of unionist memories and identities in Ulster. Recollections of the exploits of regiments like the Royal Irish Rifles (Plate 3.1) in the Battle of the Somme served to copper-fasten loyalties to the British Crown. As one of its soldiers, Willie Lynas, remarked in a letter to his wife on 15 July 1916: 'the gallantry of our boys [at the Somme] ... did not disgrace the name of Ulster or their forefathers'. They 'made a name for Ulster that will never die in the annals of history', he continued, noting with pride that 'they fell doing their duty for King and country'.[85] Similar sentiments were also evident in an old loyalist poem entitled 'Irish and Irish', an updated version of which was recited to the 6th North Staffords at Ballykinlar towards the end of 1916: 'There isn't an Irishman worthy the name who wouldn't lend a hand/To crush any foe who attempted to injure this Empire grand.'[86]

Altogether, of the *c.* 210,000 Irishmen who served in Irish units of the British Army during the Great War (including existing regulars, reservists and new recruits), a total of 49,435 '"known dead" [were later] listed in the eight volumes of *Ireland's Memorial Records*, "being the names of Irishmen who fell in the Great European War, 1914–1918, compiled by the Committee of the Irish National War Memorial"'.[87] Once most of the Diaspora and those not unmistakably Irish are excluded, a more accurate estimate of the number of Irishmen (that is, Irish-born men) who died in these units from 1914–1918 may be *c.* 35,000.

[83]   Transcript Account of C. A. Brett's Service in France with the 6th Battalion Connaught Rangers, September 1916–November 1918, Imperial War Museum (hereafter IWM), MS. 76/134/1, The Papers of Captain C. A. Brett.

[84]   Major Arthur C. Murray to A. L. Brown, 2 January 1917, National Library of Scotland, Edinburgh, MS. 8805, Correspondence of Major Arthur C. Murray (afterwards Viscount Elibank).

[85]   Willie Lynas to his Wife, Maria, 15 July 1916, IWM, MS. 89/7/1, The Papers of W. L. Lynas.

[86]   Poem Entitled 'Irish and Irish', 19 December 1916, National Maritime Museum, WH1/95, Arnold White Papers.

[87]   K. Jeffery, *Ireland and the Great War* (Cambridge University Press, Cambridge, 2000), pp. 5–7, 33.

Plate 3.1   Ration party of the Royal Irish Rifles resting in a communication trench, Battle of the Somme, 1 July 1916

Source: Imperial War Museum, Q 1.

This figure, which includes both nationalists and unionists, is still noteworthy, especially when compared to the fact that American deaths in the Great War amounted to around 48,000 (although it must be remembered that the USA fought in the conflict for only 19 months, and that its soldiers fought on the front lines for only about six months).[88] Whereas the 'immortal souls' of the 1916 Rising went on to assume the status of martyrs for the cause of the physical force tradition of nationalism in Ireland, Jonathan Githens-Mazer has pointed out that the Great War recruit was later portrayed by nationalists as 'a restless ghost', while graves and skeletons tended to dominate popular representations of their fate in certain types of literature.[89]

[88]   Ibid., p. 35; P. Fussell, *The Great War and Modern Memory*, 25th Anniversary Edition (Oxford University Press, Oxford, 2000), p. 337.

[89]   J. Githens-Mazer, *Myths and Memories of the Easter Rising: Cultural and Political Nationalism in Ireland* (Irish Academic Press, Dublin, 2006), p. 143.

'A Terrible Beauty is Born': Politics and Society after the Rising

The weeks, months and years that followed Easter 1916 proved to be a time of immense change in Irish politics and society. Soon after the executions, it became apparent that memories of the dead rebels would live on in the new Ireland. This was hinted at in the poetry that Yeats composed during the summer of 1916 in Normandy, France. In his fêted poem, 'Easter 1916' (which was eventually published in 1920), Yeats penned the unforgettable line: 'All changed, changed utterly: A terrible beauty is born.'[90] Although the poem, as John Wilson Foster has observed, reflected Yeats's own personal reservations about the use 'of physical force' for political ends, it also displayed 'a canonical image of the Rising'. This acknowledged 'the rebels' self sacrifice' and dramatically portrayed 'the guilt many mockers began to feel as the executions followed their horrifying course'.[91] Although Yeats was not in Ireland when the Rising broke out (he was staying at a farmhouse overlooking a Cotswold valley), he did manage to pay a visit to Dublin during the first week of June 1916. According to Roy Foster, he managed to survey the ruins of a lot of the city centre from his base in the Stephen's Green Club, while he needed to obtain a pass from the Dublin Metropolitan Police to embark upon a trip to Greystones in County Wicklow. The whole visit proved 'sobering' for the poet and planted 'the idea of irrevocable change' in his mind, which in turn became 'a subject for his own poetic commentary'.[92]

In spite of the ruthless clampdown on republican activities that immediately followed the Rising, there was little that the military authorities could do to deflect the gusty winds of change. In addition to the executions, the swing in public opinion towards support for the rebel cause was further exacerbated by the imposition of Martial Law and the severely hostile response to the widespread arrests, deportations and imprisonments. By the year's end, the inaugural issue of *The Phoenix* vividly captured the change that had taken place in Irish politics, noting that there had been a move towards 'a totally different political

---

[90]   Yeats, cited in F. X. Martin, 'Foreword', in F. X. Martin (ed.), *Leaders and Men of the Easter Rising: Dublin 1916* (Methuen & Co. Ltd, London, 1967), p. ix.

[91]   J. Wilson Foster, *Colonial Consequences: Essays in Irish Literature and Culture* (The Lilliput Press, Dublin, 1991), pp. 133–34.

[92]   R. F. Foster, *W. B. Yeats: A Life. II. The Arch-Poet 1915–1939* (Oxford University Press, Oxford, 2005), pp. 45, 54. Also see Zimmermann, *Songs of Irish Rebellion*, p. 72, who notes that in 1933, Yeats wrote a letter in which he alluded to 'the cult of sacrifice planted in the [Irish] nation by the executions of 1916'. For rare surviving specimens of the police passes that were issued in Dublin city as a result of the Rising, see 1916 Police Pass of Wm. Houston, Royal Irish Academy, MS. 12/Y/14 or 1916 Police Pass of Reverend E. Byrne, DDA, Archbishop William J. Walsh Papers (Laity), 385/6.

atmosphere from that which obtained ... previous to the events of Easter Week'. Moreover, it proclaimed that 'a new Ireland of strength of character and honesty of purpose' had arisen from 'the smouldering ashes of recent rebellion'.[93]

The backlash against the British government during the last eight months of 1916 made things increasingly difficult for the officers of both the Dublin Metropolitan Police and the Royal Irish Constabulary, along with the British Army troops stationed in Ireland. Tensions remained very high not only on the streets of Dublin, but across the entire island. A vivid portrayal of the extent to which anti-British sentiment escalated can be seen from the reminiscences of Captain John Lowe. As a British soldier who had helped end the Rising, he was in no doubt about the enmity that some pedestrians walking the streets of Dublin city held towards him as news of the executions spread:

> Early in May [1916] I received a telegram ordering me to join my Regiment [the 15th Hussars] in France [at the Somme]. It was a lovely sunny day and I went out for a last walk through the streets of Dublin. I had not gone far when I met a group of Irish women. Three of them immediately linked arms so I had to step off the pavement to make room for them. They looked me straight in the face with the most venomous hatred in their eyes and all three spat on the ground. I was glad when I started my journey to France ... for the Front.[94]

Down in the south of the country, soldiers also faced considerable hostility. Having arrived in the village of Macroom, County Cork on 8 May 1916, Lieutenant-Colonel James Melville Galloway reported that he was not allowed to venture outside his camp in the grounds of Macroom Castle as armed men had visited it the previous night. Macroom 'is a big dirty village', he wrote in his diary, observing that the village people were 'very unfriendly' following the arrest of several suspects.[95]

In what must have seemed like a recurring personal nightmare, Maxwell displayed considerable apprehension about the ghosts of 1916 when he wrote to Archbishop Walsh on 19 June about a 'delicate question' that needed to be addressed in order to prevent the possibility of further 'disorder and perhaps bloodshed'. 'There is a section of the people', he wrote, 'who are taking advantage of Requiem Masses said for the repose of the souls of those unfortunates who suffered death for the leading part they took in the late deplorable rebellion',

93   *The Phoenix*, 9 December 1916.
94   First World War Narrative of Captain John Lowe, IWM, MS. 75/80/1, The Papers of Captain John Lowe.
95   Diary, 8 May 1916, IWM, MS. 87/45/1, The Papers of Lieutenant-Colonel James Melville Galloway.

by staging 'political demonstrations outside the churches and chapels in which these Masses are said'. 'Yesterday', he added, 'there was a procession of perhaps 2,000 people marching along the quays and streets waving Sinn Féin flags, booing at officers ... and soldiers'. Rather courteously, he enquired whether 'the priests conducting these Masses might be asked to advise their congregations to disperse quietly after they have been said and take no part in such demonstrations'.[96] Five days later, Maxwell again conveyed his anxiety about Ireland's transformation in a confidential report to the Cabinet. He highlighted the degree to which Irish public sentiment had changed, and remarked how this was noticeable through the proliferation of souvenir objects and changes in people's behaviour:

> The wearing of what is called the ... mourning badge, display of Irish flags, the sale of photographs of the late leaders, the booing of soldiers and police by people who openly show their sympathy with the organisation that was professedly separatist and revolutionary are all signs that the causes which led up to the rebellion are still existent, and the moment new leaders are found it will become dangerous.[97]

Long-term concerns about law and order again weighed heavily upon Maxwell's mind in a letter that he wrote to Walter Long of the Colonial Office on 2 July. Although he expressed the view that it was doubtful 'that another armed outbreak is likely to recur for some considerable time', given that 'all the leaders [of Easter Week] have gone', Maxwell was particularly eager to warn against government complacency in the times that lay ahead. 'If things are allowed to again drift and new leaders come forward', he added, 'more trouble is possible'.[98]

By the month's end, Asquith was keen to urge the military to adopt a softer touch throughout Ireland, particularly in light of the commencement of negotiations concerning a resolution of the Irish question. On 27 July, he wrote to Maxwell, calling for 'as calm an atmosphere as can be secured'. Furthermore,

---

[96] General John Grenfell Maxwell to Archbishop Walsh, 19 June 1916, DDA, Archbishop William J. Walsh Papers (Laity), 385/5. Walsh's rather standoffish reply, which was sent a week later, was by no means reassuring to those who wanted to maintain law and order. Whilst explaining that he 'had to leave Dublin under medical orders', he confirmed that he had travelled back to the capital and that 'informal' measures had 'been taken to apply a remedy'. Additionally, he advised that he could not guarantee a wholly 'satisfactory result', because of 'a very widespread feeling of discontent ... growing stronger every day'. See Archbishop Walsh to General Maxwell, 26 June 1916, DDA, Archbishop William J. Walsh Papers (Laity), 385/7.

[97] The Irish Rebellion, Cabinet Print, Report on the State of Ireland Since the Rebellion, 24 June 1916, PAHPL, The Bonar Law Papers, BL/63/C61.

[98] General Maxwell to Walter Long, 2 July 1916, Princeton University Library, Department of Rare Books and Special Collections (hereafter PULDRBSC), Sir John Grenfell Maxwell Papers, Box 4, Folder 12.

he expressed 'hope that the arrests ... [and] searches are now practically over, and that you may find it possible to "go slowly" for the next week or two'. On the other hand, though, the Prime Minister did make it clear that a certain degree of vigilance and alertness was still needed. Demonstrating his knowledge and memory of Ireland's rebellious past, he ended his letter by warning Maxwell about the threat of 'reprisals' by secret societies and small clubs. The danger of retaliation, he cautioned, 'should not be lost sight of by the police', as this had 'happened before in Irish history'.[99] In the weeks that followed, the police kept a close watch on the activities of militants. On 2 August, a confidential report submitted to the Cabinet from the Inspector General of the Royal Irish Constabulary raised concerns about powerful anti-English feelings that had materialised in various parts of Ireland: 'There is no doubt that there is a strong feeling of disloyalty in the country, but it is kept in check by the existence of Martial Law.'[100]

At all times during the rest of 1916, the police remained on a high state of alert and sometimes acted with a heavy hand, as can be seen from an incident in Dublin on 6 August, when three 17-year-old girl scouts – May O'Kelly, Eileen Conroy and Margaret Fagan – were arrested by Sergeant Farrelly for carrying a flag (emblazoned with the words 'Clan na Gael Girl Scouts') on a tram, while returning from a commemoration ceremony in honour of O'Donovan Rossa at Glasnevin Cemetery. The girls were brought into custody in Mountjoy Police Barracks and later transferred to the Brideswell at Inn's Quay, where they produced an affidavit. In this, they complained of being 'detained without any trial of any description or any charge of any nature made against or communicated to us'.[101] The girls were eventually released after four days 'in view of [their] youth', but were warned 'as to their future behaviour'.[102] Tensions also remained high outside of the capital city. Royal Irish Constabulary reports on the condition of different districts typically had two common denominators – fear and anxiety. One such report, dated 7 September, detailed the arrest of a 'dangerous' teacher,

[99]   Herbert Asquith to General Maxwell, 27 July 1916, PULDRBSC, Sir John Grenfell Maxwell Papers, Box 2, Folder 8.

[100]   The Irish Rebellion, Cabinet Print on the State of Public Feeling in Ireland, 2 August 1916, PAHPL, The Bonar Law Papers, BL/63/C35.

[101]   Joint Affidavit of May O'Kelly, Eileen Conroy and Margaret Fagan, 10 August 1916, NA, Kew, CO 904/23/3.

[102]   Colonel of the Commanding Dublin District to Headquarters, Irish Command, Parkgate, 8 August 1916, NA, Kew, CO 904/23/3.

Michael Thornton, and alleged that he was 'teaching disloyalty and sedition' to the children attending Furbo National School in County Galway.[103]

The worsening state of Anglo-Irish relations was also evident on the other side the Irish Sea. In a secret and personal letter that he wrote to H. E. Duke on 11 September, in which he raised the thorny question of conscription for Ireland, Bonar Law alluded to the fact that a 'feeling of hostility to the Irish' had arisen in England, and was 'being shown by the way they are being boycotted'.[104] On a regular basis, the Cabinet in London remained fully briefed by the intelligence network on what was going on back in Ireland, but the news was not good. In a confidential Cabinet memorandum, dated 26 September, Duke delivered a very bleak assessment of the security outlook for most of the country. In his assessment of 'the political conditions of the western and south-western counties', he worryingly pointed to 'the intimate relation which the people of these districts have with hostile associations in the United States'. 'Incitement and assistance for revolutionary purposes', he noted, 'appear never to be lacking'. Even though the Rising's 'most energetic and conspicuous leaders' were 'dead, imprisoned ... or interned', he observed that 'the reaction in popular feeling upon the repression of the rebellion has altered the relation of the extremists to the general population'.[105]

In the last few months of 1916, dark clouds remained on the horizon for the police. A Royal Irish Constabulary report described a Gaelic League bazaar held in the Market House at Caherciveen, County Kerry on 1 October, and detailed how an Irish teacher, Michael Walsh, had proclaimed: 'We are Irishmen and will fight for Ireland and die in a noble cause, the men who were shot in Dublin are martyrs and their names will go down in the history of Ireland – the people by whose hands they met their deaths will fall yet.'[106] After Walsh's speech, the local butcher, Jeremiah Riordan, stood up and sang the rebel song 'Easter Week' – for which he was prosecuted at the Caherciveen Petty Sessions and fined £10 (or, in default, two months in prison). While the song was being sung, there was shouting and cheering of 'Up the rebels', 'To hell with England', 'Down with George', and 'Up the Kaiser'.[107] At another Gaelic League meeting held in

---

[103] Anonymous to County Galway Inspector's Office, 7 September 1916, NA, Kew, CO 904/216/425; *Galway Express*, 8 December 1917.

[104] Bonar Law to H. E. Duke, 11 September 1916, PAHPL, The Bonar Law Papers, BL/63/C41.

[105] The Irish Rebellion, Cabinet Print on the Condition of Ireland, 26 September 1916, PAHPL, The Bonar Law Papers, BL/63/C39.

[106] Royal Irish Constabulary Office Statement by N. Tymy, 1 October 1916, NA, Kew, CO 904/216/433.

[107] Ibid.

St. Mary's Hall, Belfast on 4 October, an officer of the Royal Irish Constabulary recalled how Father Ward from Limerick had praised the Irish Volunteers for being 'enthusiastic students of the [Irish] language even in the Convict Prisons' and declared that 'we would have no Irish nation if the language be permitted to die'.[108] Gatherings of radical nationalists at GAA matches were also a source of concern to the police. On 15 October, for example, over 2,000 spectators gathered at a hurling tournament held at Gort in County Galway. Political activists mingled with the crowd and the County Inspector, in his monthly report, commented that 'Sinn Féin colours were freely displayed on the fields where the matches took place'.[109] The mounting sedition throughout Ireland was plain and clear for all to see as 1916 eventually drew to a close, even as far away as Delhi in India. As one officer serving in the Indian Army noted in a letter that he sent to Maxwell on 20 November: 'the fruits of years of indecision ... and *laissez faire* [in Ireland] have developed a really serious situation'.[110]

In the long run, the Rising came to be seen retrospectively as right in the minds a large proportion of the general Irish public. This phenomenon was attested to by the huge public celebrations held in honour of several released internees upon their return to Ireland throughout 1917. A taste of what was in store was evident when the first anniversary of the Rising was commemorated in April of that year. Unsurprisingly, the occasion was greeted with some degree of anxiety by the authorities. As John O'Connor has shown, it 'was felt by many people to be a suitable occasion to demonstrate in commemoration of the dead patriots and their cause'. In Dublin city, a small group of women attached to the Irish Citizen Army marked the event by arranging for multiple copies of the Proclamation to be reprinted by the printer, Joseph Stanley. The reissued versions, which were done on better quality paper and which were slightly larger than the originals, were then posted on various public buildings and vantage points across the city on Easter Sunday 8 April 1917.[111] By this time, the Russian Revolution had taken place. But for the time being at least, the security situation in Ireland remained stable enough. *The Irish Times* reported that there was no conspicuous 'display of military or police in any of the ... streets' of Dublin on Easter Sunday and that no efforts had been made 'to hold a procession or

---

[108] Notes taken from Speeches Delivered at the AGM of the Gaelic League in St. Mary's Hall, Belfast, 4 October 1916, by Const. Terence Keely, NA, Kew, CO 904/23/3.

[109] County Inspector, cited in Ú. Newell, 'The Rising of the Moon in Galway', *Journal of the Galway Archaeological and Historical Society*, Vol. 58 (2006), p. 121.

[110] A. Bingley to General Maxwell, 20 November 1916, PULDRBSC, Sir John Grenfell Maxwell Papers, Box 2, Folder 11.

[111] J. O'Connor, *The 1916 Proclamation*, Revised Edition (Anvil Books, Dublin, 1999), pp. 88–89.

meeting', as a result of a Proclamation that had been issued on Good Friday by the Lieutenant General, Sir Bryan Mahon. 'Indeed', it added, 'the numbers of people in the principal thoroughfare were, if anything, below the average for a Sunday afternoon', with the cold and showery weather conditions driving large numbers of people 'to their homes somewhat earlier than usual', following attendance at various places of religious worship. One of the few gatherings of note was at the Mansion House, where the GAA held its yearly Convention in private, electing officers for the year ahead. On the north side of the city, 'exceptionally large' numbers of people attended Glasnevin Cemetery throughout the day, placing wreaths on numerous graves, 'including those of some of the men who lost their lives during the disturbances of last Easter Week'.[112]

The tranquil atmosphere that had prevailed in Dublin city centre was finally broken on Easter Monday by what *The Irish Times* described as 'a good deal of excitement', including the hoisting of a Tricolour at half mast on top of the south-east corner of the roof of the GPO shortly before 9.00 am. By 12.00 noon, the crowds on O'Connell Street swelled and another Tricolour was raised from the top of Nelson's Pillar. Throughout the day, many people turned up wearing 'black bands, surmounted with [Sinn Féin] ribbons ... on their arms, while groups of girls, with paper flags and coloured ribbons in their hair, paraded'. Republican colours were also displayed from many of the buildings that had been occupied by the rebels during the Rising, 'but in most instances the flags were quickly removed by the police'. On a few occasions in the afternoon, matters deteriorated as the police came under attack from stone-throwing youths (despite appeals for calm from men wearing republican badges), while several windows were smashed in both Lower Abbey Street (where the Methodist church was attacked) and Middle Abbey Street. In the vicinity of Nelson's Pillar, signboards were torn from a number of tramcars, while the windows of a number of businesses were broken. As a consequence of the day's disturbances, 12 people suffering from 'scalp wounds' had to be treated in Jervis Street Hospital. Four of these were civilians, while the other eight were policemen. In the end, only two arrests were made in the afternoon. Most of the crowds dispersed after sundown, when the weather worsened, bringing a 'sharp fall of hail and snow'.[113]

In Cork city on Easter Monday, the Rising's first anniversary proved slightly less volatile than in the capital. After an 11.00 am High Mass in remembrance of the executed leaders at St. Mary's Catholic cathedral, around 300 Sinn Féin supporters 'of both sexes' marched to the City Hall, where a Tricolour had been hoisted in place of a flag bearing the City Arms. This was quickly seized by

---

[112] *The Irish Times*, 9 April 1917.
[113] Ibid., 10 April 1917.

the police. The crowd then congregated at the National Monument on Grand Parade, whereupon they were ordered to disperse by baton-wielding police, under the command of District Inspector Walsh. Although one policeman was struck in the back by a woman with an umbrella, the crowd eventually dispersed in a peaceful fashion. On the same day, another Tricolour was hoisted up the flagstaff on top of the headquarters of Westmeath County Council in Mullingar, but this was also removed by the police.[114] In the days and weeks that followed, various radical newspapers passed comment on the significance of the first anniversary. *The Factionist* of 12 April, for example, declared that the bloodshed of Easter Week had not been futile and called upon people to intensify their demands for independence:

> The blood of our Martyrs has not been shed in vain. Every drop they spilled has produced men who now think as they did. They gave their lives for us. They could do no more. But we live and every one of us should swear to God by their memory that we shall redouble our efforts until we are in a position to write their epitaph in letters of gold, on the unsullied pages of the History of our country's struggle for freedom.[115]

On 28 April, *The Harp* expressed its delight that the first anniversary had enthused 'brave and unselfish men to the work of bringing the truth of a great and invincible cause to success'.[116]

Throughout the course of 1917, the political implications of the executions manifested themselves in the results of a number of by-elections. On 17 February, for example, Count Plunkett (the father of the executed Joseph Plunkett), was elected as an independent and followed a policy of abstinence from Westminster. The following month, when Lloyd George announced that he was willing to offer Home Rule 'to that part of Ireland that wants it' (but would never hand over Ulster to the rest of Ireland against its wishes), the fate of Redmond's Irish Parliamentary Party was effectively doomed. It walked out of the House of Commons to show that it would never settle for the partition of Ireland. The party never again returned to Westminster and as Mary Kenny has shown, several of its membership 'ended up living in very reduced circumstances – some in near penury – in bedsitters in Pimlico and the Elephant and Castle' in London.[117] By contrast, Sinn Féin's star was on the rise. Throughout 1917, most of those imprisoned following the Rising were released from jail under an

---

[114]   Ibid.

[115]   *The Factionist*, 12 April 1917.

[116]   *The Harp*, 28 April 1917.

[117]   M. Kenny, *Crown and Shamrock: Love and Hate Between Ireland and the British Monarchy* (New Island, Dublin, 2009), pp. 148–49.

amnesty and they subsequently joined the party en masse. It quickly coalesced into 'shorthand for all the radical nationalist movements, including the republicans', who were disenchanted with the Irish Parliamentary Party and who were 'broadly sympathetic to the Rising'.[118] In May, Sinn Féin again achieved success in another by-election.

For those supporting the troops on the Western Front, the rapid rise of Sinn Féin proved rather disconcerting. With no let-up in the death toll on the continent and the Home Rule movement in a complete state of disarray back in Ireland, profound disillusionment set in amongst many Redmondites. On 2 June, however, their morale received a temporary boost when Private Thomas Hughes of the Connaught Rangers, who hailed from Castleblaney in County Monaghan, received a Victoria Cross from King George V at London's Hyde Park for 'conspicuous bravery and determination'.[119] Redmondites also took pride in the accomplishments of the Perth miner with the ruddy skin tone, Thomas Carberry. Born in Offaly in 1884, he served in Gallipoli and France and was awarded the Military Medal in 1917 for distinguishing himself in the Australian Infantry Force.[120] Sinn Féin, however, had little appetite for the pomp and ceremony associated with the British Empire's militaristic spectacles. In the weeks after Hughes received his award from the King, tensions grew between its supporters and the authorities, as republican demonstrations began to be repressed.

This rivalry as John Borgonovo has demonstrated, eventually exploded in places such as Cork city, where a two-day disturbance occurred from 23–24 June, which the locals called 'The Battle of Patrick Street'. All seemed well initially when an elaborate (but unofficial) reception was held for some of the city's returning rebel prisoners on 23 June – among them J. J. Walsh and Diarmuid Lynch (who had fought in Dublin) and David Kent (who had survived the shoot-out at Bawnard House). After the prisoners arrived in Glanmire Rail Station, they were then escorted into the city centre by a large crowd of jubilant republican supporters. Later on that night, some protestors went on the rampage, stoning the Army Recruitment Office on Patrick Street (where two windows were broken) and removing a bronze replica of the Scales

---

[118] R. Killeen, *A Short History of the Irish Revolution 1912–1927* (Gill and Macmillan, Dublin, 2007), p. 82.

[119] Appeal for Funds to Support Pte. Thomas Hughes, VC, Connaught Rangers, National Army Museum (hereafter NAM), MS. 6106/64/1. Commenting on Hughes's award of 'the coveted decoration', *The Northern Standard*, 9 June 1917, noted that although he needed to use crutches and 'seemed very weak' at the ceremony, 'the hero was radiant'.

[120] Australian Infantry Force, Service Dossier, AMWA, Box 32, PD242, Thomas Carberry Papers.

of Justice from the top of the Court House on Great George's Street (later renamed Washington Street). Disturbances continued on Patrick Street at 9.00 pm the night afterwards, when a scuffle broke out between Separation Allowance Women and Sinn Féin supporters, the latter of whom attacked the Army Recruitment Office. A Union Jack, which had flown from the building since the outbreak of the Great War, was cut down from its flagpole and tossed into the River Lee. A Tricolour was then fastened to the flagpole, watched by a crowd of around 5,000. A riot then ensued, with the police coming under attack from stone-throwing rioters. By the time order was restored by the military, at least 12 people had been hospitalised, many with bayonet wounds, while the windows of 34 local businesses had been smashed.[121]

Despite drawing criticism from both the authorities and the Volunteer leaders (who pleaded for calm in a letter printed in *The Cork Examiner*, *The Cork Constitution* and *The Evening Echo*), violent incidents like the 'Battle of Patrick Street' did not halt the acceleration of the Sinn Féin political machine. In July, success came again when Eamon de Valera won a by-election in East Clare after defeating a Home Rule candidate. His victory came as no surprise, given both his prominence in the Rising and his 'assured performance' as leader of the imprisoned rebels at Dartmoor, Maidstone and Lewes. As David Fitzpatrick has shown, de Valera managed to develop 'an impressive portfolio of devices for niggling and goading the authorities into counter-productive severity', thus 'generating invaluable propaganda and inflaming Irish opinion'.[122] In the month following de Valera's electoral success, John Dillon invidiously observed that Sinn Fein 'has been going through the country like a prairie fire'.[123]

Unsurprisingly, the party's rapid ascent was also the source of much commentary in British press censorship reports, one of which worryingly referred to 'defiant speeches' delivered by de Valera and Countess Markievicz, who had made 'wild appeals to the principle of physical force'.[124] Then, on 25 September, tensions were further heightened following the death of Thomas Ashe, who had been arrested following a defiant speech in County Longford. Following his imprisonment in Mountjoy Prison, Ashe went on hunger strike and died from lung impairment following an effort to force-feed him at the Mater Hospital. It was at this time, as Ian Kenneally has noted, that the republicans became

---

[121] J. Borgonovo, "'Thoughtless Young People" and "The Battle of Patrick Street"': The Cork City Riots of June 1917', *Journal of the Cork Historical and Archaeological Society*, Vol. 114 (2009), pp. 12–17.

[122] D. Fitzpatrick, "'Decidedly a Personality': De Valera's Performance as a Convict, 1916–1917', *History Ireland*, Vol. 10, No. 2 (2002), p. 40.

[123] John Dillon to T. P., 11 August 1917, LCMD, Charles Edward Russell Papers, Container 8.

[124] Press Censorship Reports (Ireland), September 1917, BLUCC, MS. U.249.

'the masters of the propaganda of martyrdom', as exemplified when Michael Collins delivered an oration on 30 September at Ashe's graveside in Glasnevin. A total of around 3,000 uniformed Volunteers attended, along with a crowd numbering tens of thousands.[125] At Sinn Féin's Ard Fheis the following month, Arthur Griffith (a pacifist who had supported the concept of an Anglo-Irish dual monarchy) resigned as its leader and President, and was replaced by de Valera. De Valera at this time, according to Patrick Murray, derived 'enormous political credit for his involvement in the Rising', but also made it clear in his Ard Fheis speech that he personally regarded any future attempt to defeat British rule by physical force as justifiable – with the proviso that it had to have a chance of success.[126] Another difference between the revamped Sinn Féin movement and the Rising, as Bew has shown, was 'that it was not so much recruited from the towns and was not entirely led by townsmen'. While it encompassed more farmers and labourers, it also 'embraced, far more than the Easter Rising had, the professional and commercial classes'.[127] Throughout 1917 and 1918, rumours that conscription would be introduced into Ireland led to a huge public outcry and further swung the pendulum in favour of the restyled and resurgent Sinn Féin party.

During these years, the significance of displaying posters in public space was not lost upon republican propagandists, who increasingly saw such displays as a way in which to combat press censorship. Some of the posters were quite explicit when it came to articulating the politics of 1916. One poster, which was issued in Cork, went so far as to publicly declare that no man, no matter how far 'he has fallen away from his national faith', had 'dared to repudiate' the Proclamation.[128] Support for the actions of the 1916 rebels finally manifested itself at the General Election of December 1918 (the first in eight years), in which Sinn Féin won 70% of Irish seats and pledged not to attend the British Parliament in London, but rather to establish a parliament in a sovereign independent Irish republic. Most of those elected in Ireland pursued a policy of abstention from Westminster and met in Dublin on 21 January 1919, where they formed themselves into the first Dáil Éireann.[129] This formally reaffirmed the 1916 Proclamation and ratified a Declaration of Independence, in which

---

[125] I. Kenneally, *The Paper Wall: Newspapers and Propaganda in Ireland 1919–1921* (The Collins Press, Cork, 2008), pp. 64, 194.

[126] P. Murray, 'Obsessive Historian: Eamon de Valera and the Policing of His Reputation', *Proceedings of the Royal Irish Academy*, Vol. 101C, No. 2 (2001), p. 53.

[127] Bew, *Ireland*, p. 385.

[128] Printed Republican Poster, Cork City and County Archives, MS. SM.619.

[129] Department of External Affairs, *Cuimhneachán 1916–1966: A Record of Ireland's Commemoration of the 1916 Rising* (An Roinn Gnóthaí, Dublin, 1966), p. 22.

the matters of 'English rule ... by military occupation against the declared will of the people' and Irish resistance 'in arms against foreign usurpation', emerged as central themes.[130]

Occasionally in 1919, dissenting opinions were registered about the Rising by writers who abhorred violence. One such critic was the Maynooth academic and Catholic priest, Father Walter McDonald. In a book entitled *Some Ethical Questions of Peace and War*, he registered his moral unease with the armed resistance of 1916. In an introduction to a later reprint of the book, Tom Garvin has remarked that McDonald took a dim view of 'the tendency of senior clerics to behave in political life as propagandists and even demagogues rather than as learned and scrupulous men versed in the science of ethics'. McDonald, according to Garvin, was also uneasy about 'the political subculture of ... ecclesiastical Catholicism' in Ireland and was particularly concerned about the pronounced episcopal swing in political allegiance following the Rising.[131] In pondering the ethical and political dimensions to Ireland's experience of the Great War, and the implications of the resistance that came with both the Rising and the conscription crisis, McDonald himself wrote: 'I presume, in all humility and reverence, to say that I cannot see my way to approve of any such active or passive resistance to a Government recognised as legitimate as would leave this exposed to be crushed by a powerful foreign enemy [Germany] with whom it was engaged in a life-and-death struggle at the time.'[132] The publication of this negative intellectual critique of the 1916 metanarrative by McDonald (who passed away in 1920) made little headway with Irish nationalists at the time, as their attitude to British rule had changed irrevocably.

In the aftermath of the War of Independence, which lasted from 1919–1921, internal divisions arose in Ireland between what Michael Hopkinson has called 'pragmatists and republican irreconcilables', following the signing of the Anglo-Irish Treaty by Michael Collins in December 1921. Although the Irish Free State eventually came into being after its legal enactment in December 1922, with W. T. Cosgrave as President of the Executive Council, the first six months of that year had seen 'unsuccessful efforts to reach compromise with an agreed election and ... constitution'. Despite the pro-Treaty election victory in June, skirmishes began to escalate between pro and anti-Treaty troops. The continued militancy of republicans such as de Valera proved to be a significant factor in the outbreak of the devastating Civil War, which lasted from 27 June 1922 to 24 May 1923. Under intense pressure from the British government, the

---

[130]   *Dáil Debates*, Vol. 1, 21 January 1919.

[131]   T. Garvin, 'Introduction', in W. McDonald, *Some Ethical Questions of Peace and War: With Special Reference to Ireland* (University College Dublin Press, Dublin, 1998), pp. xiii, xv.

[132]   McDonald, *Some Ethical Questions of Peace and War*, p. 109.

pro-Treaty side initially set about ending the occupation of prominent buildings in Dublin and elsewhere by anti-Treaty IRA troops. Michael Collins, for example, ordered an attack on republican forces occupying the Four Courts on 27 June 1922 (which sadly culminated with a fire that caused the destruction of most of the archives housed in the Public Record Office on 30 June).[133] Although the Civil War resulted in defeat for the anti-Treaty contingent, bitter memories of the conflict persisted for years.

In spite of this, the dawning of a more stable era in Irish political affairs came into effect following the May 1923 elections, after Cosgrove's Cumann na nGaedheal attained a majority when Sinn Féin refused to take their seats (after refusing to take the Oath of Allegiance to the British monarchy). Throughout the first decade of the newly-independent state's existence, the machinery of a stable civilian government was carefully put into place, while an unarmed police force, An Garda Síochána, was created. Notwithstanding the bitterness that lingered on from the distressing events of 1922–1923, a range of other deeply-held feelings also managed to exert a formidable impact upon public life in the years that followed the ending of British rule. As Micheál Martin has noted, the era 'was one of powerful ideologies, nationalism and socialism being among the more potent, while in more organised forms, Catholicism, the Gaelic League and the Gaelic Athletic Association were also powerful forces within the Ireland of the day'.[134]

It was also during this time that the cult-like status that surrounded the rebels of 1916 grew stronger. De Valera, the sole surviving commandant of the Rising, once again managed to escape harm during the bloodletting of 1922–1923 (unlike other veterans of 1916 – such as Michael Collins, Cathal Brugha, Liam Mellows, and Liam Lynch). After emerging from hiding in August 1923, de Valera was duly arrested by the Free State government and imprisoned in Kilmainham (where he was the last inmate) and Arbour Hill until July of the following year. According to T. G. Fraser, his role in the Civil War enabled him to consolidate his reputation as 'the great survivor of republican politics', while his subsequent detention 'enabled him to add a new dimension to his already impressive revolutionary credentials'.[135] In the face of their military defeat, however, most of those who had opposed the Treaty gradually began to see the merits of constitutional politics. In January 1926, for example, de Valera resigned from Sinn Féin due to his dissatisfaction with its abstentionist policy from Dáil

---

[133] M. Hopkinson, 'Civil War', in B. Lalor (ed.), *The Encyclopaedia of Ireland* (Gill and Macmillan, Dublin, 2003), p. 202.

[134] M. Martin, *Freedom to Choose: Cork & Party Politics in Ireland 1918–1932* (The Collins Press, Cork, 2009), p. vii.

[135] T. G. Fraser, *Ireland in Conflict 1922–1998* (Routledge, London, 2000), p. 9.

Éireann. Soon afterwards, he founded the Fianna Fáil republican party, which rapidly became a prominent force in the political life of the fledgling state, by building up a robust and dedicated grassroots organisation. To die-hard Sinn Féin followers like Mary MacSwiney in Cork, however, de Valera's departure from the party was a source of much regret. Writing to Charles Edward Russell on 12 April, she protested that de Valera's 'new policy ... is wrong in principle'. 'I need not tell you', she added, 'that the defections in our ranks have made us very sad'.[136]

For de Valera and his followers, there was no going back on this new departure in Irish political life. In an early memorandum on Fianna Fáil's republican policy, printed in Dublin and circulated to people in County Louth in June 1926, Frank Aiken outlined his wish for 'international recognition of ... [Irish] sovereignty and unity' and highlighted how 'success for our cause' could be achieved 'by peaceful means or by arms'. 'For myself', he added, 'I should think it criminal, if we, by our neglect to use every available peaceful means, rendered war inevitable'.[137] After September 1927, as Eunan O'Halpin has noted, Fianna Fáil started 'making a coherent impact in the Oireachtas' by placing emphasis 'not only on maintaining a numerical presence, but on speaking with one voice, on advancing credible alternatives to government proposals ... and on behaving in a dignified manner befitting a government-in-waiting'.[138] De Valera's constitutional credentials were finally confirmed when he brought Fianna Fáil to office in 1932, in a moment that represented both a personal political triumph and a defining moment in Irish political democracy. Altogether, de Valera (who earned the nickname 'The Chief') served as President of the Executive Council in Dáil Éireann from 1932–1937 and then Taoiseach from 1937–1948, 1951–1954 and 1957–1959.

Although he had many political detractors, de Valera also earned plaudits from a lot of people. Michael J. O'Higgins, for example, once remarked in the Seanad that he was 'an Irish statesman of towering stature', who gave 'selfless devotion to the service of his country and its people'.[139] One of de Valera's key strengths, according to Brian Girvin, 'was his fortitude in the face of adversity'. To his loyal followers in Fianna Fáil, he symbolised 'an unbroken tradition of honesty, selflessness and integrity', while his actions 'represented

---

[136] Mary MacSwiney to Charles Edward Russell, 12 April 1926, LCMD, Charles Edward Russell Papers, Container 14.

[137] Frank Aiken, TD, 'On Fianna Fáil: A Call to Unity', Pamphlet, 19 June 1926, Dublin City Library and Archive (hereafter DCLA), Birth of the Republic, BOR F15/08.

[138] E. O'Halpin, 'Parliamentary Party Discipline and Tactics: The Fianna Fáil Archives', *Irish Historical Studies*, Vol. 30, No. 120 (1997), pp. 589–90.

[139] *Seanad Debates*, Vol. 83, 26 November 1975.

the forward march of Irish nationalism in its historic journey to a ... sovereign Ireland'. However, as Garret Fitzgerald has noted, de Valera was also a rather 'controversial figure at home', who 'venerated the past and wished to keep "the old ways"'. Nonetheless, he also garnered a reputation as a 'remarkable Irishman' who 'in the world outside Ireland ... added to his country's stature' and who became well-known around the world 'as an apostle of nationalism'.[140] In many respects, as Pauric Travers has shown, de Valera's name evoked 'mixed emotions', and he remained a contentious individual with a 'somewhat remote, schoolmasterish manner' who tended to generate 'feelings of either intense loyalty or distrust'. However, 'the central fact of his political life was survival'. This view has been echoed by Diarmuid Ferriter. De Valera, he notes, 'was an extraordinary survivor', but the national fixation 'with the events of the War of Independence and Civil War explains why he remained so divisive a figure for those who grew to adulthood in the early decades of the twentieth century'. Despite his critics, especially those who pointed to policy shortcomings on the economic front, de Valera remained ever-popular with the electorate and managed to make a significant contribution to state-building efforts through his concentration on constitutional affairs. In this regard, he can be seen as both 'a man who was of international significance' and 'a role model in the struggle of small nations to challenge and defeat imperialism in the twentieth century'.[141] Significantly, de Valera also exerted a huge influence on the nature and scope of 1916 commemorations.

In addition to de Valera, many more of the men who were 'out' in 1916 and survived the experience, went on to play prominent roles in seminal events such as the War of Independence and the Civil War. After they relinquished their guns, some of the most enterprising amongst them managed to secure prominent positions in the government or civil service during the early decades of independence. Several of Fianna Fáil's most prominent figures during this time were veterans of the Rising, including Seán Lemass, who was born in Dublin in 1899. Having fought in the GPO in 1916 (where he spent time shooting from the roof), Lemass went on to fight in the War of Independence, after which he was imprisoned for a year in Ballykinlar in County Down. During the Civil War, he fought with the anti-Treaty side and was part of the force that occupied the Four Courts, along with Rory O'Connor and Ernie O'Malley. After being

---

[140] B. Girvin, *The Emergency: Neutral Ireland 1939–45* (Macmillan, London, 2006), p. 31; G. Fitzgerald, 'Eamon de Valera: The Price of His Achievement', in G. Doherty and D. Keogh (eds), *De Valera's Irelands* (Mercier Press, Cork, 2003), pp. 202, 204.

[141] P. Travers, *Eamon de Valera* (Dundalgan Press Ltd., Dundalk, 1994), pp. 5, 51; D. Ferriter, *Judging Dev: A Reassessment of the Life and Legacy of Eamon de Valera* (Royal Irish Academy, Dublin, 2007), pp. 4, 6–8.

interned in Mountjoy and the Curragh, he subsequently turned to politics and was elected as a TD for Sinn Féin in 1924. Two years afterwards, he became a founding member of Fianna Fáil with de Valera, while from 1932 onwards, he exerted a profound influence on the economic affairs of the state. Following stints as Minister for Industry and Commerce, Minister for Supplies and Tánaiste, Lemass then served a successful period as a reforming Taoiseach from 1959–1966.

Yet another of the 1916 rebels who managed to forge a successful political career within the party was Seán MacEntee – who was once described by Baron Monteagle as 'an exceptionally high-minded man ... of education, refinement and character'.[142] Born in Belfast in 1889, MacEntee worked as an electrical engineer with Dundalk Town Council at the time of the outbreak of the Rising and played an active role in the fighting at the GPO – a time he later referred to as '"my wild week" ... the week best worth living of all my life'.[143] MacEntee's death sentence was eventually commuted to life imprisonment and he was released from jail in the general amnesty of June 1917. In the 1918 General Election he was elected Sinn Féin MP for South Monaghan, while in later years he managed to secure a senior ministerial position in every Fianna Fáil government from 1932–1965. For those aspiring to a political career in the early decades of the Free State's existence, it was clear that association with the 1916 Rising was to remain perennially popular with the electorate.

Commemorations and Nation-Building in the 1920s and 1930s

Alongside the creation of the institutions and machinery of the Free State from 1922 onwards, attention was also devoted to the symbolism needed for the construction of a postcolonial national identity. The new state seal, for example, was adorned with an image of an old Irish harp, which later featured on stamps, coins and passports. In time, a close alliance was forged with the Catholic Church, while great prominence was also accorded to both the heritage and contemporary usage of the Irish language (notwithstanding the decline in speakers). Despite being 'bedevilled' by economic difficulties, it has been observed by Síghle Bhreathnach-Lynch that the new state was especially 'anxious to establish as soon as possible a distinctive national character, one as different as possible from that of its erstwhile ruler'. While Britain 'was perceived

---

[142] Baron Monteagle to Henry Duke, 8 June 1917, University College Dublin Archives (hereafter UCDA), MS. P.67/4/3, Archives of the Fianna Fáil Party, Seán MacEntee Papers.

[143] Notebook, August–September 1917, UCDA, MS. P.67/4/3, Archives of the Fianna Fáil Party, Seán MacEntee Papers.

as urban', the Free State went to 'endless lengths' so as to prove its rurality – 'rooted in a Golden Age, that of the ancient Celtic past'.[144] Even though a rural image of Irish society was powerfully evoked in the years after independence, urban landscapes, as Yvonne Whelan has noted, also played an important part in marking Ireland's transition from colonial to postcolonial. Imperial monuments were removed (or destroyed), new ones were erected, streets were renamed, and red-coloured post-boxes were repainted in green, 'covering over although not obliterating the Royal insignia of Victoria Regina or Edwardus Rex'.[145] The Free State's elimination of the symbols associated with the British Empire was 'understandable', according to Piers Brendon, 'in view of its urgent need to fashion a separate identity, to efface the stigma of being John Bull's other island, to rip the harp from the crown'.[146]

Changes also occurred within the legal sphere, with successive governments seeking 'to secure the recognition of an independent Irish citizenship', or in other words, 'to establish that Irish citizenship was not merely a local variant of the status of British subject'. This culminated with the passing of the Irish Nationality and Citizenship Act in 1935, which was eventually revised in 1956 (so as to make citizenship more available to the descendants of Irish emigrants and to persons born in Northern Ireland).[147] One of the most conspicuous signs of the Free State's severance from the British connection was the disbandment in 1922 of long-established regiments such as the Royal Irish Regiment, the Royal Dublin Fusiliers, the Leinster Regiment, the Royal Munster Fusiliers, and the Connaught Rangers. Even though the Connaught Rangers had mutinied during the War of Independence, it seemed that the Crown was willing to forgive and forget such recent sedition. Thus in his last address to his troops on 12 June, King George V warmly commended it for 129 years of dedicated service, describing it as 'a fine regiment ... of great memories and great traditions'.[148] But as memory of the Great War became increasingly associated with unionists' deep-seated loyalty to the Crown, Remembrance Sunday began

---

[144] S. Bhreathnach-Lynch, 'Commemorating the Hero in Newly Independent Ireland: Expressions of Nationhood in Bronze and Stone', in L. W. McBride (ed.), *Images, Icons and the Irish Nationalist Imagination* (Four Courts Press, Dublin, 1999), p. 148.

[145] Y. Whelan, 'Symbolising the State: The Iconography of O'Connell Street after Independence (1922)', *Irish Geography*, Vol. 34, No. 2 (2001), pp. 135–37.

[146] P. Brendon, *The Decline and Fall of the British Empire 1781–1997* (Alfred A. Knopf, New York, 2008), p. 295.

[147] M. Daly, 'Irish Nationality and Citizenship since 1922', *Irish Historical Studies*, Vol. 32, No. 127 (2001), p. 377, 390–91, 400, 402.

[148] Last Address to the Connaught Rangers by King George V, 12 June 1922, NAM, MS. 5111/52.

to prove a contentious issue during the early years of the Free State's existence. As Tom Burke has shown, an Irish Cenotaph (in the form of a wooden Celtic cross known as the Ginchy Cross) was erected at College Green in the centre of Dublin for remembrance services in November 1924, but two years later the ceremony was moved to a peripheral location in the Phoenix Park beside the Duke of Wellington's memorial.[149] It was not until 1931 that work finally commenced on the classically-inspired National War Memorial Gardens at Islandbridge in Dublin, which were designed by Sir Edwin Lutyens.

From the outset of the Free State's foundation, the Rising continued to figure very prominently in innumerable expressions of Irish heritage, culture and identity. One anonymous contributor to *An Talam*, published on 28 April 1923, went so far as to remark that those who displayed 'heroism' in 1916 were 'the most lovable men that Irish mothers ever reared'.[150] As rebel veterans attained the status of highly regarded (and besuited) statesmen, the chronicle of their violent struggle for nationhood was accorded great prominence in Irish life. As Wills has observed, the 1916 Rising, 'which was already part of national mythology, became respectable' and 'the GPO stood as the foundation of the new polity'.[151] According to David Lloyd, the history of the Irish nation, as presented within nationalist historiography after independence, was one in which 'sporadic movements of popular resistance, sparked off by one or other colonial excess, give rise finally to a mature nationalist consciousness'.[152] As part of this chronicle, myth and myth-making were mixed in equivalent sizes in nationalist history, and in the state-centred ideology which this generated.[153] From the 1920s onwards, the story of 1916 increasingly featured in the type of history that was taught to children, especially those in schools run by the Christian Brothers or operated under the patronage of the Catholic Church. This pedagogic trend impacted significantly upon the forging of cultural identity in the Free State. As children grew into adults, many of them carried with them a lasting memory of what they had been taught about 1916 in the classroom, ranging from respect to downright reverence.

However, the 1916 metanarrative also proved politically contentious on many an occasion following the granting of independence, most especially at Eastertime. The issue of how to commemorate the Rising, as Ferriter has

---

[149]   T. Burke, "'Poppy Day" in the Irish Free State', *Studies: An Irish Quarterly Review*, Vol. 92 (2003), p. 349.

[150]   *An Talam*, 28 April 1923.

[151]   Wills, *Dublin 1916*, p. 134.

[152]   D. Lloyd, *Ireland after History* (Cork University Press, Cork, 1999), p. 44.

[153]   J. MacLaughlin, *Reimagining the Nation-State: The Contested Terrains of Nation-Building* (Pluto Press, London, 2001), p. 247.

observed, 'confronted both government and opposition' and 'frequently led to political disagreement, cantankerous debate and uncertainty'. Instead of 'being a question of solemn remembrance', commemorations also 'provided an opportunity to seek to create political capital out of the contested republican legacy and to emphasise the divisions that existed within the Irish body politic'. This sometimes led to 'a growing resentment about commemoration', with the result that 'remembrance of 1916 encouraged political confrontation' and its legacy proved 'divisive', most especially in the 'rows that commemorating ... caused'.[154] The memory of the dead of the 11-month Civil War also caused much political quarrelling – both at home and abroad. In Ireland, the conflict tore apart communities and proved to be a source of acrimony in several family homesteads. According to Anne Dolan, 'the winners of a war no one wished to fight' commonly had to contend with the dilemmas and tensions of memory by expressing whatever there was 'of pride, sorrow, bitterness, triumphalism, shame'.[155] Furthermore, as Michael Hopkinson has pointed out, the conflict had a 'depressing effect on international attitudes' toward the new state, as the fighting had given rise to grave 'disorder ... argument and violence'.[156]

As bitterness from the Civil War split in Irish politics manifested itself throughout the Free State, various factions began to squabble over the legacy of Easter Week in the mid-1920s. One quarrel occurred in the lead-up to May 1924, when the Cumann na nGaedheal government hosted the first ever official state commemoration ceremony in remembrance of the Rising. This was staged at the Garrison Church at Arbour Hill (later known as the Military Chapel of the Sacred Heart) and in the adjoining graveyard. Such were the simmering tensions in the months beforehand that a number of steadfast republicans were actually blacklisted from attending – a decision that was taken in the interests of the security of the fledgling state. At a later stage, an official attached to the President's Department conceded that there had been an 'omission to send invitations to certain persons who, by virtue of their official position or otherwise, should have been regarded as entitled to be present at any such celebrations'. In other cases, specific types of invitees stayed away from Arbour Hill in 1924. Even though invitations to the event were sent to the relatives of those executed after the Rising, only Michael Mallin's widow, Agnes, bothered to show up on

---

[154] D. Ferriter, 'Commemorating the Rising, 1922–65: "A Figurative Scramble of the Bones of the Patriot Dead"?', in M. Daly and M. O'Callaghan (eds), *1916 in 1966: Commemorating the Easter Rising* (Royal Irish Academy, Dublin, 2007), pp. 198–99.

[155] A. Dolan, *Commemorating the Irish Civil War: History and Memory, 1923–2000* (Cambridge University Press, Cambridge, 2003), p. 3.

[156] M. Hopkinson, *Green Against Green: The Irish Civil War* (Gill and Macmillan, Dublin, 2004), p. 276.

the day. While letters of thanks were returned from most of them, a Department of Defence memorandum later indicated that 'a refusal to participate' had been received in 'one or two instances'.[157] Nonetheless, the government pressed ahead with the tribute, which passed off without any incidents. Another significant development in 1924 was the renaming of Sackville Street, which officially became known as O'Connell Street, thus taking on a distinctive meaning as part of what Whelan has called 'the nation-building agenda of the independent administration'.[158]

Although the names of 25 relatives were put on the official invitation list, ongoing concerns with security in the Free State meant that a politically-inspired blacklisting policy remained in place for the Rising's ninth anniversary commemoration at Arbour Hill on 4 May 1925, when the government took the decision to specifically restrict invites 'to each Deputy and Senator who is not Irregular'. Ceremonial matters also weighed on the minds of the ceremony's organisers, including the issue of who was to be given the task of placing wreaths on the graves of the executed leaders – members of the Executive Council or Army Chiefs?[159] Once again, the day's proceedings passed off without any major incidents, save for a small organisational problem with the actual running of the Requiem Mass at the Garrison Church. As a Department of Defence memorandum later recorded, an effort had been made 'to allot seating accommodation [at the church] to the various groups of persons invited in accordance with an order of precedence', but this 'was not very successful'. For the tenth anniversary ceremony in 1926, therefore, the Department of Defence recommended that 'only some 60 seats nearest the altar' be reserved 'to accommodate a special list of visitors'. Those added to the list included members of the Executive Council, relatives of the executed 1916 leaders, the Ceann Comhairle of the Dáil, the Chairman of the Seanad, the Chief Justice, the President of the High Court, the Attorney General, parliamentary secretaries, senior Army officers from the rank of Major General upwards, and senior officers

---

[157] Roinn an Uachtaráin to the Minister of Defence, 7 April 1925, NAI, Department of the Taoiseach, S9815/A, Easter Week Commemorations; Department of Defence to Each Member of the Executive Council, 25 April 1925, NAI, Department of the Taoiseach, S9815/A, Easter Week Commemorations. Also see Ferriter, 'Commemorating the Rising, 1922–65', p. 200. The absence of some of the relatives, he notes, can be explained by their 'political affiliations', most especially the fact that they 'were avowedly republican and hostile to the ruling government'.

[158] Y. Whelan, 'Monuments, Power and Contested Space: The Iconography of Sackville Street (O'Connell Street) before Independence (1922)', *Irish Geography*, Vol. 34, No. 1 (2001), p. 32.

[159] Department of Defence to Each Member of the Executive Council, 25 April 1925, NAI, Department of the Taoiseach, S9815/A, Easter Week Commemorations; Ferriter, 'Commemorating the Rising, 1922–65', p. 200.

of An Garda Síochána. In order to enable members of the general public to pay their respects to the Rising's leaders, arrangements were also put in place to allow the public to visit the graves after 12.00 noon on the day of remembrance.[160]

Even though the relatives of the executed 1916 leaders were well-liked and highly regarded by many people, they by no means had it easy in the years that followed the Rising. As some struggled to make ends meet, their supporters had no choice but to tap into commercial opportunities that arose from the nostalgia associated with the story of 1916. Whilst the days of 1916-related items being sold for vast amounts at prestigious auction houses were still a long way away, the monetary value of heritage objects associated with the Rising did inflate during the 1920s and 1930s. This price inflation was especially discernable in the case of the personal belongings of some of the senior-ranking rebels. In some instances, fundraisers were held to support the dependents of recently-deceased republicans. On 7 October 1925, for example, a draw was held for a five-seater Ford touring car, which had been the property of Liam Mellows. Having led the rebels in County Galway during the Rising, Mellows sided with the anti-Treaty forces during the Civil War and was one of 77 republicans who were executed by the Free State. Tickets for the draw for the car cost six pence and the proceeds were collected for the benefit of his mother. Two politicians, Austin Stack and Frank Fahy, acted as Honorary Treasurers for the fundraiser, which sold over 19,000 tickets.[161]

Having purchased St. Enda's in 1920, Patrick Pearse's mother also struggled to make ends meet, with the school encountering persistent financial pressures. 'I never get one hour free', she wrote on St. Patrick's Day 1927, lamenting that 'the work and worry are greater than ever'.[162] Despite the pressures she faced, there were times during which Pearse's mother rejoiced in the reminiscences that St. Enda's held for her. As one obituary later recorded, it was a space in which a mourning mother could 'be merry and laugh at herself', as she recalled to visitors what she remembered of her executed sons, Patrick and Willie:

> What memories she had of her own children ... Margaret Pearse was a born storyteller
> ... And Margaret Pearse needed all the jokes she could gather. In one sense it was always
> Nineteen Sixteen with her ... There is little doubt that St. Enda's kept her alive after
> 1916. It softened the tragedy of the loss of her sons ... Every path in the Hermitage

---

[160] Department of Defence to Each Member of the Executive Council, 29 April 1926, NAI, Department of the Taoiseach, S9815/A, Easter Week Commemorations.

[161] Mellows' Testimonial, Ticket No. 19,334, 1925, DCLA, Birth of the Republic, BOR F20/05.

[162] Margaret Pearse to Mr. and Mrs. McGrath, 17 March 1927, Pearse Museum, Dublin, SM. 293.

had memories for her and every room. Here Patrick had walked brooding or teaching, there at that table he had dropped his glove with the revolver inside ... or the gates of the Hermitage where she had said goodbye to her sons on Easter Sunday. Or the room where one night he [Patrick] had told her, sadly and calmly, that he and Willie and the others would soon have to go and leave all the beauty of the world.[163]

Following the death of Pearse's mother in 1932 the fate of the school was sealed and it finally closed in 1935 due to a lack of funding.

Other structures associated with the Rising's built heritage fared better. On 1 July 1929, for example, the GPO reopened its doors to the public – just over four months before the 50th anniversary of Pearse's birth. The building's original Francis Johnston-designed façade was retained, while a Greek key motif was used for its double-height ceiling, in a fashion similar to the spacious design for the original. The building was significantly extended behind the façade, with the architect behind the design, T. J. Byrne, including shops at the ground floor level of the Henry Street elevation. He also introduced a sheltered shopping arcade that ran between Henry Street and Prince's Street. W. T. Cosgrave commended the quality of the Irish craftsmanship that had gone into 13 years of rebuilding work, along with the fineness of the building materials, which included Irish marble and sandstone. Cosgrave also pointed to the radically altered political landscape that had emerged. The GPO, he stated, 'has come back to us renewed and beautiful', while he also remarked that that the Irish nation was 'progressing in the path of prosperity and peace'.[164] The timing of the GPO's reopening was convenient in other ways too, for 1929 marked the 100th anniversary of Catholic Emancipation.

As was often the case during the late 1920s, the ways in which the Rising was remembered around Eastertime tended to reflect the sharp divisions that persisted between members of the Cumann na nGaedheal government and the Fianna Fáil opposition. On some occasions, it was not unusual for competing commemorations to be held on the same day. However, Fianna Fáil's ascension to power in 1932 brought with it hope for a new era of stability in Irish constitutional politics. Hardened soldiers, who had once espoused the merits of politically-motivated physical force, now acted as badges of respectability for the machinery of the legitimate government. In the same year, the Catholic Church held the 31st International Eucharistic Congress in Dublin from 22–26 June. Timed to coincide with the 1,500th anniversary of the commencement

---

[163] Typescript Obituary Notice of Mrs. Margaret Pearse, National Library of Ireland, MS. 21092, Pearse Papers.

[164] Cosgrave, cited in S. Ferguson, *At the Heart of Events: Dublin's General Post Office* (An Post, Dublin, 2007), pp. 33, 35.

of St. Patrick's mission to Ireland, the entire spectacle was witnessed by over a million people at venues such as O'Connell Bridge/O'Connell Street and the Phoenix Park. De Valera kept up a high profile throughout the entire event, and as Rory O'Dwyer has observed, 'there was no doubting his Catholic loyalties' when it ended. In due course, the event helped to copper-fasten de Valera's 'remarkable political appeal, with his distinct blend of traditional Gaelic Catholic nationalism attracting many voters'.[165] In broader terms, the International Eucharistic Congress also highlighted the broadening cultural gap between the increasingly hegemonic identities of the Free State and Northern Ireland. As Gillian McIntosh has observed, the North's Protestant identity was thrown 'into sharper relief' after the Congress, exemplified a couple of years later by Lord Craigavon's remark 'that he was "an Orangeman first and a politician ... afterwards" and that "we are a Protestant parliament and a Protestant state"'.[166]

In the South, by contrast, it has been suggested by Tom Garvin that the Catholic Church operated 'like a second government or a state within a state', while the Rising intermittently 'seemed to be confused with the Resurrection of Christ' by certain clerics.[167] For minorities living on both sides of the Border, life often proved difficult from the 1920s onwards. In her work on the lifeworlds of Protestants in independent Ireland, Heather Crawford has contended that many of them 'saw the maintenance of their ethos, culture and social practices threatened', given the domineering 'construction of national identity as Catholic, nationalist ... [and] Gaelic' and the Free State's added emphasis on the Irish language. Consequently, 'from independence to the end of the 1940s the [Protestant] community kept to itself'. One of the root causes of the community's isolationism, she adds, was its 'need to recover from its experiences during the struggle for independence and to assess where it stood in the new scheme of things'.[168] Another ingredient to the alienation was the fact that some within the minority still saw themselves as unionists to the end, and continued to campaign at a number of political and social levels to resist the forging of an independent Irish nation. Unsurprisingly therefore, the story of 1916 did not feature much in the storylines of Protestant heritage in the years after independence, nor indeed

---

[165] R. O'Dwyer, 'On Show to the World: The Eucharistic Congress, 1932', *History Ireland*, Vol. 15, No. 6 (2007), p. 47.

[166] G. McIntosh, 'Symbolic Mirrors: Commemorations of Edward Carson in the 1930s', *Irish Historical Studies*, Vol. 32, No. 125 (2000), p. 100.

[167] T. Garvin, *Preventing the Future: Why was Ireland So Poor for So Long?* (Gill and Macmillan, Dublin, 2005), pp. 3, 186.

[168] H. C. Crawford, *Outside the Glow: Protestants and Irishness in Independent Ireland* (University College Dublin Press, Dublin, 2010), p. 23.

did the War of Independence (which had proven to be a far more traumatic experience for the community as a whole).

Shortly after becoming President of the Executive Council in 1932, de Valera sought to further reaffirm his republican inheritance by proceeding to abolish the contentious Oath of Allegiance and by refusing to pay land annuities that were owed to the British Exchequer. However, as Catherine O'Donnell has pointed out, Fianna Fáil's position in the next couple of decades of the Free State's existence 'inherently meant an acceptance ... of a compromise to the ethos of 1916'. As the dream of a united Ireland proved ever more unrealistic, and the sense of a 'homogenous Catholic identity' was consolidated ever more so in the 26 counties, de Valera's commitment to unity became more 'ideological' in nature.[169] Often, it was within the arena of public commemorations of Rising that this philosophical dedication to a 32-county Irish Republic was most graphically demonstrated, as de Valera immersed himself in a populist attempt to connect the 1916 metanarrative more closely to his own party's political heritage. The enterprise was also partly designed to keep republican dissidents and other movements at bay. Membership of the IRA swelled from around 4,000–5,000 in 1931 to over 30,000 in 1933, and with growing disorder throughout the state (exacerbated by the rise in popularity of the Blueshirts, who were branded as fascists by Fianna Fáil), the government was eventually forced to do a rethink on using Emergency Powers. Having suspended military tribunals in March 1932, it decided to reintroduce them with effect from 22 August 1933.[170]

From then onwards, the relationship between the government and the IRA became rockier. By the start of 1935, according to Joseph O'Neill, it was characterised 'by uneasy mutual tolerance', but this 'worsened' considerably as a result of two incidents. In February, a group of IRA men killed the son of a land agent in Edgeworthstown, County Longford, while in the month afterwards, IRA men, who had sided with striking tram and bus workers in Dublin city, sniped at Army lorries that were being used as substitutes for public transport. From then onwards, there was a marked increase in 'arrests, surveillance and harassment of the IRA'.[171] Such were the range of domestic pressures facing de Valera at this time that he was forced to rule out the possibility of a visit to the United States, even though he had been anxious to secure a trade agreement with it for some time. In a letter to Washington-based Charles Edward Russell on 16 April, de Valera wrote that 'things change very rapidly here as elsewhere' and

---

[169] C. O'Donnell, *Fianna Fáil, Irish Republicanism and the Northern Ireland Troubles 1968–2005* (Irish Academic Press, Dublin, 2007), pp. 9–10.

[170] B. Kissane, 'Defending Democracy? The Legislative Response to Political Extremism in the Irish Free State, 1922–39', *Irish Historical Studies*, Vol. 34, No. 134 (2004), pp. 162, 168.

[171] J. O'Neill, *Blood-Dark Track: A Family History* (Granta Books, London, 2000), p. 131.

noted that there was no chance 'of being able to get away for any considerable time'.[172] Besides having to cope with the omnipresent hassle from the IRA (and criticisms of his handling of political prisoners), de Valera's woes were further exacerbated at this time by the devastating impact that British tariffs were having on the Irish agricultural population. De Valera was also battling against his own health difficulties at the time – his eyes were in poor condition and he was having trouble reading.

The upcoming 19th anniversary of the Rising provided a welcome diversion for de Valera, who purposefully used the commemoration to preemptively assert Fianna Fáil's claim to the legacy of 1916. After much preparation and strategising, a major public commemoration was held outside the GPO on Easter Sunday 21 April 1935, along with an open-air Mass. Having watched the Defence Forces parade through O'Connell Street, de Valera then unveiled Oliver Sheppard's realist-style bronze statue of the legendary figure, Cúchulainn (Plate 3.2), inside the GPO as a tribute to the sacrifices made by those who had died during the Rising. Some invitees were unable to attend the unveiling due to ill health, including Sheppard and the Chief Justice, while the President of the High Court sent his regrets that he would 'not be in Dublin on Easter Sunday'. Deputy Vincent Rice also sent his apologies, 'owing to absence from Dublin'. Maud Gonne McBride sent a note to the government in advance of Easter Sunday, stating that she could not 'be present' because of reasons that it was already 'aware of'. Those in attendance at the GPO, however, could not escape the fact that some high-profile invitees had deliberately boycotted the ceremony. One of these was the esteemed writer, Oliver St. John Gogarty, who let it be known that he considered his invitation 'an impertinence' and that he was going to refuse to support the organisers 'in playing Hamlet'.[173]

To some of his political opponents at the time, it appeared quite palpable that de Valera's crafty initiation of such grandiose celebrations for the 19th anniversary was nothing more than a shrewd tactical manoeuvre to lay down some kind of early claim to the contested heritage of 1916. Consequently, members of the Opposition, including W. T. Cosgrave, John A. Costello and a range of other Fine Gael Deputies also boycotted the ceremonies, while republican splinter groups accused de Valera of playing politics through memorialisation.[174]

---

[172] Eamon de Valera to Charles Edward Russell, 16 April 1935, LCMD, Charles Edward Russell Papers, Container 18.

[173] Easter Week Memorial Unveiling Ceremony, 1935, Miscellaneous Acknowledgements, NAI, Department of the Taoiseach, S6405/C.

[174] For details on the row that developed, see Ferriter, 'Commemorating the Rising, 1922–65', pp. 203–205.

Plate 3.2   Oliver Sheppard's bronze statue of Cúchulainn at the GPO, unveiled on 21 April 1935 by the President of the Executive Council, Eamon de Valera

Source: Author.

Accordingly, it has been argued by Bhreathnach-Lynch that through holding 'propagandistic events' a full year before the much-anticipated 20th anniversary, de Valera 'was stealing a march on his opponents and appropriating the occasion as his own'.[175] Besides the unveiling of the Cúchulainn statue, the year of the 19th anniversary was also marked by the publication of *Nineteen-Sixteen: An Anthology*, a collection of poems inspired by the Rising and other episodes in Irish history, composed by the likes of Patrick Pearse, Thomas MacDonagh, William Butler Yeats, Dermot O'Byrne, Francis Ledwidge, and Eva Gore-Booth. This was compiled by Edna C. Fitzhenry from Dundrum, whose stated aim was to 'make a book that could be read through at one sitting', thus reconstructing 'in unbroken sequence the moods in which people saw the Rising approach, take place...and end – to be followed by a resurrection in which the nation, even as it mourned its dead, strove to establish the ideal for which they died'.[176]

Although many republican dreams remained unfulfilled in 1935, de Valera was certainly eager to reassure his international contacts that he was working towards fulfilling the ideals enshrined in the Proclamation. In a private letter to Charles Edward Russell on 24 June, he admitted that he was 'just a bit anxious' to ensure that 'our friends in America' were not being 'affected by certain propaganda' about the economy. Furthermore, he stressed that important strides had already been taken regarding 'our constitutional development' and noted that he envisaged 'further progress ... in the near future'. Rather revealingly, de Valera also alleged that he had found it both 'troublesome and annoying' that 'opponents on the left wing' had been persistently 'misrepresenting both our actions and our aims'.[177] In another letter that he sent to Russell on the same day, de Valera sought to alleviate foreign criticism of the use of military tribunals in the Free State. This had come from the International Committee for Political Prisoners, of which Russell was a member. In drawing attention to the fact that he had been 'compelled' to reintroduce the military tribunals in the summer of 1933 'for maintaining order and protecting the rights of the people', de Valera argued that by successfully preventing 'grave disorder', the state's democratic institutions had been able to withstand the spread of fascism throughout many parts of Europe.[178] Nonetheless, the use of military tribunals was still proving controversial by the time of the 20th anniversary of the Rising in 1936. Between

---

[175] Bhreathnach-Lynch, 'Commemorating the Hero', p. 158.

[176] E. C. Fitzhenry, *Nineteen-Sixteen: An Anthology* (Browne and Nolan Ltd., Dublin, 1935), p. 6.

[177] Eamon de Valera to Charles Edward Russell, 24 June 1935, LCMD, Charles Edward Russell Papers, Container 18.

[178] Eamon de Valera to Charles Edward Russell (Separate Letter), 24 June 1935, LCMD, Charles Edward Russell Papers, Container 18.

1934 and 1936, extremist activity had escalated and a total of 341 IRA men were convicted by the military tribunals.[179] Having only been legalised in 1932, the IRA was outlawed by the government in the summer of 1936. Before this decision was reached, however, Fianna Fáil had other business to take care of in the months beforehand. In a broadcast on St. Patrick's Day, de Valera reasserted his vision of Irish cultural identity, declaring that Ireland had been a Catholic nation since St. Patrick's time and still remained 'a Catholic nation'.[180]

At Easter 1936, he was again closely involved in a range of activities that were organised to mark the Rising's 20th anniversary. Besides seeking out every opportunity to copper-fasten Fianna Fáil's status as the most significant power in Irish republicanism (and consolidate its growing credentials as both the gatekeeper of democracy and the spiritual partner of the Catholic Church), de Valera remained concerned with deflecting attention away from the ever-present dissidents within the IRA. Special badges in the shape of lit torches were produced by Fianna Fáil for the 20th anniversary, with proceeds going to a National Benevolent Fund. These featured a logo bearing the letters 'FF' and were advertised for sale in *The Irish Press*.[181] On Easter Sunday 12 April, a mix of commemorative politics and religion was the order of the day, as Masses and small parades of Army units in commemoration of the 1916 rebels were held on the state's day of national commemoration, at Cork, Galway, Limerick, Athlone, Letterkenny, Carlow, and Dundalk. These were reviewed by various members of the Executive Council. Elsewhere, members of the Old IRA arranged for parades in Sligo, Macroom, Letterkenny, and Islandeady.[182]

An Easter Sunday parade was also held in Dublin city by members of Fianna Fáil cumainn in the capital city (including the Vice President of the Executive Council, Seán T. O'Kelly), 1916 veterans, emigrant groups, and members of Cumann na mBan. This commenced at Lower Abbey Street at 11.45 am and was headed by an advance guard of 1916 men. When they marched past the GPO, flags were lowered in salute. The parade then proceeded through Henry Street and Capel Street, along the quays of the River Liffey and then onwards to the ending point at Arbour Hill. After prayers were recited by the graves of the executed leaders, the 'Last Post' was sounded and three volleys were fired. Easter Sunday ceremonies were also held at a variety of locations overseas. In Scotland, for example, Fianna Fáil staged a 20th anniversary memorial ceremony at St. Mungo's Hall, while in London a big meeting was organised in Hyde Park by the Casement Commemoration Committee. Back in Ireland,

---

[179] Kissane, 'The Legislative Response', p. 167.
[180] De Valera, cited in McIntosh, 'Symbolic Mirrors', p. 100.
[181] *The Irish Press*, 10 and 11 April 1936.
[182] Ibid., 14 April 1936.

more commemorative events were staged on Easter Monday, including the commencement of a new series of talks from Radio Athlone on the theme of 'Easter Week and its Headquarters'.[183]

By the conclusion of the events marking the Rising's 20th anniversary, it seemed ever more likely that Fianna Fáil's subtle republican enterprise would continue unimpeded. Nonetheless, members and supporters of the party seemed particularly anxious to reassure organisations in the United States (such as the American Association for the Recognition of the Irish Republic) about the merits of its political strategy. Four days after the Easter Sunday commemoration, for example, Tadhg MacArdee, a resident of Dundrum in Dublin, wrote to Charles Edward Russell, seeking to highlight the benefits of de Valera's strategy for constitutional change. De Valera, he argued, 'is trying ... [and] hoping to give us what Collins ... [and] Griffith promised but failed to give us, "a Republic in all but name"'.[184] As 1936 drew to a close, Seán T. Ó Ceallaig, the Vice President of the Executive Council, reflected upon what Fianna Fáil had achieved since coming to power: 'Fianna Fail policy ... has been to take hold of the governmental machine ... and to use it to the fullest possible extent to clear away all the obstacles that remain on the road to complete independence for the whole country.' 'It has been tedious uphill work', he added, 'but we have big gains to our credit'.[185] By the time that the Rising's 21st anniversary was commemorated in 1937, IRA activity had waned considerably and it seemed that elements of de Valera's strategy were paying off after a new Constitution was passed through the Dáil on 14 June. This finally came into effect on 29 December, under which the Free State moved away from its dominion status when its name was changed to Éire/Ireland (while the office of President of the Executive Council was renamed Taoiseach). Then, in the following year, the Treaty ports were finally handed back to Ireland.

The 1930s also proved significant for the way in which the story of 1916 made its mark on how the nationalist view of history was presented to pupils in primary schools. As John Coakley has shown, the introduction of a set of *Notes for Teachers* formed the cornerstone of Irish educational policy for decades afterwards. The hegemonic view of the events of the age-old story of Ireland, as presented in the instructions, was that the contents of Ireland's past could be summarised as 'the study of the Gaelic race and Gaelic civilisation, and of the resistance of that race and civilisation for a thousand years to foreign

---

[183]   Ibid.

[184]   Tadgh MacArdee to Charles Edward Russell, 16 April 1936, LCMD, Charles Edward Russell Papers, Container 19.

[185]   Seán T. Ó Ceallaig to Charles Edward Russell, 14 December 1936, LCMD, Charles Edward Russell Papers, Container 18.

domination, whether Norse, Norman or English'. Unsurprisingly, the Rising and its leaders were represented in heroic terms in history books that were sanctioned by the Department of Education.[186] Articles in newspapers also reiterated the importance accorded to the 1916 metanarrative. With the passing of each year, anniversaries typically generated an assortment of first-hand reminiscences of 1916 in local, regional and national newspapers.

It has been observed by John Tosh that 'the readership of history and the extent of public support for its study will depend on the impact which historians make on the wider public'.[187] In the case of Ireland, there was certainly no shortage of budding historians during the early decades of the twentieth century. Unsurprisingly, the number of books published on the events of Easter Week steadily increased over time. W. B. Wells and N. Marlowe's *A History of the Irish Rebellion of 1916* was first published in 1916, as was Mary Lousia Hamilton Norway's *The Sinn Féin Rebellion As I Saw It*. These were followed the year afterwards by the *Sinn Féin Rebellion Handbook*, published by the Weekly Irish Times, along with a biographical tome, entitled *The Sinn Féin Leaders of 1916*.[188] By the 1930s, there was no shortage of glowing commentaries about the Rising's place in Irish history or the lives of its leaders. In a biography of Pearse that was published in French in 1932 (and translated into English by Desmond Ryan), Louis Le Roux cast the leader in a heroic light as a saint-like figure and even suggested that he would be canonised one day.[189] When he published the officially-sanctioned *Bibliography of Irish History 1912–1921* in 1936, librarian James Carty recorded that the National Library of Ireland's catalogue of published literature contained a total of 163 listings relating to the Rising. Although many of these were simply press narratives, his alphabetical list did include a range of important books, reports, articles in periodicals, and pamphlets.[190]

At university level, however, the nationalist paradigm came under increased scrutiny from academic historians, particularly at Trinity College Dublin and

---

[186] *The Cork Examiner*, 4 April 1991.

[187] J. Tosh, 'Introduction', in J. Tosh (ed.), *Historians on History*, 2nd Edition (Pearson Education Ltd., Harlow, 2009), p. 15.

[188] See Wells and Marlowe, *A History of the Irish Rebellion of 1916*; M. L. Hamilton Norway, *The Sinn Féin Rebellion As I Saw It* (Smith, Elder & Co., London, 1916); Weekly Irish Times, *Sinn Féin Rebellion Handbook, Easter 1916* (Fred Hanna Ltd., Dublin, 1917); and Anon., *The Sinn Féin Leaders of 1916*.

[189] See L. N. Le Roux, *La Vie de Patrice Pearse* (Imprimerie Commerciale de Bretagne, Rennes, 1932).

[190] See the listings in J. Carty, *Bibliography of Irish History 1912–1921* (The Stationery Office, Dublin, 1936), pp. 74–91.

University College Dublin, where a series of historiographical adjustments led to efforts aimed at fostering a more scientific attitude to approaching and interpreting the Irish past. The main instigators in this regard were Theo Moody and Robert Dudley Edwards (both graduates of the Institute of Historical Research in London), who founded the flagship journal, *Irish Historical Studies*, in 1938. From the outset, the approach taken by the editors sought to place emphasis on the principle of the scientific research technique, the use of primary sources and the importance of 'value-free' historical interpretation posed by hermeneutics. *Irish Historical Studies* was also significant in that it had a special section entitled 'Historical Revisions', with articles containing new research that was intended to refute received wisdom concerning events, people or processes. Methodologically, this 'new' history was very much part of a transnational disciplinary mainstream, but it also displayed strong links with the orthodoxies of the English historical profession (as exemplified by journals such as *History* and *Bulletin of the Institute of Historical Research*).[191]

According to Mary Daly, one of the major consequences of the 'value-free' approach adopted by *Irish Historical Studies* was the shift 'away from a simplistic narrative of Irish history as "800 years of English misrule" towards a much more nuanced story that emphasised contingency, incongruity and complexity'.[192] Although Moody and Dudley Edwards were subsequently 'praised as the founders of modern Irish historiography', Hugh Kearney has noted that that they later came 'under criticism for constructing a bloodless model for Irish history lacking any tragic sense'.[193] Even though he concedes that the journal's accomplishments were 'significant', Joe Lee has argued that 'they tended to relate more to the administration of research than research itself'. In this regard, he feels that that the 'constant insistence' on the scientific approach 'tended to denigrate earlier achievements', along with contemporaries publishing beyond the journal's

[191] B. Bradshaw, 'Nationalism and Historical Scholarship in Ireland', *Irish Historical Studies*, Vol. 26, No. 104 (1989), pp. 334–36; C. Brady, '"Constructive and Instrumental": The Dilemma of Ireland's First "New Historians"', in C. Brady (ed.), *Interpreting Irish History: The Debate on Historical Revisionism, 1938–1994* (Irish Academic Press, Dublin, 1994), p. 4; L. P. Curtis, 'The Greening of Irish History', *Éire-Ireland*, Vol. 29 (1994), p. 7; N. J. Curtin, '"Varieties of Irishness": Historical Revisionism, Irish style', *Journal of British Studies*, Vol. 35, (1994), p. 195; S. Howe, *Ireland and Empire: Colonial Legacies in Irish History and Culture* (Oxford University Press, Oxford, 2000), p. 84; H. F. Kearney, 'Visions and Revisions: Views of Irish History', *The Irish Review*, No. 27 (2001), p. 113.

[192] M. Daly, 'Forty Shades of Grey?: Irish Historiography and the Challenges of Multidisciplinarity', in L. Harte and Y. Whelan (eds), *Ireland Beyond Boundaries: Mapping Irish Studies in the Twenty-First Century* (Pluto Press, London, 2007), p. 94.

[193] Kearney, 'Visions and Revisions', p. 113.

'orbit'.[194] During the early years of the journal's existence, the absence of articles on Ireland's revolutionary years was rather striking. Although it initially set out to enhance collaboration 'between the historian and the teacher', it could be argued that the complete avoidance of any debate about topics such as the Rising somewhat inhibited the editors' mission of preventing the divorce of 'the teaching of history ... from the results of historical research'.[195] Although there was no mention of Easter Week in the printed pages bound between the eye-catching green covers of the early volumes of *Irish Historical Studies*, memory of 1916 was by no means absent from other dimensions of the public realm.

The Silver Jubilee Anniversary, 1941

Falling as it did in the midst of World War II, the occasion of the Silver Jubilee of the 1916 Rising in 1941 was not entirely unproblematic for the Fianna Fáil government, which had adopted a policy of neutrality in the international conflict. Faced with both internal and external threats to its sovereignty in the years and months leading up to Easter Sunday, the government felt that it was best to proceed with caution on matters such as Partition. By the early 1940s, according to Dolan, 'it was increasingly clear that the [Irish] public wanted a tamer version of republicanism, one that ... indulged in a more cosy reminiscence about "being out in 1916" and doing our bit in the War of Independence'. Many of the more small-scale commemorations, however, which tended to attract 'particularly zealous Gaels', did not have 'the same appeal' for the government, and were heavily policed by the Gardaí as a consequence.[196] This, of course, was a time in which many republican dissidents intent on obtaining a 32-county united Ireland by force were interned without trial by the government. Almost two decades had passed since the ending of the Civil War of 1922–1923 and de Valera was now intent on using whatever measures were necessary to face

---

[194] J. J. Lee, 'Irish History', in N. Buttimer, C. Rynne and H. Guerin (eds), *The Heritage of Ireland* (The Collins Press, Cork, 2000), p. 126. Also see A. Leonard (ed.), *The Junior Dean R. B. McDowell: Encounters with a Legend* (The Lilliput Press, Dublin, 2003), pp. 94–95, in which R. B. McDowell downplays the notion that it revolutionised Irish history writing, and points instead to how it represented 'an updating' of historiographical techniques. This, he added, was especially the case in regards to its highlighting of 'the value and attainability of objectivity' and its emphasis upon 'unflagging industry, precision and bibliographical thoroughness'.

[195] The Editors, 'Preface', *Irish Historical Studies*, Vol. 1, No. 1 (1938), p. 1.

[196] A. Dolan, 'An Army of Our Fenian Dead: Republicanism, Monuments and Commemoration', in F. McGarry (ed.), *Republicanism in Modern Ireland* (University College Dublin Press, Dublin, 2003), p. 140.

down subversive activity within the state that he himself had originally refused to recognise after the signing of the Treaty.

On the other side of the Border, the government of Northern Ireland also grew increasingly nervous about the threat from dissidents in the years that followed the outbreak of World War II in 1939. Besides having to cope with German air assaults, the Stormont administration was fearful of republican disturbances in the lead-up to the Rising's 25th anniversary. In a secret letter sent to Major General R. V. Pollok, General Officer Commanding the Troops in Northern Ireland on 11 March 1940, the Prime Minister, Lord Craigavon, expressed fears which he and his unionist colleagues had of 'any serious disturbances that may arise as a result of republican organisations endeavouring to take advantage of any intensification of the European hostilities ... by engaging in terrorist activities in Northern Ireland'. Craigavon was also intent on avoiding history repeating itself in the event of a replication of the situation in 1916, when Maxwell had responded to the actions of the rebels with a very heavy fist. In a stark warning to Pollok, he stated that his government was 'fully alive to the objection to the use, if it can be avoided, of the armed forces of the Crown for the suppression of civil disturbances, especially when such disturbances are organised in, and originate from, Éire'. Craigavon then went so far as to express his objection to increasing expenditure on the Special Constabulary, so as to make it 'effective in a military sense', as it had been in the early 1920s. Such an action, he cautioned, 'might be regarded even in Great Britain as being provocative and might have very embarrassing results since we should probably be pressed to explain publicly why we considered the expenditure necessary'.[197]

Although tensions and rivalries remained exceptionally strong between the different varieties of republicanism across the entire island of Ireland, the Fianna Fáil government of Éire/Ireland remained unwavering (for the most part) in its aspiration to make some sort of significant gesture to commemorate the Rising's 25th anniversary. Preparations for the event lasted for more than a year. In February 1940, de Valera directed M. Ó Muimhneacháin, the Secretary of the Department of the Taoiseach, to write to the Department of Defence, seeking its views 'at an early date' about 'the question of organising a suitable commemoration'.[198] After a wait of nearly three months (caused in part by the Chief of Staff's absence on tours of inspection), the Department of Defence finally outlined its position on the matter, in a letter sent from its Secretary to

---

[197] IRA Activities in Northern Ireland: Defence of Northern Ireland, Lord Craigavon to Major General R. V. Pollok, 11 March 1940, Public Record Office of Northern Ireland, Cabinet Minutes, CAB 9G/73/14.

[198] Department of the Taoiseach to Department of Defence, 12 February 1940, NAI, Department of the Taoiseach, S11409, Easter Week Commemorations.

the Department of the Taoiseach on 6 May. 'I am directed by the Minister for Defence', wrote the Secretary, 'to state it is considered that there should be a commemoration on as large a scale as possible having regard to the events commemorated'.[199]

A detailed memorandum running to four pages was also attached to the letter. As an overarching theme, the Department of Defence recommended that a 'definite motif' should characterise all of the Silver Jubilee ceremonies – namely one centred on the idea that the Rising had been responsible for 'the national resurgence'. Most of the memorandum was preoccupied with outlining a total of 19 suggestions – from the minimal to the grandiose. The main idea that was mooted was for the state commemoration to 'extend over the entire Easter Week', entailing both 'main items' (of an official nature) and 'subsidiary items'. A total of nine 'main items' were suggested, including: religious services, memorial services at the graves of those who died in the Rising, a military review in Phoenix Park, a presentation of campaign medals (with associated concerts), a pageant of history in the Phoenix Park (culminating with an 'Easter Week episode'), a Children's Day in Dublin (involving 'community singing'), a Gaelic festival (centred on the theme: 'Not free merely but Gaelic as well'), a massive 'Victory Parade' past the GPO, and a state céilidhe. Under the heading 'subsidiary items', a total of 10 additional suggestions were put forward, including the following: an Olympic-like athletics function (with runners conveying 'Torches of Freedom' from the four provinces to the GPO for the 'Victory Parade'), an exhibition at the National Museum, a military exhibition (stressing 'the development of the Army since 1916'), an industrial pageant ('to stress the industrial development of Ireland since 1916'), a commemorative stamp, a souvenir handbook (flagged as 'an essential'), support for the publication of special newspaper supplements (emphasising the 'outstanding Border question'), the staging of 'distinctively national programmes' at all theatres and picture houses (including selected plays by Pearse and pictures such as *Irish Destiny* and *Beloved Enemy*), the floodlighting of the GPO, and the facilitation of special arrangements to transport people from the countryside to the capital throughout the week of remembrance.[200]

[199] Department of Defence to Department of the Taoiseach, 6 May 1940, NAI, Department of the Taoiseach, S11409, Easter Week Commemorations.

[200] Department of Defence Memorandum on the Proposed Commemoration of the 25th Anniversary of Easter Week, May 1940, NAI, Department of the Taoiseach, S11409, Easter Week Commemorations.

Many of these proposals never materialised in 1941.[201] Those that did manifest themselves during the Silver Jubilee included the church services, the memorial service at the graves of the executed leaders, the presentation of medals, the céilidhe, the museum exhibition, and the commemorative stamp. Significantly, it was also decided to stage a parade of military personnel past the GPO on Easter Sunday 13 April 1941. However, the 'Victory'-themed element was a source of concern to de Valera, and was the subject of a protracted set of interdepartmental correspondence during the previous year. In August 1940, de Valera sought to tone down the Department of Defence's plans. As an interdepartmental memorandum from the Department of the Taoiseach to the Department of Defence made clear, the staging of an elaborate commemoration was deemed to be unsuitable for the times. Instead, it was suggested that the Department of Defence needed to outline 'preparations ... on a scale much less ambitious than that envisaged'.[202] Following an appeal from M. J. Beary of the Department of Defence, de Valera's cautious stance was reiterated on 25 October in another Department of the Taoiseach memorandum. This stated that the Taoiseach only wanted a commemoration 'on a very much more restricted scale than that originally contemplated and that it should only take place if circumstances generally are favourable'. Furthermore, it directed that the Department of Defence should put forward revised suggestions for the Taoiseach's contemplation, such as a one-day commemorative occasion that would be confined exclusively to a 'military review'.[203] In a reply dated 19 December, the Department of Defence produced a set of watered-down plans, consisting simply of a presentation of medals, one or two religious services and a 'Victory Parade past GPO by troops, Old IRA and 1916 participants'.[204] All of these proposals were agreed upon 'in principle' by de Valera the following month.[205] However, when the Cabinet met on 24 January 1941 to firm up the proposals, plans for a triumphalist-style parade were scrapped.

---

[201] Some of the proposals were eventually put to use when the Golden Jubilee was commemorated in 1966 (including the ideas for a historic pageant, a Children's Day and an industrial exhibition).

[202] Department of the Taoiseach to the Department of Defence, 9 August 1940, NAI, Department of the Taoiseach, S11409, Easter Week Commemorations.

[203] Department of the Taoiseach to the Department of Defence, 25 October 1940, NAI, Department of the Taoiseach, S11409, Easter Week Commemorations.

[204] M. J. Beary to P. Cinnéide, 19 December 1940, NAI, Department of the Taoiseach, S11409, Easter Week Commemorations.

[205] Department of Defence Memorandum, 11 January 1941, NAI, Department of the Taoiseach, S11409, Easter Week Commemorations.

Instead, it was decided that the Easter Sunday event should 'be called the 1916, 25th Anniversary Commemoration Parade'.[206]

There were certainly justifiable reasons for such a display of vigilance during World War II. As Ferriter has argued, de Valera's cautious approach to the Silver Jubilee may be attributed to 'the sensitivities of the government appearing to overdo ... republican celebrations' at a time in which the IRA was engaged in a bombing campaign in Britain, which itself was at war with Germany. Other factors also weighed heavily on the Taoiseach's mind in the months leading up to Easter 1941. According to Wills, de Valera's political opponents in the Dáil had urged him not to be overzealous in the celebrations, 'in acknowledgement of the disastrous situation facing many of the poor and unemployed', following two decades of self government. The failure of the government to create a united Ireland and to revive the Irish language also proved a source of much frustration and hinted at both the 'erosion of the republican vision' and the abandonment of 'the dreams of a generation earlier'.[207] Nonetheless, in spite of his own cautious stance over the 'Victory' theme, and notwithstanding pressure from the Opposition on various shortcomings since the achievement of independence, de Valera's backing of a large-scale military parade on Easter Sunday was still a significant commemorative gesture.

Despite the fact that both Partition and poverty remained as major bones of contention to some parts of the population in 1941, the government still felt confident enough to put together a programme of commemorative events that signified the importance of the anniversary being remembered. In addition to the parade, it also staged a small number of Silver Jubilee events in the week beforehand. One of these was held at the main square of Collins Barracks in Dublin on Tuesday 8 April, when around 700 veterans of the Rising received commemoration medals that had been specially minted for the occasion. The first of these was presented to a sombre-looking de Valera by the Minister for Defence, Oscar Traynor. This was followed by 'a great burst of applause'.[208] For over an hour, the Taoiseach then handed out the remainder of the medals to the assembled veterans, who were each presented to him by Colonel Liam Archer. Many of the recipients wore civilian dress, but some were in the uniform of the Army and the Gardaí. Around 50 women who had been members of Cumann na mBan also received medals, as did some of the Taoiseach's closest political

---

[206] Extract from Cabinet Minutes, 24 January 1941, NAI, Department of the Taoiseach, S11409, Easter Week Commemorations.

[207] Ferriter, 'Commemorating the Rising, 1922–65', pp. 206–207; C. Wills, *That Neutral Island: A Cultural History of Ireland During the Second World War* (Faber and Faber, London, 2007), p. 311.

[208] *The Irish Press*, 9 April 1941.

allies.[209] Two additional presentation ceremonies were held at Collins Barracks on 10 and 12 April. Medal recipients at these ceremonies included Domhnall Ó Buachalla, Senator M. Stafford and J. J. Walsh.[210] Months earlier, when preparations were being made in January for the striking of the medals, de Valera had let it be known that he favoured a design featuring 'a representation of the Cúchullain [sic] statue ... in the GPO'. Apart from this suggestion (which was approved at a Cabinet meeting on 7 February), he did not have any more 'views on points of detail'.[211] In the end, a total of 2,411 of the medals were issued. Although the manufacturers, the Jewel and Metal Manufacturing Company of Ireland, issued them unnamed, some of the recipients had their own names engraved on the reverse side afterwards, thus giving them an 'enhanced value from the collector's point of view'.[212]

Although the presentation of the medals at Collins Barracks passed off without any major glitches, a slight quarrel ensued afterwards, when the government was taken to task by a long-established supporter of Michael Collins. The official circular for the presentations at Collins Barracks had described de Valera as 'the senior surviving officer' of the 1916 Rising, as had a range of press reports. This claim, however, was contested by the author and historian, W. J. Brennan-Whitmore, from Gorey in County Wexford (who had fought in the GPO with the Irish Citizen Army and was promoted to the rank of Commandant by James Connolly, as he lay injured). In a short letter to the Editor of *The Irish Independent* – which offered proof yet again of the political divisions that had lingered on since the Civil War – Brennan-Whitmore (whose book, *With the Irish in Frongoch*, was published in 1917) argued that the Fianna Fáil leader had been opportunistic in his interpretation of history and went on to accuse him of monopolising the legacy of the Rising. He wrote: 'I ... challenge the accuracy of the latest attempt to nobble a position of super-eminence. There would seem to be no limit to the vanity of politicians.'[213] Clearly, even though 25 years had passed since Pearse had read the Proclamation, claims to the legacy of 1916 remained highly contested, and certain contrarians were disinclined to give the government party an easy ride in the commemorative stakes.

[209] *The Irish Times*, 9 April 1941.

[210] *The Irish Press*, 12 and 14 April 1941.

[211] Department of Defence Memorandum, 11 January 1941 and Extract from Cabinet Minutes, 7 February 1941, NAI, Department of the Taoiseach, S11409, Easter Week Commemorations.

[212] E. O'Toole, 'The 1916 Medal', *An Cosantóir: The Irish Defence Journal*, Vol. 26, No. 2 (1966), p. 66.

[213] *The Irish Independent*, 11 and 12 April 1941.

A few days after the gathering at Collins Barracks, a specially-commissioned exhibition on the heritage of the struggle for independence was opened, on Saturday 12 April, at the ground floor level of the National Museum at Kildare Street by the Minister for Education, Thomas Derrig. In his speech, Derrig stressed that while the showcase was designed to fulfil an educational purpose, he also hoped that it would encourage pupils to express affection for those who fought for Irish independence. The exhibition was curated by Liam Gogan and featured a range of artefacts belonging to well-known personalities, including documents, uniforms, books, and assorted personal belongings. An assortment of firearms was also put on display, along with paintings by Irish artists depicting people associated with the Rising. A copy of the Proclamation was displayed near the entrance, along with a bust of Patrick Pearse, which was set on a white pedestal. Among those who attended the opening was Senator Margaret Pearse, who had donated a volume of poems and a play written by her deceased brother. In its coverage of the opening of the exhibition, *The Irish Times* reported that the display cases contained 'a remarkably wide range of interesting souvenirs attractively displayed', which occupied a position in the museum that had previously been used for the display of priceless artefacts such as the Tara Brooch and the Ardagh Chalice.[214] *The Irish Press* drew attention to how the exhibition featured 'a number of notable mementoes' handed over by de Valera. These included his Irish Volunteers membership card and a 'somewhat incomplete' flag from Boland's Mill.[215]

The major highlight of the official programme of events for the 25th anniversary was a grandiose parade, lasting two and a half hours, of some 25,000 military personnel past the GPO on Easter Sunday 13 April (Plate 3.3). Although the 'Victory' theme had been cast aside more than three months beforehand, what transpired was certainly no watered-down affair. The sight of the large numbers of marching soldiers certainly conveyed an impressive sight to the thousands of citizens who turned out to watch the spectacle. Dublin's day of commemorative activities commenced shortly before 3.00 pm, when a special Guard of Honour (composed of 1916 veterans drawn from the 26th Battalion Old IRA) was drawn up in a formation opposite the GPO, while a firing party and six buglers took up positions on its roof. De Valera and members of his Cabinet (including the Minister for Supplies, Seán Lemass and the Minister for Industry, Seán MacEntee) arrived at 3.00 pm. De Valera proceeded to inspect the Guard of Honour whilst escorted by the Chief of Staff of the Army, Major General Daniel McKenna.

---

[214] *The Irish Times*, 14 April 1941.
[215] *The Irish Press*, 15 April 1941.

Plate 3.3   Soldiers marching past the GPO for the Rising's
               25th anniversary commemoration on 13 April 1941
Source: Time & Life Pictures/Getty Images.

After the Taoiseach finished inspecting the Guard of Honour, he then proceeded
to the front of the GPO. Proinnsias Ó Cearnaigh of the Easter Week Men's
Association then presented him with a gold replica of the badge of Cumann
Óglaigh na Cásga – featuring an outline of the map of Ireland with the GPO
and an Irish Volunteer juxtaposed on it – as a mark of the high regard that his
old comrades-in-arms had of him. The Taoiseach then took his place in the
reviewing platform and spectators' eyes turned to the roof of the GPO where
the Tricolour was lowered to half mast by Gearóid O'Sullivan. The firing party
on the GPO's roof then discharged three volleys and this was followed by the
buglers sounding the 'Last Post'.[216]

A short address in Irish and English was then delivered by the 81-year-old
President, Douglas Hyde, from his sickbed in Áras an Uachtaráin. Those behind
the idea for the President to deliver the broadcast were the Minister for Defence

[216] *The Cork Examiner*, 14 April 1941; *The Irish Independent*, 14 April 1941; *The Irish Press*,
14 April 1941; *The Irish Times*, 14 April 1941.

and the Taoiseach. Almost two months beforehand, when preparations for Easter Sunday were being discussed, Traynor wrote to the Department of the Taoiseach and conveyed the view that the 'taking of any active part' by Hyde (for example, 'taking the salute'), had the potential to impose upon him 'too severe a strain ... having regard to his age and infirmity'.[217] Having considered the proposal, de Valera was in agreement that the delivery of a broadcast was the 'most effectual' means of associating Hyde with the commemoration. According to a memorandum that was drafted on the issue, the Taoiseach was also keen to have an input into the content of the speech. It was noted that he was of the view that the general substance of the Presidential broadcast 'should be an exhortation to the people to be worthy of the sacrifices made in the past and to resolve that the independence which has been gained will be maintained'.[218]

Hyde promptly agreed to get involved in the commemoration and when he eventually delivered his speech on Easter Sunday, it was broadcast on national radio and also relayed over loudspeakers to the crowds gathered on O'Connell Street. The speech, which lasted little more than five minutes, was scripted by a Mr. McDunphy following consultation with Hyde and subsequent approval from de Valera. In it, the President resorted to religious imagery as he endeavoured to summon up a collective reverence for the rebels of 1916:

> Up to twenty-five years ago we were a people without power in our own land. Twenty-five years ago, in Easter Week the chains that bound us began to be broken at last, and gradually they were thrown off, so that we here today are a free people. We are assembled in this great gathering to pay honour and homage to the men who went before us. We shall never lose the freedom which we have gained so long as we remain united, faithful to ourselves, and loyal to the spirit of our ancestors. We pray to God that He will help us in the future as He has helped us in the past.[219]

In addition to the broadcast of the President's speech, the Department of Defence had initially put forward the idea of 'a talk on 1916', to be broadcast from 11.30 am on Easter Sunday. This suggestion was recorded in the minutes of a preparatory meeting held by the Department of Defence at Parkgate on 27 March. But when the Department of the Taoiseach received its copy of the

---

[217] Department of Defence to Department of the Taoiseach, 18 February 1941, NAI, Department of the Taoiseach, S11409, Easter Week Commemorations.

[218] Department of the Taoiseach Memorandum, 19 March 1941, NAI, Department of the Taoiseach, S11409, Easter Week Commemorations.

[219] *The Irish Times*, 14 April 1941; Text of Broadcast by President Hyde on the Occasion of the 25th Anniversary of the Rising of 1916, 13 April 1941, NAI, Department of the Taoiseach, S11409, Easter Week Commemorations.

minutes, it registered its unease with the idea by underscoring the following words: 'a talk on 1916'.[220] While the additional broadcast did not go ahead, the President was not alone in addressing the crowds at O'Connell Street on Easter Sunday. The words of an announcer, who spoke about the deeds of the executed leaders, were also relayed over the loudspeakers.

After the President delivered his Easter Sunday speech, the 'Reveille' was then sounded on O'Connell Street and the Tricolour was returned to full mast. As this happened, a battery of Field Artillery on the east side of O'Connell Bridge began firing a 21-gun salute at intervals of 10 seconds. Following a flyover of three squadrons of Air Corps planes, the spectacle then proceeded at around 3.30 pm with the parade of military personnel, who were accompanied by a range of brass and pipe bands. They were greeted with enthusiasm by cheering crowds. The parade started at College Green and followed a route over O'Connell Bridge and onto O'Connell Street, where the marchers filed past the saluting platform at the GPO and then proceeded to the finishing point on Parnell Street. Those who paraded included around 10,000 regular troops from the Army and roughly 15,000 members of the Volunteer Force, members of the Local Defence Force, the Local Security Force, the Red Cross, St. John's Ambulance Brigade, the ARP Services, the Fire Services, Demolition and Rescue Squads, and Decontamination Squads. The Marine Service also participated, dressed in elegant blue uniforms. A range of the Army's equipment also featured in the parade, including field artillery and motorised vehicles. In addition to the Taoiseach, many members of the Dáil and Seanad viewed the parade from the saluting platform, as did a selection of relatives of the executed leaders of the Rising. These included Senator Margaret Pearse, Rory Connolly, Grace Plunkett, and Donagh MacDonagh. The Lord Mayor of Dublin, however, was absent from the platform and was represented instead by Alderman T. Kelly. Following the ending of the parade, the commemorative spectacle was finally drawn to a conclusion by the playing of the National Anthem at around 6.00 pm.[221]

A number of newspapers were taken aback by the scale of the parade, which had amounted to the biggest gathering of people on the streets of Dublin since the Eucharistic Congress of 1932. Given its strong links with Fianna Fáil, *The Irish Press* was unsurprisingly unabashed in its praise of the events on O'Connell Street. Its Easter Monday edition spoke glowingly of the previous day's events and went so far as to proclaim that the country's 'reborn generations' were able 'to salute the glorious week that thrilled the soul of the people' and

---

[220]   Minutes of a Meeting Held in the Department of Defence at Parkgate on 27 March 1941, NAI, Department of the Taoiseach, S11409, Easter Week Commemorations.

[221]   *The Irish Independent*, 14 April 1941; *The Irish Press*, 14 April 1941; *The Irish Times*, 14 April 1941.

'tugged at the heart of the world'.[222] *The Cork Examiner* remarked that the parade represented a fitting 'tribute to those who gave or offered their lives … [so] that Ireland might be free' and stated that it 'was the most impressive military display ever seen in the capital city'.[223] Likewise, *The Irish Independent* stated that the military parade had been 'the largest and most spectacular … the city has seen'. It also observed that several thousands of spectators 'had waited for hours' to see it at different points along its route. 'Every point of vantage', it noted, 'was occupied', including 'windows, roofs, railings', along with 'the tops of air raid shelters'. As the parade passed through O'Connell Street, it made for an impressive sight, especially 'a silver line of bayonets', which were 'glittering where the spring sun caught the blades'.[224]

The Irish Times remarked how the march of infantry in an eight-deep formation 'made an impressive display', but also drew attention to problems with overcrowding on the day. It noted that the Gardaí 'had a difficult task … maintaining a clear path' for the marchers, while the St. John's Ambulance experienced 'a busy time attending to the numerous women and children who fainted or were injured in the great crush'.[225] *The Leader* focused on problems of a different kind, by reporting a 'rumour' that Fine Gael's front bench had 'boycotted' the ceremony by not appearing at the GPO. Such a factional attitude, it alleged, was 'truly deplorable' at a time 'of crisis' in the world. Furthermore, it determined that it was 'bad citizenship' on behalf of the Opposition 'to do anything that would tend to encourage the man-in-the-street to treat Ministers with churlish contempt'.[226] An exasperated correspondent for *The Tuam Herald* made reference to problems of a different kind. Although conceding that the parade 'made us all young again as we delved into the past and brought our thoughts up to the present', he complained that the anniversary also 'had its pathetic side'. Some of those who paraded with 1916 medals, he lamented, 'are so far without a pension or any other reward from an ungrateful nation'.[227] This was a reference to the plight of the Old IRA veterans who had given 'active military service' up to 1921. The matter had been raised in the Dáil the month beforehand, when the Minister of Defence had to deal with a query concerning the operation of the Military Service Pensions Act of 1934, which had established

---

[222]   *The Irish Press*, 14 April 1941. To mark the occasion, the newspaper also published a piece by Thomas MacDonagh's son, Donagh, entitled 'The Epic Poem of Easter Week'.

[223]   *The Cork Examiner*, 14 April 1941.

[224]   *The Irish Independent*, 14 April 1941.

[225]   *The Irish Times*, 14 April 1941.

[226]   *The Leader*, 19 April 1941.

[227]   *The Tuam Herald*, 19 April 1941.

the criteria for determining entitlements to military service pensions.[228] As a consequence of prolonged delays in the processing of some cases, members of the Opposition were sometimes critical of the government's fiscal rectitude. In the weeks before and after the parade, they sought answers to when decisions would be forthcoming from referees on both claims and appeals lodged by veterans from counties such as Cork, Meath, Waterford, and Westmeath.[229]

Besides the major events staged for the Silver Jubilee, a range of additional commemorative activities – both official and unofficial – were organised by the government and different organisations. At the republican plot at Glasnevin Cemetery in Dublin, for example, republicans assembled on Easter Sunday 'under the auspices of the Easter Week Commemoration Committee' and placed wreaths on the graves. Prayers were also recited, while an oration was delivered by the poet, Brian O'Higgins.[230] Similar scenes were replicated at many graves and historic sites throughout Éire/Ireland on Easter Sunday. In the centre of Cork city, members of the Old IRA marked the anniversary by attending a Mass at St. Augustine's church on Easter Sunday. Dignitaries who attended included the Chairman of the Cork Harbour Commissioners. Various republican organisations also marked the occasion by parading from the Grand Parade to St. Finbarr's Cemetery. On the way they stopped at Gaol Cross, where the 'Last Post' and 'Reveille' were sounded by buglers from Fianna Éireann. The ritual was then repeated at the republican plot in the cemetery, which was covered with wreaths, while an oration was delivered by Eileen Crowley. Another commemoration was held at Rath Cemetery in Tralee, County Kerry, where an oration was delivered by Annie MacSwiney from Cork (the sister of the deceased Lord Mayor, Terence) and a Decade of the Rosary was recited by the attendees.[231]

In the town of Athenry in County Galway, the 25th anniversary was marked by a special Easter Sunday parade of 350 members of the Local Defence Force and 150 members of the Local Security Force. Around 40 of the marchers wore 1916 medals, including the TD, Seán Broderick. Religion formed an essential ingredient of the day's events. Prior to the parade, the marchers attended a Mass celebrated by the Reverend Father Cassin, while later on in the day they 'attended Benediction of the Most Blessed Sacrament given by the Very Rev. Vanon Lavelle'.[232] A jovial event was held in Galway city the night afterwards, when the Galway branch of the Gaelic League ran a special commemorative

228 *Dáil Debates*, Vol. 82, 20 March 1941.
229 See, for example, ibid., Vol. 82, 3 April 1941, 23 April 1941 and 8 May 1941.
230 *The Irish Independent*, 14 April 1941.
231 *The Cork Examiner*, 14 April 1941.
232 *The Connacht Tribune*, 19 April 1941.

céilidhe. This was held in the Commercial Boat Club, which was 'specially decorated' beforehand.[233] The following day, the Irish Diaspora in England made its own contribution to the Silver Jubilee anniversary when nearly 1,000 people attended a meeting of the Connolly Club at London's Holborn Hall. One of the attendees was P. J. Musgrove, the author of a biography on the life of James Connolly. A resolution was passed unanimously at the gathering which called for the release of Irish republican prisoners and pledged 'continuance of the struggle for the achievement of the Republic envisaged by Connolly and Pearse'.[234]

Back in Dublin city, another official act of remembrance was staged by the government on Thursday 24 April, when the annual 'Solemn Requiem Mass for the repose of the souls of the men who died for Ireland in the Insurrection in 1916' was held at the Military Chapel of the Sacred Heart, Arbour Hill.[235] This was followed by a ceremony at the graves of the executed leaders, which had been 'suitably tended' to in advance by the Office of Public Works.[236] Those watching included de Valera, several other high-ranking politicians, Army personnel, Garda officers, members of the clergy, and relatives of the 1916 rebels. Three volleys were fired at their graves and the 'Last Post' was sounded by buglers. After the Tricolour was raised to full mast, the 'Reveille' and the National Anthem were each played.[237] For those individuals with close family connections to 1916, it was evident that the 25th anniversary represented a date of immense historical and emotional significance. This became apparent in the lead-up to the Arbour Hill ceremony, when it appeared that some of the younger generation of relatives were going to miss out on the proceedings. One such individual was Michael Mallin's youngest son, Séamus, who wrote a letter in Irish to the Taoiseach in the month beforehand, conveying his disappointment about his absence from the next-of-kin invitation list. In a written reply, the Secretary of the Department of the Taoiseach stated that during the previous eight years, 'no alterations have been made in the list of the next-of-kin except by way of addition and ... this is the first occasion on which representations have been made on the subject of the omission of your name'. The Secretary then confirmed that his name would be

---

[233] Ibid., 12 April 1941.

[234] *The Irish Press*, 16 April 1941.

[235] Invitation Card to Solemn Requiem Mass at Arbour Hill on 24 April 1941, Military Archives, Cathal Brugha Barracks (hereafter MACBB), Department of Defence, 2/68475, Ceremonial, 1916 Commemoration, Arbour Hill Barracks, 1941.

[236] Office of Public Works to Department of Defence, 4 April 1941, MACBB, Department of Defence, 2/68475, Ceremonial, 1916 Commemoration, Arbour Hill Barracks, 1941.

[237] *The Irish Press*, 25 April 1941.

added to the list for future ceremonies.[238] The final version of the Department of Defence's 38-page invitation list for the ceremony included the names and addresses 38 'Immediate Relatives of Dead Leaders' (listed as Group B) and 95 'Relatives of Men Killed in Action in 1916' (listed as Group G).[239]

So as to be as inclusive as possible, the government also permitted the public to visit the graves of the executed leaders on four designated days – namely 13–15 and 24 April. Additional commemorations of the Silver Jubilee were held in the months that followed. One of the biggest of these was held in Dublin on Sunday 18 May, which played host to a commemoration organised by members of the Dublin Trades Council. Mass was said 'for all who died for Ireland' at St. Mary of the Angels on Church Street and then an estimated 2,000 marchers, representing 58 trade unions, marched in the Connolly Commemoration Parade from St. Stephen's Green to O'Connell Street. An address was delivered at the GPO by the President of the Dublin Trades Council, who had earlier placed a wreath at Connolly's grave at Arbour Hill. Old comrades of the Irish Citizen Army also marched in the parade, wearing their 1916 medals and carrying the Stars and Plough banner.[240] Later on in the year, everyone in Éire/Ireland got their chance to remember the story of 1916 once again, when an official two and a half pence commemorative stamp went on sale. Designed by Victor Brown, it featured an image of an Irish Volunteer in front of the GPO along with the opening words of the Proclamation. In advance of its release on 27 October, the stamp's uniqueness was highlighted by *The Irish Independent*, which reported that it was the first stamp to be 'produced in its entirety in this country' through specialised manufacturing and printing processes.[241]

Besides remembering, exhibiting and even commodifying the storylines surrounding the deeds of the rebel past, it was abundantly clear that other matters had also weighed heavily on the Irish government's mind when the main events to mark the 25th anniversary were held at Easter. In one of its leading articles on the Easter Sunday spectacle, entitled 'Dublin's Biggest Defence Parade', *The Irish Press* remarked how a particular feature of the parade 'was the mechanisation of the expanded Defence Forces'. Another article, entitled 'A Trust to Defend', suggested that the parade was 'much more than a mere

---

[238] P. Ó Cinnéide to Séamus Ó Mealláin, 20 March 1941, NAI, Department of the Taoiseach, S4393, Easter Week Commemorations.

[239] Invitation List to Arbour Hill Ceremony on 24 April 1941, MACBB, Department of Defence, 2/68475, Ceremonial, 1916 Commemoration, Arbour Hill Barracks, 1941. According to a handwritten note at the end of the list, the invitations were sent in the post on 18 April.

[240] *The Irish Press*, 19 May 1941.

[241] *The Irish Independent*, 9 October 1941.

commemoration. It was the case of Ireland stated'.[242] Although the authors of the two articles in question did not really elaborate on the bigger picture, it was clear for all to see at the time that the Taoiseach and President had been at one in sending a robust message to the powers of Europe about Ireland's neutral position in the wider world. There were genuine reasons behind this diplomatic stance. On many occasions following the outbreak of World War II, the British Prime Minister, Winston Churchill, had urged de Valera (both privately and publicly) to join the fight against Hitler's Nazi Germany by making Ireland's ports and airfields available to British forces. Only a few days before Easter Sunday 1941, these strategic facilities were again mentioned by Churchill in an address to the House of Commons. On Wednesday 9 April, he stated: 'I hope that eventually the inhabitants of the sister isle will come to realise that it is as much in their interests as it is in our own that their ports and airfields should be available for the [British] naval and air forces, which must operate farther into the Atlantic.'[243]

The Taoiseach, however, was not for turning. It was well-known, as John A. Murphy has noted, that de Valera vehemently believed that 'small states must not become the tools of any great powers'.[244] In many respects, therefore, the militaristic nature of the 25th anniversary parade on Easter Sunday functioned as far more than just a mere act of paying respect to the Rising's memory. The display of military strength served a vital pragmatic purpose in that it delivered a timely reminder and political affirmation to the wider international community of Irish military neutrality and independence – at a time in which the latter was seriously threatened. This particular angle to the ethos of the Silver Jubilee commemoration can be seen from the comments made by a correspondent writing for *The Tuam Herald*, who stated that 'the threat of invasion' was a major factor behind the decision to stage such a large military parade. It was a parade, he suggested, 'which should make any nation proud and any intending invader think twice before the "wild" Irish would get their tempers ruffled again'.[245]

In addition to Hyde's Easter Sunday message, de Valera delivered his own radio broadcast from the GPO on Easter Saturday. In this, he dwelled not only on the theme of 'sacrifice', but also spoke of the personal hope that he took from 'the heroism which was required to face the effort of 1916'. The men of 1916, he argued, had displayed 'a courage far beyond the ordinary' and 'had steeled themselves to face all that death involved for themselves and those dear to them',

---

[242] *The Irish Press*, 14 April 1941.
[243] *The Irish Times*, 10 April 1941.
[244] J. A. Murphy, *Ireland in the Twentieth Century* (Gill and Macmillan, Dublin, 1989), p. 99.
[245] *The Tuam Herald*, 19 April 1941.

so that Ireland could escape from 'the mastery of the strong hand of the stranger'.[246] De Valera then delivered a strongly-worded warning about how the population of Éire/Ireland faced grave danger because of the so-called 'Emergency'. Most pointedly, he expressed confidence that the armed forces would not be found wanting in the event that the country needed to be defended from an attack by anyone:

> Today, in a warring world, the freedom of nations is everywhere in peril ... As a state we
> have proclaimed our neutrality ... Still, every day while this war continues our dangers
> will increase ... Death and suffering we shall not be able to avoid if we are attacked,
> but we can make certain that they will not be in vain by completing and perfecting the
> organisation and training of our manhood and womanhood in the several national
> services, such as were so splendidly represented in the parade today ... Thanks to
> the men whose memory we have been honouring today, we, the first generation in
> centuries, have a freedom which is worth defending.[247]

Whilst many people across the country were already well acquainted with the economic repercussions of the war – having experienced its harsh consequences on everyday lifestyles and endured the rationing of commodities such as tea, coal and petrol (as a result of the scarcity of shipping and the consequent decline in imports from Britain) – the external dangers highlighted by de Valera were clear to see in the days and weeks both before and after the Silver Jubilee parade, as the physical impact of the war was observed at a host of locations throughout the island of Ireland.

In the four weeks leading up to 3 April, for example, de Valera confirmed to the Dáil that there had been machine-gun attacks by German aircraft on three ships sailing under the Irish flag, namely the Glencullen, Glencree and Edenvale.[248] Then, on Easter Monday, German aircraft were spotted 'flying at a great height' over Northern Ireland, with the result that 'the ground defences were in action twice'.[249] Worse still was to follow for the people living across the Border. Over a seven-hour period that stretched from 10.00 pm on Tuesday 15 April to 5.00 am on Wednesday 16 April, the docklands of Belfast city, along with other parts of Northern Ireland, were heavily bombed by almost 200 German aircraft. Many fire brigades and ambulances were sent across the Border to the North from various towns in the South, while trainloads packed with refugees from Belfast (some with wounds) arrived in Amiens Street Station in

[246] *The Irish Press*, 14 April 1941.
[247] *The Irish Times*, 14 April 1941.
[248] *Dáil Debates*, Vol. 82, 3 April 1941.
[249] *Belfast Telegraph*, 15 April 1941.

Dublin, where they were attended to by the Red Cross.[250] In total, the attack against Belfast by the Luftwaffe caused around 745 deaths, which was more 'than in any single air raid on any British city outside of London'.[251] After hearing that Churchill was once again considering the application of conscription to Northern Ireland, de Valera summoned a special meeting of the Dáil on 26 May, where he highlighted 'the seriousness of the situation which confronts us'. Spurred on by the success of the 1916 commemoration, he asserted that the continuation of Partition represented 'a deadly wound inflicted upon the body of this nation'. By regarding it as no more than 'a passing phase', he stated that it would be outrageous to conscript those in the North 'who have vehemently protested against being cut off from the main body of the nation ... against their own will'. Indicating that he wanted to maintain cordial ties with the British, he proclaimed that Irish neutrality 'has all the time been a friendly neutrality'.[252]

The dangers of the time manifested themselves again five days later, when German bombs were dropped at various locations around Dublin city early in the morning of Saturday 31 May, including the Phoenix Park, North Strand Road and Summerhill Parade. Most of the damage was caused at North Strand Road, where 34 people were killed while around 90 people were injured. A bomb was also dropped near Arklow in County Wicklow on the same date, but fortunately nobody was killed.[253] Commenting on the psychological impact of the German bombing raids, Kevin Kearns has noted that they left many Dubliners feeling 'emotionally drained', while an attitude of contentment was 'replaced with a more realistic, disquieting outlook'.[254] Although he again faced escalated pressure from certain quarters to alter his foreign policy in the aftermath of the bombings

---

[250] *The Irish Times*, 17 April 1941.

[251] T. Ryle Dwyer, *Behind the Green Curtain: Ireland's Phoney Neutrality During World War II* (Gill and Macmillan, Dublin, 2009), p. 157.

[252] *Dáil Debates*, Vol. 83, 26 May 1941.

[253] *The Irish Press*, 2 June 1941. Historians have been somewhat divided on the reasons behind the bombing incidents. Girvin, *The Emergency*, p. 180, has suggested that the German air attacks were intended to terrorise the Irish people and to highlight 'the costs of giving up neutrality'. Alternatively, S. McMahon, *Bombs over Dublin* (Currach Press, Dublin, 2009), p. 124, has speculated that the German pilots bombed Dublin by mistake, with possible reasons including 'navigational error, equipment malfunction and ... weather'. From an analysis of the records of the lookout posts along Éire/Ireland's coastline, M. Kennedy, *Guarding Neutral Ireland: The Coast Watching Service and Military Intelligence, 1939–1945* (Four Courts Press, Dublin, 2009), p. 199, has contended that the bombings were unintentional and that scattered German aircraft dropped excess weight so as 'to improve flight time, speed and altitude', thus allowing them to get back to their bases.

[254] K. C. Kearns, *The Bombing of Dublin's North Strand, 1941: The Untold Story* (Gill and Macmillan, Dublin, 2009), p. 286.

in Belfast and Dublin during 1941, de Valera steadfastly resisted and ultimately managed to keep the South out of the war by refusing to budge an inch from his stance regarding Irish military neutrality. As was clear from the Silver Jubilee parade, the Taoiseach 'had cloaked himself in the mantle of the Republic'.[255] By repeatedly campaigning for the population to cut more turf and by passing Emergency Powers legislation to increase the amount of land under tillage and thus maintain food supplies, the Irish government ultimately prevailed in its quest for survival under de Valera's leadership and the unwavering support of his Cabinet – whose most prominent members were hardened veterans of either the 1916 Rising or the War of Independence (or both). Whilst there were many human and technical skills that made neutrality a possibility, the luck of the Irish was also on de Valera's side, in a geographical sense.[256] Altogether, in spite of the hardships endured during the course of World War II, incidents such as the Lutwaffe's bombing of Belfast (only 28 hours after the culmination of the military parade in Dublin), certainly had the effect of making many people in Éire/Ireland feel 'thankful for ... independence'.[257]

Memorialisation Agendas, 1945–1965

After the eventual defeat of the Nazis in May 1945, London was the scene of blissful revelry as the British people celebrated Victory in Europe and accorded a heroes' welcome to its returning troops. Although close to 60,000 citizens of Éire/Ireland had served with the Allies during the war, Churchill still had harsh words for de Valera in his victory broadcast, noting that Britain had shown moderation and composure in not invading Ireland. Not that this stance was in any way surprising. It was a well-known fact, as Norman Davies has stated, that Éire's refusal to support Britain in its hour of need had rankled heavily with the Prime Minister, who formed the opinion that 'it was little better than a traitor'.[258] Despite being heavily criticised, de Valera dug into his revolutionary reserves and delivered an unwavering reply. In a broadcast on Irish radio, the Taoiseach calmly outlined his feelings about Ireland's long history of opposition to foreign

---

[255] Wills, *That Neutral Island*, p. 311.

[256] See T. Desmond Williams, 'Ireland and the War', in K. B. Nowlan and T. Desmond Williams (eds), *Ireland in the War Years and After, 1939–51* (Gill and Macmillan, Dublin, 1969), p. 25, who points to the 'good fortune that the occupation of Irish territory never really became absolutely vital or was thought to become vital to the security of any of the belligerents'.

[257] Ryle Dwyer, *Behind the Green Curtain*, p. 157.

[258] N. Davies, *Europe at War 1939–1945: No Simple Victory* (Macmillan, London, 2006), p. 287.

aggression. Furthermore, he resolutely defended Ireland's policy of neutrality, by casting his doubt on whether, if the Nazis had invaded and partitioned England, Churchill would have then joined with them 'in a crusade'.[259] The positive reaction by the Irish public to the Taoiseach's broadcast, according to Ferriter, highlighted the fact 'that neutrality was seen as something to be cherished and defended as the ultimate expression of Irish independence'. Furthermore, it 'did much to secure his national "father figure" status'.[260]

With Anglo-Irish relations in a perilous state at the end of World War II, de Valera's government veered away from any significant public acknowledgement of the sacrifices and intrepidity of the Irishmen who had served with the Allies. Like the men who had fought against the Germans in the Great War, there was no heroes' welcome in Dublin for these veterans. In a move that aroused the ire of some of de Valera's political opponents, 4,983 men who had deserted the Irish Army during the Emergency (with many opting to fight for the Allies), were reprimanded when they returned home to Éire/Ireland.[261] In the meantime, the primary focus of Irish commemorationist policy in the rest of the 1940s centred on the recurrent mission of perpetuating the memory of the revolutionary era. A government attempt at lifting the subdued national mood was evident at the end of 1945, when the Department of Defence made arrangements for the sale (to *bona fide* applicants) of a special batch of miniature service medals in bronze for those who had fought for Irish freedom between 1916 and 1921. According to a memorandum prepared by the Minister for Defence on 20 December 1945, a decision was taken to mount a publicity campaign to make veterans of the revolutionary era aware 'that miniature medals are available to eligible applicants', at a price of three shillings and three pence 'to cover the cost'.[262] Even though hundreds of 1916 medals had already been awarded for free during the three 1916 remembrance ceremonies at Collins Barracks back in 1941, the decision to acknowledge the active service of the veterans of the whole revolutionary era with a new set of medals certainly had its advantages for the Fianna Fáil leader.

---

[259] De Valera, cited in Ferriter, *Judging Dev*, p. 258.

[260] Ibid., pp. 255, 258.

[261] Under Section 13 of the Defence Forces (Temporary Provisions) Act, 1946, these men were blacklisted 'for desertion in time of national emergency'. For seven years from 1 April 1946, they were debarred from 'any office or employment' salaried by the government and denied payment of pensions or gratuities 'in respect of ... service in the Defence Forces'. For the full text of the legislation, see http://www.irishstatutebook.ie/1946/ (accessed on 17 January 2012). In July 2012, a Fine Gael-led government announced its intention to introduce legislation to grant these men an official pardon and amnesty.

[262] Department of Defence Memorandum on the Provision of Miniature Service, Etc. Medals, 20 December 1945, NAI, Department of the Taoiseach, S11409, Easter Week Commemorations.

Having taken a hardline stance by interning republican dissidents during the Emergency (and even allowing deaths on hunger strike), de Valera was able to outmanoeuvre some of his most outspoken political opponents (and reassert his own republican credentials in the process), by sanctioning a commemorationist policy that satisfied the demands of those who did not want to forget the revolutionary era.

Throughout the latter half of the 1940s, public interest in the story of the 1916 Rising remained undiminished. To mark the occasion of its 30th anniversary in 1946, the Parkside Press on Grafton Street in Dublin published S. P. Kelly's *Pictorial Review of 1916*, which claimed to offer a *bona fide* and exact illustration of the events of Easter Week by amassing a range of illuminating contemporary photographs. These were taken from newspapers such as *The Irish Press*, private collections (supplied by individuals such as Richard Deegan) and institutions such as the National Museum and the National Library.[263] In the year that followed, those with first-hand knowledge of the history of Ireland's revolutionary years finally got the chance to tell their side of the story to the government's Bureau of Military History – an oral history project that was set up by a committee of historians and former Irish Volunteers. Over the period 1947–1957, a total of 1,773 Witness Statements were obtained from those who had personal recollections of the 1913–1921 era, including the Rising itself. Around 36,000 pages of evidence were accumulated by the official investigators, who were predominantly Army officers. However, this 'raw material' of 'imperfect memory' was subsequently placed in 83 steel boxes in a strongroom in government buildings, where it remained off limits to historians until the death of the last witness several years later.[264]

By no means, however, did inaccessibility to this archive stall the production of new histories of 1916. One individual who tapped into the public appetite for 1916-related material in the late 1940s was Desmond Ryan, a 'highly regarded journalist and writer', who had been a student of Pearse at St. Enda's and whose 'devotion' to his mentor was 'absolute'.[265] In 1949, Ryan published *The Rising*. This significant tome ran to around 100,000 words and offered up a general account of the events that had taken place in Dublin and elsewhere some 33 years beforehand. For the author himself, 1916 represented 'one of the most

[263] See S. P. Kelly, *Pictorial Review of 1916: A Complete and Historically Accurate Account of the Events which Occurred in Dublin in Easter Week, Fully Illustrated* (The Parkside Press Ltd., Dublin, 1946).

[264] F. McGarry, *Rebels: Voices from the Easter Rising* (Penguin Ireland, Dublin, 2011), pp. xii–xiv.

[265] D. Ferriter, *Occasions of Sin: Sex and Society in Modern Ireland* (Profile Books, London, 2009), pp. 97–98.

arresting and indubitable examples in all history of the triumph of failure'.[266] Having participated in the Rising and been interned afterwards, his commentary about the fighting naturally tended to reflect his own personal biases. But according to Keith Jeffery, Ryan's work also exhibited 'particular authority' on the happenings at the GPO, because he was part of the garrison there and had also been appointed as Pearse's literary executor.[267]

At a political level, the late 1940s were also notable for the way in which Fianna Fáil faced competing claims to its much-vaunted republican credentials. Having served as leader of his country for an unbroken 16-year run and steered it clear of mass bloodshed during World War II, the sole surviving leader of the Rising was finally defeated on home turf in the 1948 General Election. Not long afterwards, it was left to de Valera's political opponents to pinch some of the republican limelight when an opportunity presented itself for Éire/Ireland to announce that it was going to become a Republic. In September 1948, at a press conference in Ottawa in Canada, the Fine Gael leader and new Taoiseach, John A. Costello (who was the compromise head of the first Interparty government that held power from 1948–1951), announced his intention to declare his country as a Republic. The unexpected announcement, as Joe Lee has noted, managed to steal the attention from 'Fianna Fáil's Sunday suit of constitutional clothes' and was justified by Costello 'on the grounds that it would take the gun out of politics'.[268] In the end, it was on Easter Monday, 18 April 1949, that Éire/Ireland formally became a Republic. To mark the occasion, a major public celebration was staged in the form of a military parade through O'Connell Street – in scenes reminiscent of the Rising's 25th anniversary some eight years earlier. Despite the attendance of various relatives of the executed 1916 leaders, the parade's main emphasis was firmly on the possibilities of the future rather than the achievements of the past.

One of the main themes of the 1949 parade, which started at Dame Street and finished at the Parnell monument, was how Ireland could improve upon its standing within the broader international community. At a press conference held inside the GPO, Costello stated that Ireland's 'international recognition as a member of the community of nations' was 'a fact of tremendous importance to us as a nation', as was the 'international recognition' of the new Republic. 'We are', he added, 'a great mother country ... making a big noise in the world,

---

[266] D. Ryan, *The Rising: The Complete Story of Easter Week* (Golden Eagle Books Ltd., Dublin, 1949), p. 257.

[267] K. Jeffery, *The GPO and the Easter Rising* (Irish Academic Press, Dublin, 2006), p. 1.

[268] J. J. Lee, *Ireland 1912–1985: Politics and Society* (Cambridge University Press, Cambridge, 1989), p. 300.

and we will make a bigger noise still'.[269] As if to prove that point, the Easter Monday celebrations included a range of noise-generating rituals – including the sounding of trumpets, the roll of drums, the playing of pipes, the firing of a *feu de joie* from the GPO's roof, and a flyover by a formation of Seafires from the 1st Fighter Squadron. That night, the ceremonies were brought to an end when vast numbers of people packed into Phoenix Park to watch a fireworks spectacle, while another crowd attended a special céilidhe in Dublin Castle. To mark the Republic's creation, parades were also held in the cities of Cork and Limerick, along with the towns of Sligo, Carlow and Thurles. No public celebrations were held in Waterford city, however, 'but ships in the port were gaily beflagged and the Tricolour was hoisted over all public buildings'.[270] In addition to becoming a Republic, Ireland also departed from the Commonwealth of Nations on Easter Monday 1949 (unlike India, which also became a Republic in the same year). Such diplomatic actions, according to Luke Gibbons, 'formally severed the imperial connection that reached its apotheosis' during the reign of Queen Victoria from 1837–1901.[271] But as Lee points out, the departure was a source of regret in some quarters, especially among those who bemoaned the abandonment of Ireland's 'possibility of playing a peacock role in the new Commonwealth, strutting on the stage in the guise of an elder statesman'.[272] Even de Valera himself had reservations about Costello's actions, as they fell short of the 32-county Republic that Fianna Fáil, just like the rebels of 1916, had desired so much.

   At the mid-point of the twentieth century, developments in the Irish political scene were monitored with keen interest by those who had emigrated, especially by individuals who had played any sort of role in the Rising. In the late 1940s, for example, Captain Robert Monteith, who had emigrated from Ireland to the USA, was a regular speaker at Easter Week republican commemorations held in New York under the auspices of Clan na Gael and IRA veterans.[273] Monteith, who spent over 25 years working for the Ford Motor Company in Detroit, held dearly to a romantic vision of Ireland whilst living in America. This, however, was somewhat shattered following a return visit to Sutton in Dublin. In a letter that he wrote on 15 May 1950 to Reverend Brother MacMahon, who was serving with the Irish Christian Brothers in California, Monteith conveyed his unease with the changing nature of Irish society – particularly with what he perceived

---

[269] *The Irish Times*, 19 April 1949.
[270] Ibid.
[271] L. Gibbons, *Transformations in Irish Culture* (Cork University Press, Cork, 1996), p. 171.
[272] Lee, *Ireland 1912–1985*, p. 300.
[273] Anon., 'Easter Week: A Distinguished Visitor', *Irish Republican Bulletin*, Vol. 5, No. 1 (April 1947), p. 1.

as its increasing Anglicisation. He also alleged that the sacrificial ethos of the Rising had become a thing of the past:

> There is a terrible change in the people since I saw Ireland last. It is not for the better. The old self-sacrificing spirit seems gone, except in a few cases. Despite our parades and protestations there is a disquieting atmosphere of West Britonism or apathy that is heartbreaking ... Half the space in our newspapers is devoted to so-called sport. Horse and dog racing, soccer, hockey, cricket, rugby ... and political claptrap that is nauseating. The better evil has reached terrific proportions.[274]

The content of Monteith's letter adds weight to the notion that disjunctions can sometimes occur in the way in which Diasporic versions of allegiance remain situated in past circumstances that 'pay little heed to the on-going evolution of identity in the metropolitan state'. As first-hand memories of the difficult years leading to Irish independence gradually faded, it is evident that the visions held of Ireland in the USA by descendants of Irish emigrants remained 'those of a nineteenth-century nationalist discourse framed as a narrative of English oppression'.[275] While Monteith may have disliked games such as rugby, it was an open secret in Ireland that de Valera was in fact a fan of the skill involved in this gentlemanly sport.

When Fianna Fáil regained power in 1951, de Valera and his closest advisors began to explore new ways of reaffirming the party's republican credentials. Not to be outdone in the commemorationist stakes by the likes of Costello, the commemoration of the 1916 Rising during the 1950s was accorded significant preference over yearly anniversaries of the formal establishment of the Republic. When asked in the Dáil by Deputy Thomas O'Higgins about whether state ceremonies would be held on Easter Monday 1952 to mark the 'international recognition of the Republic of Ireland', de Valera responded by saying that no ceremonies would take place. Instead, he added, military parades would be held in Dublin and Cork on Easter Sunday 'to commemorate the Rising of 1916'.[276] A more revealing insight into the decision of de Valera's government to overlook the third anniversary of the Republic's instigation is offered by a draft reply to O'Higgins's question (but not used in the end), which was prepared by N. S. Ó Nualláin, the Secretary of the Department of the Taoiseach: 'they [the

---

[274] Captain Robert Monteith to Rev. Brother MacMahon, 15 May 1950, AL, Box 184/ Casement Box 2, Folder 1, Monteith/Casement Papers.

[275] B. Graham, G. J. Ashworth and J. E. Tunbridge, *A Geography of Heritage* (Arnold, London, 2002), p. 182.

[276] *Dáil Debates*, Vol. 130, 25 March 1952.

government] ... do not consider that the anniversary, any more than the event, merits its [*sic*] being marked by national celebrations'.[277]

Clearly, the fact that Partition still existed weighed heavily on the mind of de Valera and the Fianna Fáil faithful. Rather significantly, the British government continued to deal with the Republic of Ireland through the Commonwealth Relations Office throughout the 1950s (a situation that persisted until the mid-1960s). As Donal Lowry has pointed out, it was Britain's policy to view Ireland 'like a Commonwealth absentee rather than an imperial escapee', a diplomatic tactic encouraged by de Valera's 'apparent willingness' to contemplate rejoining the Commonwealth during the 1950s if the question of Partition was dealt with.[278] But within the playing field of Irish commemorationist politics in general, the divisiveness that had characterised debates over how to remember the Rising during the 1930s and 1940s began to dissipate as the 1950s progressed. The realisation that the newly-established 26-county Republic was not going to expand northwards in any hurry seems to have ushered in a change in hearts and minds. For many people struggling to make ends meet, it was the need to address the dire state of the economy (and the associated problems of poverty and emigration) that mattered more than the question of Partition.[279] The scourge of emigration was felt in many communities. In Killasser in County Mayo, for example, GAA teams and cultural societies ceased to exist, so much so that the area became 'a place of apathy, cynicism and despair, with the social fabric of the parish on the verge of collapse'.[280]

As time moved on, political discordance over rival claims to the Rising's legacy became more wearisome for some elements of the Irish population. Nonetheless, nationalist sentiment still remained strong and media interest in the story of 1916 did not wane. Between 1951 and 1956, for example, around 300 articles on the events of Easter Week were contributed to *The Irish Independent* by Piaras

---

[277] Department of the Taoiseach Memorandum, March 1952, NAI, Department of the Taoiseach, S9815/B, Easter Week Commemorations.

[278] D. Lowry, 'The Captive Dominion: Imperial Realities behind Irish Diplomacy, 1922–49', *Irish Historical Studies*, Vol. 36, No. 142 (2008), p. 225.

[279] Ireland's striking population decline since the Great Famine was discussed in a collection of essays published in 1954, entitled *The Vanishing Irish*. For the 26 counties constituting the Republic, it showed that the population had shrunk from 6,529,000 in 1841 to 2,960,593 in 1951. See J. A. O'Brien, 'The Irish Enigma', in J. A. O'Brien (ed.), *The Vanishing Irish: The Enigma of the Modern World* (W. H. Allen, London, 1954), p. 7, in which he laments 'the fading away of the once great and populous nation of Ireland' and warns that 'the Irish will virtually disappear as a nation and will be found only as an enervated remnant in a land occupied by foreigners'.

[280] B. O'Hara, *Killasser: Heritage of a Mayo Parish* (Killasser/Callow Heritage Society, Swinford, 2011), p. 85.

Béaslai, a veteran of the 1st Battalion Dublin Brigade.[281] From time to time, new memorials were unveiled in the provinces. In 1951, for example, a monument in the form of a Celtic cross resting on a granite plinth (sourced from a statue of Lord Dunkellin that had previously stood in Eyre Square) was unveiled beside the Tuam Road at Castlegar in Galway, dedicated to the men from the area who fought in the Rising (and in the War of Independence and Civil War). Among those who attended were the 1916 veterans, Brian Molloy and Michael Newell.[282] In the following year, the nearby Renmore Barracks 'found itself on the threshold of ... change', when it was renamed Dún Uí Mhaoilíosa, as a tribute to the leader of Galway's Rising, Liam Mellows.[283] Mellows was honoured again in 1957, when a limestone statue of him (designed by Diarmuid Ó Tuathail and sculpted by Domhnall Ó Murchadha) was unveiled in the north-eastern corner of Eyre Square.

Throughout the 1950s, sporting organisations such as the GAA also excelled in preaching the virtues of 1916-style nationalism to communities. On both sides of the Border, its membership continued to exert a strong influence at a local level on the making of Irish cultural identity, especially in years leading up to 1959 – which marked the 75th anniversary of its foundation. At both the parish and county level, several hurling and football teams took the story of 1916 to heart by naming themselves or their grounds after personalities associated with it. On 14 June 1953, for example, the Antrim County Board opened a stadium named Roger Casement Park at Anderstown in Belfast. As was usual for events of this nature, Catholic clergy were intimately involved in the rituals of remembrance. In the souvenir programme, Cardinal D'Alton conveyed his congratulations, and expressed his hope that 'this Gaelic headquarters long keep fresh the memory' of an individual 'who gave his life and had his good name and reputation besmirched because of his love of country'.[284]

For the various governments of the new Republic, many of the questions concerning the Rising's memory during the 1950s tended to focus on heritage conservation matters, including the condition of the graves of the 1916 leaders at Arbour Hill. The worsening condition of the graves proved to be a cause of concern to a range of government departments and agencies, and various proposals were made to transform the space into a more fitting memorial site.

---

[281] Townshend, *Easter 1916*, pp. xvi–xvii.

[282] P. Ó Laoi, *History of Castlegar Parish* (The Connacht Tribune Ltd., Galway, Undated), pp. 133–35.

[283] N. Sheerin, *Renmore and its Environs: An Historical Perspective* (Renmore Residents Association, Galway, 2000), p. 21.

[284] Anon., *Souvenir of the Official Opening of Roger Casement Park, Anderstown, Belfast, From 14th till 21st June 1953* (Conor Publications, Belfast, 1953), p. 2.

Back in 1948, the Department of Defence had proposed that a plaque 'bearing in raised letters the names of the leaders, should be fitted to the wall over the graves'. At a later stage, the Commissioners of Public Works suggested removing 'the east wall [between the prison yard containing the graves and the adjoining old playground] ... and the moving back of a section of the north wall, in order to leave the graves free standing'. It was proposed that the new part of the north wall would be 'in curved ashlar masonry and suitably inscribed', while the rest of the plan 'would comprise a simple stone surround to the graves and a series of paved terraces enclosed by dwarf walls'. Finally, at a Cabinet meeting on 26 August 1952, the government adopted the memorial scheme that the Commissioners of Public Works had suggested for the graves at Arbour Hill. After requesting 'an estimate of the cost', it was informed that it would be £12,500.[285] In December 1953, additional modifications suggested by the Commissioners were 'embodied in the revised sketch plans'. These included the addition of a path from the head of the graves to the first terrace, the inclusion of a gravelled area near the second terrace and the relocation of the flag-staff to an alternative position beside the graves. Although the Minister for Defence, Oscar Traynor, gave his approval for the revised scheme, he cast doubt on a proposal to include 14 bronze plaques 'in pairs on the tombstone', as he felt it would 'lead to confusion as to the actual position of the remains'.[286]

Following a meeting between de Valera and Traynor on 5 April 1954, further revisions were incorporated into the scheme. Most of these were of a technical nature, but some were intrinsically partisan in spirit. This was especially the case with the decision to postpone the appointment of Michael Biggs to the project (who was the preferred choice to design and supervise the inscribing of names on the margin surrounding the graves), until it could be determined that he was 'an Irishman'. Whilst the decision that 'ordinary green variety of Irish yew should be planted in preference to the golden form' seemed like an obvious design solution for a highly-specialised project in which every precise detail mattered, the decision to use stone from the four provinces during construction work was particularly revealing.[287]

---

[285] Department of Defence Memorandum on the Graves of the 1916 Leaders at Arbour Hill, 19 August 1952, NAI, Department of the Taoiseach, S9815/B, Easter Week Commemorations; Memorandum on the Decision of a Government Meeting, 26 August 1952, NAI, Department of the Taoiseach, S9815/B, Easter Week Commemorations; Ferriter, 'Commemorating the Rising, 1922–65', pp. 210–11.

[286] Department of Defence to the Government, 9 December 1953, NAI, Department of the Taoiseach, S9815/B, Easter Week Commemorations.

[287] Department of Defence Memorandum on the Graves of the 1916 Leaders at Arbour Hill, April 1954, NAI, Department of the Taoiseach, S9815/B, Easter Week Commemorations.

Plate 3.4   Burial plot of the executed 1916 leaders at Arbour Hill, Dublin
Source: Author.

This modification to the Arbour Hill scheme demonstrated the lengths to which the government was prepared to go to in order to accentuate the geopolitical aspirations inherent in the Republic's memorialisation policy – one that reflected the ambitions of the Proclamation. Further work at the site continued in the years that followed. In 1959, for example, the carving of a cross on the ashlar wall of the new memorial was finished.[288] At a later stage, the text of the Proclamation was added on in a meticulous fashion by an accomplished stonemason. Additions such as these gave Arbour Hill a more sanctified and antique appearance (Plate 3.4). Christian imagery also featured in another memorial that was unveiled in 1959 in Ashbourne, in memory of Thomas Ashe. The design, according to Bhreathnach-Lynch, featured 'both human and Christ figures fused into one and carrying a cross', and was directly inspired from a poem entitled 'Let Me Carry Your Cross for Ireland', which Ashe had written in 1916.[289]

The late 1950s also witnessed the initiation of a heritage campaign aimed at restoring Kilmainham Gaol. The Inchicore site, which first opened

[288] Ferriter, 'Commemorating the Rising, 1922–65', p. 212.
[289] Bhreathnach-Lynch, 'Commemorating the Hero', p. 156.

as a prison in 1796, was a site of major cultural and historical importance, owing to the fact that it was the location where the 1916 leaders had been imprisoned and executed. But the site also contained unpleasant memories of the Civil War, as four republicans were executed there by the Free State Army in 1922. According to Rory O'Dwyer, this may elucidate 'why the building was allowed to fall into ruin after its closure in 1924 – a legacy too difficult to remember'.[290] With the passage of time, however, the space gradually acquired a more nostalgic resonance for nationalists. The brainchild behind the restoration project was a Dublin-born citizen, Lorcan C. G. Leonard, whose interest in the site was first aroused whilst waiting for a bus at the junction of South Circular Road and Emmet Road in 1942. He later recalled: 'I fell to contemplate the old jail which was citadelled before me on the rising ground.' In 1952, Leonard's interest in Kilmainham was rekindled after he was reliably informed that the Office of Public Works was contemplating demolishing the building. Following a number of meetings with a friend, Paddy Stephenson, Leonard finally took more concrete action during the summer of 1958, by drafting a plan which proposed 'restoration based on voluntary labour' and the creation of 'a historical museum'. The plan was discussed further at a meeting in Jury's Hotel in September. Besides Leonard and Stephenson, those present included Stephen Murphy, Jimmy Brennan and John Dowling.[291]

In 1960, the first issue of 'Kilmainham News' confirmed that from late May onwards, three weeks of 'hard work' had taken place to 'clear a way through what amounted to a jungle of shrubbery, fallen masonry, old and stubborn ivy', and 'trees some 30 and 40 feet high, sprung from seed presumably dropped by birds'. Afterwards, work commenced on reroofing, reglazing and reflooring different parts of the building.[292] Against all the odds, Leonard's dream gradually reached fruition, thanks in no small part to the trojan efforts of the Kilmainham Gaol Restoration Society's volunteer labour force of over 200, which included Old IRA veterans, tradesmen, skilled workers, and local children (Plate 3.5). However, due to a lack of financial support from the government, a range of fundraising activities had to be organised in the years that followed, including public collections and dances in venues such as Dublin's Hibernian Ballroom.

[290] R. O'Dwyer, *The Bastille of Ireland. Kilmainham Gaol: From Ruin to Restoration* (The History Press Ireland, Dublin, 2010), p. 9.

[291] Typescript by Lorcan C. G. Leonard, Entitled 'The Kilmainham Project as I Dreamt It and Lived It', Undated, Kilmainham Gaol Archives (hereafter KGA), 21MS-1L41-02, Lorcan C. G. Leonard Papers.

[292] 'Kilmainham News: Bulletin of Work in Progress', 1960, KGA, 21RE-1G12-07.

Plate 3.5   Lorcan C. G. Leonard (in the right background, with
            outstretched arm) overseeing restoration work at
            Kilmainham Gaol

Source: Kilmainham Gaol Archives.

Donations of building materials were also given free of charge in a significant philanthropic gesture by businesses.[293] A year before he passed away, Leonard memorably highlighted what he saw as the essence of the heritage project: 'if Kilmainham is saved Ireland is saved, and out of our poor efforts at least the children of the future will say we preserved the history of Ireland'. 'Kilmainham', he added, 'is the Calvary of republicanism in Ireland'.[294]

For the most part, state commemorations of the Rising at O'Connell Street and Arbour Hill during the 1950s and into the first half of the next decade proved to be rather small-scale affairs. The 40th anniversary of the Rising in 1956, for example, fell at a time of serious economic hardship, and did little to capture the public's imagination. The main state commemoration, which was held on Easter Sunday 1 April, was watched by John A. Costello as Taoiseach and featured a parade through O'Connell Street by 2,000 members of the Army and FCA, many of whom carried Swedish-manufactured weapons. In its report on the event, *The Irish Times* noted that those who attended had enjoyed 'brilliant sunshine' in the city centre, along with an impressive fly-past of a squadron of Spitfires over O'Connell Street, which 'left an indignant flutter of pigeons behind them'. For the first time ever, 'rope cordons' were utilised to control the watching crowd and thus 'curb the natural tendency of the average Dubliner to get as close as possible to whatever may be happening'.[295] The following day, the Rising's anniversary was again marked in the capital. At a ceremony held in St. Stephen's Green, the President, Seán T. O'Kelly, unveiled a bronze bust that featured Countess Markievicz in the uniform of the Irish Citizen Army (Plate 3.6). The memorial was produced by the Cork sculptor, Séamus Murphy, following a commission from the Madame Markievicz Memorial Committee. It was inscribed with the following words: 'A valiant woman who fought for Ireland in 1916.'

---

[293] According to P. Cooke, *A History of Kilmainham Gaol 1796-1924* (Dúchas The Heritage Service, Dublin, 2001), p. 40, the 'personal memories' of the veterans 'were gradually subsumed into the idea of a nationalist tradition greater than their individual sufferings', combined with an 'impulse to restore'. Further commentary on the restoration campaign, including its tourism aspect, can be found in E. Zuelow, *Making Ireland Irish: Tourism and National Identity since the Irish Civil War* (Syracuse University Press, Syracuse, 2009), pp. 155–65. For an assessment of the philanthropic dimension to the renovations, see S. McCoole, 'Philanthropy, History and Heritage', *History Ireland*, Vol. 17, No. 3 (2009), pp. 10–11.

[294] Lorcan C. G. Leonard to John Dowling, 12 November 1965, KGA, 21MS-1L41-02, Lorcan C. G. Leonard Papers.

[295] *The Irish Times*, 2 April 1956.

Plate 3.6   Séamus Murphy's bronze bust of Countess Markievicz in the
             uniform of the Irish Citizen Army at St. Stephen's Green, Dublin,
             unveiled on 2 April 1956 by the President, Seán T. O'Kelly
Source: Author.

In his speech at the ceremony, the President seemed rather upbeat as he fondly
recalled, 'with joy and pride', Markievicz's journey 40 years beforehand 'from the
big house to a dwelling-place in the hearts of the Irish people', where she and her
memory have ever since abided and will continue to abide'.[296]

But as Bew has maintained, much of the 40th anniversary commemorations
were 'tinged with embarrassment'. The 1916 rebels, he notes, 'had argued that a
mix of political sovereignty, nationalist economic policy, radical agrarian policy,
and compulsory Irish in the schools' would lead to sustained population growth,
but the continued use of these policies by 1956 'had produced near-collapse
and, ironically, higher levels of emigration than in the decades of British rule'.[297]
Unsurprisingly therefore, there was certainly no shortage of bitterness about the
colossal problems that the Republic faced at the time of the 40th anniversary.

---

[296] Ibid., 3 April 1956.
[297] P. Bew, "'Why Did Jimmie Die?'", *History Ireland*, Vol. 14, No. 2 (2006), p. 38.

In its Easter Monday Editorial, entitled '1916–1956', *The Irish Times* caustically reflected that 'the achievement ... of partial nationhood ... has not yet come to full flower', adding that the 'men of 1916' would not have welcomed 'the sundered and still embittered Ireland of today', one that was 'so ridden with "politics"'. Lamenting the lack of 'unified action' by politicians to tackle the scourge of emigration and growing class divisions, it concluded by urging a 'return to the sense of comradeship and common purpose that animated the leaders of that fantastic Rising'.[298]

Such was the 'over-proliferation' of 1916 commemorative parades by the early 1960s that complaints were made to the government by Old IRA veterans, who alleged that there had been a 'lack of centralisation evident in the commemoration of 1916'.[299] But the government's mind was ultimately preoccupied with the imminent Golden Jubilee anniversary, set for 1966. For many years, this momentous anniversary had been much talked about, and by the late 1950s it was evident that it weighed heavily on the minds of many of Ireland's leading personalities. Not least among them was Patrick Pearse's ageing sister, Senator Margaret M. Pearse. 'Let us keep their memory green', she wrote about the 1916 rebels in 1958, 'as an inspiration to the present and future generations'.[300] While it is a truism, as Ferriter has pointed out, that 'many were turning their attention to ... 1966', it was 'also the case that in light of these preparations, the government was conscious of two things – the general indifference of many people and the danger of reopening old wounds'.[301]

Having lost power to another Interparty government led by Costello from 1954–1957, de Valera served one last stint as Taoiseach from 1957– 1959. De Valera's last year in power coincided with the 43rd anniversary of the Rising. 'Curiously', as Tom Garvin has written, the year witnessed 'a new self-confidence'. This was exemplified by the release by Gael Linn of George Morrison's documentary on 1916 and its aftermath, *Mise Éire* ('I am Ireland'), which featured an inspirational musical score by an 'up-and-coming musician', Seán Ó Riada.[302] De Valera was in turn succeeded by a new leader of Fianna Fáil, Seán Lemass (another 1916 veteran) – who served as Taoiseach from 1959– 1966. Under Lemass's leadership, the country's confidence levels escalated.

[298] *The Irish Times*, 2 April 1956.

[299] Ferriter, 'Commemorating the Rising, 1922–65', p. 213.

[300] M. M. Pearse, 'Foreword', in The Brothers Pearse Commemoration Committee, *Cuimní na bPiarrac: Memories of the Brothers Pearse* (Mount Salus Press Ltd., Dublin, 1958), p. 3.

[301] Ferriter, 'Commemorating the Rising, 1922–65', p. 213.

[302] T. Garvin, *News from a New Republic: Ireland in the 1950s* (Gill and Macmillan Ltd., Dublin, 2010), p. 212. The same year also saw the release by Topic Records of Dominic Beehan's *The Patriot Game: Easter and After*.

The economy was steered in a new outward-looking direction, which resulted in industrial growth. With the question of Partition being sidelined by the economic and social priorities of the new era, and the departure from politics of many of the veterans of the old revolutionary era (for example, Seán MacEntee, who left the Cabinet in 1965), the times were certainly changing. Furthermore, with the state of Anglo-Irish relations steadily improving, Ferriter has observed that even prior to 1966, memory of the Rising 'was already becoming a sensitive and somewhat delicate subject'. The new Taoiseach 'was not keen on vigorous debate on the Rising or Pearse in the 1960s', with the result that he was unwilling 'to get involved in a scramble for the bones of the patriot dead'.[303]

Early in May 1965, when the issue of the following year's Golden Jubilee was debated in the Dáil, Lemass was keen to emphasise that the making of preparations for 'the form of ... ceremonies and celebrations' had been delegated to 'an informal committee, which has already held three meetings under my chairmanship'. After being questioned by the Opposition about whether the committee was 'unrepresentative' of different political viewpoints by the nature of its composition, the Taoiseach was anxious to stress that it was not. 'What happened', he remarked, 'was that I invited a number of people to come together and asked them to suggest other people who ... would be helpful and likely to be interested'. 'The present committee', he stressed, 'is as a result very much larger than when it was first instituted'.[304] Although this answer may have seemed unsatisfactory or evasive to the Opposition at the time, Lemass did at times seek to make a case for the holding of an inclusive celebration of the Rising's Golden Jubilee. At the end of 1965, for example, in an address to the Fianna Fáil Ard Fheis at the Mansion House, he asked the people of Ireland to put aside their differences during the period of the ceremonies so that 'nothing will find expression except our desire to honour the men of 1916 and our determination that the ideals which inspired them will continue to flourish in all the years ahead'.[305]

To a certain degree, it could be argued that Lemass's first-hand experience of fighting in the Rising impinged upon the nature of the pronouncements that he made in public regarding the heritage of 1916. Even though he had been inside the GPO in 1916 when he was a youth (and had also fired some shots while posted on its roof), it was a fact, as Tom Garvin has highlighted, that 'Lemass made little of his exploits during Easter Week'. The risk of dying in the fighting

---

[303] Ferriter, 'Commemorating the Rising, 1922–65', p. 215.

[304] *Dáil Debates*, Vol. 215, 6 May 1965.

[305] Presidential Address by the Taoiseach, Seán F. Lemass, TD, to the Fianna Fáil Ard Fheis, Mansion House, Dublin, 16 November 1965, UCDA, MS. P.176/772/3, Archives of the Fianna Fáil Party.

or even execution afterwards 'matured him very rapidly', while at a later stage, he expressed hope that he had not killed anybody.[306] On some occasions, however, Lemass demonstrated a capacity to be quite selective when making decisions about the individuals who would ultimately be chosen to play an intimate part in both the planning and celebration of the upcoming Golden Jubilee. As Lemass's biographer, John Horgan, has argued, Fianna Fáil was determined from the outset 'to exploit the occasion for maximum political advantage'. As the Golden Jubilee year finally dawned, he notes that Lemass acted to minimise 'the twin dangers of raw irredentism and potential revisionism'. He courteously rejected, for example, Kathleen Clarke's claim to a position on the government-appointed 'Coiste Cuimhneacháin' (Irish for 'Commemoration Committee'), conscious of the unabashed character of her republicanism.[307] On the other hand, he also justified the decision of the 'Coiste' not to issue an invitation to the Golden Jubilee ceremonies to the veteran republican, Bulmer Hobson (who had tried to prevent the Rising), in his typically curt style: 'He wasn't here for the fighting.'[308]

[306] T. Garvin, *Judging Lemass: The Measure of the Man* (Royal Irish Academy, Dublin, 2009), pp. 44, 47.

[307] J. Horgan, *Seán Lemass: The Enigmatic Patriot* (Gill and Macmillan, Dublin, 1999), p. 284.

[308] Lemass, cited in Ibid.

Chapter 4

The Golden Jubilee Anniversary, 1966

To mark the occasion of the Golden Jubilee of the Easter Rising in 1966, the government of the Republic of Ireland staged a public commemoration of grandiose proportions. This memorialisation of the revolutionary past proved to be a defining moment in the emergence of modern Ireland, with issues of national heritage, collective identity and political ideology all coming to the fore in the cultural politics of everyday life. On the evidence of what was served up for the anniversary, it is clear that the story of 1916 remained of utmost importance to the political establishment – half a century after Pearse had read the Proclamation outside the GPO. The Rising's memory was likewise cherished by the Opposition. In a Dáil debate on 22 March, for example, Fine Gael's Richie Ryan went so far as to suggest that the GPO represented a 'national shrine'.[1] He was not alone in such thoughts, for all throughout the state in the month that followed, a major effort was made by politicians and citizens alike to remember and commemorate 1916, guided as they were by the policy decisions taken by the members of the 'Coiste Cuimhneacháin'. Under the watchful eye of its secretary, Piaras MacLochlainn, the committee's members put considerable effort into ensuring that the event would not be forgotten by participants and spectators. Despite a few mishaps, they ensured that the 50th anniversary was marked in both a historic and patriotic fashion. During a succession of action-packed days in April, a series of lavish official celebrations were staged all over the country, thus stirring the national imagination (and opinion) like never before. Unsurprisingly, much anticipation had surrounded preparations for the event, and many citizens and politicians alike were eager to put on record what they felt about the position of 1916 in 1966.

At the Crossroads of History

In many senses, Ireland was at a crossroads in its history in 1966. The Golden Jubilee proved to be an occasion that evoked mixed feelings, for it offered citizens not only the chance to reflect upon the achievements of politicians since

---

[1]   *Dáil Debates*, Vol. 221, 22 March 1966.

independence, but also to ponder their shortcomings. A lot of nostalgia was also in the air, as people sensed that there was a passing of the torch from elderly statesmen who had been 'out' in 1916, to a younger generation of politicians who had not participated directly in the revolutionary era. Naturally enough, portions of the commemorative programme had a rather subdued aspect at times, given the significant loss of life that had ensued 50 years earlier. On some occasions, a degree of scepticism emanated from various critics, who questioned the track records of successive governments since independence. While the Golden Jubilee ceremonies passed off peacefully for the most part, the celebrations were marred on a few occasions by the violent actions of individuals and organisations disillusioned with Partition. Damage was caused to a range of monuments associated with the era of British rule, while the public was occasionally inconvenienced by a sabotage campaign that caused interruptions to electricity supplies and led to interference with public transport routes.

While such incidents were caused by the actions of a small minority of disillusioned extremists, a bit of further gloss was taken off the proceedings by outpourings of emotion, indifference and despair about the state of the nation's affairs from hardened sceptics – who felt that the Proclamation's ideals and objectives had still not been completely fulfilled. Partition remained a big bone of contention for a number of strong-minded republicans, while issues such as the decline of the Irish language worried a number of educationalists and Irish speakers. Others, however, were more concerned with the perilous state of the Irish economy and the pressing need to generate more jobs. For the farming community, the depopulation of the countryside and hay scarcities added to a sense of crisis, while the proposed abolition of free fishing rights angered maritime communities. Industrial unrest also affected the economy during the first half of 1966, with strikes disturbing the operations of banks, paper mills and the port of Dublin. The rising cost of food and motor tax added to people's disenchantment at a time of mounting uncertainty over the direction in which the country was being steered by Fianna Fáil – factors which provided ample ammunition for the salvos that were fired at the government from the Opposition in Dáil Éireann.

Yet another matter that surfaced during the year of the Golden Jubilee was the perennial question of Military Service Pensions. Over the course of nearly a decade, the number of people in receipt of these had fallen steadily from 13,925 on 31 March 1956 to 11,196 on 31 March 1965. Because of inflation, however, expenditure on the pensions had risen by 5.68% from £625,016 to £660,535 in the fiscal years from 1955–1956 to 1964–1965.[2] In light of the

---

[2]   Ibid., Vol. 220, 8 February 1966.

anniversary, Fine Gael's South Tipperary TD, Patrick Hogan, asked in the Dáil on 25 January 1966 whether the Minister for Finance, Jack Lynch, would consider increasing the Military Service Pensions of the rank-and-file veterans who had given service in 1916 and whose actions were now being remembered. Lynch replied: 'The matter will be kept under consideration in relation to Exchequer resources and other relevant factors.'[3] Six days later, the Taoiseach wrote to Lynch, seeking to persuade him of the merits of doubling the pensions of the 1916 veterans. By the middle of the following month, however, the proposal was shot down by the Department of Finance.[4] High inflation and the emergence of a major deficit in the government's balance of payments proved to be a cause of grave concern early on in 1966, with the result that Lynch had to introduce an extremely harsh supplementary budget on 9 March, which increased the rate of income tax by 10.5%. However, as T. Ryle Dwyer has noted, 'much of the media attention was diverted from the economy by the celebrations commemorating the Easter Rising'.[5]

At certain times during the Golden Jubilee, the prevailing mood proved to be one of merriment, as people temporarily cast aside their day-to-day worries of making ends meet. Those with patriotic tendencies were more than happy to get into the spirit of the occasion – taking satisfaction from the way in which the Rising had initiated the drive towards independence from British rule. Other spirits were also involved (namely those of a liquid nature), for an air of conviviality was added to the Golden Jubilee by the granting of court exemptions which permitted public houses to stay open and serve drink for longer periods during the week-long official celebrations at Eastertime. The children of the nation were not forgotten either, as a school holiday was arranged for 22 April. In John Horgan's view, the overriding atmosphere throughout parts of the Republic 'was of solid self-congratulation, mixed with not a little triumphalism'. Gerard O'Brien has even gone so far as to suggest that 1966 represented Ireland's 'year of national self-righteousness'. On the other hand, Rory O'Dwyer has expressed reservations about commentaries that have presented the commemoration in 'one-dimensional' terms, and given the impression 'that historical understanding in the Republic largely reflected a simplistic monolithic view of Ireland'. 'The high level of nationalist feeling in the period', he reckons, 'was generally harnessed in a very positive fashion'. Additionally, he contends that the Golden Jubilee represented 'a sincere, meaningful ... and well-organised commemoration

---

[3]   Ibid., Vol. 220, 25 January 1966.

[4]   J. Horgan, *Seán Lemass: The Enigmatic Patriot* (Gill and Macmillan, Dublin, 1999), p. 285.

[5]   T. Ryle Dwyer, *Nice Fellow: A Biography of Jack Lynch* (Mercier Press, Cork, 2001), p. 116.

with considerable educational potential for anyone with an interest in the 1916 Rising'.[6]

Many eager citizens throughout the Republic – from schoolchildren to senior citizens – turned out in large numbers for the ceremonies and exhibitions that were staged for the Golden Jubilee. There was far more to the occasion, however, than simply the case of the political establishment preaching the virtues of 1916 to a younger generation through the spectacle of historical remembrance. As a sort of coming of age that acted as a contemporaneous bridge between the past and the future, certain parts of the anniversary were marked by a mixture of maturity, integrity, piousness, harmony, earnestness, hope, and pride. Notwithstanding concerns about the economy, the Golden Jubilee did overlap with a pivotal time in Ireland's historical progression. Although there were a number of occasions in 1966 when blatant interventions were made to protect the interests of Irish manufacturing industries, the decade as a whole witnessed a rejection of economic protectionism and an embracing of both free market principles and greater efforts to attract foreign investment.

There was certainly strong justification for opening up the economy. For years, it was clear that the achievement of independence had been somewhat undermined by the sacred orthodoxy of monetary nationalism. The lack of job opportunities was evident from the emigration of over one million people between the foundation of the state and 1961. Results of the 1961 Census (the sixth since the state's foundation), recorded that the enumerated population of the Republic had fallen to just 2,818,341 persons (the lowest ever recorded). The 1966 Census, however, recorded 2,884,002 persons, representing a small increase of 2.33% in five years, with natural increase exceeding declining net outward migration.[7] The seeds of this recovery had been sown a decade earlier during the financial crisis of 1956, when T. K. Whitaker (a high-flying and London-educated economic expert at the Department of Finance), prepared a study that formed the basis of the government's flexible *First Programme for Economic Expansion*.[8] This dismissed the protectionist policy and replaced

---

[6]   Horgan, *Seán Lemass*, p. 285; G. O'Brien, *Irish Governments and the Guardianship of Historical Records, 1922–72* (Four Courts Press, Dublin, 2004), p. 100; R. O'Dwyer, 'The Golden Jubilee of the 1916 Easter Rising', in G. Doherty and D. Keogh (eds), *1916: The Long Revolution* (Mercier Press, Cork, 2007), pp. 366, 375.

[7]   Figures derived from Central Statistics Office, *Census 2006: Principal Demographic Results* (The Stationery Office, Dublin, 2007), pp. 12, 37.

[8]   M. J. Bannon, 'Development Planning and the Neglect of the Critical Regional Dimension', in M. J. Bannon (ed.), *Planning: The Irish Experience 1920–1988* (Wolfhound Press, Dublin 1989), p. 127; J. J. Lee, *Ireland 1912–1985: Politics and Society* (Cambridge University Press, Cambridge, 1989), p. 342.

it instead with one aimed at attracting foreign industry and investment over the period 1958–1963. A second plan followed from 1963–1968. Lemass's philosophy of economic expansion, as one political opponent put it at the time, was based on a 'philosophy of the survival of the fittest'.[9] Thus in many senses, as Joe Joyce and Peter Murtagh have noted, the 1960s marked a time in which 'something akin to the American dream hit Ireland', with protectionism dismantled so as 'to make way for a rawer brand of capitalism'.[10]

While Lemass, who served as Taoiseach from 1959–1966, was a politician who was said to have had the same 'head' as de Valera, his popular image was also one of a 'practical, realistic businessman quietly and efficiently leading the country to prosperity'.[11] In abandoning its protectionist strategy, the Republic made a bid in 1963 to join the European Economic Community, but this failed after Britain's application was vetoed. After revising the legislation that covered industrial investment, Lemass's government then negotiated a trade agreement with the British government in December 1965. This provided a stepping stone to membership of the European Economic Community by providing for the phasing out of all tariffs with Britain by 1970. On 7 January 1966, the Dáil ratified the Anglo-Irish Free Trade Agreement by 66 to 19 votes.[12] While critics such as the Wolfe Tone Society and Sinn Féin saw the agreement as a sellout and a diminution of economic sovereignty, the ground-breaking decision to ditch protectionism proved vital to the modernisation and reinvigoration of the Republic. The warming of Anglo-Irish relations was by no means confined to just the fiscal sphere. As will be seen later on, a number of diplomatic gestures of friendship were made by the British government and its people both before and during the Golden Jubilee commemorations.

It has been argued by Roisín Higgins and others that Lemass's emphasis on modernisation during the Golden Jubilee meant that it 'was as much about the act of looking forwards as backwards', especially in relation his dual concern of both improving the Republic's relations with Britain and Northern Ireland and securing its future membership of the European Economic Community.

---

[9]   Transcript Copy of Election Speech by Senator Gerry L'Estrange at a Fine Gael Branch Meeting in Shaw Murrays, Mullingar, 13 March 1965, University College Dublin Archives (hereafter UCDA), MS. P.39/GE/134/15-17, Archives of the Fine Gael Party, General Election Records.

[10]   J. Joyce and P. Murtagh, *The Boss* (Poolbeg Press Ltd., Dublin, 1997), pp. 85–86.

[11]   Philip Feddes to Fine Gael, 30 March 1965, UCDA, MS. P.39/GE/133/12, Archives of the Fine Gael Party, General Election Records.

[12]   M. Treacy, 'Rethinking the Republic: The Republican Movement and 1966', in R. O'Donnell (ed.), *The Impact of the 1916 Rising: Among the Nations* (Irish Academic Press, Dublin, 2007), p. 221.

These aspirations required 'a delicate negotiation between tradition and change', and nowhere was the theme of modernity more evident 'than in how the commemoration was communicated to the youth of Ireland, a group which represented the nation's future but which had mixed reactions to the lessons of the past'. By investing its future hopes in a 'dynamic younger generation', they note that the state also sought 'to positively engage the patriotism of the younger generation' so as to instigate 'a new national project' that would instil into them 'the "spirit" of the men of 1916'. In the Golden Jubilee year, 'this oft-cited "spirit" represented personal achievement, excellence ... and unity of purpose'.[13]

Besides the youth, other sections of Irish society were also caught up in the euphoria that occasionally manifested itself during the Golden Jubilee. Claremorris-born David Fahey, for example, travelled back from New York for the Easter Monday celebrations in Dublin, and was making his first visit to his native country in nine years. Asked by a reporter from *The Connacht Telegraph* about his impressions of the country that he had returned to, he replied: 'This country has improved tremendously since 1957 ... and I can see no reason why it shouldn't improve with the present generation who have grown up in freedom – a chance our generation did not get.'[14] Such optimism was also evident among senior citizens. John Flynn, from Ballymacredmond in Ballina, County Mayo, was especially upbeat after celebrating his 100th birthday on 13 April. Reflecting on the circumstances of the country, he told the *Western People*: 'Well ... we never had it so good. I remember the time when there were only a few horse carts at a funeral, nowadays these are replaced by lines of cars.'[15] A correspondent for *The Sligo Champion* also reflected on how much the country had changed, and reported that in Dublin, 'many elderly men have been telling younger men of how different it really was' in 1916, especially 'if you were a mere worker on a weekly wage' or lived amongst the thousands of slum dwellings in which 'families lived in accommodation that consisted of one room' and 'endured ... starvation'. 'The time space between 1916 and 1966 is short', he concluded, 'but the social changes have been enormous'.[16]

The Golden Jubilee also proved to be a poignant occasion for the stalwarts of the political establishment, who wished to place the memory of 1916 within a broader context that not only celebrated patriotic notions of freedom and sacrifice, but also bolstered their own personal reputations. In an article that appeared in the spring edition of *Studies*, entitled 'I Remember 1916', Lemass

[13]   R. Higgins, C. Holohan, and C. O'Donnell, '1966 and All That: The 50th Anniversary Commemorations', *History Ireland*, Vol. 14, No. 2 (2006), pp. 31, 33.
[14]   *The Connaught Telegraph*, 21 April 1966.
[15]   *Western People*, 16 April 1966.
[16]   *The Sligo Champion*, 22 April 1966.

recounted his own role in the Rising. At the time he was only 17 years old and one of his strongest memories was of having assisted in carrying the wounded Connolly from the GPO as it was being evacuated. 'Personally', he wrote, 'I assisted to carry Connolly's stretcher for a short distance to a small door opening on Henry Street.'[17] While other prominent members of Fianna Fáil were able to talk about their experiences of being 'out' in 1916, it was clear that the Golden Jubilee spotlight was reserved for one enigmatic individual, namely Eamon de Valera. By 1966, he was an 83-year-old statesman with deteriorating eyesight and was reaching the end of his first stint as President, which had commenced in 1959. Content with his present political circumstances and at one with his rebellious past, he took a very active and authoritative role in the Golden Jubilee celebrations, sensing that the occasion was ripe to put his own rhetorical stamp on how the sacrifices of the executed leaders of the Rising would be recorded in the chronicles of Irish heritage, culture and identity.

The Golden Jubilee conveniently fell during the midst of de Valera's campaign for reelection as President. Press advertisements issued by Fianna Fáil extolled his virtues, including one containing a quotation by the Taoiseach, which proclaimed that the President's reelection would 'be an affirmation by the people that the aims of our country which have directed him throughout his life remain constant, and that patriotism is as relevant in the Ireland of 1966 as it was in 1916'.[18] The implication that de Valera was akin to the spiritual father of the nation and the living embodiment of the spirit of 1916 was clear. When the man talked, people listened. With a 'magnetic' personality which he displayed throughout his political career, a close friend and party colleague of de Valera once remarked that he 'exercised a special influence and authority over public opinion ... his were the words that were listened to by the greatest number of voters'.[19] The cult of personality which de Valera forged over the years as the solitary living commandant of the Rising was as strong as ever by the time of the Golden Jubilee. Central to an understanding of his hold over his followers in 1966, as Tim Pat Coogan has remarked, was the fact 'that, insofar as the political "Mass" of 1916 was concerned, de Valera was viewed as the Chief Celebrant and occupant of the *Sedes Gestetoria* in which the memories of Pearse and the others were carried up and down the corridors of Irish political power for decades after their deaths'.[20]

---

[17]   S. Lemass, 'I Remember 1916', *Studies: An Irish Quarterly Review*, Vol. 55, No. 217 (1966), pp. 7, 10.

[18]   *The Connacht Tribune*, 26 May 1966.

[19]   Seán MacEntee to Captain Jack O'Carroll, 28 April 1976, UCDA, MS. P.176/262/4, Archives of the Fianna Fáil Party, Captain Jack O'Carroll Papers.

[20]   *The Cork Examiner*, 3 April 1991.

An indication of de Valera's hegemonic influence on public discourse can be seen from his Easter 1966 message to the people of Ireland:

> This Easter we are bringing back to our minds the Easter of fifty years ago, and are seeking to honour the men who at that time gave or risked their lives that Ireland might be free. We wish to honour, in particular, the seven brave men who, despite all the deterrents, made the decision to arrest, once more, in arms our nation's right to sovereign independence. It was a fateful decision which we know to have been one of the boldest and most-far reaching in our history ... we of this generation must see to it that our language lives. That would be the resolve of the men and women of 1916. Will it not be the resolve of the young men and women of 1966?[21]

In ending his statement, de Valera returned to the time-honoured tradition of equating the Rising to a Holy cause, asking the Irish people to place the nation's destiny 'anew under God's High Providence, and, like the men of 1916, implore His blessing upon our aims and upon our work'.[22]

Besides delivering his own pronouncements, de Valera was also the recipient of many notable messages of congratulations from the leaders of foreign powers. On 7 April, for example, the President of India, Sarvepalli Radhakrishnan, sent him the following message: 'Ireland's brave struggle for independence ... evoked the sympathy and admiration of the entire Indian nation ... we continue to watch the fortunes of the Irish people and wish them peace, happiness and prosperity.'[23] The President of the USA, Lyndon Johnson, sent de Valera a telegram, conveying his best wishes and those of the American people. 'On this Jubilee Anniversary of Ireland's Declaration of Independence', he wrote, 'the people of America join me in a salute to you and your countrymen.'[24] Another message was received from British Army Captain Edo John Hitzen, who took the surrender of the 3rd Battalion of Irish Volunteers near Boland's Mill. De Valera also received messages of congratulations from Spain's General Francisco Franco and from the President of the USSR, Nikolai Podgorny.[25] In reply to Podgomy's cablegram, which had conveyed 'cordial congratulations and sincere wishes of happiness

---

21   Message from the President to the People of Ireland, Easter, 1966, National Archives of Ireland (hereafter NAI), Department of the Taoiseach, 97/6/163. The full text of this speech was also published in *The Irish Times*, 11 April 1966.

22   Message from the President to the People of Ireland, Easter, 1966, NAI, Department of the Taoiseach, 97/6/163.

23   *The Irish Times*, 8 and 9 April 1966.

24   Ibid., 12 April 1966.

25   *The Irish Independent*, 11 April 1966, 14 April 1966 and 18 April 1966.

... to the people of Ireland from the peoples of the Soviet Union', de Valera sent back a message of appreciation through the Soviet Embassy in London.[26]

Ultimately, de Valera was able to use the tidal wave of nostalgia that emanated from the Golden Jubilee to his political advantage. Although the events that were staged to mark the Republic's commemoration of the 50th anniversary were officially 'non-party-political', Noel Whelan has claimed that they also served in a roundabout way as 'a useful series of quasi-campaign events', acting as the showpieces of de Valera's campaign for reelection.[27] As expected, he went on to defeat Fine Gael's Thomas O'Higgins in the race for the Presidency in early June. However, as Fergal Tobin has commented, de Valera's margin of victory 'was embarrassingly slight considering the year that was in it', namely one 'for the backward glance'. Out of a total poll of around 1,100,000, de Valera only won by 10,500 or so votes, compared to approximately 120,000 during the previous election in 1959.[28] In the end, the victory enabled him to enjoy a second seven-year stint that lasted until 1973. The Golden Jubilee, however, went down in history as the most significant commemorative milestone that he presided over during his tenure as President, and he did not pass over the opportunity that the anniversary afforded to reaffirm both his desire for Irish unity and his (increasingly unrealistic) aspirations for the restoration of the Irish language. It was a highly choreographed presidential spectacle of sorts, laden with traditional speech-making, pomp and ceremony.

Dublin Commemorations

The city of Dublin was the scene of an emotion-laden public display of remembrance when vast crowds of up to 200,000 turned out on Easter Sunday 10 April 1966 to watch the main parade marking the official Golden Jubilee commemoration. Those attending the spectacle on O'Connell Street included politicians, *c.* 600 veterans of the 1916 Rising (some from overseas), representatives of the Army and Civil Defence, and a range of other groups who represented national ex-servicemen, language, sporting, labour, and cultural organisations.[29] Careful planning went into the organisation of the parade and

---

[26]   *Dáil Debates*, Vol. 222, 28 April 1966.

[27]   N. Whelan, *Fianna Fáil: A Biography of the Party* (Gill and Macmillan, Dublin, 2011), p. 132.

[28]   F. Tobin, *The Best of Decades: Ireland in the 1960s* (Gill and Macmillan, Dublin, 1984), pp. 138, 143.

[29]   Department of External Affairs, *Cuimhneachán 1916–1966: A Record of Ireland's Commemoration of the 1916 Rising* (An Roinn Gnóthaí, Dublin, 1966), p. 24.

efforts were made by the 'Coiste Cuimhneacháin' to ensure that it was as inclusive as possible. Press notices had been issued five months earlier, inviting 'organised groups of Irish people' to participate in the parade, while direct invitations to groups 'had purposely not been issued lest groups not specifically invited might take offence'. In order to avoid complications, it was also decided that 'no banners other than identification be permitted'.[30] A ban was also imposed on the selling of Easter lilies, out of a fear that the proceeds would be used for illegal purposes. In the end, over 5,000 people participated in the parade, including 13 bands and groups representing Conradh na Gaeilge, Gael Linn, An Comhchaidreamh, Na Teaghlaigh Ghaelacha, old Army and Local Defence Force (LDF) comrades associations, Fianna Fáil, Fine Gael, the Bricklayers Trade Union, Comhaltas Cheoltóirí Éireann, Walkinstown Residents Association, and exiles. Others who participated included diverse groups from Tullamore, Waterford, Limerick, Wexford, and Banna Strand.[31]

The ceremonies in Dublin did not pass without controversy, however, as at least 20 seats reserved for special guests in the reviewing stand on O'Connell Street were unoccupied (and the number would have been higher had last-minute replacements not been found). Notable absentees included the Fine Gael leader, Liam Cosgrave, and the Labour leader, Brendan Corish. Another notable absentee was Fine Gael's Presidential candidate, Thomas O'Higgins. Although a list given to the press by the Department of Defence during the ceremonies showed these names as being among the anticipated platform party, official invitations were not actually posted to them. The Fine Gael leader, according to *The Irish Times*, stated 'that he had not been invited to attend and had not done so for that reason', while other Fine Gael sources 'suggested quietly ... that the lack of invitations might not altogether have been divorced from the business of politics'. The Labour chairman, James Tully, also verified that his party's leader had not received an invite, and went so far as to accuse Fianna Fáil of highjacking the event. He complained that even though there had been an accord that every political party should keep in the background during the commemoration, this appeared to mean all political parties except Fianna Fáil. He added that he was saddened by the fact that people had not been 'big enough to forget previous occasions so that all could commemorate'. A spokesman for the Department of Defence said that while the names of the Opposition leaders were on the

---

[30] Cuimhneachán 1916, Commemoration (1966) of the Rising of Easter Week 1916, Minutes of the Sixth Meeting of the Committee Set Up by the Government to Organise and Plan the Commemoration Ceremonies and Celebrations, 4 March 1966, NAI, Department of the Taoiseach, 97/6/163.

[31] Memorandum on Cuimhneachán 1966, NAI, Department of the Taoiseach, 97/6/163.

invitation list for the platform party, it appeared 'that there may have been some mess-up'.[32]

By contrast, *The Irish Press* speculated that the feeling in Fine Gael 'was that to participate as a party might be somewhat unwieldy and that it was better to join the parade as individual members of the various marching bodies'.[33] Reflecting on the hiccup, *The Irish Independent* estimated that the blunder was to blame for 'the non-invitation of about 30 prominent people'.[34] *The Connacht Telegraph* attributed the missing invitations to 'some very unfortunate misunderstandings' and 'a couple of lapses in the work of organisation'.[35] It eventually transpired that two government ministers, Joseph Brennan and Neil Blaney, had not received official invitations either, although the latter did manage to take his place in the reviewing stand outside the GPO. *The Irish Times* of 12 April reported that a formal apology had been issued by the Department of Defence for not sending invitations to 'certain prominent persons' and concluded that the whole incident appeared to be the unfortunate result of 'bureaucratic muddling rather than political direction'. The newspaper, however, did point out that there had been much criticism of the Taoiseach's chairmanship of the 'Coiste Cuimhneacháin' and commented: 'he should have paid more attention to the detail of the organisation'.[36] The government also came under fire from the Dublin Council of Trade Unions, with Donal O'Reilly of the Operatives and Plasterers Trade saying that 'it was a scandal' that they had not been invited to any of the main ceremonies.[37]

In a volatile Dáil debate over two weeks after Easter Sunday, the Taoiseach conveyed the government's 'regret' for the debacle over the review stands. Approximately 1,000 invitation cards, he noted, had been issued between 1–4 April, 'but, through oversight, they were sent only to the surviving participants [in the Rising]'. The Labour leader was not amused and remarked that the situation was both 'scandalous' and 'a shocking mess up', while Fine Gael's Gerard Sweetman accused Lemass of being 'grossly at fault' by not appointing members of the Opposition to the 'Coiste Cuimhneacháin' the previous October.[38] Further discontent amongst the Opposition was caused by the publication during the Golden Jubilee of a booklet by the Department of External Affairs, entitled *Facts about Ireland*. As a disgruntled Ruairi

32   *The Irish Times*, 11 April 1966.
33   *The Irish Press*, 11 April 1966.
34   *The Irish Independent*, 12 April 1966.
35   *The Connacht Tribune*, 16 April 1966.
36   *The Irish Times*, 12 April 1966.
37   Ibid., 13 April 1966.
38   *Dáil Debates*, Vol. 222, 27 April 1966.

Quinn pointed out years later in his political memoir, *Straight Left: A Journey into Politics*, the nature of the booklet's content had resulted in a chorus of disapproval in 1966, 'because it was written from an exclusively Fianna Fáil point of view'. 'Its near-Stalinist exclusion of Free State heroes, such as Collins and Griffith', he claimed, 'caused uproar', as did its ignorance of the Labour party's 'role in ensuring democracy through the provision of a constructive opposition in the Dáil'.[39]

Political controversies aside, the military parade past the GPO on Easter Sunday turned out to be an impressive spectacle, with dreams of the future and memories of the past strutting in cadence under bright sunshine. Much anticipation surrounded the build-up to the proceedings, culminating with the appearance of de Valera (with an escort from the 2nd Motor Squadron). The luminary President cut an eye-catching figure when he emerged from his car, bedecked with shining service medals and a rather imposing top-hat, which seemed *de rigueur* for the big occasion (Plate 4.1). After the playing of the 'Presidential Salute' by the No. 1 Army Band, he then inspected an FCA Guard of Honour. Thousands of spectators lined the streets to watch the parade, with the majority congregating along the length of O'Connell Street from the GPO southwards to O'Connell Bridge, where they watched groups who had paraded from St. Stephen's Green, down Grafton Street and through Westmoreland Street into O'Connell Street. The military parade included troops and armoured cars which converged on O'Connell Street and filed past de Valera, who took the salute from the reviewing stand outside the GPO – which was positioned in front of a large maroon drape adorned with an image of the iconic Sword of Light. High in the skies a flight of four Vampire aircraft swept past, and then returned, flying low down the length of O'Connell Street.[40] After 12.00 noon a 21-gun salute was fired, the National Anthem was played outside the GPO, the Tricolour was hoisted, and the words of the Proclamation then rang out from loudspeakers.[41]

[39]   R. Quinn, *Straight Left: A Journey in Politics* (Hodder Headline Ireland, Dublin, 2005), p. 95.

[40]   Department of External Affairs, *Cuimhneachán 1916–1966*, p. 28; *The Irish Times*, 11 April 1966.

[41]   Department of External Affairs, *Cuimhneachán 1916–1966*, pp. 26–28.

Plate 4.1   The President, Eamon de Valera, arriving at the GPO for the
Rising's Golden Jubilee anniversary commemoration on
10 April 1966

Source: National Library of Ireland, Independent Newspapers (Ireland) Collection, R 3975.

Amidst all of the pomp and ceremony, *The Irish Times* reported that there was much raw human emotion on display amongst the assembled masses, as the magnitude of the events of years earlier dawned upon their minds and memories:

> Captain Patrick Coakley, a Corkman, hoisted the National Flag. As it rose, it caught momentarily in the halyards; then the wind caught and freed it and it streamed out strongly, defiantly almost, in a fresh spring wind from the Irish Sea. The crowd cheered. They were still cheering as an Army band, with a slow roll of drums, sounded the call, 'Sunrise'. From the distance, the muffled boom of six 25-pounder guns in the grounds of Trinity College was a dull counterpoint as a battery began firing a 21-gun salute. A handful of ominous clouds which had drifted a curtain across the sunshine now went away. As the clouds dispersed and the echoes of the 21-gun salute died away,

the Irish Army of 1966 crossed over O'Connell Bridge to parade past the soldiers of another era in their salute to the men of 1916. By now the crush was beginning to take its toll of [*sic*] the crowd. A number of women, half-fainting, were lifted gently over crush-barriers by police and first-aid attendants and taken to an improvised first-aid station for treatment. Some small children, in danger of being trampled underfoot, also were lifted over the barriers. More noticeably, among the ranks of the veterans of the War of Independence, there was a sudden flurry of activity as an elderly man in one of the rear ranks collapsed. After a few moments, he was taken away on a stretcher, his hat shielding his face ... The end of the affair now was almost nigh. The parade over ... the No. 1 Army Band played the National Anthem ... itself very little older than the Rising.[42]

The Irish Times also solemnly acknowledged that despite the parade's grandeur, it 'was not merely a celebration', but was also 'the commemoration of a bloodbath where men had died in an hour of failure which, too late for them, had become a "Glorious Beauty"'.[43]

The Irish Press also commented on the poignancy of the occasion, noting that 'emotion most rightfully claimed those who sat, "soldier with equal soldier", on the platform of honour. Tears streamed unchecked down many a face lined with age'. Conversely, it also painted a more positive picture by observing how the main thoroughfare had resembled 'a glittering lake of colour' and how the spirit of 1916 had been 'rekindled with pride' by the parade. Altogether, it surmised that the spectacle had provoked 'a jubilant morning of celebration' for many citizens throughout the state, and that the significance of the anniversary had awakened 'a fierce and resurgent pride in those who watched' from various vantage points. So caught up in the grandeur of the occasion was *The Irish Press*, that it went so far as to hint that Mother Nature was in harmony with the day's events. 'A warming sun burst forth', it proclaimed, 'and glistened on the bayonets of the 100-strong guard of honour facing the GPO'. 'Two pigeons', it added, 'fluttered to a fine vantage point in the GPO portico', and towards the end of the proceedings, 'disturbed seagulls added their wailing cry' in between intervals of the 21-gun salute that 'crashed out from the grounds of Trinity College

---

[42] *The Irish Times*, 11 April 1966. According to *The Irish Independent*, 11 April 1966, a total of four Old IRA veterans collapsed during the ceremonies. Three were discharged from hospital following treatment, but one in his mid-60s was detained overnight in the Mater Hospital for observation. Fifteen other people collapsed – 12 of whom were brought to the Jervis Street Hospital, from where they were later discharged. The others were treated on-the-spot by ambulance crews on O'Connell Street.

[43] *The Irish Times*, 11 April 1966.

Dublin'.[44] *The Irish Independent* was equally impressed by the spectacle at the GPO, noting that 'Dublin was the scene of one of the greatest gatherings in the capital's history' and pointing out that the atmosphere on O'Connell Street was considerably enhanced 'by bands playing stirring national airs' and a swell of 'applause and cheering ... down to the Liffey's side'. The national, provincial and papal bunting on the flag masts, which had looked 'drab and lifeless' during cold and wet weather the day beforehand, transcended into 'a dancing riot of colour' – enhanced by brilliant sunshine and a fresh breeze. It also remarked that the demeanour of each one of the Rising's survivors had suggested 'a quiet joy to be living again something of the old exultation which launched them into a great adventure 50 years ago'.[45]

In regional newspapers circulating in the Republic, many positive Editorial views were rendered about the Dublin parade. Despite unfounded 'fears of discreditable things happening' on Easter Sunday, *The Connacht Tribune* pronounced that 'Ireland can be proud ... [of what] was a great occasion' in a city that had been 'the scene of death and great destruction in that historic Easter Week of 1916'.[46] *The Connaught Telegraph* noted that the commemoration was a 'touching tribute' that highlighted 'what can be done by goodwill and cooperation towards a desirable common goal'.[47] While the Editorial in the *Western People* used the opportunity to pour scorn on the government's decision to abolish free fishing on the country's lakes and rivers, it did concede that aside 'from political polemic, those missing invitations ... and an irritating proliferation of National Flags of the wrong hues', the Dublin ceremonies 'passed off in dignified and mature fashion'. 'After the rather heady oratory', it concluded, 'it is, nevertheless a better and revitalised nation that returns to the daily grind of grappling with the unfinished work of 1916 in the achievements of the full aims of the Proclamation'.[48]

The view from England of the Dublin parade was somewhat more subdued. Despite observing that it had taken place on a pleasant spring day, *The Times* soberly declared that 'the celebration was a sombre affair, although the streets were bedecked with ... bunting'.[49] In Northern Ireland, the *Belfast Telegraph* painted a similar picture: 'Most of those who thronged O'Connell Street ... seemed interested in the event as a spectacle and the mood was subdued, rather

44   *The Irish Press*, 11 April 1966.
45   *The Irish Independent*, 11 April 1966.
46   *The Connacht Tribune*, 16 April 1966.
47   *The Connaught Telegraph*, 14 April 1966.
48   *Western People*, 16 April 1966.
49   *The Times*, 11 April 1966.

than demonstrative of patriotic fervour.'[50] The Editorial of the nationalist newspaper, *The Derry Journal*, also commented on the sombreness of Dublin's commemoration, noting that it had been distinguished by 'a fitting solemnity'. But for the most part, it conveyed an upbeat assessment of the occasion, stating that Dublin had risen 'in memorable style' for the 'organisation of a ceremonial occasion on the grand scale'. It also commended the city for demonstrating 'pomp and panoply' that fitted the 'epic event' that was being commemorated.[51]

A similar impression of the Golden Jubilee parade surfaced years later in Hugo Hamilton's memoir of his childhood days, *The Speckled People*, which recounted his experiences of being raised by a strict nationalist father. All the lampposts in Dublin, he recalled, 'had flags so that everybody would remember how great it was that the Irish were free to walk down any street in the world, including their own'. The Irish people by 1966, he added, were reminded by the bunting that they 'were not the saddest people in the world any more, they were laughing now and nobody could stop them'.[52] Rather ironically, it seems that the flags and bunting used in the Golden Jubilee celebrations had been manufactured in England. Responsibility for adorning the capital's lampposts had rested with Dublin Corporation, who purchased the decorations from two Dublin firms at a cost of £4,953 16s. 3d. In a passionate contribution to Dáil proceedings in 1966, one exceptionally patriotic Labour politician, Michael O'Leary, expressed concern that 'the suppliers may have bought them in England'.[53] Not to be outdone in the quest to protect the interests of home manufacturing during the Golden Jubilee, Fine Gael's Patrick Hogan relayed his concerns to the Minister of Finance over 'whether imported wood or wood products' had been used in the building of the review stand outside the GPO. In reply, the Minister's Parliamentary Secretary, Jim Gibbons, noted that while imported chipboard and plywood had indeed been used, all of the remaining wood-products utilised 'were of Irish manufacture', in accordance with 'the usual practice' of using 'materials, appliances and fittings ... of home manufacture'.[54]

On the night of Easter Sunday 1966, another reminder of the legacy of the Rising was evident in Dublin city, when the Tánaiste and Minister for External Affairs, Frank Aiken, attended the premier at the Savoy Cinema of *An Tine Bheo* ('The Living Flame'), a 45-minute film which retrospectively traced the events of the Rising, using the recollections of veterans. The film was specially commissioned by the 'Coiste Cuimhneacháin' from Gael Linn and was directed

50   *Belfast Telegraph*, 12 April 1966.
51   *The Derry Journal*, 12 April 1966.
52   H. Hamilton, *The Speckled People* (Harper Perennial, London, 2004), p. 237.
53   *Dáil Debates*, Vol. 222, 5 May 1966.
54   Ibid., Vol. 222, 11 May 1966.

by the Cork-born documentary-maker, Louis Marcus (who had worked as the Assistant Editor on *Mise Éire*).[55] A commemorative concert was also held that night in the Gaiety Theatre. The following day, religious ceremonies marking the Golden Jubilee were held around Ireland in churches of all denominations. Solemn Masses were held in all cathedral churches, while services were also held in Jewish synagogues. At St. Patrick's Cathedral in Dublin, the Archbishop of Dublin, Dr. George Otto Simms, preached a special Easter Monday sermon of thanksgiving, commemoration and dedication at a united service held under the auspices of the Dublin Council of Churches. 'There is much', he preached, 'for which to give thanks on our commemoration occasion', including 'the sort of reconciliation and goodwill that has been evident in recent times'.[56] Before the service commenced, Major Ronnie Bunting of Belfast, who was in the congregation, walked out onto the aisle and laid a symbolic wreath dedicated 'in memory of the British officers who died in the line of duty in Dublin in Easter 1916'. Afterwards, the Dean of St. Patrick's, the Very Rev. J. W. Armstrong, told *The Irish Times* that the Major's wreath 'was laid without authority and was no part of the commemoration service organised by the Dublin Council of Churches'.[57]

Additional commemorative arrangements for the Golden Jubilee included the decoration and illumination of municipal and state buildings in Dublin city, while special '1916 Tours' by bus were offered by Coras Iompair Éireann (CIÉ), the national bus and rail company.[58] At Croke Park, five performances took place of *Aiséirí-réim na Cásca* ('Resurrection, the Easter Pageant'), containing a cast of 800 who retold in actions and words the story of Ireland's struggle for national independence before huge portraits of the executed leaders of the Rising. The events were interpreted through the three symbolic figures of 'Ireland', 'The Poet' and 'The Soldier'.[59] A set of gigantic prints of each of the deceased 1916 leaders, which would not have been out of place on the set of a Leni Riefenstahl film, were used in the pageant. Produced by Tony O'Malley Studios in Dublin, they measured some 30 feet by 16 feet and were known at the time to have been part of 'one of the largest photo printing jobs ever carried out in Ireland or Britain'.[60]

---

[55] Department of External Affairs, *Cuimhneachán 1916–1966*, p. 33; *The Irish Times*, 11 April 1966.

[56] *The Irish Times*, 12 April 1966.

[57] Ibid.

[58] Minute Books of Meetings of the National Executive, Loose Item Entitled 'Cuimhneachán 1916: Programme of Public Ceremonies and Events', UCDA, MS. P.176/348, Archives of the Fianna Fáil Party.

[59] Department of External Affairs, *Cuimhneachán 1916–1966*, p. 54.

[60] *Western People*, 16 April 1966.

Despite its propagandist aura, the pageant passed as light entertainment for some of those who cared to watch. In her report about the performance on the opening night on 12 April, Eileen O'Brien of *The Irish Times* conveyed mixed feelings about the whole spectacle, which featured a lot of razzmatazz and reenacted other events from Irish history besides 1916:

> Undoubtedly every child in the city who is hardy enough to withstand polar conditions should see this show – the torches, the rockets, the canons spitting fire, the Redcoats biting the dust on all sides. But for an adult it was all a little too much like a Christmas pantomime. To enjoy the pageant one must be able to enter into the spirit of the thing, to booh [*sic*] General Lake, to weep for Robert Emmet, to cheer unrestrainedly of [*sic*] the French at the Races of Castlebar. The strip neon explaining what was going on was an unnecessary touch of vulgarity.[61]

Notwithstanding the lukewarm reception from critics like O'Brien, the pageant proved to be a hit with the public and it also formed part of a special Children's Day on 17 April, when 20,000 primary and secondary school students marched to Croke Park to watch it from the Hogan Stand, headed by an FCA Colour Party and the Artane Boys' Band.[62]

One of the cast members, Ferdia Mac Anna (son of the pageant's director, Tomás Mac Anna), recalled many years later that *Aiséirí-réim na Cásca* had been 'a colourful, highly organised and completely over-the-top extravaganza that resulted in glorified confusion', but also one that conveyed to the impressionable youngsters who participated, 'a sense that we owed our freedom to Pearse and company's "blood sacrifice"'. During the months afterwards, the leaders of the Rising had 'seemed like the Magnificent Seven' to both himself and his old friends, as they playfully 'staged re-enactments of 1916 up the back fields in Howth'.[63] The staging of *Aiséirí-réim na Cásca* also had a profound influence on a future Taoiseach, Bertie Ahern, who was only 14 years old at the time. The pageant, as he later recalled in his autobiography, impacted profoundly upon his political beliefs and proved 'the real highlight of 1966 for me'. It 'left a lasting impression' and 'was when my interest in the Rising really took off'.[64]

---

61   *The Irish Times*, 13 April 1966.
62   Ibid., 18 April 1966.
63   Ibid., 13 May 2006. It was only when the Troubles started, he added, that 'it became less of a head-wreck to seek role models in the modern safety zones of rock music and movie icons'.
64   B. Ahern with R. Aldous, *Bertie Ahern: The Autobiography* (Hutchinson, London, 2009), p. 18. In elaborating on his childhood recollections of the event, Ahern (on p. 18) recalled the following: 'There were explosions and everything. James Connolly was carried in for his execution. It was a real drama. I had never seen anything like it in all my life. We were in the Hogan Stand,

Another popular cultural event organised by the 'Coiste Cuimhneacháin' was the 'Cultural and Artistic Tribute', which entailed the sponsorship of a series of 18 competitions in music, art and literature 'to enable the writers and artists of today to participate in the commemoration of a Rising the leaders of which were themselves gifted in learning and art'. These competitions attracted entries from Irish citizens all over Europe.[65] Efforts were also made to involve all elements of the education sector in the Golden Jubilee celebrations. Plans for a new scheme of 'Easter Week Commemoration Scholarships' had already been put in place by the government by the end of 1965. Based on the Leaving Certificate examination results, seven scholarships were made available in the Golden Jubilee year 'to provide a continuing commemoration of the signatories of the 1916 Proclamation'.[66] Some proposals for educational commemoration, however, never came to fruition. Within the Fianna Fáil party, for example, notice of a motion was given to a meeting of its National Executive, held on 10 May 1965, to request the government to mark the Golden Jubilee by establishing an Irish-speaking university at the site of the Royal Hospital at Kilmainham in Dublin. This overly-ambitious proposal was eventually withdrawn at a National Executive meeting held two weeks later.[67] A less costly gesture materialised on 14 April 1966 when the National University of Ireland awarded honorary doctorates at St. Patrick's Hall in Dublin Castle to the nearest surviving relatives of most of the signatories of the Proclamation. These were presented by de Valera. One of the recipients was Senator Margaret Pearse (sister of Patrick), who attended the event in a wheelchair, having been brought from Linden Convalescent Home in Blackrock, where she was a patient.[68]

While most staff and students of the National University of Ireland's constituent colleges eagerly embraced the spirit of the Golden Jubilee, the occasion proved to be rather more problematic for the esteemed dons of Trinity College Dublin. Some of them were reluctant to partake, because they remained unionists at heart. One of the senior (and most colourful)

---

mostly schoolchildren, and when they read out the names of the executed there was absolute silence. It was like everyone was holding their breath. A stadium of thousands of people that is suddenly quiet is always moving, but this was something else.'

[65]   Department of External Affairs, *Cuimhneachán 1916–1966*, p. 82. See pp. 82–84 of this source for the results of the competitions, detailing the prize-winners' names, their entries and the amounts of money that they won.

[66]   *The Irish Independent*, 4 December 1965.

[67]   Minute Books of Meetings of the National Executive, 10 and 24 May 1965, UCDA, MS. P.176/349, Archives of the Fianna Fáil Party. In May 1991, the Irish Museum of Modern Art was opened at the site.

[68]   Department of External Affairs, *Cuimhneachán 1916–1966*, p. 60; *The Irish Independent*, 15 April 1966; *The Irish Times*, 15 April 1966.

members of its staff at the time, historian R. B. McDowell, later recalled in his engaging memoir, *McDowell on McDowell*, how his political position as a southern unionist had complicated matters for him. When Trinity College apprehensively commemorated the anniversary with a lecture on Thomas Davis and a reception afterwards in the Provost's House, McDowell 'attended the lecture but declined the invitation to the reception – several colleagues whom I respected did likewise'.[69] On the other hand, many of Trinity College's younger scholars got into the spirit of the occasion. The student magazine, *TCD*, for example, produced a 42-page supplement for the anniversary, featuring writers such as Seán O'Faolain, John Horgan, Bruce Arnold, Marie Comerford, Senator Owen Lancelot Sheehy-Skeffington (son of Francis), and Sir Graham Larmour. Under the theme 'What Has Happened?', the articles revisited the aims of the Rising and questioned whether any of them had been implemented in the social sphere.[70]

Besides the universities, schoolchildren in Dublin and throughout the rest of the Republic celebrated the Golden Jubilee on 22 April with a special holiday. A Mass was held at the Pro-Cathedral in Dublin, while a special service was held in St. Patrick's Cathedral. At the Vocational School for Girls at Crumlin Road in Dublin, a special Mass in Irish was celebrated and this was followed by the opening of a historical exhibition featuring letters written by Patrick Pearse. The Minister for Education, George Colley, attended ceremonies at the Chanel College at Coolock and at the Convent of Mercy secondary school. Veterans of the Rising, members of the government and officers of An Garda Síochána attended ceremonies in many other schools, during which framed copies of the Proclamation were unveiled and Tricolours were raised.[71] Overall, it was estimated in the Dáil that the total cost of providing copies of the Proclamation to all of the country's schools was around £6,300.[72] But what did the children themselves make of it all?

Years later, after they had grown into men and women, a number of Irish writers recorded mixed impressions of their experiences during the period of the

---

[69]   R. B. McDowell, *McDowell on McDowell: A Memoir* (The Lilliput Press, Dublin, 2008), p. 158. Although he worked for nearly 40 years in Trinity College, McDowell freely admitted as a 94-year-old man (on pp. 155–56): 'I never felt completely at home in the Irish Republic ... I regarded myself as an expatriate [of the United Kingdom] living and working in the Free State or the Republic ... The nationalist tradition, the political and cultural orthodoxy accepted by the Republic, failed to attract me. A fervent nationalist would consider me purblind or tone-deaf.'

[70]   *The Irish Times*, 19 April 1966.

[71]   Ibid., 23 April 1966.

[72]   *Dáil Debates*, Vol. 222, 28 April 1966.

Golden Jubilee, in a collection of essays dedicated to the Rising. The Dublin-born writer, Dermot Bolger, wrote:

> I was a seven year old schoolboy in 1966 when the fiftieth anniversary of the Easter Rising was celebrated. Looking back now it is extraordinary how many vivid recollections I still retain of those events ... Even more extraordinary is how those same recollections are frequently shared by so many of my generation. Memorials were unveiled, speeches made beseeching God that in her hour of need Ireland would find such brave young sons again. I doubt if the dignitaries making those speeches believed a word of it ... but we, in short trousers and cropped hair, did.[73]

In the same collection, Irish language poet Nuala Ní Dhomhnaill reminisced about the 'acute embarrassment' that she had felt whilst reading out aloud her prize-winning essay on the life of Seán MacDiarmada from the stage of a concert hall. The executed 1916 leader, she noted, was like an elusive 'dark horse' to her, even though she had gathered together the basic facts of his life from her library.[74] Another contributor to the volume was the journalist, Fintan O'Toole. Born in 1958, he argued that 1966 had been a year in which fidelity to the past proved the order of the day, with 'images of 1916 ... renewed and refashioned for a new generation just at the time when they had finally lost all reality'.[75]

In *The Speckled People*, Hamilton also passed comment on how exhilarating the story of 1916 had become during the Golden Jubilee. Pictures of Pearse were displayed throughout Dublin city, 'in the windows of shoe shops and sweet shops'. Such premises also had Tricolours and reproductions of the Proclamation, which he and his young friends 'all learned off by heart'. Hamilton and his friends also got acting parts 'as croppy boys or redcoats' in a pageant staged at the Abbey Theatre, where they 'died every night'.[76] For years after 1966, the iconic image of the illuminated copy of the Proclamation on school walls, according to Declan Lynch, also captivated the imaginations of schoolchildren in the Republic, proving 'as stunningly effective as the Che Guevara poster or any of the classical images of rock 'n' roll'.[77] So too, of course, did the prominence accorded to the 1916 metanarrative in history schoolbooks and many a school tour to 1916-related sites such as Kilmainham Gaol.

---

[73] D. Bolger, 'Introduction', in D. Bolger (ed.), *Letters from the New Island: 16 on 16. Irish Writers on the Easter Rising* (Raven Arts Press, Dublin, 1988), p. 7.

[74] N. Ní Dhomhnaill, 'The Black Box', in Bolger (ed.), *Letters from the New Island*, p. 30.

[75] F. O'Toole, '1916: The Failure of Failure', in Bolger (ed.), *Letters from the New Island*, p. 41.

[76] Hamilton, *The Speckled People*, pp. 236–37.

[77] *The Sunday Independent*, 23 April 2006.

Transformations in the Cultural Landscape

Around the world, a popular method of commemorating a sense of national identity (and a sense of place) has been through the construction of built memorials in the cultural landscape. The notion of the cultural landscape, as Yvonne Whelan has noted, can act as a powerful symbol of national identity and thus be regarded 'as a social and cultural production which both represents and is constitutive of past, present and future political ideologies and power relationships'.[78] When defined in terms of landmarks of historical or personal validity, landscape can therefore serve as a widely-shared *aide-mémoire* of a culture's knowledge and understanding of its past and present.[79] Ireland has been no exception in this regard. With great imagination and flair, William J. Smyth has put forward the metaphorical notion that the Irish cultural landscape represents 'a multi-layered, eroded and rewritten text' that can be decoded by untangling a 'series of inscriptions, signs and symbols which reflect the complicated ... history of ... culture'. Rather significantly, he also notes that 'behind all these texts and artefacts are the voices, the stories and the actions of ... people who ... loved, fought and died' in Ireland.[80] The part played by 1916 commemorations in the forging of Irish cultural landscapes of remembrance was especially noticeable during the course of the Golden Jubilee year. On a number of occasions, the government did its utmost to ensure that memory of the Rising did not fade away from public consciousness. At pivotal geographical points around Dublin city centre, for example, a number of locations were given over to the creation of memorials to the Rising (and other revolutionary events), thereby constructing a commemorative heritage, culture and identity within the arena of public space. As they mushroomed from invisible into visible form, the Rising's memory was perpetuated amongst the masses.

---

[78]   Y. Whelan, *Reinventing Modern Dublin: Streetscape, Iconography and the Politics of Identity* (University College Dublin Press, Dublin, 2003), p. 207. For an illuminating commentary on the representation of the past in monuments, see C. Withers, 'Monuments', in R. J. Johnston, D. Gregory, G. Pratt, and M. Watts (eds), *The Dictionary of Human Geography*, 4th Edition (Blackwell Publishers, Oxford, 2000), pp. 521–22.

[79]   S. Küchler, 'Landscape as Memory', in B. Bender (ed.), *Landscape Politics and Perspectives* (Berg Publishers Ltd., Oxford, 1993), p. 85.

[80]   W. J. Smyth, 'The Making of Ireland: Agendas and Perspectives in Cultural Geography', in B. J. Graham and L. J. Proudfoot (eds), *An Historical Geography of Ireland* (Academic Press, London, 1993), p. 399.

Plate 4.2   A view of the remaining stump of Nelson's Pillar following the
explosion of an IRA bomb at O'Connell Street in the early hours
of 8 March 1966

Source: National Library of Ireland, Independent Newspapers (Ireland) Collection, R 4014.

Memories of Ireland's colonial past fared less well in 1966. In the early hours
of 8 March, the IRA blew up Nelson's Pillar on O'Connell Street (Plate 4.2). By
'assassinating somebody who had been dead for 160 years', as David Limond has
written, the 'remnant IRA' seemed keen on proving 'its continuing relevance'.[81]
In the aftermath of the explosion, which was set off in a small ventilation slit
about half way up the pillar, the top portion capsized under its own weight.
Fortunately, nobody was injured. The eradication of the memory of the British
naval hero, who had helped defeat Napoleon, proved extremely upsetting for

[81]   D. Limond, '[Re]moving Statues', *History Ireland*, Vol. 18, No. 2 (2010), p. 10.

the monument's trustees, who owned the private property upon which the site was situated. For years, they had taken great pride in both the aesthetic and architectural qualities of what had become one of the most significant landmarks within the capital city – acting variously as a viewing point for visiting tourists, a bus stop for locals and a meeting point for friends. However, in commenting on 'the powerful role of public statuary in cultivating narratives of identity', Whelan has noted that the monument 'had become a jarring symbol of colonial rule' since its erection in 1809. This resentment gathered pace after independence, when many groups and individuals (including Dublin Corporation councillors) campaigned for its removal. It was republican dissidents, however, who 'seized upon its political symbolism' and who 'in an iconoclastic gesture', irrevocably altered the built heritage and iconography of the principal thoroughfare.[82]

What remained of the badly damaged pillar, according to Coogan, resembled 'a giant amputated phallus', ironically serving as a powerful reminder to the Irish people that the 1916 rebels 'had fought for a united Ireland, not a partitioned one'. Afterwards, in their hurried effort to clear up the mess by blowing up the rest of the pillar in time for the Golden Jubilee parade, the Irish Army set off an explosion which 'turned out to be the Dublin version of the Bikini Atoll', and sent glass flying from numerous windows on O'Connell Street.[83] In their contributions to Dáil proceedings throughout March 1966, a number of politicians addressed the subject of the Nelson's Pillar site. A week after the first explosion, the Minister for Local Government, Neil Blaney, told the Dáil that he had 'already initiated discussions with the Dublin Corporation on the question of improving pedestrian facilities in the area'.[84] On 22 March, however, Richie Ryan expressed concern over the second explosion that had been set off to demolish the stump of the pillar. The monument, he alleged, had been 'a source of embarrassment' for the government, who had shown 'utter irresponsibility' in blowing up its remains, 'without lawful authority'. However, the Fine Gael politician's theory that there had been some sort of government-backed conspiracy to rid the city of the pillar once and for all was vigorously dismissed by senior members of the Cabinet. After Ryan suggested that law-and-order had broken down, the Minister for Justice, Brian Lenihan, contemptuously replied: 'The Deputy is blown up by his own importance.'[85] The government eventually accepted liability for the damage caused to properties by the second explosion, and confirmed that it would pay out compensation to

---

[82]   Whelan, *Reinventing Modern Dublin*, p. 207.
[83]   T. P. Coogan, *A Memoir* (Weidenfeld & Nicolson, London, 2008), p. 169.
[84]   *Dáil Debates*, Vol. 221, 15 March 1966.
[85]   Ibid., Vol. 221, 22 March 1966.

the owners of these premises.[86] The issue, however, dragged on for some time and by the end of April, 19 claims totalling £1,988 10s. 7d. were lodged as a result of the second explosion, while an additional nine claims were made 'for unspecified sums'.[87]

Another episode in the expunging of Nelson's memory from the capital city during the Golden Jubilee year was instigated by the government itself, which secured a special agreement with CIÉ to remove the name 'The Pillar' from timetables and bus scrolls and to replace it with Árd Oifig an Phoist (the Irish for 'General Post Office').[88] In private, Lemass himself may not have been too bothered at the time by the furore surrounding the unforeseen loss of Nelson's Pillar from the capital's built heritage attractions, as he had previously put forward a suggestion 'that an appropriate replacement for the figure of Nelson would be that of St. Patrick, coinciding with the Patrician Year of 1961'.[89] Furthermore, in 1964, he had informed the Minister of Finance of his desire to look at the question of erecting 'a new monument' in the pillar's place 'by 1966', noting that compulsory acquisition of the site was possible in the event of legislation being drawn up and compensation being paid to the trustees. Given the complications involved, he concluded that he did 'not think we will find it easy to reach the right answers'.[90]

---

[86]   Ibid., Vol. 222, 29 March 1966.

[87]   Ibid., Vol. 222, 27 April 1966. It was not until three years later, however, that the legal complications surrounding the saga were finally resolved. Under the terms of the Nelson Pillar Act of 1969, the monument's trustees received £26,938 18s. 3d. in compensation from Dublin Corporation, which in turn recouped its costs from the Exchequer. For details on these matters, see the following: The Nelson Pillar, Dublin, NAI, Department of the Taoiseach, 99/1/22; *Dáil Debates*, Vol. 242, 18 November 1969. Also see Government of Ireland, *Nelson Pillar Act, 1969* (The Stationery Office, Dublin, 1969), p. 7, which shows that ownership of the site was transferred to the corporation, while the trustees themselves were granted indemnity from any legal proceedings taken 'by any person in respect of their actions in ... not restoring a monument to Nelson'.

[88]   R. Higgins, '"The Constant Reality Running through Our Lives": Commemorating Easter 1916', in L. Harte, Y. Whelan and P. Crotty (eds), *Ireland: Space, Text, Time* (The Liffey Press, Dublin, 2005), p. 48.

[89]   Whelan, *Reinventing Modern Dublin*, p. 204.

[90]   Seán Lemass to Jim Ryan, 7 July 1964, NAI, Department of the Taoiseach, 99/1/22. In the end, even though the IRA explosion signalled the end of Nelson's time on O'Connell Street, it was by no means the last that Dubliners would see of him. Years later, the severed head of the statue ended up on display in Dublin Civic Museum. Thereafter, it was transferred to the reading room of the Dublin City Library and Archive on Pearse Street (located, in an ironic twist of fate, just across the street from the Padraig Pearse pub).

Plate 4.3   The Garden of Remembrance at Parnell Place, Dublin,
                 opened on 11 April 1966 by the President, Eamon de Valera
Source: Author.

Alongside the unofficial and illegal destruction of imperial monuments came the official construction of landscapes that symbolised Ireland's postcolonial status as an independent nation. The most significant public space unveiled in memory of the Rising in 1966 was the Garden of Remembrance at Parnell Square, Dublin (Plate 4.3). Dedicated to those 'who had died for the cause of Irish freedom', it was opened by President de Valera on Easter Monday, 11 April, after being blessed by the Archbishop of Dublin, Dr. Charles McQuaid. In contrast to the elation that was evident amongst some in the crowd who watched the parade past the GPO the previous day, *The Irish Press* reported that 'tones of sadness and drama were perhaps more keenly felt'.[91] At the ceremony, de Valera informed the invited guests (including the Taoiseach, Lord Mayor of Dublin Alderman Eugene Timmons, other politicians, clergy, 1916 veterans, and a choir from Clonliffe College) that it had been 30 or more years since Oscar Traynor, on behalf of the Old Dublin Brigade, IRA, had sent a memorandum to the Free

---

[91]   *The Irish Press*, 12 April 1966.

State government, urging acquisition of the Rotunda Gardens for the purpose of constructing a memorial garden. Because of its geographical proximity to key historic sites associated with the cause of Irish nationalism, the President noted that the government at the time readily accepted this suggestion. The sites within its vicinity included the GPO, the rink where the Irish Volunteers were founded in 1913, the Rotunda buildings, the old headquarters of the Gaelic League, and various houses where the executive of the Irish Volunteers and the headquarters staff of the Irish Citizen Army had met. Continuing, he stated that the purpose of the garden was 'to remind us of the sacrifices of the past, the struggle and suffering over the centuries to secure independence'.[92]

During the course of the unveiling ceremony, wreaths were laid by de Valera, Lemass, the Papal Nuncio, the Dean of the Diplomatic Corps, the Italian Ambassador, the Lord Mayor of Dublin, and veterans' organisations. To add to the spectacle, a special cadet Guard of Honour of 24 men from the Military College, under the command of Lieutenant Finbarr Studdert, held rigid positions for more than an hour on the granite semi-circular platform located on top of the marble end-wall at the western side of the garden. At 12.00 noon the Angelus was sounded and then, speaking in Irish, de Valera formally declared the garden open. In commenting on the iconography embellished within the design of the garden, de Valera noted that it was highly symbolic, for it represented 'Christian sacrifice and suffering, faith and hope, resurgence and peace', while it also challenged present and future generations to maintain Ireland as a separate nation, and to prove themselves worthy of all the sacrifices that had been made in the past. It was the government's task, he continued, 'to lay the future of the country' and 'to see to it that the dreams and ideals of the men who suffered in the past were fully realised'.[93]

In layout, the spatial design of the Garden of Remembrance consisted of 'a sunken garden in the shape of a Latin Cross with a central reflecting pool' along an east-west axis. Depicted on the floor of the pool was 'a mosaic pattern of blue-green waves bearing a design of six groups of weapons from the early Iron Age, a theme derived from the custom of the Celtic people throughout Europe of placing weapons in lakes or pools after battle'.[94] One of the iron swords depicted was a likeness of one that had been found during the excavation of a crannóg at Ballinderry, County Westmeath during the 1930s. A 'peace motif' was evident in the railings surrounding the pool, featuring a gold-coloured Irish harp and an olive branch.[95] On the day of the opening, de Valera was presented with a key

---

[92]   *The Irish Times*, 12 April 1966.

[93]   Ibid.

[94]   Department of External Affairs, *Cuimhneachán 1916–1966*, p. 45.

[95]   Higgins, 'Commemorating Easter 1916', p. 49.

of the garden's main gate by Daithí Hanley, the architect who had designed the garden 20 years previously. The 10-inch-long bronze key (with an Irish language inscription) was a three times enlarged copy of the oldest known Irish key at the time. This had been found in 1936 during archaeological excavations by Hugh O'Neill Hencken of the Harvard Expedition at the early medieval period Royal Crannóg of Lagore, seat of the Kings of Deisceirt. When de Valera opened the gate of the Garden of Remembrance with the copy of the Lagore key, a fanfare was sounded by trumpeters.[96]

Soon after its opening, local schoolchildren took it upon themselves to add their own patriotic touch to the Garden of Remembrance. As Hugo Hamilton later recalled, one of his friends in 1966 was a boy in an all-Irish Christian Brothers school in the city centre, whose brother worked in a gardening shop. Following the opening of the memorial on Easter Monday, this friend turned up at school one day with a bag of green-coloured dye that was used for mixing with fertiliser, 'so that everybody would know it was not to be eaten'. At lunchtime, he and his friends carried the fertiliser over to 'the new garden', whereupon he 'had the idea to throw the dye into the fountain for Ireland'. This soon 'turned green before we even got a chance to get back out ... again and the guards were sent for'. Having tried to evade detection by washing excess dye from his stained hands and face in the public toilets near to the GPO, Hamilton found that 'every time I put water on my face it turned even more green'. After returning late to school with green all over his face, he thought that he would face expulsion. To his relief, however, 'nothing happened because they said it was the right colour at least'. Upon returning home on the train, Hamilton's fellow passengers 'thought it was part of the Easter commemorations and that every boy in Ireland was turning green'.[97]

Other eras and personalities of Irish nationalism were also commemorated in the cultural landscape of Dublin city during the Golden Jubilee commemorations. On Saturday 16 April, some 21 years after the foundation stone was laid by the then President, Seán T. O'Kelly, de Valera unveiled a 10-foot-high bronze statue of Thomas Davis (sculpted by Edward Delaney) at

---

[96] Department of External Affairs, *Cuimhneachán 1916–1966*, p. 45; *The Irish Times*, 12 April 1966. A crannóg is an artificially-created island or lake dwelling (held within a palisade). Located six miles south-east of the Hill of Tara in County Meath, Lagore was known in the Annals as 'Loch Gabair'. Its archaeological deposits survived up to almost 10 feet and included luxury goods (produced by craftsmen), fragmentary buildings, hearths, piles, log platforms, woven wattles, peat, and vegetation. For details, see E. Evans, *Prehistoric and Early Christian Ireland: A Guide* (B. T. Batsford Ltd., London, 1966), p. 170 and N. Edwards, *The Archaeology of Early Medieval Ireland* (B. T. Batsford Ltd., London, 1990), pp. 9, 34, 38–39, 68.

[97] Hamilton, *The Speckled People*, p. 238.

College Green. To his admirers, Davis had for long been regarded as a 'glorious apostle of nationalism'.[98] At the ceremony, the President spoke of Davis's efforts to attain a parliament for the people and of his quest to make people love their country. 'To love our country' and 'and to make our country the grand old nation that it can be', stated de Valera, 'we must know her past'. He also spoke of the importance of reading the writings of Davis, because the rebels of 1916 'were his spiritual children'.[99] When he unveiled the formalistic statue amidst a steady pour of rain, the fountains in front of it came to life. Trumpeters and drummers sounded a salute, and verses from the Young Irelanders' song, 'A Nation Once Again', were played by the musicians of the No. 1 Army Band (who were accompanied by a choir of schoolchildren directed by Proinnsias Ó Ceallaigh). The ceremony ended with the playing of the National Anthem.[100] Like the Garden of Remembrance, the geographical location for the unveiling of the Davis statue was also historically significant. Firstly, it was near to the grounds of Trinity College Dublin, where Davis had told students: 'Gentlemen, you have a country'.[101] Secondly, the statue was also situated beside the old Irish Parliament House. While the parliament had originated in the middle ages as part of the Anglo-Norman's efforts at centralising government, it did not have a permanent meeting place as a 'component of the machinery of government' until a stately building was constructed at College Green, from 1729–1739. It was at this venue where Henry Grattan's Irish Patriot Party won legislative independence, from 1782–1800.[102]

The unveiling of the Davis statue did not pass off without controversy, however, and within a day it was vandalised. It also came in for stinging criticism from an architect appearing on the *Late Late Show* – the flagship chatshow of the fledgling state broadcaster, Radio Telefís Éireann (RTÉ) – who complained that Delaney's work of art did not bear a physical likeness to the Young Irelander and that the clothes did not resemble Victorian dress.[103] Despite the historical connections with events in the area, others felt that the location of the statue did not respect the built heritage of the area.

---

[98] Catalogue of an Exhibition of Objects Associated with the Young Ireland Movement, Cork City and County Archives (hereafter CCCA), MS. U.313/2.

[99] Department of External Affairs, *Cuimhneachán 1916–1966*, p. 62.

[100] Ibid., p. 62; *The Irish Times*, 18 April 1966.

[101] Department of External Affairs, *Cuimhneachán 1916–1966*, p. 62.

[102] J. J. McCracken, *The Irish Parliament in the Eighteenth Century*, Irish History Series No. 9 (Dundalgan Press, Dundalk, 1971), p. 5. For an overview of parliamentary activity at this location, see E. M. Johnston-Liik, *MPs in Dublin: Companion to the History of the Irish Parliament 1692–1800* (Ulster Historical Foundation, Belfast, 2006).

[103] *The Irish Times*, 29 April 1966.

Plate 4.4   Monument in memory of Na Fianna Éireann (the Irish National
               Boy Scouts) at St. Stephen's Green, Dublin, unveiled in 1966
Source: Author.

This point was expressed by an alarmed devotee of neoclassical architecture
from Long Island in New York, who penned a venomous letter to the Editor of
The Irish Times, conveying his disbelief that the memory of Davis had been
enshrined in what he felt was a 'colossal eccentricity' amidst the 'noble proportions'
that characterised the façades of some of the Italianate buildings on College
Green. 'It is sad', he bemoaned, 'to see the ruination of this charming place by the
intrusion of this crude figure towering over a crenulated water pond'.[104]

Another public space south of the River Liffey that figured in the Golden
Jubilee celebrations was St. Stephen's Green, which had witnessed military
action during the Rising. Near to where the bust of Countess Markievicz had
been erected in 1957, a new monument in memory of Na Fianna Éireann was
completed in 1966, consisting of a rectangular pillar on a sloping granite base. The
upper portion of the pillar was sculpted in the form of a Celtic Cross (Plate 4.4).
On the Liffey's northside, the revamped final resting place of the 1916 leaders at
Arbour Hill was the focus of attention on Sunday 24 April. Following the annual
Mass in the Military Church of the Sacred Heart for the executed leaders of the

---

[104] Ibid., 30 April 1966.

Rising, the ceremonies then moved to their grave plot. A brochure produced for the guests highlighted the important historical associations between Arbour Hill and the Rising, detailing how it was a heritage space full of republican symbolism:

> The plot in which the leaders are buried is in an open terrace of Wicklow granite, situated in what was originally the prison yard. The area containing the graves, a low grassy mound, is defined by a limestone surround in which the names of the leaders are incised, in Irish at the head and in English at the foot. The background to the graves is a curved screen wall of Ardbraccan limestone with a gilded cross in the centre. The Proclamation of the Republic is incised on this wall in Irish and English. The upper granite terrace is linked by a broad slight of steps with a series of paved areas which form an approach to the graves and an assembly area for commemoration ceremonies.[105]

For the Golden Jubilee, it was decided that a new memorial at Arbour Hill would be unveiled by the President. After delivering a short speech 'devoid of emotion', a roll of drums was sounded and de Valera then unveiled a plaque (on the wall facing the curved screen wall) 'to the memory of others, apart from the ... leaders, who were executed and are buried at Arbour Hill, who gave their lives for Ireland in 1916'.[106] The large limestone plaque bore the names of 64 members of the Irish Volunteers and Irish Citizen Army (62 killed in action, and two executed men – Thomas Kent and Roger Casement).[107] Prayers were rendered at the memorial by Father P. Duffy and the 'Benedictus' was chanted by the Clonliffe College Choir. The 'Last Post' was also sounded while the raising of the Tricolour was followed by the 'Reveille' and the National Anthem.[108] Following the conclusion of the ceremony, which had left some spectators visibly 'shaken', the Army then placed a special guard of honour around the grave, and 'remained there all day while thousands of people filed past, most to pay their respects, some to refresh old memories'.[109]

---

[105] Brochure Entitled 'Cnoc an Earbair: Arbour Hill', NAI, Department of the Taoiseach, 2000/6/112, Easter Week Commemorations at Arbour Hill.

[106] Memorandum on Ceremonial Order at 1916 Commemoration Ceremony, 18 April 1966, NAI, Department of the Taoiseach, 98/6/80, Easter Week Commemorations at Arbour Hill.

[107] Department of External Affairs, *Cuimhneachán 1916–1966*, p. 72.

[108] *The Irish Independent*, 25 April 1966.

[109] *The Irish Times*, 25 April 1966. Due to the perennial problem of visitors stepping on the grass plot over the graves, plans were later put in place to erect railings around it. For details, see

In addition to the monuments, gardens and graves that physically perpetuated the memory of 1916 within the cultural landscape, invisible forms such as placenames were also used to cultivate the Rising's memory during the Golden Jubilee year.[110] For the most part, it was the surnames of the 1916 leaders that took precedence in a range of new placenames that materialised. In what was meant to have been one of the ultimate expressions of the Lemassian era's modernisation and urbanisation, construction work commenced in 1966 on seven 15-storey apartment tower blocks in Dublin's northside suburb of Ballymun – each one named after a signatory of the Proclamation. Although six of the seven towers were later demolished (between 2004–2008), other placenames dating from 1966 fared much better in the long-run. To mark the Golden Jubilee, CIÉ renamed 15 of the Republic's principal railway stations in honour of the executed rebels, with Dublin city's Kingsbridge Station changed to Heuston Station, Amiens Street Station converted to Connolly Station and Westland Row Station turned into Pearse Street Station. The other 12 renamed stations were as follows: Casement Station in Tralee, Ceannt Station in Galway, Clarke Station in Dundalk, Colbert Station in Limerick, Daly Station in Bray, Kent Station in Cork, MacBride Station in Drogheda, MacDiarmada Station in Sligo, MacDonagh Station in Kilkenny, Mallin Station in Dún Laoghaire, Plunkett Station in Waterford, and O'Hanrahan Station in Wexford. To mark the renaming of the stations, a series of plaques were unveiled inside them as a permanent reminder of the 1916 leaders' actions.

Heritage Exhibitions, Broadcasts and Memorabilia

During the course of the Golden Jubilee, the government and other organisations did their utmost to stir the national imagination by utilising art galleries, museums and libraries both as repositories of cultural memory and forums in which to educate and inspire citizens about the storylines behind the country's

---

Michael Hilliard to Jack Lynch, 28 August 1968, NAI, Department of the Taoiseach, 2000/6/112, Easter Week Commemorations at Arbour Hill.

[110] P. O'Flanagan, 'Colonisation and County Cork's Changing Cultural Landscape: The Evidence from Placenames', *Journal of the Cork Historical and Archaeological Society*, Second Series, Vol. 84, No. 239 (1979), p. 1, notes that placenames can be classified as multifunctional signposts or 'invisible' elements within cultural landscapes, and that they can be a useful source for indicating broader changes taking place in society. Also see S. Pender, 'How to Study Local History', *Journal of the Cork Historical and Archaeological Society*, Second Series, Vol. 46, No. 164 (1941), p. 118, who notes that placenames are valuable for revealing the character of a place and enhancing an understanding of cultures, social systems, institutions, and manners of thought.

rebellious heritage. On Tuesday 12 April 1966, for example, the Lord Mayor of Dublin opened an exhibition of artworks at the Municipal Gallery of Modern Art. Among the items put on display were oil paintings and bronze sculptures that conjured up various images of the Rising.[111] In a short speech to the several hundred who attended the opening (including the Taoiseach and the British Ambassador), the Lord Mayor stated that even though most of the exhibitors had no personal connections with the Rising, they had looked into the past and brought it to life by producing their own modern impressions, 'some mournful, some defiant'. He further stated: 'To an older generation a new look at the Rising may shock or startle, but no one can doubt their sincerity or their genuine effort to commemorate the event.'[112]

The following day, a small exhibition of mementoes of the Rising (including photographs, flags, rifles, uniforms, and equipment) went on display at the building of Messrs. P. J. Carrolls at the Grand Parade in Dublin. The exhibition was curated by Cinema Arts and featured extracts from *Mise Éire*, which were shown several times each day. A novel aspect of the display was an electrically-operated map, which allowed visitors to illuminate positions depicting the garrisons, barracks and outposts located in Dublin during the Rising. Evening lectures were also run, and these were delivered by Officer P. J. Hally, Colonel J. V. Lawless (a veteran of the Rising), Dr. Lyons Thornton, and Professor Liam Ó Briain.[113] Throughout Easter, a commemorative exhibition was also held in the National Gallery of Ireland, featuring works of visual art that had come into existence under conditions 'where contemplation was possible'.[114] Among the items exhibited were historical paintings, portraits and sculptures depicting personalities and events of the Rising, along with other events in Irish history.[115]

The heritage of the Rising was also encapsulated in a special 1916 exhibition at the National Museum in Dublin, which was run in conjunction with the National Library. The exhibition was opened on 12 April by Dr. Patrick Hillery, the Minister for Industry and Commerce, in the presence of the Director of the Museum, Dr. A. T. Lucas. Also in attendance were members of the Diplomatic Corps, the government, the judiciary, the Council of State, church leaders, and relatives of the executed 1916 leaders.[116] The items on display covered a whole range of events from the Gaelic League's foundation in 1893 to the close of the

---

[111] Department of External Affairs, *Cuimhneachán 1916–1966*, pp. 50–51.

[112] *The Irish Times*, 13 April 1966.

[113] Ibid., 14 April 1966.

[114] National Gallery of Ireland, *Cuimhneachán 1916: A Commemorative Exhibition of the Irish Rebellion 1916* (Dolmen Press Ltd., Dublin, 1966), p. 83.

[115] Department of External Affairs, *Cuimhneachán 1916–1966*, p. 52.

[116] Ibid., pp. 48–49.

War of Independence in 1921, including uniforms and regalia of the fighting forces, and specimens of weapons that were in use during the period. Among the exhibits were some of the rifles landed during the Howth gun-running of July 1914, along with Russian rifles salvaged from the cargo of the Aud – a German arms ship scuttled off Cork Harbour in 1916. Other items on display included manuscript and printed items of the period (including an original copy of the Proclamation), handwritten correspondence of some of the leaders (including Patrick Pearse's last letter) and a poem written by Pearse for his mother. A number of personal relics were also included in the exhibition, including Pearse's barrister's wig and gown, the bag pipes used by Eamonn Ceannt when he played for Pope Pius X and items that belonged to Roger Casement. The prison life endured by many of the rebels who were jailed following the Rising was also represented in the exhibition by various articles which they had made during their detention.[117] Dr. Hillery, who began and ended his speech in Irish, spoke about the importance of preserving the memorabilia of those who had sacrificed themselves in that 'Easter watch and resurrection'. Furthermore, he thanked those benefactors who had parted with 'personal relics or written and pictorial documents', so as 'to make the exhibition a mirror of the men of 1916 and of the time and world in which they lived'.[118]

Before he formally opened the exhibition, two flags were presented to Dr. Hillery by Lemass for display in the museum. One of these was the Irish Citizen Army flag which depicted the Plough and the Stars. This had flown over the Imperial Hotel on O'Connell Street during the Rising. The second one included in the exhibition was the tattered remains of a green flag bearing the words 'Irish Republic'. This heritage object, which formed the centrepiece of the exhibition and remained on permanent display at the museum, had flown over the GPO during the Rising.[119] *The Irish Independent* described its condition as 'bullet-torn', with the last letter of the word 'Republic' missing, having been 'half shot away'.[120] In a significant diplomatic gesture, this flag had been handed over by the Trustees of the Imperial War Museum in London, after a request for its return had been made in writing by the Irish government to the British Prime Minister, Harold Wilson. Speaking at the exhibition, the Taoiseach spoke of his wish that the returned flag would be 'preserved as one of the important relics of that important event of Irish history and as a source of inspiration for all

117 Ibid., p. 48.
118 *The Irish Times*, 13 April 1966.
119 Department of External Affairs, *Cuimhneachán 1916–1966*, p. 48.
120 *The Irish Independent*, 13 April 1966.

who come to this museum'.[121] Historic material belonging to Captain Henry de Courcy Wheeler of the British Army – who had coordinated the surrender of the rebels (with Nurse Elizabeth O'Farrell) – was also presented to the state at a reception held in the National Museum on 19 April. The 1916-related material was handed over by Wigstrom de Courcy Wheeler in the presence of other relatives, members of the 'Coiste Cuimhneacháin', government officials, and staff from the National Library and the National Museum.[122]

Besides the GPO, the other heritage site in Dublin city that was chosen as a principal venue for the Golden Jubilee commemorations on Easter Sunday was Kilmainham Gaol (Plate 4.5). As the so-called 'haunt of the ghosts of so many Irish patriots', the building had come to be seen as representing 'a history of modern Ireland in stone'.[123] It was an obvious choice of location for the holding of an official event, as it was the prison in which 14 of the Rising's leaders had been executed. Besides the 1916 rebels, a long line of other Irish patriots had also been imprisoned there – including Robert Emmet, the Young Irelanders, the Fenian leaders, O'Donovan Rossa, the Invincibles, and Michael Davitt.[124] Having completed his presidential duties outside the GPO on Easter Sunday, de Valera's motorcar, flanked by a Garda motorcycle escort, drove slowly past streets lined with people waving Tricolours and arrived at the main entrance of Kilmainham shortly before 3.00 pm. Given the fact that this was the spot where he had been condemned to die half a century earlier, its sight obviously brought back poignant memories. As the 'Presidential Salute' was sounded, de Valera (who also held the distinction of being Kilmainham's final republican prisoner in 1924) walked through a small doorway and proceeded directly into the Stonebreakers' Yard. One report of the event recounted the following: 'Following a roll of drums and the presenting of arms, the National Flag was lowered to half mast. Trumpeters sounded the "Last Post" and honours were rendered by the Special Guard.' De Valera then advanced, 'facing the plaque on the wall of the ... yard, which bears the names of the executed leaders'. As he placed an olive wreath beneath it, 'a roll of drums was sounded', honours were given by the No. 1 Army Band and the Tricolour was raised to full mast.[125]

---

[121] Speech by Seán Lemass, Taoiseach at the National Museum, 12 April 1966, NAI, Department of the Taoiseach, 97/6/163.

[122] Seán G. Ronan to Roinn an Taoisigh, 6 May 1966, NAI, Department of the Taoiseach, 99/1/189.

[123] Kilmainham Jail Restoration Society, *Ghosts of Kilmainham* (Kilmainham Jail Restoration Society, Dublin, 1963), p. 54.

[124] Department of External Affairs, *Cuimhneachán 1916–1966*, p. 29.

[125] Ibid.

Plate 4.5   President de Valera laying a wreath in the execution yard at
                   Kilmainham Gaol, Dublin on 10 April 1966
Source: National Library of Ireland, Independent Newspapers (Ireland) Collection, R 3993.

The Irish Press reported that 'the scene in the prison yard' during the ceremony
'was one of silent reverence broken only by the hard commands of the troops
and the sad, then exultant notes of the buglers'.[126]

After the ceremony ended, de Valera moved inside the prison building,
which *The Irish Independent* described as 'now half-derelict, half-restored, an
Irish Bastille that held within its grim walls rebels against English rule'.[127] Inside,
he visited the cell block where he had been imprisoned in 1916 and also opened

---

[126] *The Irish Press*, 11 April 1966.
[127] *The Irish Independent*, 11 April 1966.

a museum exhibition.[128] Among the items on display were weapons, items of equipment and plans of military actions from various moments in the struggle for independence. After mounting a rostrum and casting a glance around the freshly-painted block, *The Irish Press* noted that he smiled as he delivered an unscripted witty remark: 'I am not strange to this place. I have been here before, but it was not as bright then as it is now.'[129] However, the poignancy of the occasion also played heavily upon de Valera. *The Irish Times* reported that even though he 'spoke strongly', it was also apparent that he 'was at the same time evidently moved'.[130]

In thanking the museum workers and the Kilmainham Gaol Restoration Society, de Valera recalled the long line of Irish patriots who had been imprisoned there:

> I do not know of any finer shrine than this old dungeon fortress in which there has not been so much suffering and courage so that Ireland should be a nation not only free but worthy of its great past. This then is a hallowed place and I hope that tens of thousands of our people will come here through the years to visit it and draw inspiration from it ... we want this place preserved ... [so] that it will inspire our people and make them remember the great efforts that were made through the centuries to preserve this nation, and encourage them to exalt it among the nations of the earth as the men of 1916 wanted it.[131]

The preservation of the Rising's heritage also weighed heavily upon de Valera's mind. Fearing that various relics associated with the fighting in 1916 might disappear as time progressed, he ended his speech by expressing his desire that Kilmainham would over time become a repository for memorabilia associated with the Rising. Whilst acknowledging that those holding artefacts in private households would be reluctant to part with them, he urged them to donate them to Kilmainham, because 'here they will be preserved without danger, and will inspire a much larger number'. Anything from a walking stick to a piece of writing that once belonged to a person who had departed, he noted, had the potential to bring the public closer to their personality, thus linking people in the present to the past.[132] By the time that he delivered this appeal, some items had already been donated to the museum, but there was a feeling by all concerned that gaps remained and that more bits and pieces could be sourced for display.

---

[128] Department of External Affairs, *Cuimhneachán 1916–1966*, pp. 29, 31.

[129] *The Irish Press*, 11 April 1966.

[130] *The Irish Times*, 11 April 1966.

[131] Department of External Affairs, *Cuimhneachán 1916–1966*, p. 32.

[132] *The Irish Times*, 11 April 1966.

One of the most unanticipated items received prior to the opening of the exhibition was a surprise package sent from an anonymous donor in England, containing a manuscript copy (probably made by a British soldier) of the letter Pearse wrote to his mother within an hour or less of his execution.[133]

To a certain degree, the spectacle that surrounded de Valera's arrival and departure from Kilmainham on Easter Sunday was more akin to scenes reminiscent of the mobbing of a major celebrity musician or movie star by adoring fans. After his motorcade arrived at Kilmainham, *The Irish Press* reported that the awaiting crowd surged forward when the President stepped out of his car, 'cheering and shouting "Good Old Dev"'. The warm welcome was duly acknowledged by de Valera, while a Garda cordon sought to control the strain from the people who had flocked to catch a glimpse of him. When he reappeared later on, he was greeted by 'renewed cheering and this time the crowd surged forward' and 'completely surrounded the smiling President'. When he eventually got back into his car, he continued to delay his departure so as to enable the assembled television crews and press photographers record the scene for posterity. While all of this was going on, members of the Fianna Phadraig Pipe Band from Wythenshawe were positioned nearby, playing 'A Nation Once Again'.[134]

Even though de Valera's personality may have seemed a little clinical at times to observers, he could not but have felt the temptation to feel a little chuffed with the genuine warmth of the public's adulation when he arrived back home at Áras an Uachtaráin in the Phoenix Park. The following Friday, de Valera seemed to acknowledge this affection with a kind gesture during the unveiling of a commemorative plaque at Boland's Mill – the building that he took over during the Rising. Under a heavy downpour of rain, he said that he had been responsible as Commandant for the welfare of his men in 1916, and that he was again taking command by ordering that the ceremony be continued under shelter. The day was again an emotional one for the President, as it was the first time in 50 years that he had met up with Grimsby-based Captain Edo John Hitzen, formerly of the Lincolnshire Regiment, to whom he had surrendered in 1916. The two old

---

[133] P. F. MacLochlainn (ed.), *Last Words: Letters and Statements of the Leaders Executed After the Rising at Easter 1916* (Office of Public Works, Dublin, 2005), p. 34.

[134] *The Irish Press*, 11 April 1966. To mark the events at Kilmainham, RTÉ broadcast a half-hour television documentary (as part of its Discovery series) on the night of Easter Monday, entitled *The Ghosts of Kilmainham*. This was edited by Michael Stoffer and narrated by Aindreas Ó Gallchoir and Chris Curran. The programme was favourably reviewed by Ken Gray in *The Irish Times*, 14 April 1966, who commended the 'brilliant camera work ... heightened by well-plotted lighting effects', which recalled the story lying 'within the walls of the old jail'.

adversaries shook hands and talked fondly about the old times.[135] Although the Boland's Mill ceremony may have seemed rather trivial when compared to some of the more grandiose events (such as the military parade and *Aiséirí-réim na Cásca*), the meeting between de Valera and Hitzen served as another important milestone in the improvement of diplomatic relations between the Irish and British governments.

In addition to the unveiling of commemorative landscapes and the holding of exhibitions, the Golden Jubilee was marked by the production of several items of commemorative memorabilia by the print media, private entrepreneurs and government agencies. Television also played a major role in the celebrations. This was a decade, as Tony Judt has demonstrated, that saw the medium catch on almost everywhere in Europe, with small black-and-white sets becoming 'an affordable and increasingly essential item of domestic furniture in even the most modest household'. Although programming throughout the continent was often 'conventional' and 'stuffy', television still impacted greatly upon impressionable minds by putting 'national politics onto the domestic hearth'.[136] This was certainly the case with RTÉ's contribution to the Golden Jubilee. Established in 1962, the confirmation of traditional values was a hallmark of much of its output in the early years of its existence. Under the watchful eye of its Director General, Kevin McCourt, the RTÉ Authority proved to be quite rigid when fashioning the slant taken in the 1916-themed material it commissioned for the 50th anniversary. As John Bowman has shown, the 1916 Programmes Committee passed a resolution 'that the overall approach to programming should be "idealistic and emotional" rather than "interpretive and analytical"'.[137]

Given the assortment of tribute programmes due for broadcast on television and radio, an item in great public demand was the 8 April 1966 edition of the *RTV Guide*, which went on sale for six pence. This bumper 48-page edition outlined the listings for (and content of) the programmes and shows dealing with the Rising. The magazine also published a revealing interview with the Listowel-based creative writer, Bryan MacMahon, who recollected his boyhood memories of 1916. Furthermore, he discussed the scripts that he had written for radio and television productions. The playwright, Hugh Leonard, also contributed an article about the making of the much-publicised *Insurrection* – a four-hour documentary-style television drama about 1916 (broadcast in eight half-hour episodes), which was shot in black-and-white. In terms of content, *Insurrection* sought to portray the day-by-day events of Easter Week 1916 as they

---

[135] *The Irish Independent*, 16 April 1966.

[136] T. Judt, *Postwar: A History of Europe since 1945* (Pimlico, London, 2007), pp. 345–46.

[137] J. Bowman, *Window and Mirror. RTÉ Television: 1961–2011* (The Collins Press, Cork, 2011), p. 82.

might have been reported had there been a televised news programme running at the time. Taking six months to film, it was scripted by Leonard, produced by Louis Lentin and directed by Michael Garvey. Kevin B. Nowlan acted as a historical adviser to the project while explosives and firearms experts from the Irish Army also assisted. The reconstruction of the Rising featured archival footage, studio action and scenes filmed on the streets of Dublin. It required 50 speaking parts (with Pearse played by Eoin Ó Súilleabháin) and 200 extras (many from the Irish Army). Leonard described *Insurrection* as 'a memorial, fashioned by people who cannot paint or sculpt' and noted that its aim 'was to be a near-as-dammit, full-scale reconstruction of the Rising... [depicting] events as if television had existed in 1916'.[138]

The first half-hour episode of *Insurrection* was broadcast on RTÉ on Easter Sunday and the remaining episodes were screened on the seven succeeding nights. The series was repeated in its entirety on 1 May, for one time only. To impressionable young minds, this was heady stuff indeed, conveying a zealous message, as Mary Raftery later recalled, that there was 'no greater glory than to die for Ireland'.[139] As filming drama was a time-consuming and costly process during the 1960s, *Insurrection*, according to Brian Lynch, relied heavily on the creativity of the staff of RTÉ's Outside Broadcast Unit, whose shot-on-location 'realism' proved to be 'a cost-effective and innovative method using electronic cameras to record drama on location'. These skills had been well honed during the making of *The Riordans* and during the coverage of President John F. Kennedy's visit to Ireland (in 1963). The half-hour format of each episode of *Insurrection*, adds Lynch, 'allowed for formal interviews, breaking stories, vox pops, and "time-shifting" of events in a modern credible TV format', while analysis of the fighting was enhanced by the use of maps and 3-D models in the studio. The series was subsequently rebroadcast in its entirety on BBC Two in Britain and ABC in Australia, while shortened versions were screened on television in Belgium, Norway, Sweden, and Canada.[140]

Notwithstanding the innovation and flair evident in the production of *Insurrection*, contemporary television critics had mixed views about the series. Brian Devenney of *The Irish Independent* called the end product a 'screen epic' that was 'superbly done'. Leonard's script, he added, 'bore the imprint of

---

[138] *RTV Guide*, 8 April 1966. Besides the listings and articles, the magazine also contained reproductions of colour and black-and-white photographs of scenes from *Insurrection*, taken by the photographer, Roy Bedell.

[139] *The Irish Times*, 9 February 2006.

[140] B. Lynch, 'TV Eye: "Through the Eyes of 1916", RTÉ, 10–17 April 1966, *Insurrection*', *History Ireland*, Vol. 14, No. 2 (2006), pp. 54–55.

the master craftsman both in the economical use of dialogue and incident'.[141] Less than impressed, however, was Ken Gray of *The Irish Times*, who initially confessed to feeling disappointed after watching the first two of eight half-hour instalments. The screening of the first episode on Easter Sunday, he remarked, 'was certainly an anti-climax to the big build-up which had been given to the series', mainly because 'the attempt to make us believe that we are watching the events of Easter Week, 1916, on television programmes of that time just doesn't work'. A particular source of ire to Gray was the 'element of extravagant theatricality' and 'playacting' about the production, which he felt robbed it 'of its historical authenticity'. 'In small ways it irritates', he continued, pointing to the false moustache that was worn by actor Maurice Doherty 'to turn him into a news reader of fifty years ago'.[142] Having watched all eight instalments, however, Gray was more positive in his television review on 21 April. Whilst noting that he never felt comfortable over the eight days with the 'live' aspect of *Insurrection*, he did concede that the series 'provided some memorable moments' and 'some of the most exciting television that has ever come out of Montrose'.[143]

Gray was less sympathetic in his verdict on another RTÉ programme, *A Sword of Steel* – Padraic Fallon's evocation of Ireland's desire for nationhood. He felt that the script – 'an imaginative, poetic view of Ireland's struggle from darkness towards the light' – was difficult 'to pin down'. Although the programme contained footage of 'magnificent landscapes and sweeping panoramic views', he felt that the footage 'at times ... began to resemble a tourist board brochure or, when the camera dwelt lingering beside a clear-flowing stream, uncomfortably like an advertisement for filter-tip cigarettes'.[144] At the 1966 Jacob's Television Awards ceremony, *Insurrection* failed to receive any citations, which came as a big surprise to many. However, an award did go to another production by RTÉ's Outside Broadcast Unit, namely a dramatic televised portrayal of Séan O'Casey's *The Plough and the Stars*, produced by Lelia Doolan.[145] A series of commemorative productions were also broadcast on radio, including Patrick Pearse's play, *The Singer*, while Pearse also featured prominently in Bryan MacMahon's *The Voice of the Rising*.

Lesser-known writers also devoted their creative energies to honouring the 1916 rebels, by putting pen to paper and producing their own commemorative verse. 'The Nineteen-Sixteen Men', a poem written by Martin Neary and

[141] *The Irish Independent*, 16 April 1966.
[142] *The Irish Times*, 14 April 1966.
[143] Ibid., 21 April 1966.
[144] Ibid., 14 April 1966.
[145] Lynch, 'TV Eye', p. 57.

published in *The Connaught Telegraph* on 14 April 1966, evokes a nostalgic respect for those who had fought in the Rising:

> We've sung them down the passing years,
> We sing them now as then;
> We'll sing them through our smiles and tears –
> The nineteen-sixteen men.
> And when we are too old too [sic] sing,
> Our children's songs will rise;
> And mountain, valley, field shall ring
> In paeans to the skies.
> For they who faced an empire's might,
> Thought not of fame to come;
> They dared, because their cause was right –
> Though death was there for some.
> So let their story long be told
> By fireside, bog and glen;
> On his'try's pages wrote in gold –
> The nineteen-sixteen men.[146]

The notion of heritage as 'something transmitted' comes across strongly in Neary's poem, which conveys a strong sense of how memory of 1916 was passed down through the generations by means of two powerful agents – songs and storytelling.

More often than not, the spatial setting for the recital and retelling of the 1916 metanarrative was within the walls of people's domestic residences. As Michael Mays has noted in a study of the cultural territory of Irish nationalism, devotion to 'personifications of the nation' can illuminate 'much about the symbolic function of the nation' and its importance in people's lives in the past. Very often, he adds, nationalism brought the nation 'into the home', thus 'making national and local culture identical' through a process which domesticated the nation and nationalised its citizenry.[147] The Golden Jubilee was no exception in this regard and the awe in which the signatories of the Proclamation were held in many a nationalist homestead can again be seen from a commemorative poem that was written by a Sandymount resident, Thomas Whelan, and subsequently forwarded to the Department of the Taoiseach. In the first quatrain of his thoughtful elegy, Whelan makes reference to pictures of the seven signatories of

---

[146] *The Connaught Telegraph*, 14 April 1966.
[147] M. Mays, *Nation States: The Cultures of Irish Nationalism* (Lexington Books, Plymouth, 2007), pp. 22–23.

the Proclamation that were on display in his house. In the second, he highlights the historic significance of their 'gallant stand' with the 'helping hand' of comrades, while in the third and fourth quatrains, he declares that the rebel dead would be duly honoured and remembered throughout 1966, for the sacrifices they made in the fight for Irish freedom.[148] The poem encapsulates a stereotypical scene inside nationalist households during the Golden Jubilee, especially in those of Catholics. On the walls of lots of these homes it was not unusual to see pictures of the executed leaders and/or the text of the Proclamation hanging in prominent positions alongside images of the Sacred Heart or the late John F. Kennedy. In some respects, as Richard Killeen has observed, the Rising had a 'rather bohemian' element to it, and it was the 'radicals and exotics' that led to its elevation as a 'great mythical event' in which key personnel 'were revered in iconic images'.[149]

Even though the younger generation had their own fair share of contemporary idols – ranging from Irish showband stars like Brendan Bowyer and Dickie Rock to the emerging icons of American and British popular culture (including singer-songwriters such as Bob Dylan and rock 'n' roll acts such as The Beatles and The Rolling Stones) – the quintessential spirit of the 'Swinging Sixties' (as exemplified by the rebellious socio-political impulses of the Counter Culture phenomenon) was met with stiff resistance from the Catholic clergy and other advocates of more traditional pastimes in Ireland.[150] Although the number of citizens who had first-hand experience or recollections of 1916 was diminishing yearly, a conservative, Catholic and nationalist ethos still reigned throughout much of the Republic of Ireland in 1966, thus ensuring that Pearse and the executed leaders had an easy ride when it came to monopolising the markets that existed for mythologising rebellious spirits. The fact that Dermot O'Brien's cover version of 'The Merry Ploughboy' (a song by Dominic Behan about the Old IRA) was at Number One in the Irish Singles Chart in Ireland at Easter 1966, said it all really. As Brian Hanley and Scott Millar have noted: 'Popular nationalism was reflected in the pop charts.' The militaristic song, 'with its cheery if anachronistic chorus', ended up spending a total of six weeks

---

[148] Thomas Whelan to Seán Lemass, 25 February 1966, NAI, Department of the Taoiseach, 97/6/163.

[149] R. Killeen, *A Short History of the 1916 Rising* (Gill and Macmillan, Dublin, 2009), pp. 119–20.

[150] According to I. MacDonald, *The People's Revolution* (Pimlico, London, 2003), pp. viii–ix, rock music 'is already part of ... the history of the Western world'. During the 1960s, he notes that it 'was ... a new, half-invented art form and ... a receptacle for rebellious social impulses'. 'The energy level of popular music', he adds, 'was ... novelty driven and pitched against a conservative social background'.

at Number One.[151] Across the Irish Sea, it was music of a different kind that had hips swaying at the time. A cover version of 'The Sun Ain't Gonna Shine Anymore', sung by The Walker Brothers, commanded the Number One position in the Easter UK Singles Chart.

Unsurprisingly, given the stratospheric proportions to which the Irish rebels' popularity had soared to in the early months of 1966, the Golden Jubilee managed to attract interest from a wide range of entrepreneurs intent on cashing in on the retailing opportunities that the big occasion afforded. Money was to be made through the retailing of the Rising's heritages and storylines, and many savvy individuals and organisations (both official and non-official) profited, as a range of ephemera went on sale or was issued in limited editions. However, there was also a cultural element to some ventures. Gael Linn released two new records to mark the Golden Jubilee – an extended play disc featuring Seán Ó Riada's soundtrack for *An Tine Bheo* and a long-playing record called *1916*. These were launched at the Royal Hibernian Hotel in Dublin on 6 April.[152] Showbands also tapped into the nostalgia emanating from the commemoration. One intrepid group, The Monarchs, used the occasion to release an LP under Ember Records, containing 14 tracks – including spoken verses and 11 ballads (sung by Tommy Drennan). In the partisan words of the *Western People*, this recaptured 'all the heroic patriotism, the tragedy and the glory of the historic days of the Easter Rising'.[153]

The Golden Jubilee also resulted in the production of a range of mementoes that were either issued in gratuity to individuals deemed worthy of honouring (which later became must-have collector's items) or issued in bulk for sale to the general public. A special commemorative medal, for example, was issued to surviving veterans of the Rising, while the Company of Goldsmiths of Dublin arranged for the striking of a special '1916 Jubilee Mark' to be placed on items of silver and gold, besides watchcases and jewellery, that were manufactured in the year 1966.[154] In Dublin, Worboys Company Ltd., a jewellers based in Grafton Arcade, issued its own 1916–1966 commemorative gold medallion 'as a private commercial undertaking'. Due to the lack of suitable presses in Ireland, however, the medal had to be struck in London.[155] Adhesive stickers, in a circular shape, were also circulated throughout Ireland to mark the Golden Jubilee. These depicted the Sword of Light in yellow, set on a black background with a white

---

[151] B. Hanley and S. Millar, *The Lost Revolution: The Story of the Official IRA and the Workers' Party* (Penguin Ireland, Dublin, 2009), p. 51.

[152] *The Irish Independent*, 7 April 1966.

[153] *Western People*, 16 April 1966.

[154] Department of External Affairs, *Cuimhneachán 1916–1966*, pp. 61, 68.

[155] *Dáil Debates*, Vol. 220, 10 February 1966.

border.[156] Special greeting cards were also produced to mark the anniversary, featuring a small ribbon of green cloth on the cover, a colour drawing of the Tricolour on a gold flag pole and the words: '1916–1966 Commemoration Year'. When opened, the card depicted an image of Pearse.[157]

The government itself launched its own pieces of commemorative memorabilia, including a set of stamps (featuring portraits of the seven signatories of the Proclamation) and a commemorative badge.[158] These stamps were designed by Raymond Kyne of Signa Ltd. and the sculptor, Edward Delaney. When eight variations went on sale in post offices on 12 April, large crowds queued outside the GPO to buy them in four basic denominations (of 3d., 5d., 7d., and 1s. 5d.).[159] The design for the badge was chosen following a competition organised by the 'Coiste Cuimhneacháin'. It consisted of an innovative stylisation of the Sword of Light. This symbol was connected in early Irish literature with the arrival of the Gaels and occurred right through later literature as an emblem of instinctive knowledge, education and progress. The symbol was later taken by scholars of the Gaelic revival, and was subsequently adopted by Irish militants to symbolise their dual objective of an armed rebellion and a cultural renaissance. Knowledge of the sword and its significance became widespread in Ireland and abroad, after it was adopted by the Gaelic League as part of the title page of *An Claidheamh Soluis*. In addition to the badges, the Sword of Light symbol was also made available to industry for the commercial production of other items such as brooches and tie-pins.[160]

Two million 10s. silver coins were also minted by the government to commemorate the Rising. These souvenirs depicted the head of Pearse on one side and Sheppard's statue of Cúchulainn on the other. Demand was also heavy for a highly polished 30s. presentation coin. A total of 20,000 of these were minted and were sold in special cases.[161] The silver 10s. coin, which was designed by the British sculptor, Thomas Hugh Paget, was manufactured by the Royal Mint in London. When the prototype scheme was considered in a Seanad debate on 16 February, many senators expressed their support for the introduction of the coinage. Professor James Dooge of Fine Gael argued that it was fitting

---

[156] Two surviving examples of the 1916 Golden Jubilee Stickers can be found in Dublin City Library and Archive (hereafter DCLA), Birth of the Republic, BOR F01/09.

[157] 1916–1966 Commemoration Year Greeting Card, 1966, DCLA, Birth of the Republic, BOR F01/01.

[158] Department of External Affairs, *Cuimhneachán 1916–1966*, pp. 92–93.

[159] *The Irish Times*, 13 April 1966.

[160] Cuimhneachán 1916 Memorandum on Design for a Commemorative Badge, NAI, Department of the Taoiseach, 97/6/163.

[161] *The Irish Times*, 13 April 1966.

'as a part of the commemoration of 1916', as fiscal independence had represented 'a symbol of the nationhood which we celebrate'. A few alterations were suggested by a number of senators. Owen Lancelot Sheehy-Skeffington spoke in favour of replacing Cúchulainn's image with a portrait of Tom Clark, while Jack Fitzgerald suggested including the names of the Proclamation's seven signatories on the coin.[162] In the end, however, the government opted to proceed with the original design. Jack Lynch noted in the Dáil on 27 April that 444,000 of the two million 10s. coins had 'been issued to the public to date', while external demand had 'amounted to a further total of 36,000 coins'.[163]

In June, however, Michael O'Leary brought it to the Dáil's attention that the special 10s. coins were being used to pay social welfare recipients, who were subsequently 'encountering difficulties in their efforts to utilise this currency'. Whilst acknowledging that the coins were legal tender, he expressed concern that 'many shopkeepers will not accept it as such'. In reply, the Minister for Finance advised that 'there should be no reluctance on the part of shopkeepers in accepting them for payment of goods sold'. 'It should not be necessary', he stressed, 'to repeat this over and over again, though I admit instances have arisen in which they have been refused'. Employment exchanges, he added, would 'facilitate those affected'.[164] The Central Bank ran into its own difficulties with the special 10s. coins. Members of the public who had taken out subscriptions for proofs of the commemorative coinage encountered long delays in receiving them. When questioned about this matter in the Dáil by Fine Gael's Stephen Barrett on 7 July, Jim Gibbons explained that the hold-up was due to 'technical difficulties ... encountered in the minting process' and hinted that the delivery process would not be completed until a few more months had passed by.[165]

Coins were by no means the only commemorative items in mass circulation in 1966, as the Golden Jubilee also inspired the print media to produce impressive 1916-themed pieces for their readers. A 16-page commemorative supplement, entitled 'A Historical Review of the Men and the Politics of the Easter Rising', was published by *The Irish Times* on 7 April. This commenced with a vivid autobiographical account by Desmond Fitzgerald (who had died in 1947) of his experiences of the Rising, entitled 'Inside the GPO'. Some of Ireland's leading historians also contributed to the supplement, including: F. S. L. Lyons (on 'Decline and Fall of the Nationalist Party'), Nicholas Mansergh (on 'The Unionist Party and the Union 1886–1916'), Robert Dudley Edwards (on 'The Achievement'), and Owen Dudley Edwards (on 'American Aspects of

162 *Seanad Debates*, Vol. 60, 16 February 1966.
163 *Dáil Debates*, Vol. 222, 27 April 1966.
164 Ibid., Vol. 223, 8 June 1966.
165 Ibid., Vol. 223, 7 July 1966.

the Rising'). Although it was an unfashionable topic in Irish historiography at the time, the newspaper also took the innovative step of commissioning Henry Harris to pen an article on Irish soldiers in the Great War (entitled 'The Other Half Million'). Another revealing aspect of the supplement was an article entitled 'A Bolshevik Viewpoint'. This was written by a Soviet journalist, L. Sedin, at the request of the newspaper. The article, which was largely based on an interview that Sedin had conducted with 70-year-old Nikolay Bogdanov, offered a curious insight into the Russian Bolsheviks' reaction to the Rising. Bogdanow, who was a member of the Soviet Union's Communist Party, recalled how Lenin had once 'studied the experience of the Dublin anti-imperialist Rising in order to make its lessons accessible to future revolutions'.[166]

Conor Cruise O'Brien also contributed to the supplement. In an article entitled 'The Embers of Easter', he offered up a stark assessment of how far Ireland had progressed by 1966, claiming that it had not entirely measured up to the political and cultural objectives enshrined in the Proclamation. Whilst conceding that Pearse and Connolly would have been satisfied 'at the appearance of the Dubliners [in 1966]; healthier, better fed, better dressed, better housed', he contended that the Golden Jubilee represented 'the funeral ceremonies of the Republic proclaimed 50 years ago'. Pearse and Connolly's 'national objective', he argued, was 'finally ... buried' (as Partition had become more of a reality), while the former's 'cultural objective of ... a fully bilingual Ireland' was 'being tacitly abandoned'. At the outset of the article, Cruise O'Brien quoted Lenin, who had once written: 'The misfortune of the Irish is that they rose prematurely, when the European revolt of the proletariat had not yet matured.' Cruise O'Brien then put forward a counterfactual hypothesis, speculating what might have happened if the Rising had taken place at the time of the conscription crisis in April 1918, when there was 'a vastly better opportunity' because of British provocation. In such a scenario, he speculated that the Rising's 'potential international significance' could have been much greater.[167] For one of Cruise O'Brien's biographers, Anthony Jordan, the pessimistic tone of 'The Embers of Easter' represented 'a devastating critique' at a time when 'only praise and triumphalism' abounded, and also offered him a forum 'to settle a few old scores with people whom, he felt, had not measured up to the historical challenge'.[168]

---

[166] *The Irish Times*, 7 April 1966.

[167] Ibid.

[168] A. J. Jordan, *To Laugh or to Weep: A Biography of Conor Cruise O'Brien* (Blackwater Press, Dublin, 1994), pp. 84, 86. Contemporary reaction to the article can be found in *The Irish Times*, 20 April 1966, in which a letter-writer commended him 'on his courageous and realistic appraisal of the progress of Ireland, politically and ideologically, since 1916'. Further discussion can be found in D. Whelan, *Conor Cruise O'Brien: Violent Notions* (Irish Academic Press, Dublin,

A lengthier supplement, running to over 30 pages, was published by *The Irish Independent* on 11 April. This featured an image of Sheppard's Cúchulainn on the front cover and commenced with a reflective article by Richard Roche. He asked whether citizens had 'forgotten the dream for which these men [of 1916] died?' and went on to suggest that people 'could do worse ... than perform now a national examination of conscience'. In addition to a poem penned by Dominic Crilly, the supplement included a range of illustrated articles, including noteworthy contributors from Carysfort College's M. Ó Dubhghaill (on 'The Plan of the Rising') and University College Galway's Liam Ó Briain (on 'The Lessons of the Rising').[169] The Independent Newspapers group also produced an A2-sized commemorative poster, which was distributed from a stand at the RDS Spring Show 1966. This fold-out souvenir contained a reproduction of *The Irish Independent* from late April/early May 1916, beneath the headline 'Drama, Terror and Death from the Files of 1916'. When folded, it depicted images of the signatories of the Proclamation.[170]

In addition to the daily broadsheets, there was no shortage of Golden Jubilee material in other sections of the print media. The April issue of *The United Irishman*, produced by Republican Publications in Dublin, devoted itself entirely to the occasion by offering 'An Evaluation of Some of the Factors which went to Make Up the 1916 Rising'.[171] Throughout the regions, many local newspapers also marked the Golden Jubilee by producing their own special supplements. The *Limerick Leader*, for example, marked the anniversary by producing a supplement on 9 April. This featured articles on 'Casement's Last Voyage', 'Edward Daly: Soldier Without Fear' (by Séamus MacConmara), 'I Walked in the Shadows of Kilmainham Jail' (by Helen Buckley), 'Labour: Foes to British Tyranny' (by Jim Kemmy), and 'West Limerick Will Remember' (by Dan Mulcahy).[172] *The Connacht Tribune* sought to flag the significance of County Galway's role in the Rising by publishing, for the first time ever, Martin

---

2009), p. 52, who suggests that the article took 'its lead' from an article that Cruise O'Brien's cousin, Owen Lancelot Sheehy-Skeffington, had published in the same newspaper in October 1962, to mark de Valera's 80th birthday. Cruise O'Brien's article, however, was more 'personal' in tone and reflected 'the views of a member of the generation which grew up in the aftermath of 1916'. His overall intention, adds Whelan, was to characterise the 'enduring political effects' of the Rising 'as ones of deception, disappointment and malaise'.

[169] See 'Supplement for the Golden Jubilee of the Insurrection, 1916–1966', in *The Irish Independent*, 11 April 1966.

[170] 1916 Commemorative Stand, RDS Spring Show 1966, Poster, 1966, DCLA, Birth of the Republic, BOR F11/10.

[171] *The United Irishman*, April 1966. Among the topics covered were reflections on John Redmond's career and commentaries on the role of the GAA and the labour movement in 1916.

[172] *Limerick Leader*, 9 April 1966.

Dolan's personal recollections of 1916, entitled 'A Dramatic Story of a Week of Rebellion'. In four lengthy instalments throughout April, Dolan attempted to convey to readers the 'tremendous importance' of Galway's Rising in the wider context. Had there been no bloodshed in the west, he felt that the British 'would have been able to foist upon the world the canard that at best a labour struggle, at worst a riot, had taken place in Dublin and that the whole affair had no national significance'. Dolan also highlighted how many of the Galway rebels were 'farmers and sons of farmers'.[173] In County Mayo, an eight-page 'Easter Supplement' was published by the Ballina-based *Western People*. This featured articles entitled 'Insurrection 1916', 'The Rising and the West', 'MacBride the Magnificent Rebel', 'The Men of Westport Were Out in 1916', 'Jacob's was a Thorn in the Side of the British Military', and 'The Poets and the Rising'.[174]

In addition to the offerings from newspapers of various kinds, the Golden Jubilee was also marked by the production of a range of commemorative publications and prints by different publishing houses. Irish Art Publications, for example, produced a small pamphlet entitled *Prelude to Freedom 1916– 1966: 50th Anniversary Review*. Through the medium of 13 colour artworks with explanatory captions beneath, it aimed to 'show and tell' readers 'a little of that great final struggle' for independence.[175] A range of printed material was also produced by the Dublin city-based publisher, Graphic Publications, including a special Golden Jubilee colour calendar with original paintings and old photographs (measuring 15.5 x 9.5 inches), full-colour prints of Patrick Pearse in two sizes (10 x 7.5 inches and 15 x 10 inches) and colour postcards featuring Pearse's image in two sizes (4.25 x 6.25 inches and 9 x 6 inches). It also produced a special '1916–1966 Golden Jubilee Issue' of *The Easter Commemoration Digest*, which ran to over 200 pages. This featured many short articles of historical interest, including one by Lemass, who made special reference to 'a new and richer meaning' to Irish-American bonds over some 300 years. The commemoration of the Rising, he proudly added, 'is more than another inscription carved on the marble scroll of Ireland's roll of honour'.[176] Several advertisements were placed in the digest by members of Dublin's business

---

[173] *The Connacht Tribune*, 2 April 1966. The other instalments of Dolan's story appeared in the paper on 9 April (on the subject of hostilities at Clarenbridge and Oranmore), 16 April (on the occupation of Moyode and the march to Limepark) and 23 April (on the arrest and imprisonment of the rebels).

[174] *Western People*, 9 April 1966.

[175] Anon., *Prelude to Freedom 1916–1966: 50th Anniversary Review* (Irish Art Publications Ltd., Dublin, 1966), p. iii.

[176] S. Lemass, 'The Meaning of the Commemoration', *Comórú na Casca Digest: The Easter Commemoration Digest*, Vol. 8 (1966), p. 12.

community, with many of these linking the geographical location of certain business premises to nearby reminders of the struggle for independence. Walden Motor Company of Parnell Street, for example, placed an advertisement which featured a new Ford motorcar in front of the Charles Stewart Parnell monument, while the one placed by the Clerys department store cheekily featured two flags flying at the top of the GPO – the Tricolour and one featuring the store's logo. From a heritage tourism perspective, the most attention-grabbing of these advertisements was the one placed by Irish Tours and Services of Parnell Street, offering four-hour bus or car tours of Dublin, with 1916 veterans as guides and part proceeds going to the Kilmainham Gaol Restoration Society.

A range of commemorative material was also produced by government departments, the Defence Forces and religious organisations. The Office of Supplies, for example, produced a bilingual pamphlet entitled *Oidhreacht 1916–1966*. Aimed at the general reader and pupils of schools, this contained a short written and pictorial history of the Rising, along with a list of the rebel dead and a reproduction of the words of the National Anthem.[177] Special issues of a number of well-known publications also went on sale in booksellers throughout the country. A notable part of *The Capuchin Annual 1966*, for example, was a distinctive presentation feature on the Rising, which ran to a total of 248 pages and featured articles by the likes of Senator James Ryan (on 'General Post Office Area'), Pádraig Ó Ceallaigh (on 'Jacob's Factory Area'), Liam Ó Briain (on 'Saint Stephen's Green Area'), Mattie Nielan (on 'The Rising in Galway'), Liam Ruiséal (on 'The Position in Cork'), and Denis McCullough (on 'The Events in Belfast'). In his 'Reamhrá' (Irish for 'Editorial'), the Editor, Father Henry, outlined his high opinion of the 'brave deeds by brave men' in 1916. Additionally, he explained the composition of the publication as follows:

> It is with pride that we present in this year of jubilee an account of their courageous fight during 1916 to make Ireland a nation once again. We have sought men who manned the posts during Easter Week to relate their memories of the stirring encounters. Their articles may suffer a trifle because of the vagaries of human memory but they make up for this in being warm, storyteller accounts, companionable, flesh and blood ... We trust our feature will be widely read in the realisation that the freedom and independence we enjoy today were purchased in a gallant stand by fellow Irishmen of whom we are proud.[178]

---

[177] Oifig an tSoláthair, *Oidhreacht 1916–1966* (Oifig an tSoláthair, Baile Átha Cliath, 1966).

[178] Father Henry, 'Reamhrá', *The Capuchin Annual 1966*, No. 33 (1966), p. 152.

The price of 27s. 6d., however, meant that purchase of the souvenir was beyond the means of many working class individuals. A more affordable option appeared in shops in May, when the Defence Forces published a commemorative volume of *An Cosantóir: The Irish Defence Journal*. This retailed at two shillings and contained four articles relating to the Rising, written by Colonel Eoghan O'Neill, Cft. J. P. Duggan, Diarmuid Lynch, and T. C. Ó hUid. Colonel O'Neill's article, which was entitled 'The Battle of Dublin 1916', sought to address what he identified as the historiographical 'deficiency' created by the lack of a 'proper military study' of the fighting in the capital during Easter Week.[179]

The Golden Jubilee also generated a number of books on the theme of 1916 by academic historians. To a certain extent, however, their capacity to bring forward fresh perspectives on the Rising was somewhat curtailed by the strict restrictions placed by both the Irish and British governments on access to a range of relevant records. Having been asked in the Dáil on 1 February 1966 about the government's intentions with regard to these, Lemass argued that it was in 'the public interest' to maintain 70-year restrictions on material in the State Paper Office, as much of it dealt 'with persons still alive or with those whose immediate relatives are alive'. He did, however, state that the Department of the Taoiseach would 'continue to offer to supply, where possible, information by way of answers to specific questions about the 1916 Rising'. Furthermore, he sought to assure '*bona fide* historians' that the government would contemplate permitting a limited amount of access 'to some ... papers ... but nothing beyond that'. When asked to define what he meant by a '*bona fide* historian', Lemass bluntly replied: 'A *bona fide* historian is identifiable.'[180]

During the course of 1966, historians frequently aired their views about the Rising in public, at a range of academic conferences and organised debates. Over the course of four evenings in February and March, a series of lectures were given to big audiences at University College Dublin, probing that institution's connections to 1916. Later in the year, these were published as a collection of six essays (five of which were written by participants in the fighting – Michael Hayes, Liam Ó Briain, James Ryan, Louise Gavan Duffy, and Joseph A. Sweeney), edited by Father F. X. Martin.[181] A 1916-themed conference, entitled 'Pearse said to Connolly', was held in April in the Wellington Park Hotel, Belfast by the National Democratic Group of Queen's University Belfast. This featured talks by a number of prominent historians, including Donal McCartney of Queens

---

[179] E. O'Neill, 'The Battle of Dublin 1916: A Military Evaluation of Easter Week', *An Cosantóir: The Irish Defence Journal*, Vol. 26, No. 5 (1966), p. 211.

[180] *Dáil Debates*, Vol. 220, 1 February 1966.

[181] F. X. Martin (ed.), *The Easter Rising, 1916 and University College, Dublin* (Browne and Nolan Ltd., Dublin, 1966).

University Belfast, who spoke on 'From Parnell to Pearse' and Art Cosgrove of University College Dublin, who spoke on 'James Connolly and 1916'.[182] In the end, some of the most significant history books that were inspired by the Golden Jubilee anniversary did not actually appear in print until after 1966 had passed. For example, Fr. Martin's edited collection, *Leaders and Men of the Easter Rising: Dublin 1916*, was published in 1967.[183] This was followed the year afterwards by *1916: The Easter Rising*. Edited by Owen Dudley Edwards and Fergus Pyle, the volume had its origins in the commemorative supplement that *The Irish Times* published on 7 April 1966.[184] Another edited book to emerge out of the Golden Jubilee commemoration was Kevin B. Nowlan's *The Making of 1916: Studies in the History of the Rising*. Commissioned by the 'Coiste Cuimhneacháin', this was not published until 1969.[185]

Special issues of academic journals were also brought out in 1966. The spring 1966 issue of *Studies*, which hit bookshops after the Easter commemoration had passed, featured a forthright endorsement of the Rising's legacy by the 40-year-old Fine Gael senator, Garret Fitzgerald (son of Desmond), entitled 'The Significance of 1916'. The actions of the rebels, he argued, proved to be 'a political event of enormous emotional power ... planned by men who feared that without a dramatic gesture of this kind the sense of national identity ... would flicker out ignominiously within their lifetime'. The alternatives that faced the rebels, he added, 'were national extinction or, by a supreme effort on their part, the possibility of another lease of life for Irish nationality, out of which a free Ireland might somehow, some day, emerge'. Delving further into the realm of counterfactual history, Fitzgerald also speculated what might have happened had Home Rule been granted instead. One possibility, he argued, was that without 'the national revival of 1916–1921', the country would not have enjoyed such a strong voice in the United Nations. From an economometric perspective, he also argued that the country might have 'shrunk like Northern Ireland into dependent provincialism, too concerned about its share of British agricultural subsidies and social welfare re-insurance provisions'.[186] The journal

---

[182]   *The Irish Independent*, 25 April 1966.

[183]   F. X. Martin (ed.), *Leaders and Men of the Easter Rising: Dublin 1916* (Methuen & Co. Ltd., London, 1967).

[184]   O. Dudley Edwards and F. Pyle (eds), *1916: The Easter Rising* (MacGibbon & Kee, London, 1968).

[185]   See K. B. Nowlan (ed.), *The Making of 1916: Studies in the History of the Rising* (The Stationery Office, Dublin, 1969), which contains contributions from historians such as F. S. L. Lyons, Robert Dudley Edwards and G. A. Hayes-McCoy.

[186]   G. Fitzgerald, 'The Significance of 1916', *Studies: An Irish Quarterly Review*, Vol. 55, No. 217 (1966), pp. 29, 35. Six years later, a more condensed version of the case for 1916 was

also featured worthwhile commentaries on 1916 by the likes of David Thornley, C. P. Curran and F. X. Martin. As already seen, Lemass also contributed an article. The journal also contained a review by Father Francis Shaw of two 1916-related books, by Leon Ó Broin and James Stephens. Although a draft of a lengthy article by Shaw himself was rejected by the journal's editor, he was not the only disappointed person at the time. The poet, John Montague, also failed to have one of his works published in the special issue. Years afterwards, he recalled that the poem in question, entitled 'Patriotic Suite', had 'disavowed' the Rising and had sought to offer up 'an answer, an antidote to the aisling of nationalism'. The Editor, he added, had refused to return the manuscript to him, having 'declared that it was "putrid"'.[187]

Regional Commemorations: A Geographical Sketch

Besides those that took place in Dublin, ceremonies sponsored by the 'Coiste Cuimhneacháin' were also held on Easter Sunday 1966 at locations such as Cloughjordan, Cork, Dundalk, Enniscorthy, Galway, Kiltyclogher, Limerick, Monaghan, Sligo, Tralee, Waterford, and Westport.[188] A parade led by 50 veterans of the Rising was held in Enniscorthy, County Wexford (and was attended by the leader of the Labour party), while a special '1916 Room' containing mementos of the Rising was opened in Enniscorthy Castle Museum.[189] In County Kildare, a number of public representatives and Old IRA veterans staged a walkout from St. Corban's Cemetery in Naas after a speaker from Dublin, Christopher Dolan, began to deliver an unscheduled oration in which he complained about censorship. After this was cut short, the Proclamation was then read and the 'Last Post' was sounded at the graveside of Volunteer Sullivan – who for nearly two decades had sounded the 'Last Post' each year outside Pentonville Prison on the anniversary of the death of Sir Roger Casement.[190] In County Westmeath, a parade featuring an element of reenactment was held in the midlands town of Athlone, where a guard of honour was formed by members of the FCA at the Old IRA memorial – flanked by schoolchildren dressed in uniforms of the Irish Citizen Army and Cumann na mBan. A wreath was then laid at the memorial by Séamus O'Meara, one of the organisers. Afterwards, the Proclamation was

---

put forward by the same author in a book dealing with the history and politics of Partition. See G. Fitzgerald, *Towards a New Ireland* (Charles Knight & Co. Ltd., London, 1972), pp. 9–12.

[187] J. Montague, 'Living for Ireland', in Bolger (ed.), *Letters from the New Island*, pp. 17–18.

[188] Department of External Affairs, *Cuimhneachán 1916–1966*, p. 11, 35.

[189] Ibid., p. 37.

[190] *The Irish Press*, 11 April 1966.

read and the Tricolour was ceremoniously hoisted in Custume Barracks at 12.00 noon.[191]

Another Easter Sunday parade was held at Portlaoise in County Laois. A guard of honour was formed by members of the Old IRA at the saluting base at the post office, where the salute was taken by a survivor of the Rising, Mr. L. Brady. In Longford over 400 people participated in a parade from Connolly Barracks to St. Mel's Cathedral, while at St. Maul's Cemetery in Kilkenny, a wreath was laid on the grave of the republican, James Morrissey.[192] A week later, the opening of a tourist information office in the restored medieval merchant's building, Rothe House, was marked by a ceremony attended by members of the Kilkenny Archaeological Society and Thomas MacDonagh's son, Donagh. Dr. T. J. O'Driscoll, the Director General of Bord Fáilte (the Irish tourist board), remarked that the restoration of the facility was done to enhance the community spirit: 'In the national stocktaking, which naturally accompanies the anniversary year of the Easter Rising, the restoration and reopening of Rothe House is a notable item on the credit side.'[193]

Throughout the province of Munster, many individuals and organisations participated in the commemorations. In County Kerry, a parade was held on Easter Sunday to the steps of Ashe Memorial Hall in Tralee, which was followed by a parade to Rath Cemetery.[194] Two days beforehand, an official ceremony was held a few miles away at a foggy Banna Strand on Good Friday, almost 50 years after Casement had tried to land arms from Germany at the same spot. Over 1,000 people gathered on the sand dunes for the ceremony, to pay tribute to Casement's memory. They applauded when Florence Monteith-Lynch, from New York, turned the first sod on the site of a memorial in honour of Casement and her own father, Captain Robert Monteith (who also took part in the historic journey from Germany to Ireland in advance of the Rising). Among the guests on the day of tribute were Captain Raimund Weisbach of the U-19 submarine and survivors of the Aud.[195] Before the ceremony commenced, the Corvette Macha vessel sailed into the bay, but fog prevented Air Corps jets from dipping their wings in salute over the memorial site, which had been donated by local farmer Eamon Stack.[196] Donogh O'Malley, the Minister for Health, was present at the event and said that he felt privileged 'to represent the government of Ireland at this ceremony, which is a very fitting prologue to the efforts of our

---

[191] *The Irish Independent*, 11 April 1966.
[192] *The Irish Times*, 11 April 1966.
[193] *The Irish Independent*, 18 April 1966.
[194] *The Irish Times*, 11 April 1966.
[195] Department of External Affairs, *Cuimhneachán 1916–1966*, p. 23.
[196] *The Irish Independent*, 8 and 9 April 1966.

nation to commemorate the 50th anniversary of a most glorious epoch in our struggle to achieve independent nationhood'.[197]

One notable absentee from the proceedings at Banna Strand was 86-year-old Joseph Zerhusen, who had spent months in Germany with Casement and Monteith. The Tánaiste and Minister for External Affairs was of the opinion that the historic Irish-German connection was a delicate matter, and that elements of it could unsettle contemporary Anglo-Irish relations, thus proving potentially 'embarrassing both for the Germans and ourselves'.[198] There were genuine grounds behind Aiken's concerns, for relations between Ireland and England had warmed 'with dramatic speed' following the election victory of the Labour Party at the British General Election in 1964.[199] This warming of diplomatic relations was certainly a welcome development for the governments of each country, as much political acrimony had been caused in the past as a result of episodes such as Irish neutrality during World War II, the economic war of the 1930s and the death toll that had resulted from the tit-for-tat killings during the most brutal phase of the War of Independence. The 'terrible waste of many lives' in the latter event, as Michael Hopkinson has remarked, managed to trigger off 'an appalling long-term embitterment in Anglo-Irish relations'.[200] Symptomatic of the new and improved Anglo-Irish relationship in the mid-1960s, however, was the decision of Labour's populist Prime Minister, Harold Wilson, following a meeting with Aiken, to allow the repatriation of Casement's remains to Ireland in the year before the Golden Jubilee.[201]

The Irish government was fortunate to encounter such an accommodating British politician as Wilson – who rather coincidentally was born in 1916. In addition to his solid 'credentials as an ordinary chap' with a self-professed fondness for HP sauce and *Coronation Street*, it has been observed that he overlooked 'no opportunity to pose on the world stage' and often displayed a tendency to seek 'solace from his domestic woes in foreign affairs'.[202] Another

---

[197] *The Irish Times*, 8 and 9 April 1966.

[198] Seán G. Ronan to An tÁrd Chonsal, Hamburg, 23 March 1966, NAI, Department of the Taoiseach, 97/6/163.

[199] Horgan, *Seán Lemass*, p. 205.

[200] M. Hopkinson, *The Irish War of Independence* (Gill and Macmillan, Dublin, 2004), p. xix.

[201] For further details on this matter, see A. Mitchell, *Casement* (Haus Publishing, London, 2003), p. 6, who notes that the Aiken-Wilson agreement was brokered at 10 Downing Street, hours after Winston Churchill's funeral. Weeks later, a representative of the Irish government witnessed the exhumation of Casement's remains from an unmarked grave in Pentonville Prison. Casement was then laid to rest in Glasnevin Cemetery on 1 March 1965, with de Valera paying a graveside tribute, in what was the Republic's first televised state funeral.

[202] Anon., 'Wilson, Harold (Baron Wilson of Rievaulx) (1916–95)', in J. Gardiner (ed.), *The History Today Who's Who in British History* (Collins & Brown Ltd., London, 2000), pp. 848–49.

bridge-building gesture made by the British, as seen earlier, was the Imperial War Museum's enlightened decision to return the 'Irish Republic' flag to the National Museum. Captain Hitzen's meeting with de Valera at Boland's Mill was of diplomatic significance too, as it was seen as a friendly reconciliation between former foes. The actual decision to allow for the repatriation of Casement's remains came as no bolt from the blue, but represented the culmination of years of intense lobbying of both governments. For many years, organisations such as the Roger Casement Committee (which was run by the likes of Frank Dunne and Kathleen Lysaght during the 1930s), had been active in perpetuating the rebel's memory at centres of Irish population around London, including Finsbury and Camden. Each year, one of its commemorative functions was to assemble people together to say prayers outside Pentonville Jail near the anniversary of Casement's death.[203] In arguing the rationale for Casement's repatriation to Irish soil, the organisation once contended that in all free countries, 'the bones of her patriots are enshrined in the soil of the land as their memory is in the hearts of the people.'[204]

Thus at Banna Strand on Good Friday 1966, it seemed to be in the best interests of the civil servants attached to the Department of External Affairs not to allow a local historical commemoration to detract from the goodwill that had been generated by Wilson's diplomacy. As the ceremony proceeded, though, Dr. Herbert Mackey exerted pressure on the British to order an investigation into the question of whether the homosexual content within Casement's so-called 'Black Diaries' was authentic or not. These diaries, as Diarmuid Ferriter has noted, had 'left behind a revealing gay geography of the many places where he spent time', but by the mid-1960s the government of Ireland 'was happy for the diaries to remain in London'.[205] As the Golden Jubilee was commemorated, however, it was apparent that Mackey and others still wanted answers to questions about the diaries, which were held at the Public Record Office at Kew (and not released for public inspection until 1994). Although the credibility of the argument in favour of the diaries being forgeries was shattered by forensic tests carried out many years after 1966, the debate over Casement's sexuality still offers a particularly illuminating window into the social mores that existed at the time of the Golden Jubilee. It demonstrates how many traditional nationalists,

[203] Minute Book of the Roger Casement Committee, London, 1935–1936, National Library of Ireland (hereafter NLI), MS. 9517.

[204] A Reasoned Statement Respecting a Nation's Desire to Honour a Patriot, by G. Allighan, Roger Casement Committee, London, Undated, Allen Library, Box 184/Casement Box 2, Folder 1, Monteith/Casement Papers.

[205] D. Ferriter, *Occasions of Sin: Sex and Society in Modern Ireland* (Profile Books, London, 2009), pp. 67, 224.

especially conservative Catholics, still held dear to the belief in 1966 that the candid accounts of promiscuous gay sex outlined in the diaries had been forged by British agents, so as to discredit Casement in the eyes of those who had been campaigning for his reprieve in 1916, including church figures and George Bernard Shaw.[206]

In addition to the delicate matter of Casement's memory, yet another issue that played on the minds of Irish and British diplomats during the early months of 1966 was the question of gaining accessibility to the British government's secret court martial records of the trials of the 1916 leaders. As Gerard O'Brien has explained, 'an apparently spurious version of an alleged statement by Thomas MacDonagh' had been circulating for many years. By the end of March, the son of the executed rebel leader, Donagh, publicly declared the 'speech' to be the work of a forger and appealed to Lemass for help in obtaining access to the original records in London. However, at the end of the following month, Breandán MacGiolla Choille of the Irish Public Record Office subsequently warned the Department of the Taoiseach about 'the inadvisability of alienating the British', out of a fear that the return of such documents to Ireland had the potential to embarrass either country.[207]

No such diplomatic delicacies impinged upon the Golden Jubilee ceremonies in Cork city on Easter Sunday, where 34 members of the Cork 1916 Men's Association gathered at St. Francis Hall on Sheares Street. The men were survivors of the 217 Volunteers who had mobilised at the exact same location in 1916, before heading westwards to Macroom in preparation for hostilities that never materialised. For the purpose of celebrating the Golden Jubilee, they were conveyed in cars to the city's northside, where they attended a Mass at

---

[206] For further commentary on this matter, see R. Sawyer (ed.), *Roger Casement's Diaries 1910: The Black and the White* (Pimlico, London, 1997), p. vii, who adds that a belief had persisted for many years that the erotic detail of the diaries had been forged 'mainly to destroy his reputation as a national hero in Ireland and also in the United States'. For a summary of the forensic investigation that was later carried in order to determine whether the diaries were authentic or not, see P. Bower, 'Appendix. Paper History and Analysis as a Research Procedure', in M. Daly (ed.), *Roger Casement in Irish and World History* (Royal Irish Academy, Dublin, 2005), p. 251, who determined that arguments for the diaries being forgeries rested 'on conjecture and circumstantial evidence together with some very specious logic'. Such reasoning, he concluded, was flawed in the sense that it took 'no account of the actual physical evidence in the diaries themselves'. For an extended commentary on how the idea initially took hold that the diaries were bogus, and how arguments on behalf of their authenticity were eventually reached, see W. J. McCormack, *Roger Casement in Death or Haunting the Free State* (University College Dublin Press, Dublin, 2002), pp. 26–43, in which the author deconstructs the credibility of W. J. Maloney's *The Forged Casement Diaries*, which set out the forgery theory in 1936.

[207] O'Brien, *Irish Governments*, p. 100.

St. Michael's Church in Collins Barracks. This was celebrated by Reverend G. Keohane and presided over by the Bishop of Cork and Ross, Cornelius Lucey.[208] The Mass was also attended by Frank Aiken, along with the Lord Mayor of Cork, Con Desmond. It was followed by a public ceremony and parade at Barrack Square, where 'awareness of ... history could be seen in the eyes of the old and in the bearing of the young'.[209] The commemoration included the raising of the Tricolour, the firing of a 21-gun salute, and the reading of the Proclamation by Captain Patrick Griffin. To add to the pageantry, the platform for the occasion consisted of a replica of the façade of the GPO, built by the Army Corps of Engineers. Wreaths were later laid on Thomas Kent's grave in the detention barracks by Aiken and by the Old IRA.[210]

On Easter Monday, a 1916 exhibition was opened at the Cork School of Art by the city's Lord Mayor. Other invited dignitaries included the city and county librarians and the two daughters of Tomás MacCurtain. Included in the exhibition were ephemera such as books, poems, paintings, sculpture, and manuscripts which reflected the artistic endeavour and ideals of the 1916 rebels.[211] In the tranquil surrounds of the Aula Maxima at University College Cork, a month-long exhibition was run on the literature of 1916. This included original letters written by Thomas Clarke's widow, Kathleen Clarke and Denis McCullagh, who was President of the IRB's Supreme Council in 1916.[212] The exhibition was opened by the 69-year-old President of the college, John J. McHenry – a distinguished experimental physicist who had a reputation as 'a man of considerable presence' and who was known to be 'dignified and courteous (if a little glacial), with considerable *gravitas*'.[213]

The biggest commemorative event in Cork city was the Special Day of Honour, held on Sunday 24 April. A crowd of 30,000 packed into the area around Daunt's Square, to watch a parade by the Army, Old IRA veterans, students, sporting bodies, and other local organisations. This was followed by an open air Mass in Irish that was celebrated by the Very Reverend Daniel Canon Connolly. Following the reading of the Proclamation in Irish by Dan Donovan and English by Michael McAuliffe, the Tricolour was hoisted to full mast over

---

208   *The Irish Press*, 11 April 1966.
209   *The Irish Independent*, 11 April 1966.
210   *The Irish Times*, 11 April 1966.
211   *The Irish Press*, 12 April 1966.
212   *The Irish Times*, 11 April 1966.
213   J. A. Murphy, *The College: A History of Queen's/University College Cork* (Cork University Press, Cork, 1995), p. 319. In his capacity as Vice-Chancellor of the National University of Ireland, McHenry also played a role in the Dublin Castle conferring ceremony on 14 April, by introducing Kathleen Clarke in Irish and English.

the altar, while buglers sounded the 'Reveille'. Those present included relatives of the 1916 rebels, including the nephew of Sean Hurley – the only Corkman killed in the Rising. Also present were the Lord Mayor, city councillors, the President of University College Cork, and members of both Houses of the Oireachtas, including Jack Lynch. Later that evening a special GAA parade of Old IRA veterans and schoolboys went to the city's Athletic Grounds to watch a special inter-county commemorative match between the Cork and Tipperary teams. The Chairman of the Munster Council, Séamus O'Riain, delivered an oration, while the Proclamation was read in Irish and English by the Vice-President of the Cork County Board, Séamus Ó Sé.[214]

Despite the abundance of Golden Jubilee-inspired nostalgia in the city, some elements of Cork's revolutionary heritage faded away in people's minds and memories, much to the regret of local old timers. *The Irish Press* reported the curious incident of the opening of 'a little ultra-modern dress shop' called Two Bare Feet at St. Augustine's Street in the city centre (formerly 13 Brunswick Street), which had been known as Wallace during the first six decades of the twentieth century. During 1916 and up until 14 May 1921, the building had secretly served as a nationalist intelligence headquarters, primarily of the Cork Brigade of the Old IRA, in addition to its function as a retailer of papers, cards and periodicals. Following the retirement of original owners Sheila and Nora Wallace around 1960, the premises then functioned as a betting office. In 1966, however, it was reopened as the Two Bare Feet clothes shop by fashion designer Florence Woods. While the new shop still retained an old Gothic window, the golden Celtic motifs over its door were painted over during renovations. 'Today', remarked *The Irish Press*, 'where leaders in the fight for freedom stood, the young girls of Cork are looking through the racks and fitting on hipster skirts, and "pop" art dresses and jewellery'.[215]

Outside of the city, around 4,000 people gathered at Castlelyons on Easter Sunday to pay tribute to the Kent brothers of Bawnard. For the most part, the day of commemorations in County Cork passed off without controversy, with the exception of an incident in the town of Midleton. Following morning Mass in the Church of the Most Holy Rosary, a melee developed after a large number of Gardaí wielded their batons and attempted to stop up to 20 young men selling Easter lilies outside the church. Three of the men were injured along with a Garda Sergeant.[216] Later on in the summer, other eras of the revolutionary past were commemorated in the county. The occasion of the Golden Jubilee spurred

---

[214] *The Irish Independent*, 25 April 1966.

[215] *The Irish Press*, 12 April 1966.

[216] *The Irish Times*, 11 April 1966.

the County Cork Old IRA Benevolent Association, chaired by Commandant-General Tom Barry, into raising funds to erect a memorial to three volunteers who died in the Kilmichael ambush in 1920 during the course of the War of Independence. Costing in the region of £2,000 and incorporating a Celtic design, the memorial was finally unveiled in July.[217]

Elsewhere in Munster, the Golden Jubilee was commemorated on Easter Sunday with a parade to the 1916 memorial at Sarsfield Bridge in Limerick city. This bronze and stone monument (which was designed during the 1930s by the sculptor, Albert Power) linked the Fenian movement with that of 1916, and featured Erin as a female form rising from bonds with an uplifted arm, and a broken shackle at the wrist.[218] The ceremony began with a prayer call and the reading of the Proclamation. Following the raising of the Tricolour and the firing of a 21-gun salute, it concluded with the playing of the National Anthem.[219] A second city parade, which included four bands and members of Limerick Corporation, proceeded to the republican plot at Mount St. Lawrence Cemetery, where a decade of the Rosary was recited in Irish and 12 wreaths were laid.[220] Another aspect to the Easter Sunday commemorations in Limerick city involved the mixing of politics and sport through the staging of the final of the rearranged 1965 Munster Senior Hurling Club Championship between Cork city's Glen Rovers and Mount Sion of Waterford. The initial match, which took place in December 1965, had been abandoned in controversial circumstances due to a fracas between 20 supporters who invaded the pitch following an altercation between two players towards the end of play. The rescheduling of the match by the GAA for Easter Sunday 1966, according to Tim Horgan, 'set many minds thinking of the seminal events that occurred in Dublin 50 years earlier'.[221]

In another commemoration in County Limerick, the Proclamation was read on Easter Sunday at the Fenian monument at Kilmallock and a wreath was then

[217] Minute Book of the Kilmichael and Crossbarry Memorial Sub-Committee of the County Cork Old IRA Benevolent Association, 30 June 1966, CCCA, MS. U.342.

[218] Correspondence and Other Papers Relating to the Limerick 1916 Memorial, 1936–40, NLI, MS. 11126, Diarmuid Lynch Papers.

[219] Department of External Affairs, *Cuimhneachán 1916–1966*, p. 38.

[220] *Limerick Leader*, 16 April 1966.

[221] T. Horgan, *Christy Ring: Hurling's Greatest* (The Collins Press, Cork, 2008), p. 304. The Glen Rovers supporters had much to celebrate when their club – which also happened to be celebrating the Golden Jubilee of its own foundation – won the replayed match by 3–7 to 1–7. Their triumph, as Paddy Downey reported in *The Irish Times*, 11 April 1966, was 'a sentimental journey' inspired by the 'magic' of the 46-year-old veteran hurler, Christy Ring (born only four years after the Rising), who reeled back the years by scoring a goal and a point. Ring's performance, added Downey, enchanted 'every soul in the 9,000 attendance', who were particularly taken by 'the whole gamut of his enormous skill; the dazzling stick-work, the lightning stroke, the dainty pass'.

laid by the President of the Kilmallock National Graves Association. In Bruff, the Proclamation was read at the Seán Wall memorial by James Moloney of the Old IRA.[222] In County Clare, the 'Coiste Cuimhneacháin' held an Easter Sunday Mass in Ennis, which was followed by a parade of 1,000 people to Drumcliffe Cemetery, where an oration was delivered at the republican plot by Commandant General Tom Maguire.[223] Day-long ceremonies were held in Cloughjordan, County Tipperary throughout Easter Sunday in honour of Thomas MacDonagh. These included the opening of a memorial park and library, and the planting of a rose tree.[224] An Easter Sunday parade of almost 2,000 people was held on the streets of Waterford city, where the salute was taken by the Minister for Education, who had earlier attended a Solemn High Mass in Waterford Cathedral celebrated by Dr. Michael Russell, the Bishop of Waterford and Lismore.[225] On 22 April, a Séamus Murphy-designed bust of Thomas Ashe was unveiled in the grounds of De La Salle College in the city – where Ashe had served as a trainee teacher. This was unveiled by the Minister for Defence, Michael Hilliard, who spoke about Ashe's role in the skirmish at Ashbourne in 1916. Two silver Thomas Ashe medals were also presented to the winners of an essay competition run by the college.[226]

In the west of Ireland, the Easter Sunday commemorations included the unveiling of a memorial dedicated to those 'who died for Irish freedom', consisting of a four-foot high Connemara granite pillar, in the grounds of the Dún Uí Mhaoilíosa military barracks at Renmore in Galway city.[227] A large parade through the city was then held in the afternoon, led by the 6th Infantry Pipe Band. As it neared its destination, around 40 survivors of the force that had mustered for the Rising in Galway formed a guard of honour at the Liam Mellows statue in Eyre Square (Plate 4.6). The Proclamation was then read in both Irish and English by 43-year-old Siobhán McKenna, an actress who grew up in the Shantalla area of the city and who, in the words of her biographer, 'won international fame without ever compromising her quintessential Irishness'.[228] During the ceremony, McKenna also recited the poem 'The Wayfarer', which Patrick Pearse had composed the night before his execution.

---

[222] *The Irish Times*, 11 April 1966.

[223] Ibid.

[224] Department of External Affairs, *Cuimhneachán 1916–1966*, p. 35.

[225] *The Irish Times*, 11 April 1966.

[226] *The Irish Independent*, 23 April 1966.

[227] *The Connacht Tribune*, 16 April 1966.

[228] M. Ó hAodha, *Siobhán: A Memoir of an Actress* (Brandon, Dingle, 1994), p. 10.

Plate 4.6   The Liam Mellows statue in Eyre Square, Galway
Source: Author.

A wreath was laid at the memorial by the city's Lord Mayor, Brendan Holland, and the ceremonies concluded with the playing of the National Anthem.[229] *The Connacht Tribune* reported that the ceremony in Eyre Square proved to be a 'dignified but stirring ceremonial'. McKenna herself spoke of her hope that 'the sun would continue to shine on Ireland spiritually and that the occasion would be celebrated with joy, peace and forgiveness'. The oration from the platform was delivered in Irish by the Right Reverend Monsignor, Thomas Fahy, a former associate of Mellows. An oration in English was then delivered by Mattie Neilan.[230] Another feature of the commemoration in Galway city was the renaming of the railway station as Ceannt Station, in honour of the executed leader. A plaque was unveiled by Josephine McNamara, a founder member of Cumann na mBan, in the presence of members of Ceannt's family and the Lord Mayor.[231]

---

229 Department of External Affairs, *Cuimhneachán 1916–1966*, p. 38.
230 *The Connacht Tribune*, 16 April 1966.
231 W. Henry, *Supreme Sacrifice: The Story of Éamonn Ceannt 1881–1916* (Mercier Press, Cork, 2005), p. 139.

Afterwards, on 8 May, a special 'Eamonn Ceannt Festival' was held at the 1916 leader's birthplace at Ballymo in the north of County Galway. A plaque in his honour was unveiled by Colonel H. Byrne at the Garda station – a former RIC barracks where Ceannt was born and spent his childhood. In attendance was de Valera, who again emphasised the need for 'the restoration of the Irish language' at a reception following the unveiling ceremony. Speaking of Ceannt, the President said that he had first met him in the Gaelic League and in Coláiste Laighin. Many of the Irish Volunteers of 1916, he added, had come to love their country through the Irish language.[232] Elsewhere in the county, a bronze bust of Mellows was unveiled under pouring rain in the grounds of the national school in the town of Athenry on Easter Sunday by Alf Monahan, who had acted as a dispatch for the Irish Volunteers during the Rising. Earlier in the day, Mass was celebrated in the parish church by Reverend L. Hennelly and a parade was held through the town.[233]

Ireland's rural traditions were underlined during the course of a commemoration in the outskirts of Athenry on 14 April, where the Minister for Agriculture, Charles Haughey, opened an agricultural college named after Mellows. He said that it was little wonder that many Irish rebels had come from farming stock and observed that it was 'on the land, so close to nature' that 'the feeling of patriotism has ever been a living thing'.[234] The building housing the Mellows Agricultural College – which specialised in year-long intensive courses in farming practice and boasted its own turkey breeding unit – had originally served as the headquarters of Mellows and a group of Irish Volunteers on Easter Monday 1916 (when it was known as the Model Farm). As already seen, they had to abandon the post the next day, due to shelling from a warship in Galway Bay.[235] A plaque commemorating this link was unveiled and then blessed by the Very Reverend Canon Heaney. Haughey also let it be known to the assembled guests that improving one's land was not only 'a duty to the nation', but 'a symptom of national health'. Present and future generations, he emphasised, 'owe it to the men of 1916 to make well organised and intelligent efforts to improve the land of Ireland and exploit its resources ... The soil is our greatest raw material and Pearse asserted the right of the Irish people to hold and control the land'.[236]

Besides its historical associations with 1916, it can be argued that the western location of the Mellows Agricultural College also made it a natural choice of venue for a significant commemorative gesture, embedded as it was

---

[232] *Western People*, 14 May 1966.

[233] *The Irish Press*, 12 April 1966.

[234] *The Irish Times*, 15 April 1966.

[235] *The Irish Independent*, 15 April 1966; *The Connacht Tribune*, 23 April 1966.

[236] *The Connacht Tribune*, 23 April 1966.

with higher cultural meanings. Geographically, the college was located within easy reach, by car, of the rugged terrain of Connemara – the incomparable heartland of the west of Ireland. For many years, the rurality of most of the west of Ireland had evoked a symbolic and internalised Otherness *vis-à-vis* the Anglicised, urbanised and industrialised eastern seaboard. During the late nineteenth century, for example, the west 'was believed to have been largely unaffected culturally and linguistically by the British presence', and was thus perceived as 'the landscape of greatest difference to ... the gently rolling green fields of the south-east'.[237] In the early part of the twentieth century, images of the uniqueness of the west of Ireland's landscape were embodied within the fabric of the nation by the writers of the literary revival and the Irish-Ireland movement. In the process, the west was fashioned into a 'Gaelic' and 'masculine' entity which 'came to stand for Ireland in general, to be representative of true Irishness. It could be seen as a ... way of life, yet also be conceived of as outside time ... as evoked ... by Seán Ó Faoláin'.[238] Pearse himself wrote vividly about the lifeworlds of the inhabitants of the west, in collections of short stories such as *Íosagán agus Sgéalta Eile* ('Little Jesus and Other Stories', published in 1907) and *An Mháthair* ('The Mother', published in 1916). These stories, according to Des Maguire, contained 'penetrating glimpses of the minds and souls of the people of the Western seaboard' and skilfully 'interpreted the inner lives of the Gaeltacht people'. Furthermore, they also offered up a psychological analysis of 'the sorrows and joys of the people of Iar-Connacht, and the tragedies of life and death from which they could never escape'.[239]

The county of Mayo also commemorated the Rising's 50th anniversary with aplomb in 1966. In what was described as 'a glittering kaleidoscope of colour', the past was remembered in Westport on Easter Sunday when thousands of people gathered 'to pay tribute to their patriotic dead and show recognition of their proud and glorious heritage'.[240] Special guests on the day

---

[237] B. Reid, 'Labouring Towards the Space to Belong: Place and Identity in Northern Ireland', *Irish Geography*, Vol. 37, No. 1 (2004), p. 105.

[238] C. Nash, '"Embodying the Nation": The West of Ireland Landscape and Irish Identity', in B. O'Connor and M. Cronin (eds), *Tourism in Ireland: A Critical Analysis* (Cork University Press, Cork, 1993), pp. 86–87.

[239] D. Maguire, 'Introduction', in D. Maguire (ed.), *Short Stories of Padraic Pearse: A Dual-Language Book* (Mercier Press, Cork, 1968), p. 7. After independence, images of the west still featured strongly in the literary output of writers such as Peadar O'Donnell. As noted by C. Travis, '"Rotting Townlands"*: Peadar O'Donnell, the West of Ireland and the Politics of Representation in Saorstát na hÉireann (Irish Free State) 1929–1933', *Historical Geography: An Annual Journal of Research, Commentary and Reviews*, Vol. 36 (2008), p. 208, the west's rural culture and landscape 'served to anchor a nationalist genealogy and iconography'.

[240] *The Connaught Telegraph*, 14 April 1966.

included five survivors of the 31 Irish Volunteers from the area who had been imprisoned after the Rising.[241] Charles Gavin, one of the veterans, laid a wreath at the plaque on the house at Westport Quay where Major John MacBride was born, while a votive Mass was celebrated in St. Mary's Church. Thousands of spectators then lined the streets to watch a parade. This was led by six bands and included official representatives, county associations, members of sporting clubs, and schoolchildren. A 'stirring ceremony' also took place at the Octagon, which was witnessed by various relatives of MacBride (including his son, Seán, who was Secretary-General of the International Commission of Jurists) and the Minister for Social Welfare, Kevin Boland. Afterwards, the guests were entertained at a dinner in Hotel Clew Bay. In an effort to boost pride in the local economy, the commemoration at Westport was also marked by a week-long exhibition of locally-manufactured goods at the Old Vocational School (featuring commodities such as meat, brushes, footwear, furniture, thread, cloth, underwear, knitwear, soft toys, and animal foods made from seaweed).[242]

Besides the events in Westport, the Golden Jubilee was also commemorated at a range of locations throughout the county. In Castlebar, an estimated crowd of 1,000 took part in an Easter Monday parade to the 1798 memorial plot on The Mall, where the Proclamation was read.[243] The local Fianna Fail Cumann marked the occasion by holding a dinner for 250 guests at the Welcome Inn on the night of 14 April, featuring speakers such as Seán Flanagan and Pádraig Flynn. Three days later, community organisations held a parade to Ballyhaunis Cemetery, led by the Foxford brass band and Old IRA veterans. A decade of the Rosary was recited in Irish at the cemetery and the Proclamation was read over the graves of two Irish Volunteers.[244] A number of guest lectures on the theme of 1916 were also held by various organisations. One of these was delivered on 26 April at the Central Cinema in Claremorris by F. X. Martin.[245] At another lecture at the Moy Hotel in Ballina on 6 May (sponsored by the Ballina Trades Council and the National Graves Association), George Gilmore from Dublin spoke on the theme of 'Labour and 1916'.[246]

Elsewhere in the west of Ireland, almost 1,000 people took part in an Easter Sunday parade to the Countess Markievicz Memorial Park in Sligo town, where the Minister for Justice inspected the Guard of Honour.[247] 'And it was only

---

241 Department of External Affairs, *Cuimhneachán 1916–1966*, p. 41.
242 *Western People*, 16 April 1966.
243 *The Irish Independent*, 12 April 1966.
244 *The Connaught Telegraph*, 21 April 1966.
245 *Western People*, 30 April 1966.
246 Ibid., 7 May 1966.
247 Department of External Affairs, *Cuimhneachán 1916–1966*, p. 40.

fitting', reported *The Sligo Champion*, 'that the focal point of the ceremonies should be in the Park, which commemorates the memory of that heroic ... woman – a Gore Booth from Lissadell'.[248] The *Western People* noted that the Sligo ceremonies 'were considerably marred by torrential rain which fell throughout the later part of the afternoon'. Later that night, Feis Shligigh presented *And They Fought*, a pageant based on the writings of the leaders of the Rising.[249] In County Roscommon, many tribute Masses were held on Easter Sunday, while thousands attended a pageant at Lanesboro.[250] On 28 April, surviving members of the Roscommon and Mayo GAA teams who had played in the Connaught Final of 1916, gathered for a function at the Abbey Hotel in Roscommon town.

Contentious Remembrance: Incidents in Ulster and Elsewhere

In the preface to *In Search of Ireland: A Cultural Geography*, Brian Graham has made the astute observation that social groupings, throughout the course of history, have habitually drawn 'upon the past to legitimate and validate both their present attitudes and their future aspirations ... within a complex geographical mosaic of locality, class and gender'.[251] This statement has particular relevance to many episodes in the making and remembrance of Irish history in modern times, including the occasion of the 1916 Rising's Golden Jubilee. Across the island of Ireland, the anniversary served as an opportune moment for various social groups to articulate a range of political viewpoints and socio-cultural ambitions. A geographical concentration of political tensions and hostilities, fuelled largely by an incisive dissatisfaction with Partition, was particularly evident throughout Easter 1966 at commemorations held in many of the Border counties.

In County Leitrim, a ceremony was held on Easter Sunday in the town of Kiltyclogher. This involved about 400 IRA veterans, Army and Civil Defence personnel, who took part in a parade to the statue of the executed 1916 leader, Sean MacDiarmada. On the surface, all seemed pretty normal on the day. The Proclamation was read in Irish by a schoolboy and the National Flag was hoisted, wreaths were laid, a decade of the Rosary was recited and the ceremonies concluded with the playing of the National Anthem.[252] The lead-up to the Kiltyclogher commemoration, however, did not pass off without incident.

248 *The Sligo Champion*, 15 April 1966.
249 *Western People*, 16 April 1966.
250 *The Irish Times*, 11 April 1966.
251 B. J. Graham, 'Preface', in B. J. Graham (ed.), *In Search of Ireland: A Cultural Geography* (Routledge, London, 1997), p. xi.
252 Department of External Affairs, *Cuimhneachán 1916–1966*, p. 38.

Two of MacDiarmada's sisters wrote a letter to *The Irish Times*, in which they complained about the preparations that were being made for the commemoration and outlined their frustration with Partition. In their acerbic letter, the elderly sisters, Margaret and Rose, registered deep unhappiness with the ban that the government had imposed on the Easter lily emblem. Furthermore, they wrote: 'The forces of the 26-County State raided our home in recent years while engaged in patrolling and maintaining the British-made border.' The letter also proclaimed that their brother had 'died for a 32-County Republic, which has yet to be achieved ... we object to commemoration ceremonies by those who have accepted less'.[253] Margaret, who was aged 89, again registered her displeasure with the establishment when she refused the National University of Ireland's offer of an honorary doctorate on 14 April.[254] Whilst de Valera may well have been disappointed at a personal level by the boycott of MacDiarmada's sister, he was well used to protests of this ilk, having encountered comparable remonstrations from diehard republicans on previous occasions of remembrance (stretching all the way back to the 1930s).

Tensions arising from Partition were also apparent in the town of Cavan. One incident of note involved the desecration of a life-sized statue of the nineteenth-century genealogist and politician, Henry Maxwell, the Seventh Baron Farnham (a representative peer for Ireland in the House of Lords, who died in the Abergele train disaster in 1868). The statue erected in his memory was disfigured on a number of occasions over the years and at once stage it was even beheaded. On the night of 9 April 1966, republicans daubed it in green paint with the slogan: 'Free Ireland, Up the Rebels'.[255] Golden Jubilee ceremonies in the town of Dundalk in County Louth included a Mass, a parade and a commemorative pageant staged in the Town Hall by the pupils of St. Louis Convent. Another parade took place in Monaghan, with 2,000 people participating.[256] This started at the North Road area and halted at the junction of Glasslough Street and the Diamond, where the Proclamation was read by a national school teacher, Seán Ó Murchadha. Writing in *Cuimhneachán Mhuineacháin, 1916–66*, a 134-page booklet that was produced for the anniversary, Ó Murchadha thanked the people of County Monaghan for their 'enthusiastic and earnest support', and observed with pride that 'the number and variety of the societies and organisations taking part testify to the healthy growth of the sapling of nationhood'.[257] Matters were less cordial

---

[253] *The Irish Times*, 8 and 9 April 1966.

[254] Ibid., 15 April 1966.

[255] Ibid., 11 April 1966.

[256] Department of External Affairs, *Cuimhneachán 1916–1966*, pp. 37, 40.

[257] Anon., *Cuimhneachán Mhuineacháin, 1916–66. Souvenir Programme* (Clogher Historical Society, Monaghan, 1966), p. 3.

in Drogheda, where disputes over representation in the commemorative parade meant that many organisations were absent – including clubs that played rugby and soccer (sports which were labelled as 'foreign games' by some nationalists), the Red Cross, the Order of Malta, and the two main political parties.[258]

In Northern Ireland, the occurrence of the Golden Jubilee managed to conjure up all of the ingredients that were necessary for a number of tempestuous encounters between nationalists and unionists. Tensions had already been simmering for years and there was little that the passing of time could do to keep these at bay. From 1929–1948, Eastertime commemorations of the Rising were prohibited in the North under the Special Powers Act, but once the ban was lifted, well-attended ceremonies were then held at republican plots in Milltown Cemetery in Belfast and in Derry City Cemetery.[259] However, in the years in the run-up to the Golden Jubilee, the Stormont administration was faced with the delicate task of determining whether or not to allow the anniversary to be commemorated in public. Given the profound demarcation between the nationalist and unionist communities, the stakes were high as politicians mulled over what to do. A genuine attempt at reconciling Catholic nationalists to the Northern Ireland regime was made by the Prime Minister, Captain Terence O'Neill. With the aim of implying their 'equality rather than inferiority within the state', Joe Lee notes that he performed a number of symbolic gestures such as visitations to Catholic schools. O'Neill's most 'courageous … initiative', according to Lee, was the invitation to Lemass to visit Stormont in January 1965, followed by his own return to visit Dublin the following month. However, the furore that ensued 'among more excitable Protestants', demonstrated 'the depths of distrust still prevailing'.[260] Afterwards, many in the Protestant community called for greater reassurances about the strength of the union with Britain.

In the months leading up to the Golden Jubilee, the Stormont administration signalled its readiness to protect Northern Ireland from any escalation in the number of IRA incidents that were likely to arise as a result of the commemoration of 1916. Harold Black, the Secretary to the Cabinet, wrote in secret to Robin North at the British Home Office in London on 18 February 1966, asking him to 'make urgent representations to the Ministry of Defence' so as to support a request from the Royal Ulster Constabulary (RUC), which had already been rejected on technical and practical grounds, for the loan of 10 Ferret Scout cars from the Army. The vehicles, stated Black, 'would be immensely valuable in dealing with the situation which may arise here during

---

[258]   *The Irish Press*, 11 April 1966.
[259]   B. M. Walker, *Past and Present: History, Identity and Politics in Ireland* (The Institute of Irish Studies, Belfast, 2000), p. 89.
[260]   Lee, *Ireland 1912–1985*, p. 416.

the celebration of the 50th anniversary of the Easter Rebellion', particularly 'if the occasion is used by the IRA to create civil commotion and strife'. Black also requested special assistance in helping to 'expedite delivery' to the police of a range of other military equipment, including 350 rifles and 12 machine guns. The security outlook, he warned, 'is now considerably worse than it was a few weeks ago'. In his reply 10 days later, North confirmed that the Army was willing to supply the weapons, while the cars were to be loaned until 1 May.[261]

Ultimately, the Cabinet decided to allow some Golden Jubilee commemorations to take place, but to prohibit others. A month and a half before Easter Sunday, O'Neill insisted that the Catholic community needed to be briefed on some of the safety measures that the government was prepared to resort to. In a letter to the Catholic Primate, Cardinal Conway of Armagh, on 22 February, he hinted that the government was ready to act tough in the event of any disturbances: 'It would certainly be my intention to come out more strongly if trouble persists or grows.'[262] O'Neill, however, left most of the toughest talking to the Minister of Home Affairs, Brian McConnell, who made the Stormont government's attitude towards parades and gatherings associated with the Golden Jubilee abundantly clear in a statement to the House of Commons on 2 March. 'The events which are being celebrated', he reminded MPs, 'do not commend themselves to the people of Northern Ireland as a whole'. The government's duty, he added, was 'to ensure that any celebrations taking place within Northern Ireland do not offend our citizens and that they should not be held in such places and in such circumstances as are likely to lead to a breach of the peace.'[263]

In the Dáil the next day, Michael O'Leary sought greater clarity on the security measures being planned in the North, asking Lemass whether he could arrange with O'Neill that those bearing Tricolours at commemoration

---

[261] Harold Black to Robin M. North, 18 February 1966, Public Record Office of Northern Ireland (hereafter PRONI), Cabinet Minutes, CAB 9G/73/14, IRA Activities in Northern Ireland: Defence of Northern Ireland; Robin M. North to Harold Black, 28 February 1966, PRONI, Cabinet Minutes, CAB 9G/73/14, IRA Activities in Northern Ireland: Defence of Northern Ireland; S. Elliott and W. D. Flackes, *Northern Ireland: A Political Directory 1968–1999*, 5th Edition (The Blackstaff Press, Belfast, 1999), p. 180; C. O'Donnell, 'Pragmatism Versus Unity: The Stormont Government and the 1966 Easter Commemoration', in M. Daly and M. O'Callaghan (eds), *1916 in 1966: Commemorating the Easter Rising* (Royal Irish Academy, Dublin, 2007), pp. 241–42.

[262] Terence O'Neill to Cardinal Conway, 22 February 1966, PRONI, Cabinet Minutes, CAB 9G/73, IRA Activities: Condemnation of Incidents by Roman Catholic Primate.

[263] Statement by the Minister of Home Affairs, Brian McConnell, MP, in the House of Commons, 2 March 1966, PRONI, Cabinet Minutes, CAB 9B/299/1, Commemoration of the 1916 Rebellion; O'Donnell, 'Pragmatism Versus Unity', pp. 249–50.

ceremonies would not be ill-treated by the authorities there. Lemass, however, was anxious not to interfere. 'I do not consider', he said, 'that representations, as suggested by the Deputy, would be helpful'. Furthermore, he said that he hoped 'that the commemoration ceremonies will take place in all parts of Ireland in a fitting and dignified manner and in an atmosphere free from dissension and animosity'.[264] However, as Brian Feeney has pointed out, news of 'extensive plans ... to celebrate' the Rising proved particularly horrifying to elements within unionism, especially to hardliners who wished to rid their state of 'all Irish imagery'. Consequently, any hopes that O'Neill had 'of a period of calm were dashed'.[265] What followed, as Coogan notes, was a backlash by hardliners against what were seen as 'celebrations of disloyalty and treachery'.[266] The IRA's surprise bombing of Nelson's Pillar in Dublin on 8 March added to the commotion, by sending shockwaves throughout the entire island of Ireland. From this moment onwards, several Protestants in the North (from moderates to hardliners) repeatedly aired their reservations about sound bites emanating from republicans bearing hawkish views about Irish unity. The Stormont government came under growing pressure by the day, and many citizens and lobby groups called on it to adopt a wholly uncompromising stance on republican attempts to observe the anniversary of 1916.

In a disparaging letter sent to O'Neill on 11 March, David Browne and 28 other signatories from the Students Union of Queen's University Belfast, registered their disapproval, 'rationally and unemotionally', at the way in which a number of Golden Jubilee commemorations had been given the go ahead by the Cabinet. The fact that supporters of the Rising were going to be 'permitted to celebrate in Belfast', they protested, ranked as 'an act of provocation and treason'. Furthermore, because of the fact that the Rising had struck a blow against British rule in Ireland, they felt that any attempts 'to celebrate a massacre of British soldiers' would be 'simply illogical' as far as the constitutional integrity of Northern Ireland was concerned. The students also lamented the fact that 'revellers' were going to be allowed 'to hold political activities and parades through the streets of Belfast on the Lord's Day'. This, they complained, 'is contrary to our heritage of Sabbath observance, on which we pride ourselves'. Many of O'Neill's colleagues also faced pressure from the grassroots of unionism. For example, the Derry MP, Robin Chichester-Clarke, took delivery of a letter from a disgruntled constituent on 18 March, who called for a ban on 1916

---

[264] *Dáil Debates*, Vol. 221, 3 March 1966.

[265] B. Feeney, *Pocket History of the Troubles* (The O'Brien Press, Dublin, 2004), p. 14.

[266] T. P. Coogan, *The Troubles: Ireland's Ordeal 1966–1995 and the Search for Peace* (Hutchinson, London, 1995), p. 49.

commemorations and who reminded him that 'our beloved Ulster is still part of the United Kingdom'.[267]

Further pressure was exerted on the Prime Minister by Captain Sir George A. Clark, the 52-year-old Grand Master of the Grand Orange Lodge of Ireland (who was well-known, according to his detractors, for having an uncompromising disposition). In a letter dated 1 April, he pleaded with O'Neill to 'curtail' public celebrations 'to be held in Northern Ireland commemorating the 1916 Rebellion'. Such demonstrations, he claimed, 'are provocative in origin, and at best must put the good name of Ulster in jeopardy since they create situations which could easily result in public disorder'. In his reply five days later, the Prime Minister stated that while he found 'the sentiments behind the 1916 celebrations' to be 'repugnant to me and my colleagues', a blanket ban on all ceremonies would not be implemented on practical grounds. While it was the government's policy not to allow 'any display or demonstration ... likely to lead to a breach of the peace', O'Neill was afraid that 'public opinion in Great Britain and ... the world' would not be favourable in the event that 'outbreaks of disorder were to occur' in the event of a complete ban being 'enforced'. Adopting a forward-looking standpoint, he speculated that were such trouble to occur, then it 'would be attributed by Ulster's enemies to repressive police action'.[268]

In the end, Stormont acted pragmatically. It decided that the most prudent solution to protecting the security of Northern Ireland was by carefully picking and choosing which celebrations would be permitted under the law and which ones would not. The government also decided that it would be in its best interests for the Prime Minister to make a significant public pronouncement on how enmity between Catholics and Protestants needed to be eradicated. As the Easter weekend finally approached, security measures were stepped up considerably throughout the North, while the climate of public fear and tension escalated too. So high were the government's worries about the possibility of

---

[267] David G. Browne and 28 Others from the Students Union, Queen's University Belfast to Terence O'Neill, 11 March 1966, PRONI, Cabinet Minutes, CAB 9B/299/1, Commemoration of the 1916 Rebellion; W. Campbell to Robin Chichester-Clarke, 18 March 1916, PRONI, Cabinet Minutes, CAB 9B/299/1, Commemoration of the 1916 Rebellion; O'Donnell, 'Pragmatism Versus Unity', p. 252.

[268] Captain Sir George A. Clark to Terence O'Neill, 1 April 1966, PRONI, Cabinet Minutes, CAB 9B/299/1, Commemoration of the 1916 Rebellion; Terence O'Neill to Captain Sir George A. Clark, 5 April 1966, PRONI, Cabinet Minutes, CAB 9B/299/1, Commemoration of the 1916 Rebellion; O'Donnell, 'Pragmatism Versus Unity', pp. 252–53; C. J. Woods, 'Clark, Sir George Anthony', in J. McGuire and J. Quinn (eds), *Dictionary of Irish Biography: From the Earliest Times to Year 2002. Volume 2: Burdy-Czira* (Cambridge University Press, Cambridge, 2009), pp. 537–38.

serious public disorder and rioting that plans for a 1916 commemoration parade in the Loup area of Derry were banned outright by the Minister of Home Affairs under the Public Order Act of 1951.[269] According to an official press release on 6 April, McConnell took the measure 'because he was convinced, from information reaching him, that the holding of the demonstration ... would undoubtedly lead to serious public disorder'.[270]

The Republic was also put on a high level of alert. Detectives from the Special Branch and uniformed Gardaí were placed on a 24-hour watch on the GPO, while both British Embassy officials and 'monuments commemorating British achievements or heroes' were also placed under armed guard.[271] O'Neill did his best to curtail Catholic resentment at Stormont's attitude to the Golden Jubilee by seeking to promote the ideals of tolerance at a conference held at the Corrymeela Centre in Corrymeela, Ballycastle on 7 April. In a speech to the assembled delegates, he pleaded for harmony in a Christian spirit between Protestant and Catholic, and a shedding of the burdens of long-established enmities. While stressing that there could be no compromise on the constitutional position of Northern Ireland, he declared: 'There is much we can do together ... It must, and – God willing – it will be done.'[272] The speech was well-received by the media in the South. *The Irish Independent*, for example, remarked that it encompassed a sincere call for religious accord, so as to put 'an end to traditional quarrels' and facilitate 'a new effort in a Christian spirit by the two communities in Northern Ireland to create better opportunities for their children and crush extremists on both sides'.[273]

In the days leading up to Easter Sunday, cross-Border cooperation between the security services was significantly heightened, with a 'hot line' established between Belfast and Dublin. On the same day as the Corrymeela conference, an RUC officer in Belfast sought to reassure the public by stating: 'There has always been a certain amount of cooperation, especially on enquiries for wanted men. Now there is a much closer link.'[274] But tensions remained high in many quarters of the RUC, whose members took the threat from extremists very seriously. A staff reporter with *The Times* noted that there was 'an atmosphere of tension and unease' all over the North and that the massive security precautions were 'contributing to a general air of anxiety'. Such was the level of security that

---

[269] *The Irish Times*, 7 April 1966.
[270] Press Release, 6 April 1966, PRONI, Cabinet Minutes, CAB 9B/299/1, Commemoration of the 1916 Rebellion.
[271] *The Irish Times*, 7 April 1966.
[272] Ibid., 8 and 9 April 1966.
[273] *The Irish Independent*, 8 and 9 April 1966.
[274] Ibid.

'admission to a police station could only be gained by knocking on the heavy front door and receiving a scrutiny through a "Judas" keyhole'.[275] On Easter Sunday, 200 police were drafted into the 'cauldron' of Derry's Loup district to enforce the ban on a republican parade that was due to be held there, but in a display of defiance, hundreds of people instead went to St. Patrick's Cemetery to lay a wreath on the republican plot containing the grave of Brigadier General Sean Larkin of the Old IRA, who was executed in 1923 by the Free State forces at Drumboe Castle, County Donegal. At the ceremony, a youth was arrested by the police for carrying a Tricolour. Adding to the already strained atmosphere was the news that a Celtic headstone in the same cemetery had been badly damaged, possibly by a crowbar, the previous night.[276] The headstone in question, according to *The Derry Journal*, marked the grave of a former Parish Priest, Father Thomas Larkin, who was an uncle of the Brigadier General. The locals, it reported, felt 'that the culprits must have taken the grave for that of Sean Larkin'.[277]

Besides the commemoration at the Loup, an Easter Sunday parade was also held in Armagh, where an attendee from Cork advocated the continuation of a physical force struggle against Britain. However, this was sharply repudiated by another attendee, who urged that 'the methods of the past cannot be those of 1966 and onwards'.[278] In Belfast, a police car led a commemorative parade on Easter Sunday from Beechmount on the Falls Road to Milltown Cemetery. Many of the estimated crowd of 10,000 wore Easter lilies, and the Tricolour flew from many houses along the Falls Road, which was ornately decorated 'with bunting and 1916 emblems'. At the republican plot in the cemetery, a Sinn Féin member delivered an oration, in which he forewarned: 'The first crack in the British Empire was made in 1916. Our next act might disintegrate it completely.'[279]

The intensive security measures that were put in place for Easter Sunday proved to be largely successful, and the only reported incidents of public disorder were the damaging of a telephone kiosk in the centre of Belfast city by a small charge of explosive and the breaking of three church windows in Milltown.[280] In the week after Easter Sunday, however, tensions remained strong throughout certain parts of the North. A number of ardent unionists responded to the occasion of the Golden Jubilee by reaffirming their own loyalty to the British Crown and by denouncing the legacy of the Rising. This was particularly evident

---

[275] *The Times*, 9 April 1966.
[276] *The Irish Times*, 11 April 1966.
[277] *The Derry Journal*, 12 April 1966.
[278] *The Irish Times*, 11 April 1966.
[279] *The Irish Press*, 11 April 1966.
[280] *The Times*, 11 April 1966.

at an Apprentice Boys of Derry demonstration at Ballynahinch in County Down on Easter Monday, when Brian Faulkner, the Northern Ireland Minister for Commerce, adopted a far sterner attitude than the more moderate-minded O'Neill, by issuing a warning to 'troublemakers' that his government would stand no nonsense. 'We will permit liberty under the law', he said, 'but Ulster is still British thank God'. In a direct attack on commemorations of the Rising in the North, he publicly criticised nationalists for 'merely celebrating an act of rank sedition – they are resurrecting an insurrection'.[281] Faulkner drew much applause from the large crowd that had gathered, especially after he announced that trade in Northern Ireland was set for an all-time record that year. A heckler, however, was also heard during the speech, but when the Minister retorted, 'the heckler gave up and disappeared'.[282]

Another prominent figure in the North who expressed grim views about the Rising was Ian Paisley, a young Free Presbyterian Church Minister with a reputation for being a fearsome street preacher. According to Ed Moloney, Paisley 'represented a virulent strain within unionism' that was resurrected in 1960s Belfast. His oratorical proficiency, he adds, 'made him a formidable opponent ... [who] could sway a mob with a few well-chosen words'.[283] Paisley's roots in Protestant fundamentalism, as T. G. Fraser has pointed out, 'ran deep'. His 'rejection of the claims of the Catholic Church' and 'distrust of what he saw as the ecumenical tendencies of the established Protestant churches' transformed into a belief 'that O'Neill and the unionist leadership were preparing to undermine the Protestant basis of Northern Ireland'.[284] Paisley came to greater public prominence in 1966, by virtue of his position as Chairman of the newly-established Ulster Constitution Defence Committee. This organisation was the brainchild of Noel Doherty, who was the devoted printer of the loyalist mouthpiece, *The Revivalist*. With Paisley on board, its membership swelled with loyalists disillusioned with O'Neill's government, who were then organised into branches and divisions of Ulster Protestant Volunteers. According to Moloney, the Ulster Constitution Defence Committee was swiftly invested with Biblical meanings by its membership:

> At the head of the committee, Christ-like, was Paisley as chairman, and underneath him were 12 committee members – 'apostles' – called together by him, pledged as a body of 'Protestant patriots' to defend the Union, the Protestant monarchy and the Williamite Settlement. Its first meeting continued the Biblical parallels. Like the Last

281 *The Irish Times*, 12 April 1966.
282 *Belfast Telegraph*, 12 April 1966.
283 E. Moloney, *A Secret History of the IRA*, 2nd Edition (Penguin Books, London, 2007), p. 61.
284 T. G. Fraser, *Ireland in Conflict 1922–1998* (Routledge, London, 2000), p. 38.

Supper, it was held in a room above a restaurant. O'Neill was the 'Judas' and Paisley would soon become the martyred saviour.[285]

In an interview with Jack Fagan of *The Irish Times*, which was published on 12 April, Paisley made clear his negative feelings about Irish republicanism. He told Fagan about his desire to 'thank God that the 1916 Rebellion was a failure and that Ulster is still free from Papal tyranny'. Paisley also explained that he had unsuccessfully sought a meeting with McConnell, so as to point out his belief that the Rising had been 'an act of treason and murder' and argue that 'Protestants could not condone the commemoration of such an event'.[286]

The strength of the Northern Protestants' resolve against their political opponents was particularly apparent on the same day, when over 10,000 junior Orangemen from over 140 lodges held their annual demonstration in Lisburn, which was attended by the McConnell. The Ulster Constitution Defence Committee erected posters at a number of vantage points in the town, warning that loyalists 'would not stand idly by' while 'disruptive and disloyal' people 'insulted the flag and reviled the Constitution'.[287] That same day, *The Times* reported with relief that that the Easter holiday period 'over which there have been so many fears ... was drawing to a close ... with fewer than the normal number of minor incidents in the radio log books at the Royal Ulster Constabulary headquarters'.[288] In the week after Easter Sunday, however, fears of a new IRA campaign were raised after a number of explosions in the Republic. These were specifically aimed at sabotaging state property. Considerable damage was caused to telephone installations in the Kilmacow and Ballykeoghan areas of County Kilkenny by two explosions on 14 April (which were rumoured to have been connected with an attempt to raise the plight of local men who were in jail in Limerick).[289] Two days later, a mysterious explosion occurred at Douglas village near Cork city, apparently aimed at wrecking ESB pylons.[290]

In the North, the security situation proved to be of major concern to the Stormont government in the lead-up to a large republican parade to Casement Park in Belfast, scheduled for Sunday 17 April. Paisley's tactical decision to hold a counter-parade on the same day also heightened fears that an ugly clash might occur. Hysterical reports circulated in some British newspapers that a force of 2,000 armed men intended to enter Northern Ireland, so as to bring a new IRA

[285] E. Moloney, *Paisley: From Demagogue to Democrat?* (Poolbeg, London, 2008), pp. 117–18.
[286] *The Irish Times*, 12 April 1966.
[287] Ibid., 13 April 1966.
[288] *The Times*, 12 April 1966.
[289] *The Irish Times*, 15 April 1966.
[290] Ibid., 18 April 1966.

campaign against the Stormont government and the security forces. Invoking his powers under the Civil Authorities (Special Powers) Acts, McConnell issued an emergency alert and banned northern-bound trains on the parts of the Dublin-Belfast line in Northern Ireland, between 9.30 pm on Saturday and 7.00 pm on Sunday. The police were also ordered to scrutinise all people arriving in the North by road and to prohibit entry to any traffic on the A1 that was deemed to be detrimental to the upholding of law and order. An official press release staunchly defended the government's emergency measures. It warned that it could 'not permit the peace of Northern Ireland to be disturbed ... by provocative incursions of hostile elements from the South', especially from those who might 'engage in subversive activities or otherwise endanger the peace'. Whilst noting that the government regretted 'the inconvenience caused', it stated that it was 'confident that the respectable citizens of Northern Ireland will appreciate that these steps are essential in the interests of public order and will cooperate in every way possible with those whose task it is to preserve the peace'.[291]

This decision was the subject of much commentary by the media in the Republic. According to *The Irish Times*, speculation had emanated from Belfast that the withdrawal of north-bound trains on the Dublin-Belfast line 'might not be due to any hope of stopping the entry of IRA men but to prevent the danger of injury to innocent passengers if trains were stoned or derailed by Protestant extremists'.[292] *The Irish Press* remarked that the ban on trains 'came as a complete surprise' to a spokesman for Lemass's government, because it had been given 'no prior notification' of the decision and had been 'very pleased' about the manner in which nationalists in the North had already marked the Golden Jubilee.[293] *The Irish Independent* reported that there was a fear in some circles in Dublin that the ban would play into the hands of extremists in the Republic, who had been causing a nuisance in places like Kilkenny. It therefore registered the opinion that 'some think the Stormont move injudicious and far too drastic'. Its Editorial was more forthright, slamming the ban as 'ridiculous' and directly accusing Stormont of panicking by giving in to pressure from Paisley and allowing 'ancient emotion to take control'.[294]

Although the republican commemoration at Casement Park on 17 April did manage to attract a significant amount of people, estimated at around 15,000 by *The Irish Independent*, rumours of the planned invasion by armed men from the

---

[291] Commemoration of the 1916 Rebellion, Press Release, 14 April 1966, PRONI, Cabinet Minutes, CAB 9B/299/1.

[292] *The Irish Times*, 15 April 1966.

[293] *The Irish Press*, 16 April 1966.

[294] *The Irish Independent*, 16 April 1966.

South turned out to be wildly exaggerated.[295] The *Belfast Telegraph* reported that only a small amount of republicans, numbering at least 100, made the journey across the Border and stated that 'police precautions had the effect of making Sunday traffic much less than normal'.[296] While *The Irish Times* also attributed the low numbers to 'the presence of heavily armed RUC' alongside the Border, it also suggested that 'the bad weather' may have been a factor. The fact that forms for the 1966 Census had to be completed by households in the Republic by 12.00 midnight that night may also have deterred people from travelling to the North. Besides a couple of mini buses from Dublin, the most conspicuous contingent to head to Belfast from the Republic was the Cork Volunteer Pipe Band – who travelled in a special bus and told of 'being treated very courteously by the police patrols' along the Border. Apart from republican contingents from all over Northern Ireland, those who took part in the parade included members of the Belfast branch of the Irish Transport and General Workers' Union, the Belfast Trades Council, the Irish National Foresters, Cumann na mBan, and the GAA.[297] In total, around 4,000 people marched in the parade, which stretched for about two-miles and was headed by a colour party and eight bands.[298] Notwithstanding the fact that the Border temporarily took on a more tangible appearance through the operation of a number of 'Checkpoints Charlie', the Belfast correspondent with *The Irish Times* wrote that there was a sense of an anti-climax following Stormont's warning about the threat of subversive activity. Despite conjuring up images of 'Biblical stories of swarms of wild locusts guzzling everything devourable', he noted that very few people in the North had actually been frightened by the notion of 'an invasion of wild and woolly men from the Republic brandishing dirks and bicycle chains or other offensive weapons'.[299]

In addition to the republican parade, a rival procession was also held to Belfast's Ulster Hall by the Ulster Constitution Defence Committee on 17 April.[300] A crowd of around 5,000 followed Paisley through the city centre to the music of bands playing 'Abide With Me' and 'Onward Christian Soldiers'. It stopped for about 10 minutes at the Cenotaph, where a wreath was laid, along with a card in remembrance 'of the members of the Ulster Volunteer Force,

[295] Ibid., 18 April 1966.

[296] *Belfast Telegraph*, 18 April 1966.

[297] *The Irish Times*, 18 April 1966.

[298] *Belfast Telegraph*, 18 April 1966.

[299] *The Irish Times*, 19 April 1966.

[300] For further commentary on this episode, see Moloney, *Paisley*, p. 125, who notes that the Ulster Constitution Defence Committee's parade, which took a route that passed by the republican gathering point, 'was the first example of Paisley's counter-march tactic, a tactic refined to an art two years later'.

RUC and civilian population who died in defence of the Ulster Constitution at the hands of the rebels during and since the 1916 uprising'.[301] Among those who paraded were malcontents from the fledgling Ulster Protestant Volunteers – whose declared aim was 'to save the Protestant heritage and to prevent rebel celebrations of the murders of 1916'.[302] Around 2,000 of the marchers filled the Ulster Hall to listen to Paisley's speech, while others listened outside the venue, where a loud speaker system relayed the words of the preacher. In his impassioned address, Paisley ridiculed O'Neill for his association with Lemass and expressed his displeasure at the holding of the republican rally. O'Neill and associates such as McConnell, he alleged, 'had made it perfectly clear' to republicans that they 'could have their own way and celebrate it [the 1916 Rising] here in Ulster'. Paisley then labelled the Rising a 'vile act' and proceeded to send out a strongly-worded message to republicans, warning 'that if Britain does not hold Ulster', then the Ulster Constitution Defence Committee 'will hold Ulster'.[303]

The 'uneasy calm' in Belfast on the day was eventually broken by a volatile incident at Denmark Street near Carlisle Circus, which was located along the geographically sensitive borderline between predominantly Protestant and Catholic streets. Three Catholic girls in their teens had to be rescued by the police when they were harassed by an angry segment of the Protestant procession returning from the Ulster Hall. The *Belfast Telegraph* reported that two of the girls 'were bundled into a parked Land Rover by police and driven off to safety', while the third was trapped (along with two policemen) by an angry mob against the home of a 76-year-old Protestant widow in Denmark Street. After the girl was taken inside by the elderly lady, it noted that 'two policemen were pinned against the wall as the crowd threw pennies and coins at them'. The mob also smashed the windows of the house using a variety of weapons. It took about 200 police reinforcements to 'restore order'. Paisley, it added, had 'tried to recall the mob and shouted over a hailer in an attempt to restore calm, but his efforts were of no avail'. In the end, a total of four people were arrested for disorder.[304] The girls, according to *The Times*, were 'badly frightened but unhurt'.[305] Trouble again flared up soon afterwards, when a youth who waved a Tricolour from the platform of a passing bus was pulled off it and also had to be rescued by the police, who remained in the area as other 'minor scuffles continued to break out'.[306]

---

301   *The Irish Independent*, 18 April 1966.
302   *The Irish Times*, 15 April 1966.
303   *Belfast Telegraph*, 18 April 1966.
304   Ibid.
305   *The Times*, 18 April 1966.
306   Ibid.

Further disturbances continued into the night when two explosions occurred in Belfast. One of them took place in the Ligoniel district in the north of the city. The other explosion was a minor one and happened near the republican plot in Milltown Cemetery.[307] Despite the high tensions on display, it has been observed by Steve Bruce that Paisley was not unaware 'of his public responsibility', and following his counter-demonstration in Belfast, he cancelled planned protest marches in Newry and Armagh, which had the potential to be 'considerably more provocative'.[308] Having endured two highly stressful weekends, the Stormont government was finally able to breathe a sigh of relief, once 1916-inspired commemorative fervour started to dwindle. In a letter that he sent to the Inspector General of the RUC on 19 April, O'Neill conveyed his 'very great admiration of the services rendered to the community by the police forces' and commended them for responding 'with exemplary efficiency, dignity and fairness ... over a most difficult and potentially dangerous period'.[309]

In the Republic, sporadic incidents of violence occurred early in the morning of 17 April, including a failed attempt to blow up a newly-erected memorial to Liam Mellows at Limepark House near Gort in County Galway. However, a gate pier near the memorial was damaged by the blast (adding to other damage a few days earlier at Emly in County Tipperary). Another incident on 17 April happened at Newland's Cross in County Dublin, when the main coaxial telephone cable was severed, thus resulting in the suspension of telephone communications in many places. Furthermore, in Dublin city, a time-bomb was placed in an automatic telephone exchange in Fitzwilliam Lane, but this failed to detonate.[310] The seriousness of the violence provoked a harsh response from some of the old stalwarts of the revolutionary era. One plea for calm came from Old IRA luminary Dan Breen, author of *My Fight For Irish Freedom*.[311]

---

[307] Ibid.

[308] S. Bruce, *Paisley: Religion and Politics in Northern Ireland* (Oxford University Press, Oxford, 2009), p. 81.

[309] Prime Minister's Tribute to Police Forces, 19 April 1966, PRONI, Cabinet Minutes, CAB 9B/299/1, Commemoration of the 1916 Rebellion.

[310] *The Irish Times*, 18 April 1966 and 19 April 1966. A full report on the attempt to blow up the Mellows memorial can be found in *The Connacht Tribune*, 23 April 1966. The incident at Emly involved an Old IRA monument, consisting of a granite plinth that was surmounted by the statue of an Irish Volunteer. This was found to have been maliciously damaged on the morning of 14 April. The *Limerick Leader*, 16 April 1966, denounced the vandalism and added that citizens of the Republic 'have good reason to feel disturbed'. *The Irish Independent*, 16 April 1966, reported that Garda investigations 'showed that the statue was smashed from the plinth, apparently by sledge hammers, and broken in two pieces when it toppled over the concrete surround'.

[311] See D. Breen, *My Fight for Irish Freedom* (Anvil Books, Dublin, 1989), which was first published in 1928 by the Dublin-based Talbot Press, with an introduction by Joseph McGarrity.

In a letter to the Editor of *The Irish Independent*, he made a passionate appeal for sensibility to prevail:

> The wrecking of war memorials to our dead must be the acts of mad men or fools. We in Ireland always honoured war memorials. They were homage to our dead and as such demanded respect. The bombing of telephone kiosks and exchanges is the act of vandals. Men do not place explosives to destroy Irish property at the risk of killing Irish citizens. You do not hurt outsiders when you blow up your own property. War is war, it is mad and horrible; the only thing worse is Civil War; I ask our youth: Don't have another. Ireland can't win. Think of Ireland; it is your duty to do so. If you love Ireland – I feel you do – do not destroy, build up Ireland.[312]

However, more trouble occurred in Dublin on 24 April, when Gardaí wielding batons made several attempts to seize a blue flag of the Dublin Battalion of the IRA (bearing the inscription in white lettering, 'Óglaigh na hÉireann').[313] This was prominently displayed during a republican parade of around 2,000 people from St. Stephen's Green to Glasnevin Cemetery – the final resting place of many notable personalities associated with the Rising (including Seán T. O'Kelly, Michael Collins, Ann Devlin, Desmond Fitzgerald, Roger Casement, and Constance Marckievicz).[314]

Following a number of running scuffles, six men were arrested and charged with disorderly behaviour and the possession of offensive weapons. A group of Belfast republicans were among the marchers and constantly taunted the Gardaí by shouting: 'You're brothers of Ian Paisley.' Having been closely protected 'by a group of young men wearing lounge suits, cloth caps and leather gloves', the marchers eventually brought the contentious flag to the gates of Glasnevin Cemetery, where they managed to rush it through the assembled Gardaí 'in close formation like rugby forwards'. The Gardaí then kept their distance as the republicans unveiled a memorial at the grave of 16 men of the Irish Volunteers

---

[312] *The Irish Independent*, 23 April 1966.

[313] Ibid., 25 April 1966.

[314] For details on the location of various 1916-related plots at Glasnevin Cemetery, see Fact Pack Travellers' Guides, *Glasnevin Cemetery* (Morrigan Books, Killala, 1997), pp. 6, 8, 10–16, 29. Some of the high-profile casualties were buried in the cemetery's main section, located on the northern side of the Finglas Road. However, many of the rank-and-file members of the Irish Volunteers and Irish Citizen Army who perished in 1916 were buried in a mass grave in the St. Paul's section of the cemetery, located on the southern side of the Finglas Road. Further details on the rebels buried in Glasnevin can be found in R. Bateson, *They Died by Pearse's Side* (Irish Graves Publications, Dublin, 2010), pp. 325–37.

and Irish Citizen Army who were killed in the Rising.[315] As a result of the earlier exchanges, at least four civilians and one Garda ended up in Jervis Street Hospital while another 12 were treated for injuries at Parnell Street.[316] Five days later, in a protest against the Gardaí's efforts to disrupt the parade, two groups of Sinn Féin members picketed the GPO and Mountjoy Prison in Dublin.[317] In a Dáil debate on 3 May, however, the Minister for Justice defended the actions of the Gardaí in a robust manner, noting that 'the [blue] flag in question purported to be the flag of a military organisation' and that it represented 'a challenge to the Oireachtas ... in which is vested the exclusive authority in regard to military force in this country'. He also stated that the parade's organisers had been 'informed in advance that the display of such a flag was contrary to the law and that it would not be tolerated'.[318]

Another incident that added to the commotion at Eastertime involved groups of Irish language enthusiasts (linked to the Cumann Chluain Ard branch of the Gaelic League), who commenced a five-day hunger strike at 12.00 noon on Easter Monday in both Dublin and Belfast. This lasted until 3.30 pm the following Saturday. The protest was coordinated by the radical Misneach language movement, who felt that a proper effort had not been made to revive the Irish language. Its time-span was 'fixed to mark the duration of the Rising' and was supplemented on Easter Monday by pickets at the Garden of Remembrance and Leinster House. In both locations they attracted large crowds as they paraded in the rain carrying placards that read: 'Celebration?' and 'Éire 1966: Mo Chlann Féin a Dhíol a Máthair' (Irish for: 'Ireland 1966: My Own Children Sold their Mother').[319] To mark the beginning of the hunger strike, a ceremony was held at the GPO, where a wreath was laid at the base of the Cúchulainn statue by Joseph Clarke, an 84-year-old veteran of the battle at Mount Street Bridge in 1916.[320] The protesters, who included a number of Irish language writers, claimed that they were on hunger strike because they did not believe that the men of 1916 gave their lives 'to have their death celebrated but rather that their aims be achieved'. Elaborating, they claimed that economic, intellectual and physical independence had not been achieved for Ireland, and that the decline of the Gaeltacht was 'the surest yardstick' of the failure of political 'huxters' of

---

[315] *The Irish Times*, 25 April 1966.

[316] *The Irish Independent*, 25 April 1966.

[317] *The Irish Times*, 30 April 1966.

[318] *Dáil Debates*, Vol. 222, 3 May 1966.

[319] *The Irish Times*, 12 April 1966.

[320] *The Irish Press*, 12 April 1966.

the day.[321] A total of 12 men and a woman were involved in the Dublin protest in a dilapidated building at Little Denmark Street, while six men participated in Belfast, where they lay on beds in a room in Hawthorn Street. The protest in Dublin was followed by a press conference conducted entirely in Irish at Dublin's Gresham Hotel. Sipping soup, the protestors told journalists that they were feeling 'a bit weak on the feet'. Their spokesman, Pádraig Ó Clereigh, said that the hunger strike 'was also concerned with economic independence, the end of Partition, and equal educational rights for all the citizens, regardless of their ability to pay'.[322]

Overall, the hunger strike failed to arouse large-scale support from the Irish public. Official language bodies refused to sympathise with the protestors, and the only financial support Misneach received was £5 from Sinn Féin's Cumann Séan Ruiséal.[323] In a letter to *The Irish Independent*, Patrick Browne of the apolitical Language Freedom Movement, called for 'a return to realism' on the language question and put forward a case for the maintenance of English as 'the normally spoken tongue'. He also called upon 'all men and women in Ireland to put an end to dramatic gestures and emotional name-calling in this matter and to seek instead a scientific investigation of the facts'.[324] Messages of support for Misneach, however, were received from the Students' Representative Council of University College Cork and from 23 secondary school teachers. The hunger strikers' plight also received international attention, when they were interviewed by English, American, German, and Italian television.[325] Another gesture of support in Ireland was sounded in the pages of the *Gaelic Weekly* by the GAA, who proclaimed that the Misneach folk were 'not cranks or mischief-makers', but rather 'intelligent, well-educated people ... [including] brilliant writers'. It concluded that the movement's protest would not be in vain if they managed to 'get some of the general public to reconsider their lackadaisical attitude towards the [language] revival'.[326] The *Irish Catholic* newspaper was also quick to point out that the hunger strike in Belfast had enabled the Irish language receive 'a bigger advertisement in the North than it has received in a long time'.[327]

In the Republic, however, the ambitions of Irish language enthusiasts received a setback towards the end of August, when the preliminary report of

---

[321] Pamphlet Entitled 'On Hunger-Strike Against the Huxters', James Hardiman Library, NUI Galway, G 26, Pádraig Ó Mathúna Papers.

[322] *The Times*, 12 April 1966; *The Sunday Independent*, 17 April 1966.

[323] *The Irish Independent*, 18 April 1966.

[324] Ibid., 25 April 1966.

[325] *The Irish Times*, 18 April 1966.

[326] *Gaelic Weekly*, 16 April 1966.

[327] *Irish Catholic*, 21 April 1966.

the 1966 Census revealed just how bad the plight of the native tongue was. Even though the state's population had risen since 1961, the number of native Irish speakers declined to less than 70,000, compared with almost 400,000 speakers in 1926. For the rest of the Golden Jubilee year, tensions remained high amongst Irish language speakers. When the Language Freedom Movement held a meeting at the Mansion House in Dublin on 21 September, around 2,000 people attended. But most of them 'were unfriendly to the organisers' and began heckling them. One attendee made an unsuccessful attempt to seize a Tricolour that was displayed on the platform and as he was being led away, 'a shower of papers was flung at the stage and a stink bomb was left off'. A fight then broke out, but calm was finally restored after the Chief Executive of Gael Linn, Dónall Ó Moráin, intervened by mounting the platform and calling for calm. This action 'probably averted a riot'.[328] Notwithstanding the arresting decline in the number of speakers, it is clear that the native tongue was still regarded as a key ingredient of Irish heritage, culture and identity by many nationalists in 1966. By vehemently exalting the language's merits at the time of the Rising's 50th anniversary, the radicals of Misneach seemed to be rearticulating the age-old traditions of resistance to Anglicisation, in a similar fashion to the Irish-Irelanders of Pearse's generation.

The Diaspora Remembers

On Good Friday, Seán MacEntee delivered a commemorative speech entitled 'The Genesis of 1916', in which he argued that the Rising had been designed 'to attract the attention and evoke the sympathy and support of the nations'. It awakened, he argued, 'the conscience of the Irish abroad' and 'gave to the Irish everywhere high cause for pride'.[329] Given its special place in the hearts and minds of the Irish Diaspora around the world, several prominent overseas organisations were unsurprisingly active in holding events to commemorate the Golden Jubilee. This was especially the case in the USA, where millions claimed Irish ancestry.[330] In the lead-up to Easter 1966, the interest shown by Irish-Americans

---

[328] Tobin, *The Best of Decades*, pp. 152–53.

[329] Text of a Talk by MacEntee on 'The Genesis of 1916', Delivered to a Cáirde Fáil Lunch, 8 April 1966, UCDA, MS. P.67/21, Archives of the Fianna Fáil Party, Seán MacEntee Papers.

[330] Many years earlier, in *The Republic*, 2 August 1919, Hanna Sheehy-Skeffington memorably captured the essence and importance of the Irish-American bond, when she remarked that the Irish question 'has been interwoven into the warp and woop of American politics'. Furthermore, she added that 'sentiment for Ireland among American citizens with any drop of Irish blood in their veins' amounted 'to a passion'.

in the Golden Jubilee was evident from the many letters that were sent from the USA to the Irish government. Typical of the mood in America at the time was a letter that was sent on 5 April by Mrs. H. T. Diehl from California, who asked the Taoiseach to 'accept congratulations and admiration for those who stayed to fight and win from a descendant of those who ran away'. Ireland 'can do no wrong', she wrote, before adding that she wished 'the USA had the same magic right now'.[331]

Much correspondence on operational matters connected with the commemoration was received by the Department of External Affairs from the top echelons of American society. This illustrated how close ties were between those wielding powers in both countries. In a letter sent from New York on 30 March, for example, Judge James J. Comerford informed Frank Aiken of his intentions to arrive in Dublin with his wife on Easter Saturday and to stay at the Gresham Hotel. Comerford also let it be known that various members of the Ancient Order of Hibernians would be travelling to Ireland for the occasion, and requested that their presence be made known to 'some responsible members' of the 'Coiste Cuimhneacháin'. The judge was also keen to emphasise in his letter that the organisation's secretary, John F. Geoghan, was a high-ranking civilian administrator in the New York City Police Department, whose parents 'were natives' of County Offaly. In a second letter to Aiken, the judge confirmed that he would be staying in Ireland for three weeks and stated that he was particularly keen 'to participate in as many of the Easter Week ... activities as possible'.[332]

Dublin and Shannon Airports were extremely busy throughout the second week of April 1966, as large numbers of Irish-Americans travelled back to Ireland for the Easter break. Many made the trip purely to experience the Rising's commemoration – which was dubbed 'the Golden Anniversary of Ireland's fight for freedom' by a New York-based exile in his 'American Newsletter' for the *Western People*.[333] However, heavy fog resulted in chaos in airports, while the travel plans of others were almost hampered by more curious incidents. One such episode involved the 130-strong American Easter Rebellion Association, who arrived at Shannon from New York on 7 April, on board a KLM jet named Sir Alexander Fleming (which had been specially chartered by Irish Airlines). The original jet that had been scheduled to bring the group to Shannon was switched at the last minute in New York by 'an alert official', who saw that the name of the aircraft was Sir Winston Churchill. Having been alerted to the

[331] Professor and Mrs. H. T. Diehl to Seán Lemass, 5 April 1966, NAI, Department of the Taoiseach, 97/6/163.

[332] Judge James J. Comerford to Frank Aiken (Two Letters), 30 March 1966, NAI, Department of External Affairs, 2000/14/72, 50th Anniversary of Easter Rising.

[333] *Western People*, 12 March 1966.

politico-linguistic sensitivities of the Shannon-bound travel party, the KLM authorities came to the decision 'that it would be more diplomatic to put it on a Montreal to Amsterdam run', and instead offer them a substitute aircraft named after the renowned scientist.[334]

The strength of the Irish-American connection was evident for all to see on 7 April, when the former Chairman of the Philadelphia Authority, Michael von Moschizisker, who had been living in Ireland for a number of months, delivered a talk. In this, he spoke about the admiration that many Americans had for Ireland and outlined some possibilities for promoting and encouraging heritage tourism. Of Americans, he noted: 'When we think of Ireland it is in terms of charm, courtesy, wit … and courage … our illusions about … the leprechauns vanished long ago.' While acknowledging that many Americans' knowledge of Irish history was mediocre, he expressed the far-sighted view that there was great potential to attract and educate visitors by making O'Connell Street 'a sculptured greenway to the Liffey', preserving its architectural heritage and offering 'guided tours of the city and its environs'. 'Ireland and Ireland's visitors', he continued, 'should be the richer for it'.[335] One of the highlights of the Golden Jubilee for Irish-Americans was a ceremony held at Iveagh House on 13 April, where de Valera accepted a seven-foot-high bronze statue of Robert Emmet from a group led by Congressman Daniel J. Flood. The statue, which had been sourced from an antique dealer's yard in Pennsylvania, was a replica of the original by Jerome O'Connor, a Kerry-born sculptor.[336] On the following day, the people of Massachusetts paid tribute to the 1916 rebels when Bishop Jeremiah F. Minihan and various officers of Irish-American societies, presented Aiken with a scroll declaring 10 April 1966 as a day of commemoration of the Rising in the State of Massachusetts. This scroll was signed by the Governor of Massachusetts, John Volpe.[337]

A range of Golden Jubilee ceremonies were also held in other states in the USA with Irish communities. In the city of San Diego in California, a commemorative dinner was organised by a committee chaired by businessman Alfred E. O'Brien and Rear Admiral Paul F. Duggan of the US Navy. Proceeds from the event went towards the funding of the Casement monument at Banna

---

[334] *The Irish Times*, 8 and 9 April 1966.

[335] Ibid.

[336] Department of External Affairs, *Cuimhneachán 1916–1966*, p. 57; *The Irish Times*, 14 April 1966. The original statue was presented to the National Gallery of Art in Washington, DC in 1917, while another statue made from the original cast was unveiled by de Valera two years later in the Golden Gate Park in San Francisco.

[337] Department of External Affairs, *Cuimhneachán 1916–1966*, p. 73.

Strand.[338] Unsurprisingly, given the year that was in it, third level students and lecturers showed a lot of interest in a nine-month Scholarship Exchange Programme between the USA and the Republic of Ireland, operated by An Bord Scolaireachtaí. A financial analysis for all the scholarships awarded during the academic year 1965–1966 indicates that the total value came to £42,551, representing an 18.5% increase on the figure of £35,901 that was awarded during the previous academic year. Of the 29 individuals who benefited from the scheme, 21 were Irish. Of the eight American citizens who were awarded scholarships to visit Ireland, half ended up in Trinity College Dublin. One historian, Professor W. S. Sanderlin, was awarded £2,681 for a nine-month stint at University College Galway.[339]

The Golden Jubilee was also marked by Irish communities living in South America. Argentines of Irish descent, for example, marked it by presenting the Irish government with a bust of General José de San Martin, a national hero in Argentina.[340] Nearer to home, the Golden Jubilee was marked in England by a range of concerts, speeches and parades. One commemoration was held by the United Irish League (previously known as the Anti-Partition League), whose Honorary Organising Secretary at the time of the Golden Jubilee was Frank McCabe. According to a report in *The Irish Independent*, he remained committed to working 'through peaceful means' for Irish unity and to awakening 'the conscience of the British people' on the issue of Partition.[341] Organisations such as the Gaelic League and the GAA also served as a forum for consolidating a community spirit amongst Irish emigrants and their descendants in 1966.

The highpoint of the Golden Jubilee celebrations in London occurred on Easter Sunday. A Mass at Southwark Cathedral was attended by the Irish Ambassador, John G. Molloy, at which Archbishop Cowderoy preached a sermon on Irish patriotism.[342] In the afternoon, up to 5,000 exiles attended a rally at Trafalgar Square. Traffic was held up by a colourful party of 500 men and women, led by the South London Girl Pipers. Among the marchers were Old IRA veterans, members of Cumann na mBan and the 1916 London Committee. The Proclamation was then read by the Abbey actor, Eddie Golden.[343] An oration

---

[338] *The Irish Independent*, 26 April 1966.

[339] Minutes of Bi-National Educational Commissions, 1966, United States National Archives and Records Administration, College Park, RG 59: 150: 84/20/6, General Records of the Department of State, Bureau of Education and Cultural Affairs, Board of Foreign Scholarships, Box 34, Ireland.

[340] Department of External Affairs, *Cuimhneachán 1916–1966*, p. 58.

[341] *Irish Independent*, 11 April 1966.

[342] *The Irish Press*, 11 April 1966.

[343] *The Irish Times*, 11 April 1966; *The Times*, 11 April 1966.

was also delivered at the event by Ruairí Brugha (son of Cathal), who appealed to emigrants to indulge in 'practical patriotism' by buying Irish-manufactured products wherever possible. His speech then proceeded to matters of geopolitics. 'Ireland', he stated, 'is in the middle of a herculean effort to rebuild herself'. 'The unity of our country ... and the solution to the problem of the remaining British forces on Irish soil', he predicted, would only 'come about when Irishmen understand that they cannot have two loyalties – claiming to be Irish on the one hand and claiming allegiance to a foreign power on the other'. In words that echoed the writings of Pearse in the lead-up to 1916, Brugha argued that future progress in Ireland could only be achieved through a 'conception of nationality' that 'must be broad and unifying, embracing what is best in the traditions that were brought together in Ireland with the heritage we have from our Gaelic past'. After outlining his dissatisfaction with Partition, he ended his speech by reflecting upon Ireland's standing in international affairs and expressing the aspiration 'that our children realise that there is more to Ireland than an endless conflict with England ... if they are ... to continue the work of putting Ireland on her feet'.[344]

The optimism behind these ambitions was not to be realised in the years that lay ahead, as political tensions in Northern Ireland's six counties escalated out of control, culminating with the outbreak of a prolonged armed struggle by republican paramilitaries against British rule. In the process, the legacy of the Rising became increasingly complicated for subsequent governments of the Republic, who found themselves having to tread far more carefully when negotiating the precarious floorboards upon which the ghosts of 1916 trod, especially during Eastertime commemorations of the revolutionary past.

---

[344] *The Irish Times*, 11 April 1966.

Chapter 5

Renegotiations of Memory, 1966–2005

On Saturday 16 April 1966, weather conditions were atrocious as the GPO played host to the closing ceremony for the week-long official commemorations organised by the 'Coiste Cuimhneacháin' for the Golden Jubilee of the 1916 Rising. 'The commemoration ceremonies', reported *The Sligo Champion*, 'came to an end in the most appalling April weather that Dubliners have experienced for many years'. 'So the people of Dublin', it added, 'looked up despairingly at the weeping skies and wondered whether the entire island of Ireland would sink beneath the waters of the Irish Sea and the Atlantic'.[1] Despite the deluge, the ceremonies went ahead as planned. After the Tricolour on the GPO was lowered by Captain Patrick Coakley, trumpeters then sounded 'Sundown'. Lights were dimmed and the 'Last Post' was sounded, and then the lights were brightened again for 'Reveille'. Under the command of Captain Maurice Downing, a firing party of 120 men lining the roof of the GPO then fired a *feu de joie* and six 25-pound guns in the grounds of Trinity College Dublin fired a 21-gun salute. The spectacle then finished with the playing of the National Anthem.[2] Undaunted by the bad weather, de Valera remained upbeat throughout the ceremony. Buoyed up as he was by his deep personal connection to the story of 1916, he enthusiastically outlined his vision of the future of the nation in an address to the crowds that had gathered along O'Connell Street. In drawing attention to the aspirations of the Proclamation, he spoke about his long-held desire to tackle what he saw as a trilogy of concerns – namely the political unification of Ireland, the creation of an all-island parliament and the restoration of the Irish language. Speaking of the future, he told the assembled masses that they could 'look forward to the people of the North wishing to be with us' in 'a representative all-Ireland Parliament'.[3]

---

[1] *The Sligo Champion*, 22 April 1966.

[2] *The Irish Times*, 18 April 1966.

[3] De Valera, cited in Department of External Affairs, *Cuimhneachán 1916–1966: A Record of Ireland's Commemoration of the 1916 Rising* (An Gnóthaí, Dublin, 1966), p. 64. De Valera's speech was by no means atypical. As noted by J. Bowman, *De Valera and the Ulster Question 1917–1973* (Clarendon Press, Oxford, 1982), pp. 299–300, his 'playing of the "Green Card" … was difficult to resist'. Although his rhetoric had the potential to alienate unionists, Bowman argues that de Valera's speechifying enabled him 'to curb republican extremism and maintain his position

Political reaction from the North to the Irish government's ambitions for an all-Ireland parliament was hostile to say the least. Two days after the GPO ceremony, the Northern Ireland Prime Minister, Captain Terence O'Neill, issued a forceful declaration in which he stated that he 'totally rejected' the views that de Valera had advocated, and went on to proclaim that the North and South were 'poles apart' in political, social and economic terms. The North's 'constitutional heritage', he continued, 'is our most precious possession'. 'Let no-one suppose', O'Neill warned, 'that, in our readiness to promote more friendly relations with the South, we are one whit less determined than our forefathers to retain our loyalty to the British Commonwealth and to prosper as an integral part of the United Kingdom'.[4] O'Neill again reaffirmed his resistance to de Valera's stance in an address to the annual conference of the Ulster Unionist Council in Belfast on the night of 30 April. Referring to the tensions that had been created by Golden Jubilee commemorations in Northern Ireland during the Easter period, the Prime Minister commended citizens for showing restraint under pressure and firmly declared: 'We will never allow anyone to separate us from Britain.' 'Our patience', he continued, 'was sorely tried by offensive demonstrations which dragged on over two weekends'. O'Neill also conveyed his belief that his own government's response to the 'nonsense' caused by the Golden Jubilee had been both fair (by permitting many republican parades to take place) and firm (as in the case of the ban on the parade at the Loup and on cross-border trains).[5]

The unionist politicians of Northern Ireland were by no means the only people who were prepared to take de Valera to task over his pronouncements about the unification of Ireland. Earlier in April, there had been a guarded reaction from certain elements of the media in the Republic to attempts at using the Golden Jubilee to stir up passions about unification. An Editorial in the *Limerick Leader* on 9 April, for example, called for great caution regarding matters of geopolitics. Whilst conceding that the 'splitting of hairs' over unity 'was certainly a burning issue in the years gone by', and recognising that a minority of the community was refusing to participate in commemorations bearing the government's 'official imprimatur', it remarked that 'today, idealistic considerations [about unity] cut very little ice with the great majority of the population, intent only on the more mundane task of earning their bread and

---

as the leader of anti-Partitionism'. Lemass was just as forthright about the national question, as can be seen from a Transcript of an Article by An Taoiseach, Seán F. Lemass, Entitled 'And Now for the Next Fifty Years', National Archives of Ireland (hereafter NAI), Department of the Taoiseach, 97/6/163. In this, Lemass argued that an Ireland 'reunited in harmony within our natural borders' depended upon 'the sense of national purpose which inspires our people'.

4   *The Irish Times*, 19 April 1966.
5   Ibid., 30 April 1966.

butter'.[6] Similar sentiments were expressed again by the press in the days after de Valera's speech at the closing ceremony. On 21 April, an Editorial comment appeared in *The Connaught Telegraph*, which stated that 'it is not natural that we should harp upon it [1916] indefinitely'. It also urged the nation to 'look forward and march forward ... by mutual understanding and cooperation' and to accept 'the simple fact that the aims and aspirations of fifty years ago ... are today fifty years behind the times'. 'The day of shouting and flag-waving', it suggested, 'is over'.[7] De Valera's speech was also criticised in the letters page of *The Irish Times*. 'Nothing could more clearly underline the real issue to be faced', wrote a disgruntled Ballsbridge resident on 27 April, 'than this total failure of such an eminent person to see, even in 1966, that these two objectives are mutually exclusive'. The letter-writer also argued that the President's dialogue had hindered rather than helped 'the cause of national unity'.[8]

Deflating 1916: The Impact of the Troubles

In the years that followed 1966, dark clouds of gloom and despondency descended upon the political climate in Northern Ireland. Due to the prolonged and bitter conflict that subsequently unfolded, the issue of the Golden Jubilee's impact upon the course of late twentieth-century Irish history was destined to become a subject of much debate amongst politicians, journalists and historians. As seen in the previous chapter, the Ulster Constitution Defence Committee's hostile reaction to the 1916 commemorations played a major role in heightening sectarian tensions in the North. With Ian Paisley at its helm, it took to the streets of Belfast after Easter 1966 to reiterate its members' determination to preserve the union with Britain. But even greater political instability was to follow in the summer. In the loyalist stronghold of the Shankill Road, a group of men led by Gusty Spence formed a new paramilitary group called the Ulster Volunteer Force or UVF (named after the Protestant force that was established in 1913). This organisation of extreme loyalists was heavily influenced by Paisley's oratory and was ultimately responsible for killing two Catholics in the month of June. In addition to these deaths, the UVF also carried out violent petrol bomb attacks on Catholic properties and businesses. Although these killings were a manifestation of the deep sectarian divisions that had lingered on since Partition, it has been suggested that other factors may have played a part too. In *Lost Lives*, a tome

---

6   *Limerick Leader*, 9 April 1966.
7   *The Connaught Telegraph*, 21 April 1966.
8   *The Irish Times*, 27 April 1966.

which has meticulously documented the stories of all those who lost their lives in Northern Ireland's Troubles, the UVF killings have been placed within the context of a 'rise in unionist and Protestant anxieties about the perceived threat posed by republicans in the wake of large-scale ceremonies marking the 50th anniversary of ... 1916 ... in Dublin'.[9] Likewise, Ed Moloney has interpreted the UVF killings as something akin to 'a pre-emptive strike', carried out in response to its belief 'that the IRA was planning a new violent campaign', inspired by the Rising's Golden Jubilee.[10]

The outbreak of more protracted violence in the North in the late 1960s had profound consequences for the way in which 1916 was remembered thereafter – both north and south of the Border. A complex set of factors and circumstances impinged upon the turmoil, and these are worth briefly recalling. For the duration of the Troubles, which lasted for around 30 years and resulted in more than 3,500 deaths, life for the citizens of Northern Ireland was blighted by the horrors of violent conflict (which ultimately resulted in 1916 commemorations being toned down considerably in the Republic). The North, as Victor Mesev, Peter Shirlow and Joni Downs have explained, became 'synonymous with conflict and ethno-religious sectarian violence', but over the course of time, the root causes of the violence remained much contested:

> it is understood that during the Troubles, republicans fought an anticolonial struggle against state repression and loyalist intimidation. In contrast, loyalists viewed the war as conditioned by protecting the Protestant people and repelling the threat from armed republicans. Less is certain about the role of the state as a defender because there are no accurate figures concerning the extent of collusion and thus state violence can only be attributed by known responsibility.[11]

---

9   D. McKittrick, S. Kelters, B. Feeney, C. Thornton, and D. McVea, *Lost Lives: The Stories of the Men, Women and Children who Died as a Result of the Northern Ireland Troubles*, 3rd Edition (Mainstream Publishing, Edinburgh, 2008), p. 25. Further details of the killings are included in ibid., pp. 25–28. The first of the UVF's victims was a 28-year-old Catholic storeman, John Patrick Scullion, from the Clonard area of the Falls district of West Belfast. Scullion was shot near his home while returning from a pub and died two weeks later on 11 June 1966. The second victim was Peter Ward, an 18-year-old Catholic barman in the International Hotel, who came from Beechmount Parade off the Falls Road. Ward died on 26 June 1966, after being gunned down in the Shankill Road area.

10   E. Moloney, *A Secret History of the IRA*, 2nd Edition (Penguin Books, London, 2007), p. 61.

11   V. Mesev, P. Shirlow and J. Downs, 'The Geography of Conflict and Death in Belfast, Northern Ireland', *Annals of the Association of American Geographers*, Vol. 99, No. 5 (2009), pp. 901–902.

So when, where and why exactly did things start to go wrong?

Although the summer months of 1966 were marred by increased violence and sectarian tensions, it was not until the end of the 1960s that some of the problems created by almost half a century of Irish Partition came to the surface and worryingly flared up out of all proportions. A wide range of factors contributed to Northern Ireland's political instability. In the lead-up to 1966, there was an infuriated reaction by a number of diehard unionists to O'Neill's policies of moderate reform, while his meetings with Lemass also provoked much consternation. Despite the conciliatory and friendly nature of the meetings between the two leaders, there were others in Stormont who were more uneasy about recognising pluralist pasts and presents. As seen in the previous chapter, Brian Faulkner's speech to the Apprentice Boys of Derry on Easter Monday 1966 (in which he infamously accused the North's nationalists of 'resurrecting an insurrection'), served as one of the most vivid indications of the polarities that still remained between elements of the governments of Northern Ireland and the Republic of Ireland. As one letter-writer to *The Irish Times* argued afterwards, the 'few words' uttered by Faulkner had 'destroyed the understanding forged by men of goodwill on both sides of the Border'.[12]

Another factor in the North's volatility in the mid to late 1960s, as Ed Moloney has demonstrated, was the emergence of a far greater 'Catholic and nationalist assertiveness'. This was exemplified by the emergence of a better educated yet more frustrated Catholic middle class. Following the O'Neill-Lemass meetings, the entry in 1965 of the Northern Nationalist Party to Stormont signified 'a step that signified recognition to the Northern Ireland state'. A further sign of the new tone 'was an increasing willingness of Catholics to stand up and demand their rights', especially in relation to public housing and employment. Influenced by the National Association for the Advancement of Coloured People's (NAACP) campaign for black civil rights in the USA, the Northern Ireland Civil Rights Association was established in Belfast in 1967 with the objective of instigating four major reforms – namely the ending of a rule that limited to property owners the right to vote in local authority elections, the ditching of gerrymandered electoral wards, the dissolution of the B Specials, and the elimination of the Special Powers Act.[13]

On 5 October 1968, a banned civil rights march in Derry was baton-charged off the streets by RUC officers, leading to the injury of 88 demonstrators. An RTÉ camera crew captured scenes of protestors being scattered by water

---

12   *The Irish Times*, 27 April 1966.
13   Moloney, *A Secret History*, pp. 61–62.

cannon and these were later broadcast in news bulletins around the world.[14] Incidents such as this sparked wider instability in Northern Ireland. This was especially the case in 1969, which was marked by an upsurge in confrontations – largely though not exclusively – between Catholic civil rights demonstrators and Protestant opponents. According to Gabriel Doherty, the clashes had the effect of producing 'an increasingly tense atmosphere which erupted into open communal violence'. In an attempt 'to restore calm', the British Army was eventually deployed onto the streets of the North on 14 August 1969. However, as Doherty notes, miscalculated policies such as the curfew imposed on Belfast's Falls Road in July 1970 and internment without trial in August 1971 backfired on the British.[15] Thereafter, as Tommy McKearney has written, the Provisional IRA 'assumed much greater influence thanks to its increased capacity as a result of its ranks being swollen with new recruits responding to internment'.[16] Thus the early years of the 1970s, as Richard English has observed, were characterised by 'appallingly high levels of killing', as a result of bombings and conflict. As the formidable forces of nationalism and violence intersected, the Provisional IRA's operations embodied 'the force of aggressive ethno-religious identity as a vehicle for historical change'.[17] On many occasions, the physical force legacy of 1916 was cited as legitimate justification by those seeking validation for a range of paramilitary actions.

One reporter who vividly chronicled the calamitous times was Kevin Myers, who worked as RTÉ News's Belfast correspondent during the early 1970s. In his memoir, *Watching the Door*, he later recalled how he had 'witnessed the bloody chaos' that had ensued after 'the tribe' had been 'exalted over the individual', and 'personal morality' had been 'abandoned to the autonomous ethos of some imagined community'.[18] For Myers, the 'fleeting entity' of the 32-county Republic that had been 'declared in the 1916 Rising in Dublin', managed to remain 'as real and as vital' to Northern nationalists at the beginning of the 1970s, 'as it had been at the moment of its declaration'. After interviewing one die-hard nationalist from the Clonard area of Belfast, the intrepid reporter was

---

[14]   Ibid., pp. 63–64.

[15]   G. Doherty, 'Modern Ireland', in S. Duffy (ed.), *Atlas of Irish History*, 2nd Edition (Gill and Macmillan, Dublin, 2000), p. 128.

[16]   T. McKearney, 'Internment, August 1971: Seven Days that Changed the North', *History Ireland*, Vol. 19, No. 6 (2011), p. 34.

[17]   R. English, *Armed Struggle: A History of the IRA* (Macmillan, London, 2003), p. xxiii–xxiv.

[18]   K. Myers, *Watching the Door: A Memoir 1971–1978* (The Lilliput Press, Dublin, 2006), p. vii.

left in no doubt as to how important the story of 1916 remained at the time to Catholics in the North:

> It [the Republic declared in 1916] lived within him, awaiting incarnation beyond his flesh ... a reverse of the Catholic sacrament of communion. Listening to him talking about the Republic, which had been momentarily seized before being lost, it was as if his Ireland had died on the cross, and we were waiting for its return on a somewhat delayed third day ... People welcomed the message that there was a cure to all their woes ... a united Ireland, achieved by the purifying flame of war ... That's what he believed ... so too did many Northern Irish nationalists.[19]

As the death toll from the Troubles increased during the 1970s, the Rising's legacy became an extraordinarily sensitive political issue for successive governments in the Republic.

A rewriting of the history of the revolutionary past also occurred. As Cal McCarthy has noted, the story of 1916 began to be reinterpreted in more negative tones as the actions 'of a group of destructive dreamers who tore a city [Dublin] asunder in a fight which could never have resulted in victory'.[20] In the long run, therefore, the Troubles had huge implications for the way in which subsequent governments in the Republic approached not only the Northern Ireland question, but also by consequence, the increasingly awkward question of how to approach Eastertime commemorations of the 1916 Rising when paramilitary violence was ongoing. Although the nationalist Social and Democratic Labour Party (SDLP), which abhorred bloodshed, managed to retain majority support from the nationalist and Catholic community in the North, support was strong in certain areas for Sinn Féin and the Provisional IRA. Throughout the course of the Troubles, it was common for supporters of republican paramilitaries to perpetuate the idea – through the use of written statements, speeches, wall murals, and other forms of symbolism – that they were the true heirs to the 1916 rebels. Firmly believing that they were engaged in a legitimate war to free Ireland from British rule and thus unite the island, spokesmen for the paramilitaries frequently pointed to the links between the geopolitical objectives of their own armed campaign and those of the 1916 rebels.

Within a few years of taking over in 1966 as Taoiseach from Sean Lemass, Jack Lynch faced the delicate task of reaffirming Fianna Fáil's republican credentials, at a time in which the Provisional IRA was waging its armed struggle for political ends. On many times throughout his stints as Taoiseach,

---

[19]   Ibid., p. 20.

[20]   C. McCarthy, *Cumann na mBan and the Irish Revolution* (The Collins Press, Cork, 2007), p. 51.

which lasted from 1966–1973 and 1977–1979, Lynch remained steadfast in dismissing the notion that Irish unity could be achieved by the same means used by the 1916 rebels – violence. In a speech which he delivered to a challenging Fianna Fáil Ard Fheis on 17 January 1970, for example, Lynch remarked that while Partition was 'a deep, throbbing weal across the land, heart and soul of Ireland ... emotionalism and the brand of impetuous action or demands that it leads cannot possibly solve, or even help in dealing with such a problem'.[21] Again and again, Lynch showed that he was particularly dubious about Fianna Fáil's 'radical republican camp' and avoided using 'traditional emotive ... rhetoric' when commenting on matters like Partition.[22] This stance owed a lot to his personal background, for unlike previous leaders, Lynch had not been alive when the 1916 Rising occurred. He was born in a house on Cork city's northside on 15 August 1917 and first rose to prominence as a result of his reputation for prowess on the sporting field with the Cork GAA team – with whom he won six All-Ireland medals in a row.

In a biography, Dermot Keogh has placed great emphasis on the significance of Lynch's personal milieu, writing that he 'was not a scion of a "great" republican family'. On the contrary, he notes that he 'lived in a very different era' to old Fianna Fáil – for he was the party's first post-revolutionary leader and was 'not a survivor of the 1916 Rising'. His political pluralism and openness, however, worked to his advantage by enabling him 'to stand out against the forces of atavism' that exposed Ireland to the risk of 'returning ... to a state of war and mayhem'. This was especially apparent during the Arms Crisis of 1970, when Lynch 'provided decisive leadership' by successfully standing 'against physical-force nationalism' and mobilising the state's institutions 'to fight the subversion of a reconstituted IRA'.[23] Following reports of a conspiracy to import arms into Dublin illegally (for the relief of Catholic nationalists in distress in the North), Lynch sacked two members of his Cabinet on 6 May, namely Charles Haughey, the Minister for Finance and Neil Blaney, the Minister for Agriculture. A third minister, Kevin Boland, resigned his position in protest over the sackings.[24] Although Lynch, according to Roy Foster, could have taken action against his hawkish colleagues 'earlier', he was most likely 'inhibited by his position within

---

[21] Presidential Address by the Taoiseach, Jack Lynch, TD, to the Fianna Fáil Ard Fheis, Mansion House, Dublin, 17 January 1970, University College Dublin Archives (hereafter UCDA), MS. P.176/776/4, Archives of the Fianna Fáil Party.

[22] D. Keogh, 'Jack Lynch and the Defence of Democracy in Ireland, August 1969–June 1970', *History Ireland*, Vol. 17, No. 4 (2009), p. 30.

[23] D. Keogh, *Jack Lynch: A Biography* (Gill and Macmillan, Dublin, 2008), pp. 286, 473.

[24] M. Mills, 'Arms Crisis (1970)', in B. Lalor (ed.), *The Encyclopaedia of Ireland* (Gill and Macmillan, Dublin, 2003), p. 46.

Fianna Fáil, which was that of the compromise candidate who had succeeded the charismatic Lemass'. Nonetheless, 'his sane and careful response', argues Foster, was 'spectacularly vindicated with time'.[25]

While the Arms Crisis created long-lasting divisions within Fianna Fáil, Lynch in the early 1970s remained steadfast in his determination not to be accused of sending out the wrong message to republicans advocating physical force. On many an occasion, therefore, he used a cautious tone when speaking about the contemporary significance of revolutionary events such as the 1916 Rising. On 22 February 1971, Lynch faced his most difficult Ard Fheis, as party divisions surfaced over both Partition and the deterioration of the security situation in the North. During the course of the event, which was held at the Royal Dublin Society (RDS) in Ballsbridge and attracted around 5,000 delegates, his leadership was persistently tested by the heckling cries from the supporters of Haughey, Blaney and Boland. When Lynch's close ally, the Minister for External Affairs, Dr. Patrick Hillery, approached the podium to speak, he was interrupted by an incensed (and finger-wagging) Boland. As scuffles broke out on the floor, the normally mild-mannered Hillery was resolute in his support for Lynch. 'You can have Boland, but you cannot have Fianna Fáil', he bellowed, in an impromptu outburst that became one of the most memorable TV images of the period.[26] Lynch was of course fortunate to have had such strong allies in the Cabinet, who were also unprepared to raise the ghosts of 1916. As Ronan Fanning has observed, Hillery's 'self-appointed role' after replacing Frank Aiken as Minister for External Affairs in June 1969 had been 'to serve as a lightning rod to soak up republican rage within Fianna Fáil', to resist pressure to invade the North and to formulate an administrative policy on it (despite opposition from certain party colleagues and senior civil servants).[27]

Even though Lynch managed to face down internal party dissent during the 1971 Ard Fheis, he had no shortage of detractors elsewhere in the days that followed. One of these was a New York-based Irish-American organisation known as the Impeach Lynch League. On 26 February, its Corresponding Secretary wrote an impassioned letter to President Richard Nixon (which was

---

[25] R. F. Foster, *Luck & the Irish: A Brief History of Change, 1970–2000* (Allen Lane, London, 2007), p. 111.

[26] Hillery, cited in Keogh, *Jack Lynch*, p. 281. Also see Hillery's obituary notice in *The Irish Times*, 14 April 2008.

[27] *The Sunday Independent*, 20 April 2008. Also see J. Walsh, *Patrick Hillery: The Official Biography* (New Island, Dublin, 2008), pp. 255–56, who argues that Hillery 'remained a crucial figure in establishing constitutional nationalism as a viable political alternative in the face of increasing paramilitary violence'.

duly dispatched to the Director of the Secret Service), in which he protested against the White House's plan to entertain the Taoiseach during the upcoming St. Patrick's Day celebrations. In explaining the organisation's opposition to the invitation, he went so far as to allege that the Taoiseach's 'action in arresting Irishmen and placing them on trial for conspiracy' had 'led to the widely held belief that Lynch acted in this instance on information received from Britain's Secret Service'. He also claimed that links could be drawn between the predicaments of Sir Roger Casement in 1916 and those who had been put on trial during the Arms Crisis. 'The forgeries of the Roger Casement trial', he contended, 'is the sort of thing these people excell [*sic*] at and when a so-called Taoiseach acts in collusion with such disgusting men in order to convict fellow Irishmen he deserves this damning title "felon setter"'.[28]

When he finally embarked upon his official three-day visit to the United States from 15–17 March, Lynch was unwavering in rebuffing criticism, especially from those who had accused him of shortcomings with regards to his policy position on the North. On a number of occasions, when opportunities presented themselves, he resolutely reaffirmed the Irish government's policy on the national question. After finally meeting President Nixon in private for an hour on 16 March, Lynch told a press conference in Blair House that he remained committed to Irish unity by peaceful means.[29] At another event in Philadelphia on St. Patrick's Day, he delivered another firm message to those who were willing to endorse armed struggle in the North or summon up the ghosts of 1916. 'Every act of violence', he argued, 'is retarding the ultimate unification of Ireland'. 'We don't like to have British troops in Northern Ireland', he explained, 'but it seems they are necessary'.[30]

Within a month of his return from the United States, the Republic of Ireland commemorated the 55th anniversary of the 1916 Rising by soberly staging a half-hour military parade of approximately 1,800 military personnel past the GPO on a warm and sunny Easter Sunday, on 11 April. Among those marching were cadets from Zambia, who had come to Ireland for training. According to *The Irish Times*, a 'frail' President de Valera (who had missed the previous year's ceremony due to a bout of laryngitis) 'arrived promptly on schedule' before the main proceedings commenced, accompanied by an escort of 29 motor-cyclists from the 2nd Motor Squadron. He then stood to attention

---

[28]   Visit of Prime Minister Lynch of Ireland, March 1971, United States National Archives and Records Administration, College Park (hereafter USNARACP), RG 59: 150: 73/16/5-6, General Records of the Department of State, Bureau of European Affairs, Office of Northern European Affairs, Records Relating to Ireland, Box 7.

[29]   *The Irish Times*, 17 March 1971.

[30]   Ibid., 19 March 1971.

in front of the GPO for the playing of the National Anthem, carrying a top hat in his right hand and 'an unseasonal umbrella' in his left hand. Following the sounding of the 'Presidential Salute' by the No. 1 Army Band, de Valera then assumed a commanding position on the saluting platform and 'scarcely moved' until the parade of soldiers and armoured cars, under the command of Colonel William Barrett, had ended. As per usual, the spectacle also included flyovers by helicopters and Vampire jets. A few tourists watched the proceedings, and despite some heckling from a small group of nationalists styling themselves as The 1971 Easter Protest Group (who carried placards protesting against issues such as the Criminal Justice Bill, Partition and membership of the European Economic Community), 'the overall impression was of a small, but good humoured crowd'.[31]

By the end of the day, the 1916 commemoration was somewhat overshadowed in the news bulletins by excitable reports of what had happened in Belfast, where the GAA's annual congress had taken the historic decision to remove the controversial Rule 27 from its handbook (which had banned its members, for over 66 years, from playing or attending 'foreign' games like soccer, rugby and cricket). The removal at the same congress of the lesser-known Rule 29, which had prevented GAA clubs from promoting foreign dances, served as another cultural signpost of the changing times. In the longer term, however, that day's military parade was of major historic significance itself, as the date of 11 April 1971 went down in history as the last occasion in the twentieth century that the Irish Army paraded past the GPO on Easter Sunday as a tribute to the memory of the 1916 Rising. Rather poignantly, the day also marked the last time in his life that the ailing de Valera was to bear witness to an Easter Sunday parade by the Army (Plate 5.1). From the perspective of memory-making in the Republic, matters were clearly in the process of becoming more problematic than ever before.

During the summer of 1971, the government availed of the opportunity to switch the focus of its commemorationist activities to something other than 1916, when it held a series of commemorations to mark the 50th anniversary of the ending of the War of Independence. At the Mansion House on Saturday 10 July, de Valera unveiled a plaque which commemorated the Truce. The following day, he went to the western end of the Garden of Remembrance, where he unveiled a 25-foot-high bronze sculpture depicting the mythical Children of Lir (Plate 5.2) – the Lord of the Sea in ancient Irish legend. This was designed by Oisín Kelly.

---

[31]   Ibid., 12 April 1971.

Plate 5.1   President de Valera watching the Irish Army parade past the
              GPO for the Rising's 55th anniversary commemoration on
              11 April 1971

Source: National Library of Ireland, Independent Newspapers (Ireland) Collection,
R 471/475.

Born in 1915, Kelly 'greatly influenced design in Ireland' through his
work as artist-in-residence at the Kilkenny Design Workshops, where he
completed many iconic works.[32] His Children of Lir sculpture was specifically
dedicated in memory of all those who had fought and died for Ireland. After
de Valera laid a wreath at it, the 'Last Post' was sounded and the National
Flag was then hoisted to full mast. In the course of the ceremony, Lynch
delivered a major policy speech on the national question, during which
he called for a greater recognition of inclusive or pluralist Irish identities.

---

[32]   M. Bolger, *Statues and Stories: Dublin's Monuments Unveiled* (Ashfield Press, Dublin,
2006), pp. 14–15, 20. For an insight into the inspiration behind the design of Kelly's sculpture,
see P. O'Brien, 'Honouring the Dead', *An Cosantóir: The Defence Forces Magazine*, Vol. 71, No. 5
(2011), p. 27, who states: 'The concept was that at certain points in history people are transformed
and the artist used the depiction of human figures transforming into swans, symbolising rebirth,
victory and resurrection, as in the mythological tale of the Children of Lir.'

Plate 5.2   Oisín Kelly's bronze sculpture of the Children of Lir at the western end of the Garden of Remembrance, unveiled on 11 July 1971 by the President, Eamon de Valera

Source: Author.

He said that there were many types of Irishmen and asked that the national majority 'examine their consciences' as regards the national minority.[33] Lynch then proceeded to make a case for a United Ireland, which he felt would lead to reconciliation between the peoples of the North and the South. Criticism immediately came from some of the usual suspects. In his assessment of this situation, Keogh has noted that the Taoiseach's 'emphasis on Irish unity eclipsed his very positive comments about the diversity of Irish identity and the insensitivity of the majority towards minority rights'.[34]

While hanging on delicately to the cherished ideals of the Proclamation, Lynch and Hillery's success in overcoming internal party dissent during the early 1970s ensured the survival of Fianna Fáil's policy line that the only road to Irish unity was through constitutional means. In order to counteract the spread of paramilitary ideology, which came to be seen as a menacing threat to the Republic, serious efforts began to be made by the government to reorientate citizens' hearts and minds and to make a distinction between the actions of the 1916 rebels and those of the Provisional IRA. As Diarmuid Ferriter has shown, 'significant reform' was instigated in the teaching of history in schools during the 1970s.[35] In addition to pedagogical issues, John Coakley has argued that political motivations were firmly behind noteworthy changes that the government made to the primary school history syllabus in 1971. A new teacher's manual emphasised that tuition in history should be 'unspoiled by special pleading of any kind' and highlighted the need to take stock of 'how to present past ill-doing without so arousing the child's emotion as to prejudice his mind in relation to existing conditions'. Prior to that, there had been a 'tendentious approach in ... schools in the four decades since independence' and too much emphasis on a certain uncompleted business had served to provide 'implicit justification for the IRA, who could present themselves as the successors to the "men of 1916"'. While the 1971 teacher's manual for national schools, according to Coakley, introduced a programme that 'was animated by the idea that the child should engage in historical discovery, rather than receiving a strongly nationalist perspective', many of the first generation of the primary school history books that appeared in 1974–1975 still 'tended to use language similar to that of their predecessors in relation to 1916'.[36]

If proof was ever needed of the fragile eggshells that Lynch and his Cabinet colleagues were treading on during these turbulent times, it came in 1972.

---

[33]   Lynch, cited in Keogh, *Jack Lynch*, p. 302.
[34]   Ibid.
[35]   D. Ferriter, *The Transformation of Modern Ireland 1900–2000* (Profile Books Ltd., London, 2004), p. 531.
[36]   *The Cork Examiner*, 4 April 1991.

During this chaotic year, the North experienced 'a spiral of violence which culminated ... with an orgy of shootings and bombings unparalleled since the violence of 1922–23'.[37] One of the most notorious incidents of the Troubles happened on 30 January, when the British Army killed 13 unarmed civilians who were marching through the streets of Derry. The event, which became known as Bloody Sunday, was followed by the burning down of the British Embassy in Dublin, after it was beset by an angry mob. The killings were also denounced around the world by a range of concerned commentators, including many in the United States. In remarks delivered before a subcommittee on Europe in the House of Representatives on 29 February, for example, Congressman Jonathan B. Bingham contended that the killings had demonstrated the 'utter bankruptcy' of Britain's policies in Northern Ireland, which seemed to have transcended into a 'bull-headed' reaction 'to demonstrations of protest'. On the same day, Martin Hillenbrand, the Assistant Secretary of State for European Affairs, gloomily observed that Ireland was 'passing through times as critical as any it has faced' from the time when the Civil War had ended. 'Emotions', he added, 'are understandably running very high', but 'the realities of life in Ireland ... must be dealt with in a spirit of patience and moderation'.[38]

After Bloody Sunday, Lynch tread even more carefully when speaking about the national question. The last thing that he wanted to see was violent recriminations against the authorities in the North or any further attempts to seek Irish unity by physical force. In a dogged speech to his party on 19 February, he noted: 'There is no peace among the exploding bombs. There is no justice in the rule of the gun and there is no unity in setting Irishmen against Irishmen.' 'On the contrary', he added, 'violence simply frustrates the creation of the conditions in which serious political discussions can take place'.[39] Lynch's denunciation of violence at this time was warmly welcomed by many citizens across the state, especially by those in favour of toning down the Rising's upcoming 56th anniversary. One of these was K. R. Wilson from Mount

[37]   Doherty, 'Modern Ireland', p. 128.

[38]   House of Foreign Affairs, 1972, USNARACP, RG 59: 150: 73/16/5-6, General Records of the Department of State, Bureau of European Affairs, Office of Northern European Affairs, Records Relating to Ireland, Box 3. The impact of Bloody Sunday was felt for years afterwards. See N. O'Faolain, *Are You Somebody?* (New Island, Dublin, 2007), p. 143, who observes that Northern Ireland's social revolution had turned into a 'life-and-death war'. Having been immersed (as a television producer for BBC's *Open Door* programme) in the city's Bogside, she wrote that she had 'met for the first time people completely alienated from the state that was supposed to claim their allegiance'.

[39]   House of Foreign Affairs, 1972, USNARACP, RG 59: 150: 73/16/5-6, General Records of the Department of State, Bureau of European Affairs, Office of Northern European Affairs, Records Relating to Ireland, Box 3.

Merrion in County Dublin, who wrote to the Department of the Taoiseach on 23 February, urging it to draw Lynch's attention to a suggestion to abandon the 'usual' military parade on Easter Sunday, 'in view of the present situation in the country'. Wilson also suggested that in light of the government's 'desire for a peaceful solution to our problems, a "Walk of Witness" be substituted'. 'This', he added, 'could be an Ecumenical affair, and would be much more in keeping with the celebration of the resurrection of our Lord and would highlight once again the true Christian character of the country'.[40] Six days afterwards, Wilson received a reply from R. Ó Foghlú of the Department of the Taoiseach, who informed him: 'that as this is a matter for consideration, in the first instance, by the Minister for Defence your letter has been forwarded to him for attention'.[41]

As Easter Sunday approached, both domestic and foreign matters impinged upon discussions regarding the Republic of Ireland's military policy. On the one hand, the government seemed keen on highlighting its military capacity to the wider world by reiterating its ongoing commitment to the contingent of troops that it had sent to Cyprus to serve with a United Nations' peace-keeping force. In a Cabinet Meeting held on 21 March, its obligation to these overseas operations was reaffirmed after 'it was decided ... to maintain the Irish contingent ... at its present strength, for a further period of up to three months', extending beyond 26 March.[42] On the domestic front, however, the pessimistic outlook with regard to the state of affairs in the North meant that change was soon afoot in the way that the government was prepared to display its military wares during commemorations of the 1916 Rising in the capital city. Calls to abandon the military parade in remembrance of the Rising seem to have been well-received by the Minister for Defence. Thus on 24 March, Diarmuid Ó Cróinín of the Department of Defence wrote to the Taoiseach in Irish, informing him that there would be no parade or marching for the upcoming commemoration on Easter Sunday 2 April: 'Mar is eol duit ní bheidh aon paráid mhíleata agus mairseáil thar bráid i mBaile Átha Cliath ar Domhnach Cásca i mbliana' (Irish for: 'As you know, there will be no military parade and march passing by in Dublin on Easter Sunday this year').[43]

---

[40] K. R. Wilson to the Department of the Taoiseach, 23 February 1972, NAI, Department of the Taoiseach, 2003/16/65, Easter Week Commemorations at Arbour Hill, etc.

[41] R. Ó Foghlú to K. R. Wilson, 29 February 1972, NAI, Department of the Taoiseach, 2003/16/65, Easter Week Commemorations at Arbour Hill, etc.

[42] Government Cabinet Minutes, 21 March 1972, NAI, Department of the Taoiseach, 2001/5/1.

[43] Diarmuid Ó Cróinín to Jack Lynch, 24 March 1972, NAI, Department of the Taoiseach, 2003/16/65, Easter Week Commemorations at Arbour Hill, etc.

When this information was relayed to Lynch on the same day, somebody in the Department of the Taoiseach appended a handwritten memorandum. This contained a suggestion that would have been wholly inconceivable only a few years beforehand (most especially to Fianna Fáil's most ardent supporters) – that the Taoiseach stay away altogether from the Easter Sunday commemorations at the GPO and the Garden of Remembrance: 'Taoiseach: Could you not give yourself a break from these ceremonies? The men are dead over 50 years now & they wouldn't mind! May I say you'll be away?'[44] Lynch ultimately rejected this idea and on 28 March, his private secretary wrote to the Department of Defence, confirming that the Taoiseach would indeed be accepting his invitation to the downsized Easter Sunday ceremonies.[45] Even though it had been customary for many years for the Irish Army to parade past the GPO on Easter Sunday, Lynch did not seem too bothered about the cancellation of this time-honoured tradition by the Department of Defence. But the abandonment of the parade did prove rather perplexing to some in the media. On Good Friday, *The Irish Times* reported that the cancellation had been explained by a need 'to release Army personnel and equipment for duty at likely trouble spots over the weekend'. An Army spokesman, it added, had 'said yesterday that no extra troops were being sent to the Border'.[46]

In Northern Ireland itself, parades of all kinds had been banned by the Prime Minister, Brian Faulkner, since the beginning of 1972. Then, following the introduction of direct rule from Westminster on 30 March, the new Secretary of State, William Whitelaw, restated the official ban. 'So, for the present', he stated, 'the law must stand'. To this statement, he then added the conciliatory remark that he would be willing to lift the ban if some kind of compromise could be found in the future between the representatives of different types of marchers.[47] As expected, however, his words were completely ignored and there was the usual widespread defiance of the ban on marches throughout the North on Easter Sunday. Republican parades in remembrance of the Rising's 56th anniversary took place in many locations, including one to Milltown Cemetery in Belfast, where members of the Provisional IRA fired a volley of shots over the republican plot. In the end, the whole day passed off peacefully, and the new

---

[44]   Department of the Taoiseach Memorandum, 24 March 1972, NAI, Department of the Taoiseach, 2003/16/65, Easter Week Commemorations at Arbour Hill, etc.

[45]   H. S. Ó Dubhda to the Department of Defence, 28 March 1972, NAI, Department of the Taoiseach, 2003/16/65, Easter Week Commemorations at Arbour Hill, etc.

[46]   *The Irish Times*, 31 March and 1 April 1972.

[47]   Ibid.

Secretary of State was reported to have watched some of the proceedings from the comfortable vantage point of a helicopter hovering high in the sky.[48]

With the Army parade cancelled for the official commemoration in the Republic, the government instead marked Easter Sunday with a simple flag-raising ceremony at the GPO and a wreath-laying formality at the Garden of Remembrance, both of which involved the participation of Lynch and de Valera. Even though security was very tight on O'Connell Street, with crash barriers in place and a helicopter circling overhead, somebody still managed to cause an inconvenience by setting off a smoke bomb opposite the GPO, shortly before the ceremony commenced. A total of 250 Old IRA men participated in the ceremony, along with a small group of Cumann na mBan women. However, the sight and sound of marching soldiers was missing. After the ceremony was brought to a conclusion with the playing of the National Anthem by the No. 1 Army Band, the political dignitaries were driven to the Garden of Remembrance, where de Valera laid a wreath on behalf of the Irish people. On the same day, Fianna Fáil held its own commemoration of the Rising at the Arbour Hill graves, with Lynch laying a wreath on behalf of the party. A Decade of the Rosary was recited by the party's Assistant General Secretary, Captain J. J. O'Carroll, while the Proclamation was read by Cathal Brugha. The 56th anniversary was also commemorated at Collins Barracks in Cork city, where the Minister for Justice, Des O'Malley, laid a wreath on behalf of the government and the Defence Forces, following a Mass in St. Michael's Garrison Chapel. Other attendees included Fianna Fáil's Pearse Wyse, Fine Gael's Peter Barry, members of the Kent family, Lord Mayor Toddy O'Sullivan, and the Bishop of Cork and Ross, Dr. Cornelius Lucey.[49]

The next day, in its front-page report on the Easter Sunday commemorations, *The Irish Times* lamented the absence of the Army parade on O'Connell Street. 'The public, and the many children who enjoyed this parade every year', it remarked, 'missed the opportunity of seeing the Army's latest Panhards and Unimogs which were bought recently and which would have been the main feature of the parade'. In an Editorial piece entitled 'Rethink', it was suggested that other factors had obviously impinged upon the abandonment of years of tradition. 'Mr. Lynch's action in scaling down the normal military parade', it wryly observed, 'is hardly explained by the need for manning the Border'. All things considered, it reckoned that 'a couple of companies with a band and a few vehicles could mount a reasonable march-past at the GPO'. The Editorial then put forward the following hypothesis to explain the parade's cessation:

48   Ibid., 3 April 1972.
49   Ibid.

'Rather, some see it as a move away from a demonstration which, in the context of current events, he feels could be read as triumphalism. Only Mr. Lynch, with his, at times, labyrinthine mind, could answer that.'[50] As the Army parade was not reinstated until many years later, it appears that *The Irish Times* was spot on in its determination of the real reason for the absence of the Army from O'Connell Street on Easter Sunday 1972 – namely the Taoiseach's fear of the Republic appearing too jubilant at a time in which the North's security situation had become horrendous (as evident from the high levels of violence that had led to the prolongation of draconian measures like internment without trial).

In doing his best not to appear too gung-ho in orchestrating the government's commemoration of the 56th anniversary, Lynch was treading on thin ice as far as some commentators were concerned. The void created by the absence of the Army from O'Connell Street was clearly one that republican paramilitaries (espousing the continuance of an armed struggle) were more than willing to fill, given the right circumstances. This was one of the main reasons for the sense of disenchantment that was conveyed by some sections of the media upon hearing of the cancellation of the Army parade. Lynch, however, was not for turning and almost certainly felt that he had acted with the best of intentions by cancelling the parade, despite continuing to espouse views about Irish unification. No matter what gestures he made, it seemed that the Taoiseach's political viewpoints were open to continual challenge, sometimes quite dramatically. It was not long before proof of the sensitivities associated with the Republic's territorial claim to the North manifested themselves from a most unforeseen source – from north of the Border. On Tuesday 4 April, two days after the GPO ceremony, nine members belonging to a Belfast-based left-wing group known as the Workers' Association for the Democratic Settlement of the National Conflict in Ireland, stormed the Department of Foreign Affairs headquarters at Iveagh House in St. Stephen's Green. After chaining themselves to radiators and statues in the hallway, they staged a two-hour sit-in as part of a protest against the Irish government's constitutional policy on Northern Ireland – namely the 'claim ... to rule over Northern Ireland' that was enshrined in Articles Two and Three of the Republic's Constitution. In explaining the reasoning behind its actions, the spokesperson for the interdenominational group, Jeff Dudgeon, outlined its opposition to a united Ireland: 'As long as the "one nation" myth exercises any significant influence in the South, there will always be those, like the Provisional IRA who will seek to follow out the logic of this assertion.'[51]

---

[50] Ibid.

[51] Ibid., 5 April 1972.

Over three months later, Sinn Féin supporters aired their own grievances against Lynch's government, by staging a protest march to the Army's Curragh Camp in County Kildare on Sunday 9 July – the day before the 51st anniversary of the conclusion of the War of Independence. The demonstrators, according to *The Irish Independent*, wanted an end to government intrusion into the affairs of RTÉ and the granting 'of full political status' to republican prisoners held in the camp. However, the protest eventually turned ugly, as the troops guarding the camp came under attack from a breakaway group of rioters armed with rocks, hurleys and steel bars. In the end, a total of three Gardaí and six soldiers were injured.[52] The Curragh incident was described by *The Irish Press* as the 'biggest' security operation that the country had witnessed in peacetime, and it went on to commend the Army for not being provoked by 'the stoning, the savage invective, the hate' on show.[53] Given the criticism that had emanated from some quarters as a result of the cancellation of the Easter Sunday military parade two months earlier, the Army itself was naturally keen on safeguarding its public image amidst the increasingly volatile conditions of the times. Although the military operation to defend the camp was deemed a success, an official report that was swiftly compiled on the incident reiterated the importance of maintaining sound relations with the media. Even though 'a good relationship' between the Army's press officers and newspaper reporters had been evident on the day of the protest, the report stressed the need for junior officers to be conscious of the media aspect in any 'future confrontations'. 'An unfavourable written comment', it prudently deduced, 'can be just as damaging as a photograph'.[54]

From a historiographical perspective, 1972 was of significance for the way in which the Rising's legacy came under attack in two seminal publications. In a lengthy article in the summer issue of *Studies*, entitled 'The Canon of Irish History: A Challenge', the late Jesuit priest, Father Francis Shaw (a former Professor of Early and Medieval Irish at University College Dublin), offered a powerful and polemical critique of the place of 1916 in Irish history. It had taken some time before Shaw's article saw the light of day. Despite being specially commissioned for the spring 1966 issue of *Studies*, it was initially withheld from publication. His revisions, as Brendan Bradshaw later reflected, 'did not reflect the uncomplicated acceptance by the Irish in 1966 of the Easter Rising as a great event'. Consequently, the Editor of *Studies*, who was also a Jesuit, deemed the article both untimely and unsuitable to the occasion of the

---

[52]   *The Irish Independent*, 10 July 1972.

[53]   *The Irish Press*, 10 July 1972.

[54]   Report: Curragh Protest, 10 July 1972, Military Archives, Cathal Brugha Barracks, Department of Defence, 2003/15/115, G2/C/1919, Demonstrations.

Golden Jubilee commemorations, and therefore 'suppressed it'.[55] Reflecting on the saga, Shaw himself conceded that it was understandable 'that a critical study of this kind might be thought to be untimely and even inappropriate in what was in effect a commemorative issue'.[56] However, he remained undaunted by the hullabaloo and managed to finish a second version of the article in August 1966. It was this longer version that was published in the journal in the summer of 1972 – one and a half years after the author's death.

From the outset of the published version of the article, Shaw objected to the simplistic reduction of Irish history to 'a straight story of black and white, of good "guys" and bad', with a 'virtuous and oppressed' Ireland pitted against 'the bloody Saxon'. 'The truth', he argued, 'is different', as 'the story of Easter Week is not as straightforward a tale as we are asked to believe'. Shaw then proceeded to lay out a challenge to 'an accepted view', namely 'a canon of history' which had 'come into being' and which had stamped 'the generation of 1916 as … in need of redemption by the shedding of blood'.[57] Following a harsh appraisal of Pearse's use of violence for political ends, he then arrived at the determination 'that the resort to arms in support of the separatist doctrine in 1916' had 'inflicted three grave wounds on the body of the unity of Ireland'. These, he noted, were 'the wound of Partition', the damage inflicted by the Civil War and the failure of the island of Ireland to 'be at one' in commemorating the thousands of Irishmen who had fought and died in the Great War.[58] Whatever Shaw's motives had been in writing such a critical piece, it has been observed by D. George Boyce that the article was ultimately successful in casting light on the predicament that the Rising 'posed for Irish political ideas, with its mixture of high romantic endeavour and destructive violence, its self-sacrifice and its infliction of death on others'.[59]

The Rising's place in history was also subjected to a harsh assessment in Conor Cruise O'Brien's book, *States of Ireland*, which was also published in 1972. Combining historical analysis with autobiographical and political perspectives, it put forward the argument that England itself was not the root cause of Ireland's problems. Disparities between nationalists and unionists, he argued, had to be accommodated in order to reach an appropriate solution that

---

[55] *The Irish Times*, 1 April 1991.

[56] F. Shaw, 'The Canon of Irish History: A Challenge', *Studies: An Irish Quarterly Review*, Vol. 61, No. 242 (1972), p. 113.

[57] Ibid., p. 117.

[58] Ibid., p. 151.

[59] D. George Boyce, '1916, Interpreting the Rising', in D. George Boyce and A. O'Day (eds), *The Making of Modern Irish History: Revisionism and the Revisionist Controversy* (Routledge, London, 1996), p. 179.

curtailed enmity. Cruise O'Brien, who was working at the time as a Labour TD for Dublin North-East, also cast a very cold eye upon the repercussions of the Golden Jubilee of the Rising, seeing it in rather pessimistic terms as a 'commemorative year ... in which ghosts were bound to walk, both North and South', not least because of the fact that the anniversary 'had to include the reminder that the object for which the men of 1916 sacrificed their lives – a free and united Ireland – had still not been achieved'. 'The general calls for rededication to the ideals of 1916', he added, 'were bound to suggest ... that the way to return to them was through the method of 1916: violence, applied by a determined minority'. Despite acknowledging the attempts of the Irish government to discourage violence in 1966, Cruise O'Brien felt that there was actually no way of effectively deterring it 'within the framework of a cult of 1916'. In addition to blaming the resurgence of the IRA movement upon the Golden Jubilee and 'numerous commemorations of similar type', he also sought to draw attention to a perception amongst Ulster Protestants that the Dublin commemorations 'seemed a celebration of treachery', 'a threat to "Ulster"' and a sure sign, if ever needed, 'that the leopard had not changed his spots'.[60]

Despite a hostile reaction to *States of Ireland* from influential Northern politicians such as the SDLP's John Hume, Cruise O'Brien's influence on the Irish political scene was destined to increase in the years that followed the book's publication. He was appointed as Minister for Posts and Telegraphs following a General Election on 28 February 1973, which returned a Fine Gael-Labour Coalition led by Liam Cosgrave. In their geopolitical analysis of the results of the General Election, Mervyn Busteed and Hugh Mason have noted that it 'produced a marked change in the Irish political scene'. Fianna Fáil (which by then had held office for all but six years since coming to power in 1932) lost out by four seats to Fine Gael and Labour, even though it had 'campaigned strongly as a national movement' and had placed great weight upon Lynch's 'personality and statesmanship'. In a major blow to those with hawkish tendencies towards the North, Frank Aiken's newly-established radical republican party, Aontacht Éireann ('Irish Unity'), had a catastrophic election. Having run 13 candidates

<hr>

[60] C. Cruise O'Brien, *States of Ireland* (Hutchinson & Co. Ltd., London, 1972), p. 150. Years later, Martin Mansergh produced a more circumspect view of the Golden Jubilee's impact upon the North's history. See M. Mansergh, *The Legacy of History for Making Peace in Ireland: Lectures and Commemorative Addresses* (Mercier Press, Cork, 2003), p. 16, who expresses concern about critics who 'claimed that it made the policy of north-south *détente* more difficult and provoked its opponents in Northern Ireland'. 'The renewed Troubles that started in the late 1960s', he adds, 'had far wider and deeper causes', before advising 'that commemorations should be handled sensitively, without triumphalism ... while not denying or denigrating the mainstream experience'.

upon a ticket that espoused 'a programme of vigorous direct intervention in the Northern Ireland situation', it failed to win a single seat (securing a minuscule .91% of total first preference votes).[61]

The vast majority of the electorate had clearly registered its revulsion for belligerent gestures. This was not forgotten by the newly-elected Coalition, which stood firm in the face of great adversity – including threats to subvert the state. Whilst in Opposition in the Republic, Lynch unsurprisingly continued to draw a line between militancy in the Irish past and present. In a speech to Fianna Fáil's National Executive in 1975, for example, he was forthright in declaring that the party's founder, de Valera, 'had fought in arms for the attainment of the aims – Éire Saor is Éire Gaelach [Irish for: 'A Free Ireland and a Gaelic Ireland']', 'but when he saw that the continuance of the struggle in arms was unwise in the national interest, he espoused peace throughout the land'.[62] At Fianna Fáil's 50th anniversary celebrations in Cork city in 1976, Lynch again reaffirmed the party stance 'that those who constitute the majority, for the time being, in the North of Ireland, cannot be coerced into joining us; we rule out force as the enemy of reconciliation'.[63]

Propped up by intellectual heavyweights such as Cruise O'Brien, the Fine Gael-Labour Coalition was equally passionate about condemning republican violence and advocating the merits of constitutional politics. By the middle of the 1970s, as Joe Joyce and Peter Murtagh have noted, he had 'made a political career out of debunking nationalism'.[64] After his appointment as Minister for Posts and Telegraphs, Cruise O'Brien was personally instrumental in the passing of the Broadcasting Authority (Amendment) Act of 1976, which explicitly banned spokespersons for Sinn Féin and the Provisional IRA from being interviewed on Irish radio and television. He also served as the Labour Party's

---

[61]   M. A. Busteed and H. Mason, 'The 1973 General Election in the Republic of Ireland', *Irish Geography*, Vol. 7 (1974), pp. 97, 99–100.

[62]   Minute Books of Meetings of the National Executive, 2 September 1975, UCDA, MS. P.176/350, Archives of the Fianna Fáil Party.

[63]   50th Anniversary Celebrations Address of Jack Lynch, TD, President of Fianna Fáil, at Grand Parade, Cork, 16 May 1976, UCDA, MS. P.176/946/7, Archives of the Fianna Fáil Party.

[64]   J. Joyce and P. Murtagh, *The Boss* (Poolbeg Press Ltd., Dublin, 1997), p. 84. Over time, Cruise O'Brien came to be regarded as one of the leading intellectuals in Irish life. For further observations on his legacy, see R. English, 'Directions in Historiography: History and Irish Nationalism', *Irish Historical Studies*, Vol. 37, No. 147 (2011), p. 450, who notes that 'his life and arguments' have 'deservedly prompted very many analyses', not least for his 'opposition ... to the supposed rightness of Irish unification'. One of Cruise O'Brien's 'lasting, ironic legacies to subsequent historians', he argues, 'was his relentless capacity to prompt ... interrogation of those of his provocative arguments that turned out, ultimately, not to be as convincing as they were stimulating'.

spokesman on Foreign Affairs and Northern Ireland and robustly opposed the Provisional IRA on a persistent basis. Years later in his memoirs, Cruise O'Brien explained the thinking behind his censorship policy, arguing that the principle behind the controversial Act had entailed 'the protection of the security of the democratic state against the broadcasting of subversive propaganda by organisations whose function was to work under the orders of the leadership of a private army for revolutionary purposes'.[65] In many ways, the censorship of militant dogma had implications for how Easter commemorations of 1916 were reported upon by the media, insofar as it curtailed physical force republicans from disseminating, on a wider basis, their firmly-held belief that their violent actions were no different to the techniques adopted back in 1916.

Unsurprisingly, the notion of using military pomp and spectacle to mark successive anniversaries of the Rising remained consigned to the dustbin of official state commemoration by Cosgrave's government. Thus on Easter Sunday 18 April 1976, there was again no Army parade along O'Connell Street to mark the occasion of the 60th anniversary of the Rising. This conspicuous absence spoke volumes, serving as a profound contrast to the excitement that had greeted the parading of the troops 10 years beforehand. By no means, however, did the media ignore the 60th anniversary. On Easter Monday, for example, *The Irish Times* carried two special articles to mark the occasion. One of these was penned by Liam de Paor, who asserted the opinion that the Rising's long-term significance lay in the fact that it had given 'self respect to Irish independence' and 'showed that the cause of the Republic was a serious one'. The other was written by Ruth Dudley Edwards (the daughter of Robert Dudley Edwards), who had uncovered a missing part of the Rising's heritage – namely a long-lost copy of a first edition of *The Irish Times* from the second day of the fighting. This censored edition, she noted, was 'rich in insurrectionary detail' and served as 'a belated tribute' to the newspaper's competence 'in those distant days'.[66]

Whilst things were relatively quiet and subdued in Dublin on Easter Sunday 1976, the opposite was the case in Northern Ireland, where rioting occurred following a commemoration held at Belfast's Milltown Cemetery. A three-story building close to the cemetery was set on fire, while four people were arrested.[67] Although mainstream political parties in the Republic largely ignored the 60th anniversary, due to 'a sense of a progressive strengthening of state security', 10,000 people still turned up for a Sinn Féin parade from St. Stephen's Green to the GPO on 25 April. Even though the event was banned by the Gardaí,

65 C. Cruise O'Brien, *Memoir: My Life and Times* (Poolbeg, Dublin, 1998), p. 356.
66 *The Irish Times*, 19 April 1976.
67 Ibid.

the maverick Labour TD for Dublin North-West, David Thornley, was amongst those who appeared on the platform. While his appearance was motivated by a personal concern with sustaining the right to protest in public, he was expelled from his party three days later.[68]

When Lynch served again as Taoiseach from 1977–1979, both his statesmanship and his attitude towards the legacy of 1916 continued to be tested by the delicate state of affairs in Northern Ireland. As preparations gathered pace in Dublin for the impending centenary commemoration of Patrick Pearse's birth, a source of worry to unflagging constitutionalists was the willingness of elements of the Irish population to articulate admiration for the Provisional IRA's motivations. Late in November 1978, for example, Conor Cruise O'Brien delivered a lecture at New York University in memory of the late British Ambassador, Christopher Ewart-Biggs (who was assassinated by the Provisional IRA on 21 July 1976, when a bomb was detonated under a road leading from his residence in Sandyford to the British Embassy in Ballsbridge). Cruise O'Brien pointed to the results of a Gallup poll that had been carried out earlier that year in the Republic for the BBC. While the poll showed that only a very small minority of 2% of those sampled actually approved of the Provisional IRA's campaign of violence, a rather large minority of 35% were actually willing to attribute idealistic motives to the Provisionals and were also willing to express respect for these. While 51% of the sample 'condemned the Provisionals absolutely', Cruise O'Brien lamented that 'it was the smallest majority possible' and went on to point to what he saw as serious discrepancies in the way in which a powerful figure within the Catholic Church in Ireland had 'in the same breath' condemned violence yet praised the ideals of those who had carried it out.[69] His pessimism did not go unnoticed, and he was by no means alone when it came to feeling concerned about the ambivalence that existed within some communities towards the dubious issue of the gun in Irish political culture. The unshakable reluctance of the state to hold military parades in remembrance of the Rising at Eastertime during the rest of the 1970s was proof of this.

By no means, however, did the Troubles (and apprehension about the Provisional IRA's claim to the Rising's legacy) manage to entirely extinguish the flame of 1916. Despite the continued absence of Irish Army parades on Easter Sunday, the late 1970s was a time in which other opportunities arose for the commemoration and exhibition of the story of 1916, especially after Fianna Fáil's return to power in the General Election of 1979. One such opening

---

[68] E. Sweeney, *Down Down Deeper and Down: Ireland in the 70s and 80s* (Gill and Macmillan, Dublin, 2010), p. 88.

[69] C. Cruise O'Brien, *Neighbours: The Ewart-Biggs Memorial Lectures 1978–1979* (Faber and Faber, London, 1980), pp. 60–61.

related to the matter of Roger Casement's memory, as represented by Oisín Kelly's bronze statue. This had originally been commissioned following the repatriation of Casement's remains in 1965, and was finally completed in 1971. The statue was then placed in storage by the Office of Public Works at Lad Lane and it was not until the staging of a travelling exhibition of the sculptor's work in 1978 (at venues in Dublin, Cork and Belfast) that the public were able to view it. After the rejection of proposals to erect the statue at locations such as Glasnevin Cemetery, Banna Strand and Murlough Bay, it was finally decided at a government meeting, held on 12 June 1979, that a new memorial to the 1916 rebel, incorporating Kelly's statue, was to 'be erected at the roundabout centre-piece at the pedestrian entrance to the car-ferry pier' at Dun Laoghaire (Casement's birthplace) in County Dublin.[70]

The Pearse Centenary, 1979

Towards the end of the 1970s, the legacy of 1916 was again subjected to an iconoclastic assessment by a revisionist historian. In *Patrick Pearse: The Triumph of Failure*, which was published in 1977, Ruth Dudley Edwards presented a far sterner biographical appraisal of the life of the 1916 leader than anything published beforehand. Although Dudley Edwards was credited for humanising Pearse as a person, she aroused the ire of some of the reading public by presenting him as an indecisive and tormented megalomaniac, who was vainly obsessed with the doctrine of blood sacrifice. In a preface to a later edition of this biography, Dudley Edwards recalled the adverse reactions that she had encountered when her work first appeared in print. Much hostility, she noted, had come from nationalists who felt that she had spoken ill of the dead. Whilst conceding that she had developed some degree of 'affection' for the rebel leader by the time that she had completed the book, Dudley Edwards stood over her conclusion that Pearse's actions and rhetoric had generated 'a posthumous Pandora's Box of horrors' (namely various latter-day expressions of the physical force tradition). Such thinking, she added, had brought her into a conflict of ideas with 'those who were genuinely upset because they could not cope with a Pearse with human failings'.[71]

---

[70]   D. Ó Súilleabháin to An Rúnaí Príobháideach, An tAire Airgeadais, 12 June 1979, NAI, Department of the Taoiseach, 2009/135/25, Sir Roger Casement, Repatriation of Remains and Reinterment in Ireland.

[71]   R. Dudley Edwards, *Patrick Pearse: The Triumph of Failure* (Irish Academic Press, Dublin, 2006), p. xx.

Plate 5.3   The Pearse Museum at St. Enda's Park in Rathfarnham, Dublin, opened on 10 November 1979 by the President, Patrick Hillery
Source: Author.

Notwithstanding the negative verdicts that were registered in Dudley Edwards's book, the government was by no means willing to ignore the legacy of the rebel past completely. This was demonstrated in 1979 by its official celebration of the centenary of Pearse's birth, which was marked by the opening of the Pearse Museum at St. Enda's Park (The Hermitage) in Rathfarnham, Dublin (Plate 5.3). A protracted set of negotiations had surrounded the fate of St. Enda's in the years beforehand and these are worth recounting. The building that housed the Pearse Museum was originally built in the 1780s. As already seen, Pearse opened it as a school called St. Enda's in 1910. At first he leased the building from William Woodbyrne, but after his execution in 1916 the property was occupied by the British Army. The school then relocated to Cullenswood House in Ranelagh, but in 1919 it moved back to Rathfarnham. In 1920 Pearse's mother, Margaret, purchased the property from its owner, Sir John Robert O'Connell, who was the trustee of the Woodbyrne Estate.[72] When she passed away on 22 April 1932, the St. Enda's property was bequeathed to her two daughters, Mary-Bridget

---

[72]   Typescript Entitled 'Brief Chronology of St. Enda's Park', Pearse Museum.

(who died on 12 November 1947) and Margaret. While St. Enda's ceased to function as a school in 1935, it was not until Margaret's death on 7 November 1968 that the property was finally left to the Irish state as a memorial to her executed brother, Patrick. After she died, a state funeral was held in her honour at Glasnevin Cemetery. Expenditure of £115 7s. 0d. was incurred in preparing the area around her grave, building a viewing platform and erecting amplification equipment.[73] Then, at a formal ceremony held on a raised platform outside St. Enda's on Easter Monday 1969, de Valera, as President, accepted the key of the property from Éamon de Barra. On 13 April 1978, Pearse Wyse, the Minister of State at the Department of Finance, informed the Dáil that £20,000 had been set aside for works at St. Enda's, and that the first phase of the building's 'internal rehabilitation' had been 'virtually completed'. He also stated that planning for the second phase, which involved 'laying out' the premises as a museum, had started.[74]

Even though the building's restoration was not finished on time, the Pearse Museum was finally opened on 10 November 1979. With Lynch away on a trip to the USA, the task of performing the opening was left to the President, Dr. Patrick Hillery (who had taken over the office following the resignation of Cearbhall Ó Dálaigh). A bust of Pearse, sculpted by John Behan (and cast at the Dublin Art Foundry), was also unveiled in the grounds of St. Enda's. Originally, the government had mooted the idea of erecting a statue in his honour at the College Green end of Pearse Street, but this never saw the light of day after stiff resistance from the Office of Public Works.[75] On the day that the Pearse Museum was opened, it was clear that events in the North (not to mention Dudley Edwards's controversial biography of the 1916 leader) were playing on the minds of many of those attending the ceremony. Hillery's formal speech reflected the political sensitivities of the time by emphasising many aspects of Pearse's educational legacy, rather than simply concentrating solely upon the memory evoked by his military activities. Although he praised the Pearse family for being 'enshrined in Irish history, remembered and revered', Hillery seemed particularly keen to remind the invited guests that Pearse was more than just a rebel leader. Pearse, he said, was a 'poet, scholar, teacher, visionary, inspirer of others, man of action, soldier ... [and] patriot'. Given the nature of Pearse's association with St. Enda's, the President then concentrated on his educational legacy: 'Here he endeavoured to educate in the fullest sense the youth in whose

---

[73] Statement of Account with Commissioners of Public Works, 14 August 1970, NAI, Department of the Taoiseach, 2003/16/30, Death of Senator Margaret Pearse.

[74] *Dáil Debates*, Vol. 305, 13 April 1978.

[75] Pearse Commemoration: Progress Report, 12 October 1978, NAI, Department of the Taoiseach, 2009/135/348, Patrick Pearse Centenary, 'One-Man Show', by Ulick O'Connor.

generosity of spirit and dedication to high ideals he had such great and abiding faith.'[76]

Although military parades marking the Rising had not taken place since 1971, the wider significance surrounding the opening of the Pearse Museum should not be underestimated. As an Office of Public Works-run heritage attraction that was somewhat hidden away inside a walled park in the capital's south suburbs, it certainly managed to escape the controversy that an officially-sanctioned Army parade through the city centre would have generated. Yet in its own way, the decision to open St. Enda's as a museum was an act of political significance, for its exhibits comprised part of a government-funded exhibition that evoked present-centred connections to the story and memory of 1916. Not that this episode was in any way unique or surprising for a state that was preoccupied at the time with copper-fastening its place in Europe and the wider world. As Elizabeth Crooke has pointed out, the creation of museums internationally has 'been linked historically to political aspirations and developments'. 'The creation and dissemination of knowledge of the past in museums', she adds, 'is an ideological process'.[77] Or as David Lowenthal has perceptively remarked, when humans domesticate the past, they 'enlist it for present causes'. In this way, legends about endurance and victory have often managed to 'project the present back, the past forward', with heritage clarifying pasts 'so as to infuse them with present purposes'.[78]

After being appointed as the first curator of the Pearse Museum, Pat Cooke set about the politically sensitive task of determining the precise nature and content of the storylines that the attraction would retell for years to come. Some years later, in an small pamphlet outlining the story of St. Enda's, he reflected upon the museum's origins, recalling that it had arisen 'out of the individual and collective impulse to admire and venerate great men and their achievements by raising up some physical, enduring monument to them'. However, he did highlight the fact that Pearse's standing had 'suffered the extremes of blind defence and debunking assault' since the intensification of the Troubles. 'Somewhere between the two', added Cooke, 'lies the more complicated truth'. Whilst acknowledging Pearse's 'courage and idealism', the curator also pointed out that the 1916 leader 'bore also the common marks of human frailty'. For visitors to the museum and its

---

[76]   *The Irish Times*, 12 November 1979.

[77]   E. Crooke, *Politics, Archaeology and the Creation of a National Museum of Ireland: An Expression of National Life* (Irish Academic Press, Dublin, 2000), pp. 8, 10.

[78]   D. Lowenthal, *The Heritage Crusade and the Spoils of History* (Cambridge University Press, Cambridge, 1998), p. xv.

grounds, Cooke expressed hope that they would 'with an effort of imagination, sense something of ... [Pearse's] courage and commitment'.[79]

Besides the opening of the Pearse Museum, many other cultural and educational projects were arranged in 1979 in order to mark the centenary of the rebel leader's birth. These were coordinated by an inter-departmental government committee, which had been established in August 1978, under the auspices of the Department of the Taoiseach. According to an official press release issued in the lead-up to the centenary celebrations, it was announced that these projects were to include the commissioning of Louis Marcus to produce a documentary film, the commissioning of Seoirse Bodley 'to write a symphonic work for full orchestra', the commissioning of a publication on the environment 'to mark Pearse's interest in nature and the landscape', and the allocation of a subvention towards the costs of publishing two volumes on *The Educational Writings of Pádraig Pearse*. The committee also announced the instigation of a series of scholarships for students of secondary schools and colleges. Furthermore, it sanctioned the holding of a three-month course in journalism through the Irish language at University College Galway, to highlight 'Pearse's work as a journalist and editor'. As with many previous anniversaries of events in Irish history, a special commemorative postage stamp was also issued to mark Pearse's centenary.[80] The committee also accepted a proposal from the writer, Ulick O'Connor, to write a One-Man Show, 'based on aspects of Pearse's life and work'.[81]

Beyond the official programme, the Pearse centenary was also marked by the staging of a special commemoration dinner on the same day as the opening of the Pearse Museum. This was run by Fianna Fáil's Cumann Bhean Uí Phiarsaigh and was addressed by the Minister for Health, Charles Haughey. Having been found not guilty years earlier in his trial by jury in the charge of conspiring to illegally import arms, Haughey's return from the political doldrums was a remarkable feat in itself. In contrast to Hillery's carefully-worded and far-ranging speech at St. Enda's, Haughey opted for more ardent rhetoric and seemed particularly keen on defending Pearse's memory from various detractors who had castigated him for being a militaristic individual. In addition to dismissing the notion that Pearse had been 'a man who glorified in war', Haughey also downplayed the

---

79   P. Cooke, *Scéal Scoil Éanna: The Story of an Educational Adventure* (Office of Public Works, Dublin, 1986), pp. 2, 4.

80   Memorandum on Pádraig Pearse Centenary Commemoration, 1979, NAI, Department of the Taoiseach, 2009/135/347, Patrick Pearse Centenary.

81   Rory MacCabe to Accounts Branch, Department of Finance, 15 October 1979, NAI, Department of the Taoiseach, 2009/135/348, Patrick Pearse Centenary, 'One-Man Show', by Ulick O'Connor.

significance of the constitutional efforts of the Irish Parliamentary Party during the 1910s and argued that Pearse had 'sought freedom by arms only because there was no other way open to him in the circumstances of his time'. Upon hearing what his arch rival had said about Pearse's resort to arms, Lynch felt that it was necessary to put a greater wedge between the past and present by adding the following words to a speech that he was about to make in the USA to the American-Irish Foundation: 'The paradox of Pearse's message for the Irish nation today is that we must work and live for Ireland, not die, and most certainly not kill for it.'[82]

Revisionism, Remembrance and Forgetfulness

A month after the opening of the Pearse Museum, the old divisions in Fianna Fáil resurfaced when Jack Lynch was ousted as party leader by his old adversary, Charles Haughey, who became Taoiseach on 11 December 1979. Given Haughey's earlier involvement in the Arms Trial, his appointment as Taoiseach was seen at the time as one of the greatest comebacks in the history of Irish politics. British politicians' unease with Haughey's more robust ideology of nationalism came to the surface half a decade later, when he opposed the signing of the Anglo-Irish Agreement between Garret Fitzgerald and Margaret Thatcher in 1985 (which gave the Republic a consultative role in the affairs of Northern Ireland). Nevertheless, while the charismatic personage of Haughey was certainly more zealous about the national question and the memory of 1916 than his predecessor, he seemed less Anglophobic than he was during his student days at University College Dublin's Earlsfort Terrace campus in 1945 – when he burned a Union Jack flag in Dublin city centre on Victory in Europe (VE) Day, as a retort to a perceived insult to the Tricolour by a group of students who were celebrating Germany's defeat in World War II. Indeed on many occasions throughout the 1980s, Haughey publicly condemned the violent deeds of the Provisional IRA and was careful not to overstate his party's much-cherished connection to the heritage of 1916.

Towards the end of his first term as Taoiseach, which lasted until June 1981, Haughey had to confront a major crisis in the North, when sympathy for the physical force tradition was bolstered across the island of Ireland by the H-Block hunger strike campaign in the Maze prison, by republican prisoners seeking the restoration of 'special category' status. This was a time, in Conor Cruise O'Brien's view, that 'the Pearsean ghosts walked to the greatest effect and "slipped between

---

[82] *The Irish Times*, 12 November 1979.

the lines" most disturbingly'.[83] The most prominent of the 10 hunger strikers who died during the campaign was Bobby Sands, who passed away on the 66th day of his hunger strike on 5 May 1981, having been elected to Westminster as an MP for the constituency of Fermanagh-South Tyrone less than a month earlier.[84] The H-Block protests sparked international attention, and the death of Sands elevated him to the status of a martyr for the Provisional IRA and its supporters – not unlike the way in which Pearse had become a martyr after his execution by the British almost 65 years previously. Such was the popularity of Sands that at least 100,000 people lined the route to his funeral at Milltown Cemetery. In many parts of the island of Ireland, the appearance of more and more militant republican iconography within the arena of public space served to raise awareness of the plight of the hunger strikers. Numerous walls were daubed with murals of Sands and slogans such as 'SMASH H-BLOCK', while countless lampposts were draped with black flags. Spur-of-the-moment street appearances by Provo supporters, wearing black masks and armbands, added to the anxiety of the times.

To a certain extent, as Cruise O'Brien later reflected, the deaths of the prisoners brought about reactions in the Republic 'which were similar to ... the reactions which had followed the execution of the leaders of the 1916 Rising'.[85] While public opinion throughout Ireland was firmly split on the emotive issue of 'special category' status for the hunger strikers, the one thing that Provo sympathisers were quick to point out on many occasions of commemoration was their belief that there was no distinction between the militant tendencies of Pearse and Sands – both of whom they idolised as martyrs for the cause of physical force republicanism. For most politicians in the Republic, however, the issue of celebrating the Rising in a triumphalist fashion remained just as much a no-go area in the 1980s as it had been in the 1970s. On some occasions, nonetheless, Haughey was particularly forthright about the cause of Irish unity. In delivering an address at the White House on St. Patrick's Day 1982 shortly after commencing his second term as Taoiseach, Haughey spoke about 'the goodwill of that worldwide Irish spiritual empire which is stronger here in the United States than anywhere else', and expressed his pride in the fact that Ireland 'holds a special place in the affections of millions of Americans'. The US President, Ronald Reagan (whose ancestry was traced to Tipperary), then heard Haughey express his hope that feelings of goodwill would continue to inform 'American policy and actions and ensure that the encouragement of

83　Cruise O'Brien, *Memoir*, p. 415.
84　J. Bardon, 'Hunger Strike', in Lalor (ed.), *The Encyclopaedia of Ireland*, p. 507.
85　Cruise O'Brien, *Memoir*, p. 417.

Irish unity ranks high among her international objectives'.[86] At a speech to the Irish-American community in New York a few months later, however, Haughey reaffirmed his government's commitment to 'the peaceful reunification by agreement of our people', and then reminded his audience that nobody in the USA 'should support or subscribe to policies which envisage violence and terror as the means of bringing about the unity of Ireland'.[87]

As the 1980s progressed, the legacy of 1916 continued to remain a delicate subject. In many respects, memory of the Rising found itself embroiled in a struggle against forgetfulness, owing to complications arising from the persistence of the conflict in Northern Ireland. The brokering of the Anglo-Irish Agreement in 1985 offered some hope for the negotiation of a roadmap towards peace. However, with the Republic suffering from a prolonged recession and high levels of unemployment and emigration, there was little to celebrate by the time of the 70th anniversary of the Rising in 1986. With Fine Gael back in power and Garret Fitzgerald serving as Taoiseach, the anniversary proved to be just as inconspicuous as the 60th anniversary a decade beforehand. On Easter Sunday 30 March 1986, Dublin city played host to a series of small-scale commemorations organised by some of the Opposition parties. At Kilmainham Gaol, for example, around 50 people turned up in the morning for a Workers' Party commemoration. A wreath was hung on the prison's gate by the party leader, Tomás MacGiolla, while an oration was delivered by Councillor Pat Rabbitte. In a sign of the challenging times, Rabbitte stated that it was tragic that the Rising was being used by some individuals 'to justify actions ranging from Provo terrorism to rampant exploitation, all in the name of freedom'. Later in the day, Arbour Hill was the scene of a modest enough Fianna Fáil commemoration, attended by a crowd of about 300 people, including Old IRA veterans and TDs such as Ray Burke, Niall Andrews and Michael Woods.[88] During the three days that followed, the 70th anniversary was also marked by the publication in *The Irish Times* of a series, fittingly entitled 'The Shadow of 1916'. This featured two contributions apiece from Dick Walsh (the newspaper's political editor) and Nollaig Ó Gadhra (a lecturer in liberal studies at Galway Regional Technical College). In his Easter Monday piece, Walsh's concluding words reflected much

---

[86] An Taoiseach, Charles Haughey's Saint Patrick's Day Address to the President of the USA, Ronald Regan, at the White House, 17 March 1982, UCDA, MS. P.176/947/25, Archives of the Fianna Fáil Party.

[87] Speech by the Taoiseach, Charles Haughey, at a Reception for the Irish-American Community in the Waldorf Astoria Hotel, New York, 10 June 1982, UCDA, MS. P.176/947/27, Archives of the Fianna Fáil Party.

[88] *The Irish Times*, 31 March 1986.

of the sentiment then prevailing in the Republic: 'it's so difficult now to take a clear-eyed view of 1916 and what it means to us'.[89]

In contrast to the inconspicuous events held in the Republic on Easter Sunday, the 70th anniversary commemorations drew larger numbers of attendees in Northern Ireland. In Derry city, for example, around 1,000 people took part in an Easter Sunday march from the Bogside to the City Cemetery. In Belfast, some 700 people participated in a march to Milltown Cemetery, where Sinn Féin's annual commemoration of the Rising was staged. A statement from the leadership of the Provisional IRA was read out, predicting that the Anglo-Irish Agreement would not succeed, 'as it is not aimed at removing the root cause of the conflict, Britain'. Elsewhere in the North, Sinn Fein's President, Gerry Adams, addressed a substantial commemoration in Carrickmore, County Tyrone, where he too criticised the Anglo-Irish Agreement and reiterated his support for the Provisional IRA's armed struggle. 'We defend', he stated, 'the use of force today against the same enemy and in the same cause as that which made the Easter Rising a necessary and morally correct form of struggle'. Furthermore, Adams accused the Irish government of 'revisionism', alleging that it had become submissive to British interests and was engaging in policies that contradicted the 1916 Proclamation.[90]

In addition to the public discourse associated with the politics of historical commemoration, the mid-1980s proved to be a time when 'revisionism' impinged heavily upon historiographical debates concerning the making of Irish history. Following on from the hard-hearted critiques of the Rising that were published in the previous decade by scholars such as Shaw, Cruise O'Brien and Dudley Edwards, comprehensive histories of the Rising proved to be a rare commodity in the 1980s. However, in some of the general histories of the modern era that were published, the story of 1916 was accorded less prominence than heretofore – as new interpretations of the past rejected some of the traditional views espoused by nationalists. Commenting on revisionism's wide-ranging impact, Tom Dunne has argued that 'Irish historians – like the population at large' became 'less preoccupied with nationalist rhetoric' and consequently felt 'less defensive about research findings' that supported rather than subverted 'elements of the traditional account'. In Stephen Ellis's view, the essence of historical revisionism represented 'a debate about the ousting of traditional faith-and-fatherland perspectives on the growth of an Irish nation and state in favour of a more broadly-based, pluralist approach to Irish history'. The Irish 'revisionist renaissance', adds Peter Cottrell, facilitated a questioning

---

89   Ibid.
90   Ibid.

of nationalist interpretations, with historians coming to terms with the past 'as a piece of historical study rather than an emotional experience'. Additionally, as Gareth Ivory has pointed out, the revisionist debate in Ireland had 'a certain edge'. One of its most controversial aspects arose from the way in which the revisionists 'challenged the established view that Irish history was to be seen as an eight-hundred-year oppression of a monocultural Irish nation by the British, ended only by revolutionary action leading to an independent Ireland'.[91]

With distaste for the spate of killings in the North impinging heavily upon hearts and mindsets, the revisionist school of thought gained a lot of ground in debates concerning Irish historiography in the 1980s. The decade also saw the emergence of Waterford-born Roy Foster, who was based at Birkbeck College in London, as one of Ireland's leading historians. In the same year as the 70th anniversary of the 1916 Rising, he published a seminal article in the first issue of *The Irish Review*, entitled 'We Are All Revisionists Now'. In commenting on the impact of revisionism, he argued that it had succeeded in eliminating 'as much as possible of the retrospectively "Whig" view of history which sees every event and process in the light of what followed it rather than what went before'. Foster also emphasised that revisionism aimed to downplay the benefit of hindsight in historical interpretation: 'To blame every unwelcome development in Irish history on British malevolence, disallowing economic, social and political forces in Ireland, is an attractively easy option; it also implies an Irish moral superiority which leads too easily to self-righteous whingeing.' Furthermore, he noted that 'history need no longer be a matter of guarding sacred mysteries', especially for 'a country that has come of age'.[92]

Although critics of the revisionist enterprise sometimes accused its adherents of being anti-nationalist or pro-unionist, Charles Townshend has disagreed with this standpoint. He argues that revisionism in Ireland (and elsewhere) involved confronting 'less flattering' facets of 'national foundation myths' and recognising 'their elisions and fabrications'. Under such a thematic approach, he feels that revisionists sought to highlight 'the complexity out of which an alternative story

---

[91]   T. Dunne, 'New Histories: Beyond Revisionism', *The Irish Review*, No. 12 (1992), pp. 10–11; S. G. Ellis, 'Writing Irish History: Revisionism, Colonialism, and the British Isles', *The Irish Review*, No. 19 (1996), p. 3; P. Cottrell, *The Anglo-Irish War: The Troubles of 1913–1922* (Osprey Publishing Ltd., Oxford, 2006), p. 11; G. Ivory, 'The Meanings of Republicanism in Contemporary Ireland', in I. Honohan (ed.), *Republicanism in Ireland: Confronting Theories and Traditions* (Manchester University Press, Manchester, 2008), pp. 87–88.

[92]   R. Foster, 'We Are All Revisionists Now', *The Irish Review*, No. 1 (1986), pp. 1, 5. For further reflections by Foster, see *Making History: The Irish Historian* (RTÉ Hidden History, 2007). 'The kind of history that was presented by the IRA in the North', he recalls, 'demanded and called forth a riposte from the Academy, which is what it got'.

could have emerged', as an alternative to 'a linear, teleological story of national liberation'.[93] The politics of the present sometimes filtered through in critiques of nationalist historiography, as is evident from the writings of scholars such as John A. Murphy, who was Professor of Irish History at University College Cork. In a short collection of essays on *The British-Irish Connection*, published in 1986, Murphy expressed his desire for an improvement in 'the climate of Anglo-Irish relations' and his distaste for an Irish tendency to display 'the prickly sensitivity of the scarred ex-colonial'. In elaborating on why he was perturbed by the lingering 'influence of nationalist mythology' and the myths 'cultivated by the propagandist purveyors of anti-English hate', he claimed that interpretations of Irish history as an 800-year-old 'liberation struggle' did not always stand up to critical scrutiny. As an alternative, he suggested that 'sources of enlightenment and reconciliation in Anglo-Irish relations' could theoretically be found in the 'truths of historical research'.[94] Although this line of argument carried little weight with hardcore nationalists at the time, there were others who saw sense in what Murphy was advocating historiographically.

Revulsion against physical force republicans reached a new crescendo on 8 November 1987, following the Provisional IRA's bombing of a World War I Remembrance Day ceremony at Enniskillen, County Fermanagh. Eleven civilians (six men and five women) were killed, while 63 people (including children) were injured or maimed. The antipathy towards republican violence that followed the bomb was graphically captured in the Phil Joanou-directed concert movie, *Rattle & Hum*, featuring U2 on the road during their *Joshua Tree* tour in the USA in 1987. On the night of the bombing, prior to singing a cathartic and moving version of 'Sunday Bloody Sunday' (one of the band's most explicit anti-violence songs), lead singer Bono spoke out against the actions of the Provisional IRA from the stage at the McNichols Arena in Denver, Colorado. 'Where's the glory?', he asked, 'in bombing a Remembrance Day parade of old-age pensioners' and leaving them 'dead under the rubble of a revolution that the majority of the people in my country don't want'.[95] As the devastating repercussions of the Enniskillen bomb lingered on in the collective consciousness of many Irish people, an ever more guarded approach was evident

93   C. Townshend, *Easter 1916: The Irish Rebellion* (Allen Lane, London, 2005), p. 353.

94   J. A. Murphy, 'A Look at the Past', in J. McLoone (ed.), *The British-Irish Connection* (The Social Study Conference, Galway, 1986), pp. 11, 15–16.

95   *Rattle and Hum* (Paramount, 1988). For further commentary on the Enniskillen atrocity, see Father Brendan Hoban's article in the *Western People*, 13 March 1991. The killings, he observed, 'had truly sickened the Irish people' and resulted in 'almost universal condemnation'. The message from the paramilitaries, he felt, 'was loud and clear'. 'In the new Ireland that the Provos hope one day to usher in', he feared that 'only the Provos will be allowed to remember'.

in matters concerned with the Rising's remembrance. Writing in 1988, Dermot Bolger reflected that a 'curious ambivalence in public attitudes' had emerged towards the events of Easter 1916, so much so that it had become 'a controversial topic and one in which people are often left in a no-win situation'.[96]

In the months that followed the bomb, there was no shortage of critics of the Provisional IRA's actions, or indeed of their territorial aspirations – the latter of which happened to be the same as those of the 1916 rebels (namely the creation of a 32-county republic). As an independent member of Seanad Éireann (representing the National University of Ireland constituency), John A. Murphy continued to live up to his reputation as a one of the most vocal intellectual detractors 'of the myths and shibboleths of physical-force republicanism'.[97] In this respect, he was a vociferous critic of the territorial aspirations outlined in Articles Two and Three of the Constitution and had much to say about the legacies of 1916. In a Seanad debate on 21 January 1988, Murphy remarked that 'the horror of Enniskillen' had 'thankfully, seen a widespread and welcome revulsion against that horror and all it stands for'. In calling upon the Irish government 'to clarify its policy on Northern Ireland', he registered his opposition to all those who were in favour of what both the 1916 rebels and the Provisional IRA wanted – namely the creation of a united Ireland. In clear-cut language, Murphy signalled that it was time to face what he saw as some uncomfortable realities about links between different strands of nationalism – in both the past and present. 'One of the pseudo historical points which the Provos make', he argued, 'is that Partition is the cause of all the trouble'. 'I suggest', he added, 'that there is no basis for ... concepts of self deception ... Certainly, the 1916 Rising made Partition even more inevitable'. Murphy then suggested that 'we should renounce not only Provo methods but Provo objectives', and thus 'liberate ourselves from the will-o'-the-wisp of territorial unity'.[98]

The revisionists' moral crusade against the continued use of the gun in contemporary Irish politics had a profound impact on the historiography of the revolutionary era. Roy Foster's bestseller, *Modern Ireland 1600–1972*, which was first published in 1988, called for 'a more relaxed and inclusive definition of

---

[96] D. Bolger, 'Introduction', in D. Bolger (ed.), *Letters from the New Island: 16 on 16. Irish Writers on the Easter Rising* (Raven Arts Press, Dublin, 1988), p. 7.

[97] T. Dunne and L. M. Geary, 'Introduction', in T. Dunne and L. M. Geary (eds), *History and the Public Sphere: Essays in Honour of John A. Murphy* (Cork University Press, Cork, 2005), p. 3.

[98] *Seanad Debates*, Vol. 118, 21 January 1988. Additional perspectives on the logic behind Murphy's call for 'constitutional adjustment' and his quest for 'good relations' between North and South, can be found in V. Power, *Voices of Cork* (Blackwater Press, Dublin, 1997), p. 182.

Irishness, and a less constricted view of Irish history' that stressed the different varieties of Irishness running through the past. In an analysis of the history of the 1916 Rising, he acknowledged the occurrence of a series 'of appalling incidents' against innocents under Martial Law. To the distaste of some nationalists, however, he sought to advance a more complex assessment of the reasons behind the British Army's reaction to the Rising, especially in relation to the execution of the rebel leaders. 'The draconian reaction' by the British authorities to 1916, he argued, 'should be understood in terms of international war and national security'. The response, he added, 'also has to be seen against the background of alienation and Anglophobia inherent in so much of the Irish experience'. After 1916, he contended, 'the IRB tactic already defined came into its own ... the garrison was forced on to the aggressive, and the Volunteers secured a moral ... advantage'.[99] A significant theme of Foster's book, as Niall Ferguson later remarked, was a veiled counterfactualism that repeatedly called into question 'the nationalist teleology of inevitable independence from "English" rule'.[100]

The 75th Anniversary, 1991

The Republic of Ireland was by no means in a mood of universal despondence during the late 1980s and early 1990s. As the deficit within the public finances began to be tackled by the cross-party Tallaght Strategy, there was a cautious optimism that the economy could be salvaged. Furthermore, a feelgood patriotism was sparked by the exploits of the Jack Charlton-managed Republic of Ireland soccer team, following its impressive performance at the 1988 European Championships in Germany. In their opening game in Stuttgart, the Irish beat the much-fancied English team 1–0, thanks to the on-pitch adroitness of Ray Houghton. Sport, it seemed, had replaced history as the forum for collective triumphalism when it came to rejoicing in successful encounters with the so-called 'Old Enemy'. This became glaringly obvious nearly three years later during the Republic's restrained commemoration of the 75th anniversary of the 1916 Rising – which also happened to fall on the 10th anniversary of the hunger strikes, the 50th anniversary of James Joyce's death, the 100th anniversary of Charles Stewart Parnell's death, the 200th anniversary of the foundation of the United Irishmen, the 300th anniversary of the Battle of Aughrim, and the 800th anniversary of the commencement of worship at the site of St. Patrick's Cathedral in Dublin. The official commemoration of the Rising failed to stir

---

[99]   R. F. Foster, *Modern Ireland 1600–1972* (Allen Lane, London, 1988), pp. 484, 596.
[100]   N. Ferguson, 'Introduction. Virtual History: Towards a "Chaotic" Theory of the Past', in N. Ferguson (ed.), *Virtual History: Alternatives and Counterfactuals* (Pan Books, London, 2003), p. 19.

the public imagination and what surfaced instead was a nationwide searching of hearts and minds. In expressing their views about the occasion of the 75th anniversary, many politicians were hesitant to say anything too excessive about the 1916 Rising's legacy.

When asked in the Dáil on 27 February about the government's plans for the upcoming anniversary, the Taoiseach, Charles Haughey, was keen to emphasise that it would 'be commemorated in a dignified and fitting manner' and that in addition, 'a wide programme of events' would also be held in the capital to celebrate Dublin's status as 'European City of Culture' for 1991. Although Haughey was anxious to highlight the fact that he personally regarded the Rising as a 'great event in our history' and that he was in favour of encouraging local authorities to commemorate it 'in their own way' in April, he was also eager to stress that a state ceremony would be staged the following July, in the Royal Hospital at Kilmainham, to remember 'all those who died in past wars and on service with the United Nations'.[101] In the same Dáil debate, Proinsias de Rossa of the Workers' Party urged the government to exercise caution in the way that it was going about the business of encouraging various local authorities to remember 1916. In a statement that reflected how sensitive attitudes to 1916 had become, he asked:

> Would the Taoiseach agree or, perhaps, undertake to convey to various councils and organisations that are being encouraged ... to commemorate 1916[,] that it is important to emphasise that this is but one part of our heritage and that there are other traditions on this island which do not see 1916 in the same light as many of us do, particularly in view of the fact that 1916 may be used by the Provisional IRA to justify their activities, and that this is one of the main concerns which many people have?[102]

De Rossa was by now means alone in asserting that the 1916 metanarrative had to be handled with care. Around the same time, many other high-profile public figures were more than happy to openly register guarded opinions about matters of remembrance. When asked for his opinion in the following month on the matter of whether to celebrate the imminent 75th anniversary or not, the captain of the Republic of Ireland soccer team (and former Dublin GAA footballer), Kevin Moran, urged that any celebrations should be done 'sensitively, so as not to offend people in the North'. 'It's nicer to look forward', he cautioned, 'than to

---

[101] *Dáil Debates*, Vol. 405, 27 February 1991.

[102] Ibid.

dwell on the past'.[103] The sea-change in public opinion since 1966 could not have been more pronounced.

This hesitance was clear for all to see as Easter 1991 finally approached. As Garret Fitzgerald later reflected, 'the appalling violence in Northern Ireland' had 'shifted the balance powerfully against the case for 1916' for lots of people at the time. The aggression, he adds, 'undoubtedly contributed to the deep-seated national reluctance' to 'celebrate'.[104] Historians have portrayed the 75th anniversary in a similar fashion. Diarmuid Ferriter, for example, has observed that government's 'low-key approach' to the official commemoration was proof that 1916 had clearly 'gone out of fashion'. Likewise, Rebecca Graff-McRae has pointed to the absence of 'glorificationist tendencies' throughout the commemoration, which she attributes to a 'changed and changing context' which 'apparently called out for a downplayed approach to the problematic past'.[105] The journalists and writer, John Waters, has suggested that the 75th anniversary amounted to no more than 'a collective staring at shoes', as reflected by the holding of 'a few earnest but desultory commemorations, from which most people maintained an agnostic detachment'.[106] Going on the evidence of what did (and did not) happen at an official level, all of these assessments have validity. For the most part, the official ceremony that was held in Dublin on Easter Sunday proved to be a damp-squib of an affair – confronted as it was by a Pandora's Box of contemporary ethical complications and uncertainties that saw forgetting take the upper-hand in a struggle against memory. Lasting a mere 15 minutes, the subdued and sensitive ceremony was a big turnaround on the week-long ceremonies that took place a quarter-century beforehand for the Golden Jubilee, when pronouncements about Irish unity and the restoration of the Irish language had been commonplace. Although the Troubles cast a huge shadow over the 75th anniversary, other factors also impinged upon the nature and shape of the commemoration. One significant factor in the waning of the 1916 metanarrative was the passing away of nearly all of the veteran rebels by 1991 and the emergence of a younger breed of politicians with separate outlooks and concerns.

The 1991 Census (the 12th since the state's foundation), revealed that the Republic's population was 3,525,719, which represented a 22.25% increase on

    103   *The Sunday Tribune*, 24 March 1991.
    104   G. Fitzgerald, *Reflections on the Irish State* (Irish Academic Press, Dublin, 2003), p. 5.
    105   D. Ferriter, *What If? Alternative Views of Twentieth-Century Ireland* (Gill and Macmillan, Dublin, 2006), p. 79; R. Graff-McRae, *Remembering and Forgetting 1916: Commemoration and Conflict in Post-Peace Process Ireland* (Irish Academic Press, Dublin, 2010), p. 44.
    106   J. Waters, *Was It For This? Why Ireland Lost the Plot* (Transworld Ireland, London, 2012), p. 8.

the total of 2,884,002 persons recorded in 1966.[107] De Valera, who had played such a prominent role in the Golden Jubilee events in his capacity as an iconic and authoritative President, had been dead for nearly a decade and a half by 1991, while on 7 November of the previous year, the country had unexpectedly elected its first female President in Mary Robinson, who was then only in her mid-40s. Her election by a rainbow coalition 'of left, centre left, liberals, Greens, and women', as Gemma Hussey noted a few years afterwards, shocked Fianna Fáil profoundly. The main government party, she wrote, had lost 'the office they had always considered their own' and the defeat of Brian Lenihan, the hitherto highly-tipped and much-admired 70-year-old candidate of the government, symbolised an intoxicating defeat of 'the older, male, conservative power-bloc of national politics' by 'a fresh, liberal and female candidate'.[108] Although Lenihan's election campaign had imploded in a controversy that arose from a personal lapse in memory, the ascent of one of the shining lights of the liberal left to the Presidency still served as a potent reminder to the old vanguard about to commemorate 1916 that other priorities were now competing with the 'sacred cow' that was the nationalist agenda.

Although many of the stalwarts of constitutional nationalism found themselves having to defend the worthiness of the 1916 Rising, they seemed up to the challenge. Having examined its own political conscience and withstood a certain amount of hullabaloo about its historical faith, the Fianna Fáil-led government remained steadfast in its desire to honour the rebels of the 1916 Rising in a dignified and patriotic manner. There was, however, an aura of measured reticence about the way that it actually handled the sombre 75th anniversary. Writing in *The Cork Examiner* about the uneasy alliance of the traditional and modern that clouded the issue of national identity in 1991, Brian Girvin commented that there had been a debate about the Rising which would have been 'impossible in 1966'. Furthermore, he noted that 'the monolithic view of Ireland' had 'disappeared and with it the unthinking devotion to a narrow nationalism'. The ultimate transformation since the Golden Jubilee, he observed, was that 'one section of public opinion now believes that the negative consequences override the positive ones and, consequently, feel uneasy about celebrating it in traditional fashion'.[109]

---

[107] Figures derived from Central Statistics Office, *Census 2006: Principal Demographic Results* (The Stationery Office, Dublin, 2007), p. 37.

[108] G. Hussey, *Ireland Today: Anatomy of a Changing State* (Town House, Dublin, 1993), p. 10. According to a report in *The Sunday Tribune*, 14 April 1991 on a new Lansdowne Market Research opinion poll, Robinson enjoyed a 74% approval rating from voters, making her more popular at the time than any of the country's political party leaders.

[109] *The Cork Examiner*, 2 April 1991.

As it transpired, the official government commemoration at Easter 1991 was designed so as not to upset the general populace or make them feel collectively uncomfortable by glorifying the deeds of the 1916 rebels at a time in which violence was being waged in the North – with the Provisional IRA citing the Rising as legitimacy for their own use of physical force. So great was the Irish government's fear in 1991 of fanning the fires of the Provo's campaign, that an attempt at the beginning of the year by the curator of the Pearse Museum to gain official support for a conference which he wanted to hold on '1916 and its Interpretations' proved unsuccessful.[110] Clear distinctions were drawn between the violence of the past and the present, with official policy dictating that decisions concerning commemoration had to be made so as not to exacerbate unionist anxiety.[111] As Girvin remarked at the time, nationalist opinion about unionists 'is now coming to realise that the traditional belief concerning them is neither realistic nor convincing ... like it or not they do not consider themselves to be Irish and that is slowly being recognised'.[112] Amidst the heightened sensitivities during what was a fragile moment in Anglo-Irish relations, a significant diplomatic predicament for the government was the wording of Articles Two and Three of the Constitution. These outlined the Republic's territorial claim to the North and were proving to be major obstacles to efforts by the Secretary of State for Northern Ireland, Peter Brooke, to initiate peace talks. Appearing on RTÉ Radio 1's *This Week* programme on Easter Sunday 1991, Peter Robinson, the deputy leader of the Democratic Unionist Party, reaffirmed his well-known view that Articles Two and Three were like a 'Sword of Damocles' hanging over the unionist community 'in the shape of an aggressive claim of jurisdiction over Northern Ireland which is seen by us to be the bar to proper, normal relationships with our neighbour'.[113]

---

[110] M. Ní Dhonnchadha and T. Dorgan, 'Preface', in M. Ní Dhonnchadha and T. Dorgan (eds), *Revising the Rising* (Field Day, Derry, 1991), p. ix.

[111] For some unionists, the 75th anniversary brought back disquieting memories of the Golden Jubilee and its aftermath. In an interview with *The Irish Times*, 5 April 1991, the Ulster Unionist MP for Upper Bann, David Trimble, spoke about how the commemorations in 1966 had exacerbated tensions in the North. At a time in which O'Neill had been 'presenting himself as capable of ushering in a new era of accommodation', Trimble felt that 'the old republican ghosts started walking and the Northern nationalists walked after them'. This, he believed, had enabled Paisley to arrange his first efficient counter-demonstration – 'a pivotal event in starting the slide which became apparent ... later'.

[112] *The Cork Examiner*, 2 April 1991.

[113] *The Irish Times*, 1 April 1991. An MRBI poll conducted from 15–16 April 1991, and published in ibid., 23 April 1991, showed that 82% out of a representative sample of 1,000 electors in the Republic aged over 18 aspired to a united Ireland, but the same figure also supported the idea of postponing efforts to secure it if it helped in bringing about a settlement in the North.

Plate 5.4   The Taoiseach, Charles Haughey, greeting the President,
             Mary Robinson, on arrival at the GPO for the Rising's
             75th anniversary commemoration on 31 March 1991
Source: National Library of Ireland, Independent Newspapers (Ireland) Collection,
391-639B/11-11a.

When the official Easter Sunday commemoration of the 75th anniversary of
the Rising eventually came to pass on 31 March 1991, there was no repeat of the
calls for Irish unity that had been made 25 years previously. A few administrative
blunders in the organisation of the event led to some media criticism, but on
the whole, the commemoration proved relatively uncontroversial. At 11.50 am,
Charles Haughey, who was now serving a third and final term as Taoiseach,
arrived in O'Connell Street, where he was greeted by senior members of the
Defence Forces outside the GPO.[114] He was followed five minutes later by the
President, Mary Robinson, who arrived in a 1948 Rolls-Royce Wraith (the same
one used by de Valera in 1966), along with an Escort of Honour commanded
by Captain Frank Lawless.[115] After being greeted by Haughey (Plate 5.4),
Robinson inspected a guard of honour and was escorted by Lieutenant
General James Parker to the central walkway, from where she watched the short

---

[114]   Ibid., 1 April 1991.
[115]   H. Bonar, 'National Commemoration: 75th Anniversary of the Easter Rising, 1916',
An Cosantóir: The Irish Defence Journal, Vol. 51, No. 4 (1991), p. 6.

military spectacle. It commenced with the lowering of the Tricolour and the reading of the Proclamation by Captain Sean Fitzpatrick of the Cadet School Military College (who was the only person to speak during the ceremony). This was followed by the hoisting of the Tricolour on the roof of the GPO and the ceremony was then brought to a swift conclusion with the playing of the National Anthem by No. 1 Army Band and the Bank of Curragh Command.[116]

The Taoiseach, as Joe Carroll remarked in his report for *The Irish Times*, had insisted that the commemoration 'would be a simple, dignified ceremony with no elaborate trimmings and that is what the nation got'. In the same newspaper, Eileen Battersby passed comment on the government's 'diffidence towards the occasion' and remarked that the ambiguous legacy of 1916 was glaringly obvious throughout the 'muted official ... celebratory gesture'.[117] The downgrading of 1916 in the commemorative calendar did not go down well with some individuals, who expressed their disillusionment in the media. Speaking to *The Cork Examiner*, Bernadette Byrne, the daughter of a dispatch carrier for the Irish Citizen Army in the GPO, complained that the ceremony 'was a bit of an excuse'.[118] Writing for the same newspaper about 'the obvious shift in emphasis which has occurred since 1966', Tim Pat Coogan remarked that 'the pendulum' had swung 'in official attitudes, from ... overkill ... to the deliberate neglect of one of the most significant Irish events of the century'.[119] Another columnist for *The Cork Examiner*, singer-songwriter Jimmy Crowley, claimed that 'the guts, blood and the very soul of Ireland is being swept under the carpet' and registered his incredulity that 'the very people who owe their existance [*sic*] to the revolution, the politicians, are tongue-tied completely' and could not 'find it in their hearts to celebrate'.[120] Others took a more light-hearted view. In a very short communication to *The Irish Times*, a Cork-based letter-writer highlighted how 'incredibly low-key' the commemoration ceremony had been and wondered whether people could 'expect a similarly muted Twelfth of July march' later in the year.[121] At a later stage, Declan Kiberd described the 75th anniversary commemoration as both 'sheepish' and 'spare'.[122]

---

[116] *The Irish Times*, 1 April 1991.

[117] Ibid.

[118] *The Cork Examiner*, 1 April 1991.

[119] Ibid., 3 April 1991.

[120] Ibid., 9 April 1991.

[121] *The Irish Times*, 6 April 1991.

[122] D. Kiberd, 'The Elephant of Revolutionary Forgetfulness', in Ní Dhonnchadha and Dorgan (eds), *Revising the Rising*, p. 1.

Plate 5.5   Veterans of the Rising (seated in the front row) in the central enclosure outside the GPO for the 75th anniversary commemoration on 31 March 1991

Source: National Library of Ireland, Independent Newspapers (Ireland) Collection, 391-639C/17-17a.

The size of the crowd who turned out to watch the 75th anniversary ceremony was estimated at 600.[123] Having gone through security screening, they then watched the formalities and offered genteel applause from behind barricaded positions at the Abbey Street and Henry Street junctions with O'Connell Street. The invited dignitaries who attended the commemoration included members of the Cabinet (including Bertie Ahern, Dr. Michael Woods, Dr. Rory O'Hanlon, Mary O'Rourke, Séamus Brennan, and Bobby Molloy), the Chairman of the Fianna Fáil Party, the Attorney General, a small number of Junior Ministers and Senators, and the Lord Mayor of Dublin, Vincent Brady. The former President, Dr. Patrick Hillery, was also present, as was the former Tánaiste, Brian Lenihan. A handful of surviving veterans of the Rising – namely 90-year-old John Andrew Flynn, 96-year-old James Henry, 93-year-old Bill Hogan, and 84-year-old William Flood – also attended the ceremony.[124] They sat down

---

[123] *The Irish Press*, 1 April 1991.
[124] Ibid.

in the only chairs provided in the central enclosure, while members of their families and a few War of Independence veterans stood behind them (Plate 5.5).

In a repeat of the 1966 debacle, however, the leader of the Fine Gael party (a position now held by John Bruton) was again absent, as were all the members of its front bench. The only representatives from the main Opposition party who did show up were the former Taoiseach, Liam Cosgrave and his son, Senator Liam Cosgrave. Sean Barrett, the Fine Gael spokesman on Justice, denied that they had boycotted the ceremony and tried to explain his own absence by stating that he had not received an invitation. Other politicians who were present included the SDLP's Séamus Mallon and Denis Haughey. Michael McDowell, the Chairman of the Progressive Democrats, was also in attendance. He and others expressed their surprise at the absence of most of Fine Gael's leading figures.[125] While the Civil War divide in Irish politics again seemed clear for all to see within the public arena of historical commemoration, deeper issues may have explained their nonappearance. In the lead-up to Easter Sunday 1991, many Fine Gael politicians had been seeking to draw attention to the role of constitutional nationalists in Irish history. One of these was Austin Deasy, who received a blank refusal when he approached Haughey for his support in erecting a monument at 'the home of the Redmondites' on Ballybricken Hill in Waterford.[126]

Overall, the contrast with 25 years earlier was striking and the stark impression left by the official commemoration at O'Connell Street was that those who did go to the trouble of showing up to watch it were simply going through the motions and merely paying lip-service to the actions of the 1916 insurgents, whose historical legacy had been tainted by over two decades of bloodshed caused by paramilitary violence in Northern Ireland. As was clearly evident on Easter Sunday, a particularly problematic issue for Haughey's government was the serious ambiguity arising from the Provisional IRA's claim that they were the rightful heirs of the 1916 rebels, and their argument that the absence in 1916 of a mandate from the Irish people for the Rising was enough justification for their own armed campaign. After the military ceremony on O'Connell Street ended, Haughey promptly moved inside the GPO and from behind closed doors he launched a joint An Post/OPW video and display about the Rising (along with five new stamps marking Dublin city's year as 'European City of Culture'). In a short speech, during which he struck an impressive pose next to Oliver Sheppard's statue of Cúchulainn, Haughey placidly emphasised that the Rising was 'an inalienable part of our heritage which belongs to all our

---

125   *The Irish Times*, 1 April 1991.
126   Ibid., 9 March 1991.

people and not to any faction'. The Proclamation, he said, was 'a noble document of great power and eloquence'. 'We take pride', he continued, 'in the courage, vision and honourable conduct of the men and women of 1916 as an integral part of our pride in nationhood'.[127]

When asked for his opinion on the ceremonies, James Henry (who served in Boland's Mill in 1916), said that the commemoration was 'as it should be, for the time that it is commemorating'.[128] In a sign of times, tight security surrounded Haughey as he spoke inside the GPO. However, this did not go down so well with the assembled press corps as a number of them, along with some invited guests and politicians, were refused entry to the building and left standing outside its doors. According to *The Irish Press*, one of those locked out was John Andrew Flynn, who also fought in Boland's Mill and handed up de Valera's rifle after the surrender. The veteran was delayed by conducting interviews with the press, and when he eventually was led through the doors of the GPO by the Gardaí, the proceedings inside were coming to an end.[129] *The Cork Examiner* alleged that the veteran had been 'insulted by uncaring An Post personnel', thus 'spoiling his day and souring the 75th anniversary official commemoration'. To make matters worse, it also reported that further embarrassment had been caused by the failure of the original Department of Defence invitations to cater for the veterans' transportation requirements to and from the ceremony.[130]

Following the abrupt GPO ceremony, the Taoiseach flew from Dublin to Wexford in an Air Corps Dauphin helicopter to attend an afternoon commemoration ceremony at Abbey Square in Enniscorthy, to mark the town's role in the 1916 Rising. He arrived there at 2.45 pm and was greeted by bands and bunting for a ceremony that proved to be both longer and more upbeat than the one held a few hours earlier in the capital.[131] It was fitting, said Haughey at Enniscorthy, to once again pay tribute to the men and women of 1916 'for the freedom and independence which we enjoy today'.[132] After he reviewed the guard of honour from the 10th Infantry Battalion, to the sound of music from the Band

---

[127] Ibid., 1 April 1991. Afterwards, in an interview that was published in *The Irish Press*, 1 April 1991, Haughey maintained that the ongoing paramilitary violence in Northern Ireland had 'nothing to do with 1916', and conveyed his belief that the 1916 rebels would 'not want their cause dishonoured' by bloodshed. The Rising, he argued, was 'inspired by a noble spirit of patriotism and sacrifice and conducted with generous chivalry. Its leaders sought to avoid unnecessary bloodshed'.

[128] *The Irish Press*, 1 April 1991.

[129] Ibid., 1 April 1991.

[130] *The Cork Examiner*, 1 April 1991.

[131] *The Irish Times*, 1 April 1991.

[132] *The Irish Press*, 1 April 1991.

of the Western Command, a five-minute Ecumenical Service was then led by Dr. Brendan Comiskey, the Bishop of Ferns and Archdeacon Ken Wilkinson, a Church of Ireland rector. In attendance were two 1916 veterans – Billy Dagg and Paddy Brennan. Under the warm sunshine, wreaths were laid by the Taoiseach and the Vice-Chairman of Enniscorthy Urban District Council under the 1798 monument in the town, and a 'crushing crowd' then followed Haughey to the Athenaeum (which had been turned into a citizens' information centre) where another wreath was laid. Following a minute's silence, the Tricolour was raised to full mast. Finally, another ceremony took place at the Séamus Rafter monument by the River Slaney, where Haughey paid tribute to the citizens of Enniscorthy and called God's blessing down on the insurgents of 1916. The ceremony then drew to a close with the reading of the Proclamation by the grandson of a 1916 veteran. The Taoiseach then lit a 'flame of freedom' and a plaque was unveiled at the monument.[133]

In addition to the Easter Sunday ceremony in Dublin, another official event organised by the government was the 'Road to Freedom' exhibition in the refurbished 1916 room at the National Museum of Ireland. Curated by Michael Kenny, the museum exhibition opened in mid-April and included a range of artefacts from the revolutionary period (including guns, flags and uniforms), a photographic record and a labelled mini-history. Due to problems 'with lighting, heating and humidity', Kenny decided to use replicas of the original flags in the display cases.[134] Outside of Dublin, Easter Sunday commemorations were held at various locations throughout the Republic. At Ennis, 500 members of the public paraded from the Catholic cathedral to the 1916 memorial in Abbey Street, where a wreath was laid by the Minister for Defence, Brendan Daly. In Cork, Junior Minister Denis Lyons and a group of fellow Fianna Fáil politicians – including Gene Fitzgerald, Dan Wallace, Micheál Martin, John Dennehy, and Dave McCarthy – visited many of the city's memorials after Mass in St. Augustine's church. The programme of events concluded with an oration by Lyons at the republican plot in St. Finbarr's Cemetery. Denouncing the ongoing violence in the North, he argued that the 1916 rebels 'gave their lives for Ireland' and stated that it was appropriate to 'recognise their sacrifice'. Additionally, he urged people 'to acknowledge that we have the responsibility to work for the improvement of the heritage they left us.'[135]

Elsewhere in County Cork, a commemoration was held in Skibbereen by the Urban District Council to honour the memory of Gearóid Ó Súilleabháin,

133 *The Irish Times*, 1 April 1991.
134 *The Sunday Tribune*, 31 March 1991.
135 *The Cork Examiner*, 1 April 1991.

who raised the Tricolour over the GPO in 1916. His nephew, Dr. Mícheál O'Sullivan, placed a wreath under a plaque outside the Town Hall that had been unveiled in 1966 to honour the volunteer. The Proclamation was read beside the Maid of Erin statue by the Fianna Fáil Councillor, Letty Baker. In Tipperary town, around 1,500 people attended a commemoration ceremony organised by South Tipperary County Council in conjunction with the Third Tipperary Brigade, Old IRA Commemoration Committee. Its Chairman, Frank McCann, stated that instead of violence what was needed in the North was peace and reconciliation. 'The bullet', he added, has changed very little, making it more difficult to reconcile the various cultures and differences that separate people in the North from the people down here'. In Waterford, members of Cumann Seán Oghlaigh Fianna Éireann paraded to the National Memorial at the Quayside, where a wreath was laid by the Mayor, Alderman Liam Curham.[136]

In the West of Ireland, Fianna Fáil arranged a Mass at St. Patrick's Church in Galway city, where readings were given by two TDs, Maire Geoghan Quinn and Frank Fahey.[137] In the county, over 400 people attended an Easter Sunday commemoration ceremony at the Pearse cottage at Rosmuc, where an elected member of the Údarás na Gaeltachta Board, Seosamh Ó Cuaig, called for 'practical patriotism' to tackle local problems such as unemployment and emigration. Writing at the time for *The Connacht Tribune*, Harry McGee noted that the mood in Galway about 1916 'is now a source of indifference and shame', in contrast to the 'great rejoicing and celebration' witnessed during the Golden Jubilee. 'The Rising', he mourned, 'has been left to the few of the marginal interests to commemorate ... only Axe-Tax protests draw the people to the streets nowadays', thus leaving the limestone statue of Liam Mellows in Eyre Square in Galway city 'as handy target-practice for the birds'.[138] At Westport in Mayo, an Easter Sunday commemoration was held by members of the Westport Historical Society at the Major John McBride memorial in the Mall. In Cong, Father Enda Howley offered prayers at an Easter Sunday commemoration, while a wreath was laid by Lelia Luskin, who was a member of Cumann na mBan.[139]

It was reported that the government's subdued Easter Sunday commemoration ceremony at Dublin was the subject of contemptuous debate amongst some of the crowd who attended the Enniscorthy tribute. The local Fianna Fáil TD, John Browne, was particularly unapologetic, stating: 'The nation that forgets its past is deliberately abandoning its own identity.'[140] He was not the only Fianna Fáil

---

[136] Ibid.

[137] *Western People*, 10 April 1991.

[138] *The Connacht Tribune*, 5 April 1991.

[139] *Western People*, 17 April 1991.

[140] *The Irish Times*, 1 April 1991.

TD who was ruffled by the downgrading of the 1916 metanarrative. A similar sentiment had been uttered weeks beforehand by Eamon de Valera's grandchild, Síle de Valera, who headed Fianna Fáil's own committee for commemorating the 75th anniversary. 'We should not be running away from our history', she said, nor 'allowing other people to hijack our history'.[141] Another party member descended from de Valera was Senator Éamon Ó Cuív – known at the time as 'Young Dev' due to his striking resemblance to his grandfather. Speaking to John Waters of *The Irish Times*, he declared that he had 'no difficulty ... about 1916, or with nationalism properly expressed'. 'I would submit', he continued, 'that, in the 1991 context, one of the problems we have with 1916 is that a lot of people do not appreciate the value of what was won ... They haven't worked it out – largely because we don't talk about it'. 'One thing that was good in my upbringing', he added, 'was that all these things were discussed'.[142] On Easter Sunday, Ó Cuív again staunchly defended the ideals of 1916 at a Fianna Fáil-organised commemoration ceremony at the republican plot at Kilcrumper Cemetery in the town of Fermoy, County Cork. 'We have our problems', he admitted, 'but in tackling them we should not run down the achievements of the past'.[143]

While a 'most pressing reason' for the astonishing differences between the commemorations of 1966 and 1991 was 'the fear of giving aid and comfort to the IRA', Charles Townshend has noted that this also signalled 'an acknowledgement ... of how successfully republicans had appropriated the 1916 legacy'.[144] As they had been doing for years, supporters of Sinn Féin and the Provisional IRA came out in force on Easter Sunday 1991 to commemorate the Rising (and bemoan Partition) at venues in the Republic such as Glasnevin Cemetery in Dublin and Mount St. Lawrence Cemetery in Limerick. Under the watchful eye of a Garda escort, a crowd of about 400 followed a Sinn Féin colour party from the GPO to Glasnevin, but there was no repeat of the heavy scuffles that had marred the fractious procession to the cemetery back in 1966. When they arrived at Glasnevin, the Sinn Féin press officer, Martha McClelland, referred to Northern Ireland as a 'bloody and failed political entity' and said that republicans did not need the 1916 Rising 'to justify today's struggle for Irish freedom'. She also proclaimed that the military occupation by the British, along with Partition, provided 'justification for the armed response', as did issues such as 'plastic bullets, systematic house raids and death squads, and the kind of justice given to the Birmingham Six, the Guilford Four and the men and women who have

---

141 Ibid., 9 March 1991.
142 Ibid., 4 April 1991.
143 *The Cork Examiner*, 1 April 1991.
144 Townshend, *Easter 1916*, p. 352.

passed through the Diplock courts'.[145] Republican Sinn Féin also held a 1916 commemoration outside the GPO on Easter Sunday, which was attended by a crowd of about 150.

In Northern Ireland, Sinn Féin commemorations of 1916 were held at Carrickmore and Moortown in County Tyrone, Milltown Cemetery in west Belfast, Crossmaglen in County Armagh, Newry, and Derry – all of which passed off peacefully. Speaking to a crowd of 2,000 who turned up to a commemoration ceremony in Carrickmore, Gerry Adams took the opportunity to engage in some political point scoring by pointing out what he believed were glaring disparities between the Irish government's commemoration of the 50th and 75th anniversaries of the Rising. Adams declared that Haughey's government was complicit in erasing the event of 1916 from the public consciousness, and argued that the Republic's lack of commitment to the vision of the Rising had become even greater than it had been in 1966, especially in its policy towards the North. Back in 1966, he noted that 'verbalised republicanism was sufficient and necessary to win electoral support in the 26 counties', yet at the same stage, 'Dublin did nothing to correct the crisis created by the British Partition of Ireland'. 'Now', he alleged, 'the Dublin establishment is clearly seen to be part of that crisis', and its fear of 1916 had a logic to it. 'How', he asked, 'could they speak of the hunger strikers MacSwiney or Thomas Ashe without conjuring up visions of hunger strikers Martin Hurson or Bobby Sands?' and how could they disapprove of 'the British occupation of that time while collaborating with the British occupation today?'[146] Adams also criticised the proposed Brooke talks, alleging that they contained a British and unionist agenda and that the SDLP were 'being wrong-footed again', just like in 1973.[147]

A statement issued by the leadership of the Provisional IRA at a 1916 commemoration attended by several thousand republicans at Milltown Cemetery on Easter Sunday 1991 did little to allay fears of further republican violence. 'We have the personnel, the material, and most importantly, the commitment to fight', said a man shielded with umbrellas from a group of RUC officers and British soldiers some 40 feet away, 'until the foreign military presence removes itself or is removed'. To loud cheers, he then told the crowd that the Provisional IRA would never be defeated.[148] At the same ceremony in Milltown Cemetery, the Sinn Féin chairman, Martin McGuinness, denied accusations that the republican movement was living in the past. He argued that in order to analyse and comprehend the present, one had to look objectively at

---

[145] *The Irish Times*, 1 April 1991.
[146] Ibid.
[147] *The Irish Press*, 1 April 1991.
[148] *The Irish Times*, 1 April 1991.

the past, learn from it and move onwards. While saying that the 1916 Rising was a remarkable historical event, he sought to elevate the status of the armed struggle of the Provos by stating: 'The republican struggle of today is not carried out because of the Easter Rising, but it is certainly carried out for the same reasons, that is to break the connection with England and establish national democracy in Ireland.'[149]

Such pronouncements were, of course, well anticipated by Haughey's government, which ultimately succeeded in its objective of delivering an extremely guarded commemoration of 1916, falling as it did less than 41 months after the Enniskillen bomb. Yet another sign of the changes that had taken place by 1991, according to Townshend, was 'a shift of priorities, from traditional nationalism to a wider Europeanism – and indeed an embrace of the materialism thought by nationalists to be so alien'.[150] Irish nationalism, as Fintan O'Toole later observed, had become 'vastly more complicated, a set of troubling questions rather than of easy answers'. In addition to the pain caused by the Troubles, both membership of the European Union and cultural globalisation were making nationalism 'a slippery and ambiguous concept'.[151] Events such as the collapse of the Berlin Wall had already started a process that was destined to change the political map of Europe beyond recognition. With the 1916 metanarrative in such a perilous position by Easter 1991, a number of Fianna Fáil politicians were more than happy to shine the spotlight on the burgeoning notion of a broader European identity. In a Seanad debate on Dublin's designation as 'European City of Culture' on 24 April, the Minister of State at the Department of Justice, Noel Treacy, welcomed the awarding of the accolade with open arms. 'We should not be parochial', he advised, 'in this matter'. 'It is right for all of us in Ireland', he added, 'to take pride in our European cultural heritage' and to better appreciate 'contemporary European culture', so as to improve communications and better understand 'the peoples of the European Community'.[152]

The Role of Artists and Writers in the 75th Anniversary

Although the combined impact of revisionist arguments and public revulsion against violence in the North led to a sea change in how the story of 1916 was commemorated at Easter 1991, there were still people at the time in the Republic

---

[149] Ibid.

[150] Townshend, *Easter 1916*, p. 352.

[151] F. O'Toole, *Enough is Enough: How to Build a New Republic* (Faber and Faber, London, 2010), p. 3.

[152] *Seanad Debates*, Vol. 128, 24 April 1991.

who remained unwavering in their determination not to let citizens completely forget the significance of the Rising's place within storylines concerning Irish heritage, culture and identity. This was especially true of a number of practitioners within the creative arts sector, who felt compelled to win over people's hearts and minds when it came to espousing what they saw as the merits of the legacy of 1916. With military parades still in abeyance, the time seemed more than ripe for to them to celebrate the 75th anniversary in an alternative fashion – namely through the staging of a more light-hearted commemoration through the guise of a festival. Under the banner of 'Reclaim the Spirit of 1916', a voluntary committee headed by the artist, Robert Ballagh (and supported by the Arms Trial trio of Neil Blaney, Kevin Boland and Captain James Kelly), was initially established in 1990 'to look anew' at the principles which provoked the Rising and 'to further constructive dialogue and to demonstrate pride in our history'.[153] By the beginning of 1991, James Stephenson, an experienced festival organiser, was also serving on the committee. Following much debate, it produced a range of ambitious proposals for a 'secular celebration of our culture', which was designed for the purpose of filling 'a vacuum on 1916' and facilitating 'the purposes of joyous self-identification'.[154]

The culmination of the committee's work came with the holding of an artistic festival on Easter Sunday 1991 called 'The Flaming Door'. In an imaginative effort aimed at tackling forgetfulness, six double-decker busloads of people participated in a day-long commemorative venture that included poetry readings and airings of political manifestoes at various sites throughout Dublin that had associations with 1916 – including Liberty Hall, Mount Street Bridge, the derelict Jacob's biscuit factory, the City Hall, the Four Courts, and the GPO. Prominent in delivering the readings were a host of well-known writers and poets, including Anthony Cronin, Ulick O'Connor, Richard Murphy, Brendan Kennelly, Seamus Deane, Michael Longley, and Máire Mhac an tSaoi. The Labour TD, Michael D. Higgins, participated in the festival in a personal capacity as a poet and conveyed his disappointment to *The Irish Times* at the 'ashamed, forgetful and subservient' attitude which he alleged was being shown towards 1916 by the government. Anthony Cronin said that the nature of the event demonstrated 'a welcome attitude of celebration mixed with a questioning'.[155] Later that night, 'The Flaming Door' festival concluded with an

---

[153] *The Irish Times*, 9 March 1991.

[154] Ibid., 29 January 1991. For an insight into the general appeal of festivals throughout history, see M. K. Smith, *Issues in Cultural Tourism Studies* (Routledge, London, 2003), p. 140, who notes that their potential to be inclusive, participatory and people-based has enabled them, over and over again, to 'become the quintessence of a region and its people'.

[155] *The Irish Times*, 1 April 1991.

Easter Sunday musical event at Kilmainham Gaol that was attended by a crowd of 400. This featured Ronnie Drew (founder of The Dubliners), Donal Lunny and members of The Hothouse Flowers.[156]

For six weeks afterwards, Kilmainham Gaol also played host to a daily 'Sixteen-Ninety-One' dramatic tour from Tuesdays to Sundays, in place of its usual guided tours and audiovisual shows. Featuring eight actors in period dress and the odd song, the 45-minute tour attempted to offer visitors a reenactment of certain parts of the story of the Rising in spaces such as the great hall, the corridor where the leaders were imprisoned and the Stonebreakers' Yard.[157] Other activities organised by 'Reclaim the Spirit of 1916' included a series of walking tours, an exhibition of paintings by Countess Markievicz at the Irish Labour History Museum on Tuesday 2 April 1991 and a debate in the Mansion House the next day on Articles Two and Three of the Constitution. A commemorative pageant watched by a crowd of 5,000 was also held by 'Reclaim the Spirit of 1916' on a stage outside the GPO the following Saturday. Featuring a showcase of artistic talent, including the Dance Theatre of Ireland and actors like Frank Kelly (who was cast in a satirical role as a politician witnessing 'the final obsequies of Éireann'), it offered an amusing and dramatic presentation of what the organisers saw as a determined attempt by the government to write 1916 out of Irish history. In an address to the crowd, Ballagh commended them for turning up and demonstrating 'that you are proud to celebrate the 75th anniversary of 1916, that you are not embarrassed to celebrate your Irishness, that you are not ashamed of your history and culture'.[158] The artist had harsher words for the government. In a short article that he wrote for the *Irish Reporter* in the lead-up to the parade, he alleged that there had been a degree of 'official prevarication' concerning the story of 1916, and that people's attempts to celebrate its anniversary had been frustrated by the rejection of projects and the withdrawal of grants. It had been illuminating, he bemoaned, 'to observe the serried ranks of the ... establishment squirm in the face of their own history'.[159]

---

[156] *The Irish Press*, 1 April 1991.

[157] *The Irish Times*, 2 April 1991.

[158] Ibid., 8 April 1991.

[159] B. Ballagh, '1916: Goodbye to All That?', *Irish Reporter*, No. 2 (1991), pp. 7–8. As a result of his prominence at the time, the artist had to put up with certain inconveniences. Years later, in *The Irish Times*, 15 April 2006, he complained of having been put under surveillance by the Special Branch in 1991: 'I was walking up Parnell Square for lunch one day, and an unmarked squad car suddenly braked beside me, out jumped two plain-clothes officers who pushed me against the railing. They asked me for ID, then jumped in the car and drove off. An amused crowd had gathered to see who this dangerous criminal was ... It was horrendous ... we were accused of giving aid and comfort to the IRA.'

Outside of the capital, the Galway committee of 'Reclaim the Spirit of 1916' organised an artistic commemoration on Easter Monday, followed later that night by what was described as 'a celebratory traditional music session' in the Raparee Bar on Dominick Street.[160] In the county, actors with the Tuam Theatre Guild also celebrated the 75th anniversary by running an imaginative theatrical performance called 'The Delirium of the Brave'. This recalled the events and aftermath of 1916 through the reading of excerpts from old newspapers and from the writings of the 1916 leaders and literary figures such as William Butler Yeats and Séan O'Casey. *The Connacht Tribune* reported that the production 'told it as it was' and managed to steer clear of the controversy over 'whether the 75th anniversary of the Rising should or should not be celebrated, whether it was right or wrong' and whether the 'country was the better or the worse for the blood sacrifices made'.[161]

Not all creative figures were as upbeat as those connected to 'Reclaim the Spirit of 1916', when it came to the issue of commemorating the 75th anniversary. Playwright Hugh Leonard felt that the Ireland of 1991 was not mature enough to celebrate the Rising, arguing that any hint of a rejoicing would resurrect the philosophy previously taught in schools 'that it was better to die for Ireland rather than live for Ireland'.[162] The acclaimed Leitrim-based novelist, John McGahern, who had by then achieved major international acclaim for works such as *The Dark* and *Amongst Women*, offered his own illuminating perspective on the prevailing scepticism about the Rising in a piece that he wrote for *The Irish Times* on 3 April 1991, entitled 'From a Glorious Dream to Wink and Nod'. In this, he declared with characteristic candour: 'I think that the 1916 Rising was not considered to be of any great importance in the country I grew up in ... it was felt secretly to have been a mistake.' In a swipe at the way de Valera had been revered by elements of the public, McGahern charged that in private terms, 'his name was always slightly tainted by the fact that he alone, of all the signatories, had escaped execution' and that 'he looked more like a lay cardinal than a revolutionary'. The writer also bewailed that the Proclamation had been undermined by 'a theocracy' within 'the Free State that grew out of that original act of self-assertion in the General Post Office'.[163]

---

[160] *Western People*, 10 April 1991.

[161] *The Connacht Tribune*, 5 April 1991.

[162] *The Sunday Tribune*, 24 March 1991.

[163] *The Irish Times*, 3 April 1991. Also see J. McGahern, *Memoir* (Faber and Faber, London, 2005), p. 210, in which the writer returns to this theme and holds dearly to his deep-seated personal convictions, arguing that a theocratic state had emerged by the middle of the twentieth century and gone 'against the whole spirit of the 1916 Proclamation'.

The prevailing awkwardness associated with the 75th anniversary resulted in a marked reluctance by historians to conduct research on the Rising in the lead-up to Easter 1991. From a sales perspective, it was a work of fiction by English-born novelist Peter Rosa that proved most successful at tapping into any lingering nostalgia. First released in March 1990, de Rosa's *Rebels: The Irish Rising of 1916* offered a dramatised account of the events leading up to and including the Rising.[164] After topping the best-seller list, it was released as a Corgi paperback in 1991, with extracts appearing in *The Cork Examiner* between 1–6 April.[165] Only a few non-fictional works on the Rising were published in 1991. One of these was *Revolutionary Woman*, an absorbing autobiography of Kathleen Clarke (wife of Thomas Clarke), edited by her grandniece, Helen Litton.[166] Secondly, the text of a lecture that had been delivered by C. Desmond Greaves during the Golden Jubilee commemorations 25 years beforehand was published by the Fulcrum Press, entitled *1916 as History: The Myth of the Blood Sacrifice*.[167] Its publication was overseen by Anthony Coughlan, a lecturer in Social Policy at Trinity College Dublin, who added extra endnotes to support the arguments of Desmond Greaves, who had been his personal friend until he passed away in 1988. A special issue of *An Cosantóir: The Defence Forces Magazine* was also produced for the 75th anniversary, retailing at £1. This contained seven articles relating to the Rising (including one by Commandant Jim Burke), adding to the 38 that the magazine had already published between April 1941 and November 1987 (with 13 or 34% of these being published between 1966–1967).[168]

The most significant publication that marked the 75th anniversary was a small book entitled *Revising the Rising*, edited by Mairín Ní Dhonnchadha, a researcher attached to the Dublin Institute for Advanced Studies, and the Cork poet, Theo Dorgan. Running to 142 pages in length and published by Field Day in both hardback and paperback, this interdisciplinary collection of

---

[164] P. De Rosa, *Rebels: The Irish Rising of 1916* (Bantham Press, London, 1990). Extracts of the novel were also published in *The Irish Times*, 17 March 1990 and 19 March 1990. The author described his work as 'history written purely for pleasure'.

[165] Interviewed by Ralph Riegel in *The Cork Examiner*, 1 April 1991, de Rosa said that 'very few people ... are aware of the real facts of the Rising' and expressed regret 'that there's such a huge number of modern connections that people seem to take totally for granted'.

[166] H. Litton (ed.), *Revolutionary Woman: Kathleen Clarke 1878–1972. An Autobiography* (The O'Brien Press, Dublin, 1991).

[167] C. Desmond Greaves, *1916 as History: The Myth of the Blood Sacrifice* (The Fulcrum Press, Dublin, 1991).

[168] Figures derived from J. White, 'Index of Articles Relating to 1916 Previously Published in *An Cosantóir*', *An Cosantóir: The Irish Defence Journal*, Vol. 51, No. 4 (1991), p. 30. These figures do not include the short biographical sketches of some of the 1916 leaders that appeared in the magazine's 'Irish Leaders of Our Time' series between 1945–1946.

essays featured contributions by three literary scholars (Declan Kiberd, Edna Longley and Seamus Deane), two historians (Micheal Laffan and Joe Lee), two political scientists (Tom Garvin and Arthur Aughey), and a solitary folklorist (Gearóid Ó Crualaoich). In their preface to the volume, the two editors noted that whilst anniversaries 'are often problematic', hardly any had 'been as loaded with ambiguities and contradictions' as the Rising's 75th anniversary. Efforts at understanding the story of 1916, they contended, had 'prompted fractures and disputes among historians, politicians and citizens – but, outside the realms of embattled historiography, open and general discussion has been curiously muted, not to say inhibited'. Seeking to address this perceived discrepancy, they claimed that 'amnesia – private or communal – is both unhealthy and dangerous', and stated that their objective was to offer 'a forum where a range of opinions on 1916, and its meanings', could be put before a broader public than the more specialised audience accustomed to 'the in-fighting of recent historiography'.[169] By no means, however, did these principled aims serve to dilute the passion of some of the book's contributors. In Declan Kiberd's chapter, memorably entitled 'The Elephant of Revolutionary Forgetfulness', the writer seemed to be on a mission designed to counter-attack the revisionist deprecation of the story of 1916. In no uncertain terms, he accused powerfully-coordinated 'cadres' in 'the Irish intelligentsia' of airbrushing 1916 from 'official history'.[170]

Unlike the Golden Jubilee, no major supplement commemorating the Rising was published by the main broadsheet newspapers in 1991. Several column inches, however, explored the legacy of the Rising. *The Cork Examiner* marked the 75th anniversary with a full page of commentaries for six days, under the heading 'The Easter Rising 1916–1991'. Two of the articles explored the differences between 1966 and 1991 and the impact of revisionist historiography on interpretations of 1916.[171] In seeking to inform its readers about all angles to the 1916 debate, ample column inches were also devoted by *The Irish Times* to a wide range of divergent opinions on the Rising. The former Taoiseach, Garret Fitzgerald, conveyed a sympathetic view about 1916 and its consequences in five articles that he wrote for *The Irish Times* from 13–18 July 1991. These were based on a lecture that he had prepared for delivery at the Carlingford Summer School.[172] To assess the impact of revisionism on the 1916 metanarrative, the newspaper also published a head-to-head debate between the distinguished

---

[169] Ní Dhonnchadha and Dorgan, 'Preface', p. ix.

[170] Kiberd, 'The Elephant of Revolutionary Forgetfulness', p. 5.

[171] See *The Cork Examiner*, 2 April 1991 and 6 April 1991.

[172] See *The Irish Times*, 13 July 1991, 15 July 1991, 16 July 1991, 17 July 1991, and 18 July 1991.

Cambridge-based nationalist historian, Father Brendan Bradshaw and Pearse's biographer, Ruth Dudley Edwards.

In his article, entitled 'Dishonouring Heroes in History's Name', Bradshaw lambasted the revisionist school's 'denigration of 1916', claiming that its philosophy was at odds with its practical application in Irish historical studies. Its foundation, he noted, 'lay in the idea of history as a science and, accordingly, in the idea of the historian interpreting the historical evidence in the same clinical, disengaged way as the scientist is supposed to examine natural phenomena'. While 'revisionists aspired to provide a "value-free" account of Irish history ... free of the bias that had coloured and, they believed, distorted earlier accounts', Bradshaw remarked that 'the practice', as he experienced it, was rather separate from 'the theory'. To his horror, the violence in the North was being 'seen as a legacy, via the IRA, of the protest in arms of 1916', and this had 'emboldened' revisionists to adopt a 'more strident tone'. Blaming the violence in the North on the Rising, concluded Bradshaw, was 'about as historically valid as to blame Jesus Christ for the anti-semitism which produced the Holocaust'. In her own article, entitled 'Is it Unpatriotic to be Honest?', Dudley Edwards accused Bradshaw of being 'one of those overwrought faction fighters who insists on belting bystanders with his hurley, and therefore forces even the most pacific of us to self-defence'. Pearse, she argued, had been a 'tortured and complex man' – who was 'obsessed with Calvary', led his 'unworldly' brother to his death, 'frequently changed his mind on ... priorities', was 'financially irresponsible', and possessed a 'vanity' that 'prevented him from ever questioning his own judgement'. Although concurring that terrorism would have appalled Pearse, she concluded that the Provisional IRA 'were logically his heirs'. Through his martyrdom and immortality, she wrote that Pearse had 'left behind him a self-justificatory political testament that turned out to be a Pandora's Box'.[173]

Besides the heated debates over the legitimacy of the Rising and the question of how to commemorate it, the 75th anniversary was also marked by a certain degree of debate that ventured into the embryonic sphere of counterfactual history – pondering what might have happened in the event of the 1916 Rising not taking place.[174] An Editorial in *The Irish Times*, for example, speculated that it 'is probably true that without the Easter Rising an independent Irish State would have come into existence anyway'. 'It is also possible', it ventured, 'to argue

---

[173] Ibid., 1 April 1991. A longer exposition of the priest's ideas can be found in B. Bradshaw, 'Nationalism and Historical Scholarship in Ireland', *Irish Historical Studies*, Vol. 26, No. 104 (1989), pp. 329–51.

[174] For an assessment of counterfactual history, see Ferguson, 'Virtual History', pp. 2, 89, who describes it as a historiographical format that seeks to enhance learning through voyages into 'imaginary time'.

that if there had been no recourse to arms, Ireland today would not be divided and the North would be at peace'. However, it also stated that that 'it ought not be forgotten that before ever a shot was fired on behalf of Irish nationalism in these years, the unionists had armed and trained and declared their intention to resist a new settlement by force'.[175] Writing for the same newspaper, John McGahern wrote that it was impossible to answer 'with any certainty' what might have occurred if the 1916 rebels had waited – 'if that freedom would have come about anyhow without violence'. While pronouncing that 'it does not matter now', he pessimistically added 'that North and South would have separated anyhow in their need to out-bigot one another'.[176] In *The Cork Examiner*, T. P. O'Mahony also pondered what might have happened had the 1916 rebels 'not acted as they did'. However, he did not go any further than merely wondering whether the state would have been other than where it was by then. 'The historical reality', he argued, 'is that the Easter Rising *did* happen. We can't change that. In Hollywood jargon, we can't get the toothpaste back into the tube'.[177]

As counterfactual techniques had made little inroads into Irish historiography by the time of the 75th anniversary, academic historians steered clear of tackling the subjunctive conditional.[178] The prominence accorded to counterfactualism by the media, however, did fracture little bits and pieces of the 1916 metanarrative's hallowed veneer. In other ways, the decay also extended to the state of the built heritage of the Rising within the capital city. When the poor physical condition of the GPO was discussed in the Dáil on 14 May, the Minister for Tourism, Transport and Communications, Séamus Brennan, sought to allay Opposition concerns by putting it on record that the government had allocated £1 million from National Lottery funds in 1991 (along with a further £500,000 the previous year) for restoration work. The Minister, however, was not able to confirm when the works would be fully completed, as this depended 'on the availability of additional financial resources'. However, this statement did not go down well with Labour's TD for Cork North Central, Toddy O'Sullivan, who said that

---

[175] *The Irish Times*, 29 and 30 March 1991.

[176] Ibid., 3 April 1991.

[177] *The Cork Examiner*, 1 April 1991.

[178] It took another fifteen years before counterfactual debates on 1916 infiltrated mainstream Irish historiography. See, for example, J. J. Lee, '1916 as Virtual History', *History Ireland*, Vol. 14, No. 2 (2006), p. 5, who acknowledges counterfactualism's merits as a 'potentially illuminating form of historical methodology', but expresses concern about historians using it 'to get the present they want, as if somehow all the intervening variables can be frozen to satisfy a current wish-list'. A more extended treatment of the theme appears in Ferriter, *What If?*, pp. 76–89. Alternatively, P. Bew, "'Why Did Jimmie Die?'", *History Ireland*, Vol. 14, No, 2 (2006), p. 39, labels it as 'little more than an amusing educational parlour game'.

it was both 'an absolute disgrace' and 'an insult to the men who sacrificed their lives in declaring the Proclamation outside the GPO', that the government was now depending upon the National Lottery to fund the restoration of a building that he considered (just like Richie Ryan had in 1966) to be 'a national shrine'.[179]

The Peace Process and the Forging of Pluralist Identities

Despite ongoing moral questioning of the 1916 Rising's legitimacy, the decade and a half that followed the awkward commemoration of the event's 75th anniversary witnessed a gradual rehabilitation of its status as a key metanarrative of Irish heritage, cultural and identity. Whilst the problem of political violence in Northern Ireland had served to deflate the status of 1916 for more than two decades in the lead-up to Easter 1991, the positive repercussions of various successes in the laborious Peace Process negotiations during the rest of the 1990s, and from the early to mid-2000s, facilitated the emergence of an increasingly fashionable post-revisionist historiography – under which memories of 1916 were cast in a more positive light. Signs first emerged in the early 1990s that a durable peace settlement was possible in the North after Charles Haughey (followed by his successor as Taoiseach, Albert Reynolds) engaged in talks with the British Prime Minister, John Major. Secret talks between Gerry Adams and the SDLP's leader, John Hume, also bore fruit. Success finally came on 31 August 1994, when the Provisional IRA announced a historic ceasefire, with loyalist paramilitaries following suit the following month. But on 9 February 1996, just a couple of months before the commemoration of the Rising's 80th anniversary, hopes and expectations were mercilessly dashed when the Provisional IRA ended its ceasefire by setting off a bomb in the up-and-coming financial district of Canary Wharf in London, which killed two individuals.

As a result of this unforeseen setback to the Peace Process, the Rising's 80th anniversary was again rather minimalist on Easter Sunday 1996. One of the most significant acts of commemoration took place weeks afterwards on 12 May, when President Robinson unveiled Eamon O'Doherty's bronze statue of James Connolly at Beresford Place in Dublin. This depicted him 'standing in front of an undulating wall' showing the flag of the Irish Citizen Army – the Starry Plough.[180] Located directly opposite the high-rise trade union headquarters of Liberty Hall in Dublin, the site of the memorial was vested with historic significance as it was an area where the father of Irish socialism

---

[179] *Dáil Debates*, Vol. 408, 14 May 1991.
[180] Bolger, *Statues and Stories*, p. 32.

had often addressed political and trade union rallies in the years leading up to the Rising. The construction of the memorial, which was the brainchild of the James Connolly Memorial Initiative, was financed through funds obtained from a plethora of sources – including the government, Dublin Corporation, labour organisations (such as the Irish Congress of Trade Unions), American groups (including the Chicago and New York chapters of the Irish-American Labour Coalition), and individual citizens. In addition to being 'dedicated to the ideals of working class solidarity', the following words from an old edition of *The Workers' Republic* (dated 8 April 1916) were added to monument: 'The cause of labour is the cause of Ireland, the cause of Ireland is the cause of labour' (Plate 5.6).

For the most part, though, it appeared that the new Fine Gael-Labour government had little interest in milking the significance of the 80th anniversary. Out of apparent frustration at the inconspicuous nature of the anniversary, a group of writers and artists established the Ireland Institute in the following year. Counting Robert Ballagh and Declan Kiberd amongst its members, it aimed to nurture writing aimed at tackling revisionism.[181] Efforts were also made in 1997 to revive memories of the personalities associated with the Rising, as exemplified by the publication of a new edition of *James Connolly: Selected Writings*, almost a quarter of a century after its initial release. In his preface to the tome, editor Peter Berresford Ellis proclaimed that Connolly's work was 'far from being dead and forgotten' and expressed a personal desire 'that the study of what Connolly actually taught will eventually create an understanding of the national question'.[182]

After a second ceasefire was announced by the Provisional IRA in July 1997, the Peace Process gathered considerable momentum and certain Fianna Fáil politicians began to reevaluate the implications of the legacy of 1916 for the politics of the present. One of these was Bertie Ahern, who was elected Taoiseach in the month beforehand and who quickly proved his worth as a peacemaker. His prowess in this regard was not surprising. As Ken Whelan and Eugene Masterson have noted, he possessed 'astute negotiating skills' and seemed 'to have time for everyone'. Ahern, according to Michael Clifford and Shane Coleman, was also known for being a notable 'practitioner of the art of pure politics' and had 'a reputation as a man who got things done'. All of these traits proved vital in his efforts to bring about a lasting peace in Ireland.

---

[181] Foster, *Luck & the Irish*, p. 177.

[182] P. Berresford Ellis (ed.), *James Connolly: Selected Writings*, New Edition (Pluto Press, London, 1997), pp. x–xi.

Plate 5.6   Eamon O'Doherty's bronze statue of James Connolly at
Beresford Place near Liberty Hall, Dublin, unveiled on 12 May
1996 by the President, Mary Robinson

Source: Author.

Important too, as Colm Keena has noted, was Ahern's 'legendary patience' and his 'extraordinary ability to project feelings of ... understanding' towards physical-force republicans.[183] In public and in private, Ahern also managed to get on extremely well with the British Prime Minister, Labour's Tony Blair, who also came to power in 1997. Blair's mother hailed from Ballyshannon in County Donegal and with an unassuming personality, he easily clicked with Ahern on a personal level after first meeting him. As Mick Temple has noted, Blair at the time possessed 'a youthfulness, charisma and broad popular appeal', truly 'appeared the *regular sort of guy* he claimed to be' and 'deserves credit for continuing the initiative for peace begun by John Major's [Conservative] government'.[184] When he wrote his autobiography, Blair confirmed that his own personality had served him well when dealing with the Irish situation. As he was 'not big on the "dignity of office" stuff', he concentrated instead 'on motivating and persuading people, not frightening them'. Of his relationship with Ahern, he confirmed that they 'got on immediately like the proverbial house on fire' when they first met, and that 'he became a true friend'. Although the family of Ahern 'had fought the British, had been part of the Easter Uprising', Blair recalled that the Taoiseach was 'free of the shackles of history ... he was a student of history, not its prisoner ... he chose repeatedly to put the future first'.[185]

The chemistry between Ahern and Blair added a significant impetus to the Peace Process negotiations. Success came with the brokering (and subsequent approval in referendums) of the Good Friday Agreement in 1998 – the terms of which reassured unionists by virtue of the Irish government's undertaking to amend the territorial claim to the North that was enshrined in its Constitution. The Good Friday Agreement also introduced the principle of 'consent', whereby a united Ireland could only come about in the future with the expressed wishes of a majority of the population in both the North and the South. This principle, as Foster has argued, allowed the British and Irish administrations 'to disengage

---

[183] K. Whelan and E. Masterson, *Bertie Ahern: Taoiseach and Peacemaker* (Blackwater Press, Dublin, 1998), p. 213; M. Clifford and S. Coleman, *Bertie Ahern and the Drumcondra Mafia* (Hachette Books Ireland, Dublin, 2009), pp. 17, 40; C. Keena, *Bertie: Money and Power* (Gill and Macmillan, Dublin, 2011), p. 137.

[184] M. Temple, *Blair* (Haus Publishing, London, 2006), pp. 50, 118–19.

[185] T. Blair, *A Journey* (Hutchinson, London, 2010), pp. 159, 163, 167–68. Jonathan Powell, who worked as Blair's political advisor during the peace negotiations, was equally impressed by Ahern's politics. See J. Powell, *Great Hatred, Little Room: Making Peace in Northern Ireland* (Vintage, London, 2009), pp. 94, 309–10, in which he writes that Ahern was 'above all a practical man'. In spite of his 'firm republican credentials', he 'did not carry the complexes of the past as his predecessors had', nor did he retain any major hang-up 'about the Brits'. Furthermore, he recalls that Ahern seemed 'prepared to override his system by rejecting traditional Irish positions and to take political risks in order to achieve peace'.

themselves from embarrassing inherited positions'.[186] As matters transpired, one of the most significant signs of the sowing of the seeds of peace in the late 1990s was the greater acceptance and recognition of pluralist heritages that unfolded across the island of Ireland. In the process, government commemorationist policies in the Republic began to be reshaped as the twentieth century came to an end. As the state gradually rid itself of many of the obstacles that had been imposed by paramilitary violence and subversive activity, it sought out new ways in which to elevate various revolutionary episodes from the past to the top of the public commemoration agenda. Although it aroused critical commentary from a number of revisionist historians, the bicentennial anniversary of the 1798 Rebellion was marked by hundreds of events across the Republic in 1998 – both official and non-official. Whilst the legacy of 1798 was debated intensely, the events organised by the government were particularly notable in that they displayed little of the embarrassment that had been so clearly evident during the ceremonies held to mark the Rising's 75th anniversary only seven years beforehand.

After the conclusion of the 1798 bicentennial, the Fianna Fáil-led government soon turned its attention to the upcoming 85th anniversary of 1916 Rising. Many in Fianna Fáil felt that the time was finally ripe for the party to reassert its republican credentials by realigning it politics far more closely once again to the story of 1916. Hopes were high that the Rising could be commemorated with more aplomb in the future, as it had been in 1941, 1966 and other occasions of remembrance. In addition to assisting in the continuing rehabilitation of 1916, the Peace Process also brought about conditions which facilitated a greater recognition of a more pluralist perspective on Irish heritage, culture and identity – as exemplified by the revival of interest, throughout the Republic and nationalist parts of the North, in the storylines, memories and traditions of Irishmen who had served in the British Army during the course of the Great War. For years, as David Murphy has noted, the men who had been part of the Southern divisions and regiments had often been 'portrayed in their own country as misguided dupes or, even worse, as traitors'.[187] However, as this subject matter was reexamined more widely, there followed, as Keith Jeffery remarked in 2000, 'a historiographical revolution involving the ways in which we regard the Great War years in Ireland'. In the process, scholars (and others) were successful in amending the so-called 'national amnesia' which had applied to nationalist Ireland's involvement in the conflict.[188]

---

[186]   Foster, *Luck & the Irish*, p. 141.
[187]   D. Murphy, *Irish Regiments in the World Wars* (Osprey Publishing Ltd., Oxford, 2007), p. 54.
[188]   K. Jeffery, *Ireland and the Great War* (Cambridge University Press, Cambridge, 2000), p. 1.

The seeds of this phenomenon can be traced to a range of factors, such as the journalistic writing of Kevin Myers in the 1980s and the restoration and reopening of the Irish National War Memorial Gardens at Islandbridge in September 1988. Until that point, as Dermot Bolger has written, the eight-hectare gardens had been 'allowed to become so overgrown that anyone stumbling upon them by the Liffey might imagine they must have belonged to some lost civilisation'.[189] Then, in April 1995, greater government recognition of the Islandbridge memorial was shown at an official ceremony presided over by Fine Gael's John Bruton, who served as Taoiseach from 1994–1997.[190] The restored space, according to Bruce Arnold, managed to gain recognition as one of Ireland's 'most peaceful open spaces, steeped in an atmosphere of sympathy and eloquence'.[191] While 1998 was a momentous year for Irish history-making, as exemplified by the signing of the Good Friday Agreement and the mature celebration of the 1798 bicentennial, the year also ended on a high note far away in the fields of Flanders.

On 11 November, a Peace Tower symbolising reconciliation was unveiled at the Island of Ireland Peace Park at Messines, Belgium by the President of Ireland, Mary McAleese, in the presence of Queen Elizabeth II of England and King Albert II and Queen Paola of Belgium. The memorial, according to Patsy McGarry, was the 'brainchild' of Paddy Harte (a former Fine Gael TD from Donegal) and Glen Barr (a former political adviser to the Ulster Defence Association, or UDA), while the opening day itself was marked by 'one of the most significant speeches' that McAleese delivered in her capacity as President and Supreme Commander of the Irish Defence Forces.[192] In her remarks at the opening, the President said that the memory of many Irishmen who fought in the Great War had 'fell victim' to the struggle for independence 'at home', and explained that the inauguration of the new memorial was aimed at forging 'a mutually respectful space for differing traditions, differing loyalties' in the island of Ireland, thus helping 'to change the landscape of our memory'. She also expressed her desire for peace and reconciliation, stating: 'None of us has the power to change what is past but we do have the power to use today well to shape a better future.'[193] By the day's end, it was clear that the heart-rending

---

[189] D. Bolger, 'Milestone to Monument: A Personal Journey in Honour of Francis Ledwidge', in D. Bolger (ed.), *The Ledwidge Treasury: Selected Poems* (New Island, Dublin, 2007), p. 101.

[190] J. Hill, *Irish Public Sculpture: A History* (Four Courts Press, Dublin, 1998), pp. 161, 269.

[191] *The Irish Independent*, 22 September 2005.

[192] P. McGarry, *First Citizen: Mary McAleese and the Irish Presidency* (The O'Brien Press, Dublin, 2008), pp. 212–13.

[193] M. McAleese, *Building Bridges: Selected Speeches and Statements* (The History Press Ireland, Dublin, 2011), pp. 255–57. A souvenir booklet was produced for the ceremony at

commemoration would go down in history as a seminal moment in McAleese's presidency, which took as its prevailing theme the notion of 'building bridges'.

At the outset of the 2000s, the emphasis in Ireland remained firmly on bridge-building between divided communities and making progress on outstanding issues in the Peace Process. After the shock of the '9/11' terrorist attacks on the USA in 2001, there followed a backlash against paramilitary violence of any kind throughout much of the Western World. In Ireland, however, historical commemoration continued to offer some degree of hope for the curtailment of long-standing enmities. As nationalists and unionists began finding more and more common ground in history and memory, they continued to make pluralist gestures in the name of peace and reconciliation. One significant development came in November 2001. After Sinn Féin's Francie Molloy was elected Mayor of Dungannon, he hosted a reception for the British Legion on Remembrance Day, in an effort to enable people to 'understand each other'.[194] Similar gestures were evident on a number of occasions in the following year. At 9.00 am on 2 July 2002, for example, the Sinn Féin Lord Mayor of Belfast, Alex Maskey, laid a laurel wreath at the granite plinth of the Belfast City Hall Cenotaph to mark the 86th Anniversary of the Battle of the Somme (two hours before the main council-led commemoration). He followed this gesture on 15 September by laying wreaths at the Ulster Memorial Tower at Thiepval in France and at the end of the peace walk at the pristine Island of Ireland Peace Park in Belgium.

Like many nationalists in the North, Maskey had a relative who had served in the British Army during the Great War, namely his maternal grandfather, Patrick McClory (who fought with a Scottish regiment). There were also wider issues at stake and much discussion had taken place amongst Maskey's team of advisors about how the wreath-laying at the Belfast Cenotaph would go down with grassroots republicans. According to Maskey's biographer, Barry McCaffrey, one of his closest advisors during his tenure as Mayor was the senior Sinn Féin party strategist, Jim Gibney, who had been interned in Long Kesh in

Messines. See Anon., *A Journey of Reconciliation: The Island of Ireland Peace Park, Messines, Flanders, Belgium* (DBA Publications, Dublin, 1998), p. 5, in which McAleese outlined how the tower had been erected by the Journey of Reconciliation Trust, in memory of 'all those from the island of Ireland who fought and died in the First World War'. One of the park's functions, she added, 'will be a place of pilgrimage for everyone from the island of Ireland to visit and reflect on that terrible time and the sacrifice of those who lost their lives'. In a short contribution to the same booklet, Queen Elizabeth II wrote (on p. 7) that it gave her 'great pleasure to be in Belgium', so as to honour 'the great contribution of all those from throughout Ireland who fought valiantly and sacrificed so much in the First World War'.

[194] Molloy, cited in A. Maillot, *New Sinn Féin: Irish Republicanism in the Twenty-First Century* (Routledge, Oxon, 2005), p. 168.

the early 1970s. During the mid-1990s, Gibney had met Protestant community leaders and churchmen 'in an effort to identify common ground with unionists on social and cultural issues'. One of the political strategies that Gibney encouraged during Maskey's Mayoralty, adds McCaffery, was 'a continuation of a confidence-building process directed at unionism', especially one that would 'show the unionist community what they could expect in a united Ireland'.[195] When asked at the time for his reaction to Maskey's involvement in the Belfast commemoration, the Northern Ireland Secretary of State, John Reid, said that it had provided a symbol of 'encouragement and hope for the future'.[196]

The notion that *lieux de mémoire* ('realm of memory') provides 'an ideal way to trace underlying continuities and discontinuities in national identity politics'[197] was again graphically illustrated towards the end of 2002 when BBC 1 Northern Ireland screened a moving documentary, entitled *Somme Journey*. The programme featured David Ervine, the leader of the Progressive Unionist Party and Councillor Tom Hartley of Sinn Féin (another advisor to Maskey) travelling on an emotion-laden journey to heritage spaces that powerfully evoked memories of a shared history – namely the World War I cemeteries in France and Belgium. Experts have for long been agreed on how the 'uniformity of standard headstones ... tend to make ... military cemeteries so impressive',[198] and this was certainly evident in the *Somme Journey* documentary, which screened moving footage of Ervine and Hartley walking amongst the many rows of simple white headstones that dominate the cemeteries maintained by the Commonwealth War Graves Commission. In doing so, they set aside old grudges and shared their feelings about the tragic experience of suffering that had been endured by Protestant and Catholic Irishmen in the 16th and 36th Irish Divisions when fighting against the Germans during the Great War.[199]

Overall, despite taking some flak from hardliners within its own grassroots, Sinn Féin's authentic outreach gestures to the unionist community during the early 2000s added considerable energy to the Peace Process. Additionally, the gestures of friendship and reconciliation by Molloy, Maskey and Hartley did

---

[195] B. McCaffery, *Alex Maskey: Man and Mayor* (The Brehon Press, Belfast, 2003), pp. 156–58, 160, 162, 164.

[196] Reid, cited in Ibid., p. 165.

[197] B. Forest and J. Johnson, 'Unravelling the Threads of History: Soviet-Era Monuments and Post-Soviet National Identity in Moscow', *Annals of the Association of American Geographers*, Vol. 92, No. 3 (2002), p. 525.

[198] Diary of a Tour of Australian Cemeteries and War Memorials in Egypt, Gallipoli, France and Belgium, Entry, 8 April 1930, J. S. Battye Library of West Australian History, State Library of Western Australia, ACC4618A, MN. 1460/12, Sir John Joseph Talbot Hobbs Papers.

[199] *Somme Journey* (BBC, 2002).

much to enhance Sinn Fein's political respectability and maturity in the eyes of some of its political opponents in unionism, as indeed did the strong work ethic displayed by Martin McGuinness during his term of office as Minister for Education at the Stormont Executive from 1999–2002 (which was marked by policy efforts aimed at ending segregation in the educational system). Sinn Féin's honouring of the Great War dead was also warmly welcomed by many nationalist families in the North, whose ancestors had played a part in the events of 1914–1918. Traditionally, many of these families had felt somewhat alienated from World War I remembrance ceremonies in Northern Ireland – due to their strong evocations of the symbolic imagery of the British Army and the Orange Order.

In the Republic, Bertie Ahern's pluralist politics found the full backing of his Cabinet colleagues in the years that followed the brokering of the Good Friday Agreement. Behind the scenes and in public, leading Ministers played an important role in fostering the conditions for better North-South relations. When it came to matters of historical commemoration in the mid-2000s, the need for give and take on both the nationalist/republican and unionist/loyalist sides seemed crystal clear to all. In an article written for *The Irish Times* on Armistice Day 2005, entitled 'Shared History Can Help Build a Shared Future', the Minister for Foreign Affairs, Dermot Ahern (who hailed from the border town of Dundalk), made it clear that 1916 was 'an iconic year in Irish history' for reasons other than the Rising. The Republic of Ireland, he conceded, needed to honour the dead of the Somme, just as it had honoured the 1916 rebels for decades. 'We can no longer have two histories', he argued, 'separate and in conflict'. 'We must acknowledge', he added, 'that the experiences of all the people on this island have shaped our present and, in some way, defined what it is for all of us to be Irish'.[200]

Repossessing 'The Spirit of 1916'

The advent of peace and the rapid warming of both North-South and Anglo-Irish relations from the late 1990s onwards had widespread political and cultural implications in the years that followed. It was in the opening years of the twenty-first century that the rehabilitation of the 1916 Rising's legacy really gathered pace throughout the island of Ireland, as the dividends of the Good Friday Agreement began to be reaped in a more fertile and less contentious field of historical remembrance. Notwithstanding occasional questioning of

---

[200] *The Irish Times*, 11 November 2005.

the Rising's legacy, the early to mid-2000s was a time in which the success of bridge-building between different traditions in the island of Ireland enabled the Fianna Fáil-led government in the Republic to make considerable progress in its desire to rehabilitate the memory of 1916. Within the wider political context, much of the good work that had been done in securing ceasefires in 1994 and 1996, followed by the securing of the peace settlement in 1998, had led to the dawning of a time in which hope replaced pessimism, which in turn facilitated the opening up of new directions in Irish historiographical interpretation – especially in relation to the history and memory of 1916. As the Peace Process took hold, so too did a new post-revisionist historiography that showed more empathy towards those who fought for Irish independence. Harsh political and historical assessments of the legacy of the Rising became less common, while anniversaries of the event were deemed to be ever more suitable for official government recognition than had been the case during the Troubles.

In the changed climate, growing numbers of historians began to reengage with research on the Rising. At a wider level, the increased popularity of the story of 1916 as a topic for investigation can also be attributed to various initiatives in archival policy. The release for the first time of many 1916-related records that had remained under lock and key in Britain and Ireland allowed researchers to conduct fresh examinations of the Rising – both from a top-down and a bottom-up perspective. Broader political developments were important in this regard, especially the Open Government Initiative that had been implemented in Britain by John Major's Conservative government during the 1990s. As Angus Mitchell has noted, the quick release of a 'vast amount of hitherto inaccessible material' in the form of hundreds of files on 1916 rebels such as Roger Casement, 'reflected a desire in some circles to break the culture of secrecy and implement a greater degree of transparency as part of democratic accountability'. Casement's 'Black Diaries' were finally released unconditionally at the Public Record Office in Kew in March 1994. Extra Home Office and Police Commission files were released in October 1995, followed by the Security Service records in 1999 and documentation relating to the Metropolitan Police Special Branch in 2001.[201]

On a number of occasions, the renewed interest in the Rising was reflected by the unveiling in public space of new monuments and by the making and screening of historical dramas. On 14 May 2000, for example, a bust of Thomas Kent (Plate 5.7), resting on a limestone plinth, was unveiled by Kathleen Kent in the grounds of Kent Station in Cork city.

---

[201] A. Mitchell, *Casement* (Haus Publishing, London, 2003), p. 7.

In Onóir
to Thomas Ceannt
Ceannport Céad Briogáid Oirthear Chorcaí
Óglaigh na h-Éireann
In Onóir
Thomas Kent
Commandant 1st East Cork Brigade
Óglaigh na h-Éireann
Executed by British armed forces
at Victoria Barracks now Cork Prison
9th May 1916
Erected by Thomas Kent Memorial Committee
Larne Óc Éireann workers
Thomas Kent Station Cork
Unveiled by Kathleen Kent 14th May 2000

"From the graves of patriot men and
women spring living nations."

Plate 5.7   Bust of Thomas Kent, unveiled by Kathleen Kent in the grounds
               of Kent Station in Cork city on 14 May 2000
Source: Author.

Then, in the early months of 2001, the lead-up to the Rising's 85th anniversary
saw the actions of the 1916 rebels dramatically resurrected for TV audiences in
both the North and the South by the screening of *Rebel Heart* – an epic BBC/
RTÉ drama series written by Ronan Bennet and directed by John Strickland.
Seven years in the making, the series was made independently by Picture Palace
Films, with significant funding from the Irish state broadcaster and An Bord
Scannán. Focusing mainly on the human story of the fictional character of
18-year-old Ernie Coyne (played by English actor James D'Arcy), *Rebel Heart*
was shot on 150 different sets in Dublin city (including the GPO). It included
180 speaking parts and the participation of over 3,000 extras (many of them
from the Reserve Defence Forces – who had previously participated in the
filming of Mel Gibson's *Braveheart*).[202]

    In the Republic, *Rebel Heart* was first screened on RTÉ television in January
2001, while BBC viewers were able to watch it the following month. Featuring a

---

[202] *The Irish Independent*, 9 December 2000.

soundtrack that included music from the Dundalk quartet, The Corrs, the series depicted scenes from the Rising and its aftermath – including the tumultuous events that occurred in the lead-up to the Treaty of 1922. The brainchild behind the four-hour series was BBC Northern Ireland's Head of Drama, Robert Cooper. In an interview published in *The Irish Independent* prior to its screening, he noted that he had wanted to use 'a northern voice' to bring a fresh perspective on the event. 'I wanted to capture', he said, 'the popular imagination with a huge epic set around the birth of the Irish nation, with the spotlight focused on personal experiences in and around the battlefield'. In the face of criticisms about the objectivity of the series from David Trimble, writer Bennett claimed that he was 'not banging the drum for either side'. He also added that his intention had not been to romanticise the rebels or to pass judgement on the later Troubles in the North: 'This is not a proselytising piece, I'm not trying to educate or persuade anyone. If it clarifies issues or provokes discussion of Irish history, that's in the hands of individuals, all I cared about was writing a moving story.'[203]

A range of Fianna Fáil politicians also jumped upon the commemorative opportunities presented by the atmosphere of peace, reconciliation and cross-border cooperation that prevailed in the year of the 85th anniversary. At a pragmatic level, they seemed intent on doing their utmost to rehabilitate the reputation of the 1916 metanarrative. Chief among them was the party leader, Bertie Ahern. At various times during 2001, pronouncements by Ahern and other Fianna Fáil politicians pointed towards a major surge in the attempt by the 'Soldiers of Destiny' to reclaim pre-eminence over 1916, in the face of proprietorial claims to its legacy by Sinn Féin and the Provisional IRA. The broadcast on 9 April of an RTÉ True Lives documentary, entitled *Pearse: Fanatic Heart* (made by Stephen Carson of Mint Productions), featured a revealing interview with Ahern, who made it abundantly clear to viewers that Patrick Pearse was his idol. The Taoiseach declared that he wanted the Irish people to see Pearse in perspective and said that the Proclamation of 1916 'cannot be, I think, altered ... he [Pearse] gave his life. We are his heirs. He has given us the right to fill our destiny. Without violence'. Footage of the Taoiseach showed him seated in his office in front of a framed picture of Pearse that was mounted on a wall beside a Tricolour. Rather revealingly, Ahern also spoke of his desire that by 2016, the Rising's centenary, things would go 'full circle'. Furthermore, he noted: 'I think people will see that Pearse ... played an extraordinary part in Irish history.'[204]

---

[203] Ibid.

[204] *Pearse: Fanatic Heart* (RTÉ True Lives, 2001); *The Irish Times*, 7 April 2001; *The Sunday Independent*, 15 April 2001.

Later on that year, after much planning and following lengthy negotiations with the families concerned, Ahern pressed ahead with his strategy of rehabilitating another aspect of Ireland's revolutionary heritage, by sponsoring the reburial of 10 Old IRA volunteers who had been executed by the British authorities during the War of Independence. The remains of the so-called 'Forgotten 10', which had lain in Mountjoy Prison for 80 years, were moved on 14 October to new graves near the entrance of Glasnevin Cemetery, while the remains of one of them (namely Patrick Maher), were taken to Ballylanders in Limerick. With uncertainty over progress with the Peace Process, some commentators expressed the view that the reburial was untimely. Others alleged that event was about political point-scoring, coinciding as it did with the end of a Fianna Fáil Ard Fheis. In *The Daily Telegraph*, for example, one letter-writer charged that that the event 'was merely a cynical manoeuvre by the ideologically bankrupt Fianna Fáil party, reaching into the ample republican necropolis for a diversion from the stench of its own political corruption'.[205] But as Catherine O'Donnell has argued, the commemoration also represented more than an attempt 'to dethrone Sinn Féin as the apparent heirs of the 1916 and War of Independence legacy'. An effort was also made to demonstrate that Fianna Fáil 'represented continuity with the republican roots of the state' and had also 'successfully settled the issue' by virtue of its role in negotiating the Good Friday Agreement.[206] Altogether, the symbolic movement of the Old IRA remains from Mountjoy Prison to the sacred space of Glasnevin, according to Nuala Johnston, represented a 'recognition of their sacrifice through the performance of a state funeral' that aroused 'the collective and personal memories of the living'.[207]

By the end of 2001, it was clear that the new political climate of peace and reconciliation had also energised a post-revisionist historiography that sought to recast the revolutionary years in new light. To coincide with the year of the 85th anniversary, the well-known nationalist historian and journalist, Tim Pat Coogan, published a new history of the event, entitled *1916: The Easter Rising*.[208] Also in 2001, Gerry Adams republished his decade-old history of Belfast republicanism from 1900–1916, entitled *Who Fears to Speak ...? The Story of Belfast and the 1916 Rising*. This work, wrote Adams, was originally published as 'a little commemorative gesture to mark the 75th anniversary of the

[205] *The Daily Telegraph*, 17 October 2001.

[206] C. O'Donnell, *Fianna Fáil, Irish Republicanism and the Northern Ireland Troubles 1968–2005* (Irish Academic Press, Dublin, 2007), p. 172.

[207] N. Johnson, *Ireland, The Great War and the Geography of Remembrance* (Cambridge University Press, Cambridge, 2003), p. 1.

[208] T. P. Coogan, *1916: The Easter Rising* (Cassell & Co., London, 2001).

Easter Rising'. Whilst conceding that 'there is a lot yet to be done', he prefaced the new edition of his book by expressing hope that 'Belfast is now becoming a shared city with a future for all its people'.[209] From an archival perspective, a decision by the British government to open up the official records held in Kew of the proceedings relating to the court-martialling of the executed leaders of the Rising, was the subject of much curiosity around the time of the 85th anniversary. Writing in *The Irish Times* in May 2001, Dara Redmond, the grandson of executed 1916 leader, Thomas MacDonagh, noted that the development came 'as quite a surprise' – especially because the prevailing consensus was that they would not be released until 2016.[210] Irish historians were quick to delve into the new materials, most especially Brian Barton, who successfully navigated the issues at stake in a book entitled *From Behind a Closed Door: Secret Court Martial Records of the 1916 Easter Rising*, which was published in 2002.[211] One of the most notable post-revisionist history books to be published in the early to mid-2000s was Diarmuid Ferriter's *The Transformation of Ireland, 1900–2000*, which sough to portray a deeper understanding of the Rising's policy objectives, especially in regard to how it was viewed by certain contemporaries.[212]

Another sign of the rehabilitation of the 1916 legacy in the early to mid-2000s was evident from the way in which people began to reconnect with the Proclamation. In his acclaimed family history, *Blood-Dark Track*, which was published in 2000, Irish-American novelist Joseph O'Neill recalled being overcome by 'intense sensations of patriotic exhilaration' when he first encountered the Proclamation during his university days in England a decade and a half previously. 'When I first came across the text', he wrote, 'the surge of emotion I felt was so strong that my scalp and cheeks and ears prickled, and it's even possible that my eyes clouded over'. 'No matter how familiar I grow with it', he added, 'I am always moved ... and grateful for it'.[213] In the years following the novel's publication, a sharp rise occurred in the prices that rare copies of the Proclamation began to fetch at auction houses. In mid January 2001, for

---

[209] G. Adams, *Who Fears to Speak ...? The Story of Belfast and the 1916 Rising*, Revised Edition (Beyond the Pale Publications, Belfast, 2001), pp. vii–viii.

[210] *The Irish Times*, 5 May 2001.

[211] B. Barton, *From Behind a Closed Door: Secret Court Martial Records of the 1916 Easter Rising* (The Blackstaff Press, Belfast, 2002). Also see M. Foy and B. Barton, *The Easter Rising* (Sutton Publishing Ltd, Stroud, 1999).

[212] See, for example, Ferriter, *The Transformation of Modern Ireland*, p. 142, in which he states: 'It is too simplistic to dismiss it [the Rising] as a hopeless blood sacrifice ... there was much more to Connolly, and indeed Pearse, than mystic notions about spilling blood for the benefit of future generations.'

[213] J. O'Neill, *Blood-Dark Track: A Family History* (Granta Books, London, 2000), p. 150.

example, a copy with 'with a bit of mud on it and a couple of pinholes' was sold by Whyte's Auctioneers in Dublin for £52,000.[214] Commenting on this transaction, Linda King remarked that the sum of money fetched indicated the Proclamation's 'rarity' and 'inherent symbolism'. As 'the most politically significant...document' of the twentieth century, she reflected that it represented 'the most widely recognised example of Irish printed ephemera'.[215] Almost three years later, at the end of 2004, another one of the 20 or so surviving original copies of the Proclamation was sold at an auction in Dublin for a massive €390,000 – more than eight times the price achieved for another original sold by Sotheby's in 1997.[216] Such rampant price inflation confirmed the Proclamation's early twenty-first century status as an iconic heritage object endowed with significant financial value and prime cultural significance.

The renewed interest in the story of 1916 during the early to mid 2000s was also reflected by the popularity during the summer months of privately-run walking tours of many of the sites associated with the Rising in Dublin city. Ongoing sales of publications like Mick O'Farrell's *A Walk through Rebel Dublin 1916* and Joseph Connell's *Where's Where in Dublin: A Directory of Historic Locations 1913–1923* were emblematic of the demand from visitors for published information regarding the location and stories behind some of the principal heritage sites associated with the Rising.[217] The final phase of the redevelopment of the GAA's Croke Park headquarters on Dublin's northside during 2003–2004 again highlighted the emotional attachment that many people, especially the citizens of the capital, retained to the tangible heritage of the 1916 Rising. In his memoir, *Rule 42 and All That*, the GAA's President from 2003–2006, Kerryman Seán Kelly, recalled that the final building works by the Sisk construction company posed a predicament for the sporting organisation. At issue was the question of what to do with the fabled Hill 16 – a terrace 'called after the 1916 Rising because the rubble from the destruction of Sackville Place was used in its construction'. The space, he added, was 'hugely sentimental for GAA supporters, none so more than the ... Boys in Blue, the Dubs, who made

---

[214] *The Irish Times*, 15 January 2001.

[215] L. King, 'Text as Image: The Proclamation of the Irish Republic', in E. Sisson (ed.), *History/Technology/Criticism: A Collection of Essays* (Dun Laoghaire Institute of Art Design and Technology, Dublin, 2001), p. 5.

[216] *The Irish Times*, 10 December 2004.

[217] See M. O'Farrell, *A Walk through Rebel Dublin 1916* (Mercier Press, Cork, 1999) and J. E. A. Connell Jnr., *Where's Where in Dublin: A Directory of Historic Locations, 1913–1923. The Great Lockout, The Easter Rising, The War of Independence, The Irish Civil War* (Dublin City Council, Dublin, 2006).

it their home whenever Dublin were playing'.[218] In the end, the GAA replaced it with a more modern terrace that became known as either the Hill 16 End or the Nally/Hill 16 Terrace, while rock from the original terrace was collected as souvenirs by some people.

Although the early to mid-2000s saw a greater acceptance of different varieties of history across the island of Ireland, this did not deter some writers from airing long-held reservations about the actions taken by the 1916 rebels. At times when the Peace Process ran into difficulties over the decommissioning of paramilitary weapons, it was not unusual for certain journalists to repeat well-rehearsed criticisms of the Rising's legacy. Writing at the end of 2003, for example, Emer O'Kelly of *The Sunday Independent* delivered an iconoclastic vilification of Pearse, denouncing him as a 'blood-lusting fanatic'. 'Any modern psychological examination of a man who called on people to die uselessly and bloodily', she wrote disparagingly, 'would find him dangerously obsessed, maybe even psychopathic'.[219] Outside of media circles, the Rising's place in history proved to be the subject of a well-publicised reappraisal by the former Taoiseach, John Bruton. This occurred in September 2004, when he delivered the opening speech at a conference in Dublin's Mansion House, held by the Reform Movement (an organisation in favour of the Republic of Ireland rejoining the Commonwealth). Reflecting upon the conference theme of 'Reforming Ireland: Towards Pluralism', Bruton (who was now serving as the European Union's Ambassador to the USA), outlined to delegates how his political roots were in the Irish Parliamentary Party. Ireland, he claimed, 'could have achieved independence without 1916'. Bruton also stated that those who had used violence in more recent times had felt that 'they were drawing on respectable traditions'. He then proceeded to use counterfactual thought to speculate that even though Partition may have been inescapable under Home Rule, the treatment of northern Catholics might have been better due to the influential position of southern MPs in Westminster.[220]

In the weeks following the Reform Movement conference, Bruton's speech was the subject of much discussion. Those with ancestral connections to the Home Rule movement were especially eager to air their views. Writing for *The Sunday Independent* on 3 October, John Dillon described the ex-Taoiseach's observations as 'most interesting'. Having outlined his own admiration of the Home Rule movement, he then criticised the actions of the rebels of Easter Week.

---

218 S. Kelly, *Rule 42 and All That* (Gill and Macmillan, Dublin, 2008), p. 140. One of Kelly's main legacies was the GAA's enlightened decision to open up Croke Park to 'foreign games' such as rugby and soccer.

219 *The Sunday Independent*, 28 December 2003.

220 *The Irish Times*, 20 September 2004.

In a reflection of the type of political discourse that was then in vogue (most especially as a result of the USA's so-called 'War on Terror'), Dillon denounced the Rising as 'a very shrewdly-planned terrorist act'. Pearse, he argued, had 'shared with modern terrorists the desire to bring down the established order by delivering a violent shock to it'.[221] An even more vitriolic attack on the legacy of 1916 appeared in *The Sunday Independent* in May 2005. This came from a letter-writer from Celbridge in County Kildare, who went so far as to describe the Rising as 'a brutal proto-fascist attack on a developed constitutional democracy, and on a nationalist political consensus ... [when] most nationalists ... were also "constitutional democrats", and wished to remain within "the Empire", on Home Rule terms'.[222]

By no means, however, did such harsh criticisms dissuade historians from pursuing fresh investigations into the story of 1916. As was typical with previous commemorations of the Rising, there was an upsurge in scholarly activity amongst historians as the 90th anniversary approached, and they did not disappoint when it came to satisfying the public's enhanced appetite for new reading material. The most noteworthy of the books published in the lead-up to the anniversary was a 442-page tome, *Easter 1916: The Irish Rebellion*, written by the English military historian, Charles Townshend. This was first published in hardback in 2005 (and was released as a Penguin paperback the year afterwards). 'There have been several excellent books', wrote Townshend in his preface, 'but it remains true that few academic historians have felt it worthwhile to fill the gaps'. 'To do so', he proclaimed, 'is my main purpose' – a task which was facilitated by the release in 2003 of the Bureau of Military History Witness Statements of 1947–1957, in both the Military Archives at Cathal Brugha Barracks and the National Archives of Ireland.[223] Another work to benefit from the release of the Witness Statements was Annie Ryan's *Witnesses: Inside the Easter Rising*, which was also published in 2005.[224]

Demand for new biographies of the 1916 leaders also escalated as 2006 drew nearer. In *James Connolly: 'A Full Life'*, published in 2005, former trade unionist Donal Nevin used Connolly's contributions to 27 journals and some 200 personal letters to trace the evolution of his political thinking as a revolutionary socialist. This work represented the ninth substantial biography of Connolly since 1924. Sir Roger Casement's place in history was also the subject of a short biography published in 2003 by Angus Dillon, while the rebel's posthumous reputation was debated in an edited collection of essays that was published in 2005, based on a

---

[221] *The Sunday Independent*, 3 October 2004.

[222] Ibid., 8 May 2005.

[223] Townshend, *Easter 1916*, p. xvii.

[224] A. Ryan, *Witnesses: Inside the Easter Rising* (Liberties Press, Dublin, 2005).

conference that was held in the Royal Irish Academy.[225] Other aspects of the lives of the 1916 leaders also came under scrutiny. In *Pearse's Patriots*, for example, Elaine Sisson strove to move beyond the standard tendency to portray Patrick Pearse in a way that either debunked 'the myths of martyrdom' or reinforced the leader 'as an icon of the past'. Instead, she sought to offer a cultural history of his school, St. Enda's, between the years 1908–1916 and to explore 'the visual and literary myth-making discourses of national identity and masculinity which the school so successfully promoted'.[226]

The publication of a new biography of the life of Eamonn Ceannt in 2005 provided Éamon Ó Cuív (the Minister for Community, Rural and Gaeltacht Affairs) with an opportunity to reassert Fianna Fáil's renewed enthusiasm for reclaiming the 'spirit' of the 1916 Rising. In his 'Foreword' to William Henry's *Supreme Sacrifice: The Story of Eamonn Ceannt 1881–1916*, Ó Cuív argued that labelling the 1916 leaders as 'bloodthirsty militarists' was a mistruth that diminished their 'incredible legacy'. The 1916 rebels, he added, were idealists striving to do the best for their country, and their actions needed to be placed against the background of their time – especially their belief that Irish freedom could not be obtained by 'no other way'. Although acknowledging that Ireland could have obtained granted Home Rule had the Rising 'not taken place', Ó Cuív speculated that its 'whole nature ... would have led to a very different Ireland than the one we enjoy today, the one of the Celtic Tiger and of world economic success'. Home Rule, he cautioned, 'was about a limited form of self rule' in the British Empire.[227] On further occasions throughout 2005, others within Fianna Fáil expressed equally congenial sentiments about the story of 1916. Writing for *The Sunday Independent* on 1 May, just three days before an official commemoration at Arbour Hill, Willie O'Dea described Pearse and Connolly's republicanism as 'an embracing philosophy ... about the inherent rights and dignity of man'. He also used the opportunity to take a swipe at Gerry Adams and others in Sinn Féin, whom he accused of peddling 'the myth that "provisionalism" and "republicanism" are synonymous'. 'One obvious way they try to further this myth', he contended, 'is by claiming that they, and they alone, are the sole true inheritors of the legacy of 1916'.[228]

---

[225] See A. Dillon, *Casement* (Haus Publishing, London, 2003); M. Daly (ed.), *Roger Casement in Irish and World History* (Royal Irish Academy, Dublin, 2005) and D. Nevin, *James Connolly: A 'Full Life'* (Gill and Macmillan, Dublin, 2005).

[226] E. Sisson, *Pearse's Patriots: St. Enda's and the Cult of Boyhood* (Cork University Press, Cork, 2004), p. 2.

[227] E. Ó Cuív, 'Foreword', in W. Henry, *Supreme Sacrifice: The Story of Éamonn Ceannt 1881–1916* (Mercier Press, Cork, 2005), pp. ix–xi.

[228] *The Sunday Independent*, 1 May 2005.

However, it was at Fianna Fáil's Ard Fheis, held in Killarney on 21 October 2005, that Bertie Ahern grasped the opportunity to play his party's trump card in its pragmatic quest to reclaim the 'spirit of 1916' from Sinn Féin. By this time, Ahern had been in power as the democratically-elected Taoiseach for almost a decade, and had built up a formidable reputation as leader of a country which had more or less reached full employment – owing to the runaway growth rates of the Celtic Tiger economy. His political negotiation skills, as already seen, had proved vital in securing the Good Friday Agreement, while his skills as a 'fixer' made it possible for him to act as a leading architect of various social partnership agreements during the boom years. By the middle of the 2000s, the Republic of Ireland was nearing the unprecedented height of a frenzied property boom – or, as economist David McWilliams later described it, a 'construction leviathan' that resembled a 'pyramid scheme' based on 'the hot air of cheap credit'.[229] During this time of peace and prosperity, the anticipation was naturally high in Killarney when Ahern stepped up to the podium to deliver the leader's address to the Ard Fheis delegates. To the surprise of many, he dramatically informed his party's grassroots and a watching television audience, that he was going to reintroduce (for the first time since 11 April 1971) the full pomp and ceremony of a state-run military parade past the GPO on Easter Sunday 2006, to mark the Rising's 90th anniversary. Such an exercise, he believed, would act as the ultimate mechanism to rehabilitate the legacy of 1916 from the severe bashing that it had taken during the height of the Troubles.

When he published his autobiography a few years afterwards, Ahern explained the rationale behind both his unexpected Ard Fheis announcement and his lifelong passion for the heritage of 1916:

> There was one event in particular that I was always determined to revive [as Taoiseach]. Looking back on my childhood, one of the great occasions in my life was the fiftieth anniversary of 1916 ... Then the Troubles came, with the violence, the bombings and the killings. The Provos tried to claim they were the heirs of the Rising. The revisionists trashed the whole 1916 legacy. People got so uncomfortable with militarism that the annual military parade was scrapped. The Good Friday Agreement had changed this context. I was determined to take 1916 back from both the IRA and the revisionists for all the people of Ireland ... I had been going to commemorations for the Rising all my life...It had hurt me when Sinn Féin hijacked these events, so I was anxious that the anniversary should be retrieved from all that ... This would be a commemoration by the state and the people for those who gave their lives for Irish freedom.[230]

---

[229] D. McWilliams, *The Generation Game* (Gill and Macmillan, Dublin, 2007), pp. 52, 87.

[230] B. Ahern with R. Aldous, *Bertie Ahern: The Autobiography* (Hutchinson, London, 2009), pp. 293–94.

According to Ahern, it was the Attorney General, Rory Brady, who personally gave him 'great encouragement' to make the decision to announce the reinstatement of the Army parade.[231] Pat Leahy, however, has produced another account of where the stimulus for the parade originally came from. In his book *Showtime: The Inside Story of Fianna Fáil in Power*, he argues that the idea to reinstate the parade 'was actually not a Fianna Fáil idea at all', but rather the 'brainwave' of Michael McDowell of the Progressive Democrats. In his capacity as Minister for Justice, McDowell had 'detested Sinn Féin' and regularly 'believed that the Irish state and its mainstream parties had to reclaim ownership of the event'. On one occasion, following a Cabinet dinner at Farmleigh House, Leahy notes that McDowell 'fell into a discussion on the rise of Sinn Féin' with Brady, Mary Harney 'and some others'. McDowell, he contends, was the first to float the idea of the 'state military commemoration', with Brady agreeing 'enthusiastically'. The Taoiseach, in this version of events, was then 'quickly persuaded of the merits of the idea'.[232]

Unsurprisingly, the Ard Fheis announcement was met with overwhelming approval by the Fianna Fáil party faithful, especially those who had no personal recollections of witnessing the Army parade through O'Connell Street. During the long hiatus that ensued after Easter Sunday 1971, the party's most ardent supporters had watched with dismay each year, as other groups put their own stamp on the Eastertime commemorations. Between the early 1970s and the mid-2000s, it was the supporters of Sinn Féin and the Provisional IRA who consistently turned out in force to participate in Easter Sunday spectacles. These unofficial events, featuring flag-bearers clad in standard black paramilitary clothing, usually involved some degree of parading to republican cemeteries. More often than not, these occasions also attracted the attention of An Garda Síochána, whose officers kept a close watch on the proceedings from the sidelines (Plate 5.8). For some observers, Ahern's declaration that he was reviving the Army parade seemed to smack of political opportunism. Less than a month before his speech at Killarney, Sinn Féin had taken to the streets of Dublin once again, staging a special rally on 24 September under the banner of 'Make Partition History'.

---

[231] Ibid., p. 293. Also see Ahern's obituary tribute to Brady in *The Sunday Independent*, 25 July 2010. In this, Ahern wrote the following tribute: 'Rory was a true republican in the best sense. He cherished the memory of those who strove to establish this nation. It was Rory Brady who first raised the idea that we should mark the 90th anniversary of the 1916 Rising with full state honours.'

[232] P. Leahy, *Showtime: The Inside Story of Fianna Fáil in Power* (Penguin Ireland, Dublin, 2009), p. 289.

Plate 5.8   A scene at the Sinn Féin 1916 commemoration parade at the
              National Monument, Grand Parade, Cork on 31 March 2002
Source: Author.

To the bemusement of some of Fianna Fáil's most senior figures, however, what
had been billed as a cultural celebration involving street theatre turned out
rather differently. As the Minister for Defence, Willie O'Dea, later wrote in
The Sunday Independent: 'In place of the street theatre and costumes were IRA
banners, paramilitary-style marching bands and children on floats brandishing
mock rifles.' Much to O'Dea's annoyance, not everybody seemed prepared
to acknowledge the existence of only once legitimate Defence Forces in the
Republic. 'This state', he noted, 'does not honour its patriots and its history with
balaclava-clad gangs and cardboard Kalashnikovs.'[233]

To a certain degree, Fianna Fáil's 2005 Ard Fheis proved to be a case
of lucky timing. Following important advances in the Peace Process, the
government was in a position to rebut critics of its plans to commemorate the
90th anniversary, by pointing to the Provisional IRA's pledge earlier on in 2005
to decommission its weapons. After he broke the news of the military parade's
reinstatement, Ahern was greeted with rapturous applause and cheers from the

---

233   *The Sunday Independent*, 26 February 2006.

assembled delegates. In echoes of the discourse used by the arts community for 'The Flaming Door' festival in 1991, he said that the people needed 'to reclaim the spirit of 1916'. This, he said, 'is not the property of those who have abused and debased the title of republicanism'. In rejecting the Provisional IRA's claims to the Rising's legacy, Ahern made it clear that the Defence Forces (Óglaigh na hÉireann) were 'the only legitimate army of the Irish people' and 'the true successors' of the Irish Volunteers. Ahern also said that the Rising formed part of 'our State's inheritance'. 'We must protect it', he said, 'from those who will abuse it and from revisionists who would seek to denigrate it'. He also sought to justify Fianna Fáil's claim to the heritage of 1916, by pointing out that it had 'rightly commemorated the heroic struggle of the men and women of 1916', since its foundation in 1926. 'But it is now time', added Ahern, 'that we suitably recognise the self-sacrifice of our forebears'. 'Many of those who fought in 1916', he stated, 'became the founding members of our party. We all know the names of de Valera and Markievicz. We are also the party of Pádraig Pearse's mother and sister'.[234] In addition to reviving the military parade, Ahern also announced that he was going to establish a new all-party 1916 Centenary Commemoration Committee, to whom he was entrusting the task of planning for the Rising's centenary.

As it transpired, the Taoiseach's decision to revive the 'spirit of 1916' generated mixed reactions, especially from an array of bewildered Opposition politicians, who voiced serious reservations over the fact that the announcement had been made from an Ard Fheis platform. Consequently, as John Cooney has written, 'the parade's preparations' became 'mired in the inevitable political squabbling engendered by inter-party rivalry'.[235] In a short letter to *The Irish Times* four days after the Taoiseach's announcement, Councillor Marie Baker of Fine Gael stated that it was 'extraordinary that Bertie Ahern should decide to re-establish a military parade', especially when communities in the North were 'trying to move away from partisan parades'.[236] Then, in a Seanad debate on 27 October, Fine Gael's Brian Hayes said that it was 'inappropriate' for Ahern to have had made 'an announcement on a national day of commemoration at what was a political rally'. The Army, he alleged, had been used 'as a pawn' in a 'partisan party political way' by Fianna Fáil, for the purpose of demonstrating 'its neo-national credentials'. The Trinity College Dublin senator, David Norris, concurred with Hayes, stating that the parade's announcement at 'a party rally' was lamentable, 'because it makes it the possession of one section rather than the

---

[234] *The Irish Times*, 22 October 2005.

[235] J. Cooney, *'Battleship Bertie': Politics in Ahern's Ireland* (Blantyremoy Publications, Dublin, 2008), p. 147.

[236] *The Irish Times*, 25 October 2005.

entire community'. Fine Gael's Paul Coughlan spoke in favour of the decision to revive the parade, but stated: 'our great republican democracy would be better served if there was consultation'. Another Trinity College senator, Shane Ross, also backed the decision to commemorate the 90th anniversary, but added: 'there is a problem with the Taoiseach announcing it at a tribal rally ... it has ... become a political football. Political parties should not compete for possession of this part of our history'. Labour's Michael McCarthy also expressed dismay at the lack of consultation, wittily bemoaning 'that the announcement was made in the wigwam in Killarney when the Taoiseach was putting on his warpaint'.[237]

On 2 November, the Taoiseach again faced harsh criticism during a Dáil debate on a Sinn Féin motion about Irish unity (which was ultimately defeated the day afterwards). During the proceedings, the Green Party leader, Trevor Sargent, accused Fianna Fáil of making 'political footballs out of serious issues such as the commemoration of 1916'. 'Fianna Fáil', he complained, 'has decided the [1916 commemoration] ceremonies are an Ard Fheis device to rally its troops and try to out-do the Sinn Féin party'. 'People', he felt, 'will see it for what it is'. A similar view was aired by Fine Gael's Bernard Allen, who said that the Taoiseach's Ard Fheis announcement 'should have come about following extensive consultation with all political parties and not as part of a political party jamboree'. 'It seems to me', he added, 'that what we are witnessing is a type of power play between Fianna Fáil and Sinn Féin', whereby each party was 'more concerned with consolidating its respective position with certain elements of the electorate'.[238] Reflecting upon the Ard Fheis announcement in *The Irish Times*, the Labour leader, Pat Rabbitte, later remarked that the manner in which Ahern announced the parade had been 'inappropriate'. Furthermore, he went on to point out his belief that 'there is much more to Ireland than Sinn Féin's tribalism or Fianna Fáil's narrow conservatism'.[239]

Many journalists and media figures provided extended coverage of the political wrangling that followed the Ard Fheis decision. In a satirical piece that appeared in *The Irish Independent* the day after the Ard Fheis, Miriam Lord quipped that 'Gerry Adams and pals must be sick'. 'No sooner do they decommission IRA weapons', she wrote, 'than Bertie Ahern announces the resumption of military parades in the Free State. He's a devious one alright'.[240] A more guarded view was aired in an Editorial in *The Sunday Times*, entitled '1916 March a Gimmick'. Urging caution, it remarked that the pressure for the creation of a united Ireland emanating from Ahern's revived brand of

---

[237] *Seanad Debates*, Vol. 181, 27 October 2005.
[238] *Dáil Debates*, Vol. 609, 2 November 2005.
[239] *The Irish Times*, 30 December 2005.
[240] *The Irish Independent*, 22 October 2005.

republicanism would 'not reassure the unionist community that the concept of consent', which had been the guiding principle of the Good Friday Agreement, was 'likely to be honoured by Irish nationalists'. However, it did concede that the decision to revive the parade was 'admirable' at 'one level', for it reflected Ahern's efforts 'to reclaim ownership of Irish nationalism from the blood-spattered hands of Sinn Féin'.[241] Some commentators felt that Fianna Fáil was clearly doing the right thing at the right time, including RTÉ broadcaster Ryan Tubridy, who wrote in *The Sunday World* about how the decommissioning of Provisional IRA weapons had changed matters: 'the rules have changed and everything's up for grabs ... Bertie wasted very little time and, as I see it, he's making the right decision'. 'The time has come', he declared, 'to take back an element of ... history that was borrowed/hijacked/stolen and if we are going to remember those who fell [in 1916] ... we should do it properly'.[242]

From a political perspective, Ahern's perceived fear of the electoral threat from Sinn Féin in the mid-2000s was by no means exaggerated. The local elections on 11 June 2004 had proved calamitous for Fianna Fáil. By winning only 302 of the 883 seats on the country's local authorities, it suffered its worst election results since 1927. While Fianna Fáil's 'dramatic loss' entailed the loss of nine per cent of its seats (representing a total of 80), Sinn Féin on the other hand more than doubled its number of seats from 21 to 54.[243] These results, according to John Downing, hit the Taoiseach 'like a hammer blow', while further bad news emanated from the European Parliament elections, when Fianna Fáil's number of MEPs dropped from six out of 15 in the previous assembly to four out of a reduced allocation of 13 in the new one.[244] Altogether, as John Cooney later remarked, Ahern's 'humiliating rebuffs' at the local elections put his 'hegemony over the Irish political landscape ... under serious strain, if not threat, not least within his own jittery Fianna Fáil party'.[245]

As matters subsequently transpired, Ahern's carefully-orchestrated response to the mid-term reverse was to attempt to reinvent Fianna Fáil, by shifting the emphasis in budgetary strategy towards both the lower paid and the elderly in society. In a Cabinet reshuffle, Brian Cowen was promoted to Minister for Finance in place of the more right-leaning Charlie McCreevy, who was

---

[241] *The Sunday Times*, 23 October 2005.

[242] *The Sunday World*, 30 October 2005.

[243] L. Weeks and A. Quinlivan, *All Politics is Local: A Guide to Local Elections in Ireland* (The Collins Press, Cork, 2009), pp. 71–72.

[244] J. Downing, *'Most Skilful, Most Devious, Most Cunning': A Political Biography of Bertie Ahern* (Blackwater Press, Dublin, 2004), pp. 225–26.

[245] J. Cooney, *'Battleship Bertie'*, p. 8.

dispatched to Brussels as the country's EU Commissioner.[246] Additionally, the Taoiseach invited the well-known campaigner for the underprivileged, Father Seán Healy, to address the Fianna Fáil party at Inchydoney in West Cork on 6 September 2004. In the same month, in a high-profile interview with *The Irish Times*, Ahern publicly declared that he saw himself as a socialist and remarked that he was a fan of Robert Putnam's acclaimed book, *Bowling Alone*, which had drawn attention to the theme of social depletion in modern societies. Thereafter, a major theme in Ahern's second term of office as Taoiseach, according to Noel Whelan, was 'the disjointed nature of ... [Irish] society and the need to encourage volunteerism'.[247]

Having recovered some ground in the opinion polls in early 2005, the narrow loss of two by-elections in Meath and Kildare North on 11 March served as 'another warning ... that Fianna Fáil still had a lot of ground to make up'.[248] This obviously forced Ahern into a rethink yet again – one that saw him look towards the party's republican roots in the lead-up to the 2005 Ard Fheis. In a way, it was as if Ahern wanted to summon the ghosts of 1916 for a new age, out of a belief that republicanism was the best way to boost morale and harmony in the party. There was by no means anything unusual about these actions. As one of Fianna Fáil's most central and influential figures noted a few years afterwards in a collection of essays on Irish political theory: 'republicanism is the most important element that unites the party, regardless of an otherwise wide left-right spectrum of views on economic, social and socio-moral questions, not to mention foreign policy'.[249] These comments may go some way to explaining why Ahern was well shielded by the party faithful from the heavy criticism that he faced after his speech in Killarney. Many high-ranking members were more than pleased at their leader's reassertion of the party's republican credentials and publicly backed his decision to reclaim the 'spirit of 1916'. In a letter to *The Irish Times*, for example, TD Barry Andrews declared that his party's 'intentions in regard to the revival of an Easter 1916 commemoration are honourable' and self-confidently stated that Fianna Fáil had 'a proud republican heritage – something to be cherished rather than tip-toed around'.[250]

---

[246] S. Collins, 'The Election Background', in S. Collins (ed.), *Nealnon's Guide to the 30th Dáil & 23rd Seanad* (Gill and Macmillan, Dublin, 2007), p. 6.

[247] N. Whelan, *Showtime or Substance? A Voter's Guide to the 2007 Election* (New Island, Dublin, 2007), pp. 192–93.

[248] Collins, 'The Election Background', p. 6.

[249] M. Mansergh, 'Fianna Fáil and Republicanism in Twentieth-Century Ireland', in I. Honohan (ed.), *Republicanism in Ireland: Confronting Theories and Traditions* (Manchester University Press, Manchester, 2008), p. 107.

[250] *The Irish Times*, 28 October 2005.

After taking a lot of flak for his Ard Fheis speech, the Taoiseach finally addressed the matter of 1916 in an address to the Dáil on 8 November 2005. In defending his actions, Ahern was eager to stress that he was prepared to engage in consultation with the Opposition leaders about plans for future commemorations of 1916. 'I am presently considering', he said, 'the most appropriate way of marking the 90th anniversary of the 1916 Rising next year, with particular emphasis on how this may present an opportunity to put in place, over the following decade, a wide-ranging programme of activities and events in the lead-up to the centenary commemorations'. After indicating that he proposed 'to enter into discussions with party leaders shortly on how these matters may be progressed', he confirmed that 'the military parade by the Defence Forces, Óglaigh na hÉireann, commemorating the 1916 Rising, traditionally organised to take place each Easter at the GPO but in abeyance since 1971, will be restored to the annual calendar'. Significantly, Ahern also stated 'that this parade will reflect the evolved role of the Defence Forces and include significant representation of their peacekeeping service abroad with the United Nations'.[251] A far lengthier debate on plans for future commemorations of 1916 took place in the Dáil on 7 December, in which the Taoiseach avowed his belief that the Rising had represented 'a defining event of modern Ireland' – one that was of 'fundamental importance' for the state's creation. He further added: 'I believe it is appropriate that commemorative events should be organised to respectfully acknowledge the achievements and sacrifices of past generations and to inculcate an awareness and appreciation in modern Ireland of the events and issues of those times.'[252] Having done his part in orchestrating the revival of the 'spirit of 1916', Ahern then momentarily passed the commemorative baton to the President, who was soon set to deliver a major speech in Cork about the perceived meanings of the Rising to the citizens of twenty-first century Ireland.

---

[251] *Dáil Debates*, Vol. 609, 8 November 2005.

[252] Ibid., Vol. 611, 7 December 2005.

Chapter 6

Recasting the Rising:
The 90th Anniversary, 2006

As they cast aside the constitutional efforts of the Redmondites and engaged in physical force for political ends, a degree of controversy was always bound to persist around the momentous actions of the rebels who fought and died for Irish freedom in 1916. On Easter Sunday 2006, however, the Rising went some way to reclaiming the primacy that it had once occupied in the canons of Irish heritage, culture and identity. Ninety eventful years had passed since the Irish Volunteers occupied the GPO, and the whole island of Ireland had changed beyond all recognition. It was a time of opulence in the Republic in the months leading up to the 90th anniversary commemoration, with many citizens reaping the benefits the Celtic Tiger economy. Furthermore, after many attempts, Ireland was fundamentally at peace, following the decommissioning of the Provisional IRA's arms the year beforehand. Anglo-Irish relations were in a better state than they had ever been, while many historians in Ireland were heralding the arrival of a new post-revisionist era that sparked a revival of interest in the events of the revolutionary era. Encouraged by the atmosphere of cross-community reconciliation that the political settlement in Northern Ireland had generated, the Fianna Fáil-led government of the Republic, with peacemaker Bertie Ahern at its helm as Taoiseach, proceeded to follow through on its Ard Fheis promise to stage a major military parade past the GPO on Easter Sunday, for the first time since 1971. Whilst the 75th anniversary in 1991 had been marked by a low-key affair designed not to offend, the vast crowd that turned out to watch the 90th anniversary commemoration in Dublin revealed how attitudes to 1916 had turned full circle.

Although the Rising still had its outspoken detractors – including unionist politicians, historians who remained faithful to revisionism and journalists such as Kevin Myers – Ahern saw the 90th anniversary as an opportune time in which to reassert Fianna Fáil's republican credentials and reclaim the Rising's legacy from the Provisional IRA. The one saving grace for Fianna Fáil politicians intent on marking the 90th anniversary commemoration was that the violence brought about by the Rising had lasted for only a week or so, compared to over 30 years in the case of the conflict in Northern Ireland.

Another notable occurrence in the aftermath of the reinstatement of the Army parade was the government's enlightened move, following years of neglect, to hold a significant public commemoration of the 90th anniversary of the Battle of the Somme at Islandbridge on 1 July. The pluralism inherent in this official gesture represented yet another major milestone in the annals of Irish history-making, commemoration and heritage. As the Great War was accorded greater priority in the collective memory of the Republic of Ireland, the commemorative trajectory of the 1916 Rising took a new direction – as the notion of paying respect to the Easter Week rebels became acceptable once again. To all intents of purposes, therefore, much of the substance of the 90th anniversary commemoration was about recovering lost memories of the revolutionary past. In what seemed like a public relations exercise of grandiose proportions, Easter 2006 was all about recasting the Rising in a new positive light and sanitising its legacy from all of the negative connotations associated with the actions of the Provisional IRA during the course of the Troubles.

The Lead-up to Easter Sunday

Soon after the dust had settled on the animated discussion generated by Ahern's surprise announcement at Killarney, the 1916 debate raged all over again, following President Mary McAleese's keynote speech, on 27 January 2006, to an academic conference organised by the History Department of University College Cork on the theme of 'The Long Revolution: The 1916 Rising in Context'. According to *The Irish Times*, the President's speech, which had been approved in advance by the government secretariat, marked 'a further move to bring the 1916 Rising back to a central place in the official establishment view of Irish history'.[1] When examined within its broader context, it is clear that each sentence of the carefully-crafted speech was aimed at debunking years of revisionist criticisms of the legacy of 1916. As she spoke in the university's Aula Maxima, with large colour portraits of the institution's past presidents (known affectionately as the Rogues' Gallery) staring down at her, McAleese solemnly reflected upon the arduous journey that Ireland had experienced since the fight for independence. Doing her utmost to vindicate the actions taken by the 1916 rebels, she put forward the suggestion that the prosperity of the Celtic Tiger-present could not have been achieved without the sacrifices that had been made in the past in the name of Irish freedom.

---

[1]   *The Irish Times*, 28 January 2006.

'With each passing year', she said, 'post-Rising Ireland reveals itself, and we who are of this strong independent and high-achieving Ireland would do well to ponder the extent to which today's freedoms, values, ambitions and success rest on that perilous and militarily doomed undertaking of nine decades ago, and on the words of that Proclamation'. The President then probed the historical context of the Rising, seeking to reassert the legitimacy of the motivations behind Irish republicanism in the early twentieth century. She argued that Ireland in 1916 had merely been 'a small nation attempting to gain its independence from one of Europe's many powerful empires'. She also castigated the administration of the country at the time, claiming that it was 'being carried on as a process of continuous conversation around the fire in the Kildare Street Club by past pupils of public schools'. This, she decried, 'was no way to run a country, even without the glass ceiling for Catholics'. McAleese then focused upon a more delicate facet of the story of 1916, namely the resort to arms to resolve a territorial dispute. Arguing that the world in 1916 was one 'of violent conflicts and armies' where the strongest did 'what they wished' while the weakest endured 'what they must', she put forward the notion that diplomacy only 'existed to regulate conflict, not to resolve it'. She felt that this was the framework within which the Rising's leaders had seen 'their investment in the assertion of Ireland's nationhood'. In drawing the speech to a conclusion, McAleese stated: 'Like every nation that had to wrench its freedom from the reluctant grip of empire, we have our idealistic and heroic founding fathers and mothers, our Davids to their Goliaths.' The rebels, she determined, had 'inhabited a sea of death, an unspeakable time of the most profligate worldwide waste of human life', but 'their deaths rise far above the clamour – their voices insistent still'. 'Enjoy the conference', she added, 'and the rows it will surely rise'.[2]

As the President predicted, the conference did indeed generate heated quarrels over the motives and methods of the 1916 rebels – thus proving that the Rising's memory remained as sentient a political issue as ever. Within three days of her speech, a major debate on the place of 1916 in Irish history was in full swing, as exemplified by the content of Kevin Myers's 'An Irishman's Diary' column in *The Irish Times* on 31 January. Myers, whose 25-year tenure at the so-called 'paper of record' was coming to an end, had for long been a dissenting voice on 1916 and was certainly in no mood to alter his opinions. In his characteristically acerbic writing style, he attacked the President's keynote speech wholeheartedly, and venomously commented that her remarks at Cork had ranked 'among the most imbecilic ever' delivered by an Irish President.

---

[2]   For the full text of the President's keynote address at University College Cork, see ibid., 28 January 2006 or M. McAleese, '1916: A View from 2006', in G. Doherty and D. Keogh (eds), *1916: The Long Revolution* (Mercier Press, Cork, 2007), pp. 24–29.

A particular source of indignation to Myers was the fact that many civilians had been killed during the Rising, contrary to the Proclamation's assurance of 'equal rights' for all citizens. The journalist also repeated his well-rehearsed view that the events of 1916 had cultivated 'catastrophic, 1916-inspired violence' in the North and had led to 'the formal inauguration of a political cult of necrophilia whose most devoted adherents over the past 36 years have been the Provisional IRA'. The President, he concluded, had deliberately avoided asking hard questions about the 'orgy of violence' in 1916, and had only rationalised upon an assortment of self-righteous notions about the Rising and the Proclamation.[3] Myers's column the next day proved equally explosive in nature. He again vituperatively slammed 'the President's dreadful speech at UCC', and expressed concern that the achievements of the Irish Parliamentary Party had been 'wiped from our public history yet again'. He also denounced the President's view of 1916 as a polygonal endeavour as 'utter rubbish', pointing to how 'attacks on rural Protestants' had subsequently taken place during the early 1920s.[4]

Unsurprisingly, debates about 1916 continued to dominate agendas when the matter of history was discussed in both Houses of the Oireachtas, in newspaper columns and letters pages, on talk radio shows, and on television programmes during the early months of 1916. On these occasions, a wide range of opinions surfaced – both positive and negative. In a Seanad debate on 1 February, David Norris remarked that the President's vilification of the Kildare Club 'was not helpful'.[5] The following day, however, many senators went on record to support the President's speech, despite reminders from the Cathaoirleach that such discussion was 'out of order', as she was 'independent of this House'. Nonetheless, John Dardis of the Progressive Democrats expressed his belief that the President had the 'loyalty and ... confidence' of the House, 'irrespective of what criticisms we might have of the utterances she might make'. Even though he had still not yet read the speech in question, Fine Gael's Paul Coghlan observed 'that however one might regard individual

---

[3]   *The Irish Times*, 31 January 2006. The full text of this column was reproduced at a later stage, under the heading 'Irish Nationalism & 1916', in K. Myers, *More Myers: An Irishman's Diary 1997–2006* (The Lilliput Press, Dublin, 2007), pp. 106–108. Myers's views came as no surprise to his readers. For an earlier commentary on the theme of 'Easter 1916' by the same journalist, see K. Myers, *From the Irish Times Column 'An Irishman's Diary'* (Four Courts Press, Dublin, 2000), pp. 35–36, in which he denounced the Rising as 'an unmitigated evil for Ireland' and contested the notion that the violence was 'historically justifiable'. 'And most enduringly of all', he lamented, 'the cult of the gun became sanctified within Irish political life'.

[4]   *The Irish Times*, 1 February 2006. The President's speech was again discussed by Myers in his 'An Irishman's Diary', in ibid., 24 February 2006, when he bemoaned that 'political blood-worship in this country has become a pathological and incurable perversion'.

[5]   *Seanad Debates*, Vol. 182, 1 February 2006.

utterances, the President's comments were of course meant in that spirit of mutual respect'. Among the senators who had read the speech by then was Fine Gael's Brian Hayes, who singled out one part for special praise, namely the element that had touched upon the forthcoming 90th anniversary of the Battle of the Somme. The President, he said, was correct 'to highlight the fact that the commemoration could be a point of reconciliation for every person on the island of Ireland'. Whilst endorsing the merits of 'looking at our history in a spirit of mutual respect and reconciliation', Fianna Fáil's Martin Mansergh also pointed out that he did not 'subscribe to efforts to effectively repudiate and criminalise parts of our history, such as the 1916 Rising'. In registering his own verdict about the Rising's place in history, Independent Joe O'Toole felt that the time was ripe to 'openly discuss 1916', but warned that it would be incorrect 'to try to apply the views of today to that time'.[6]

As the Seanad's impromptu debate about 1916 progressed, some senators decided that it was time to be more explicit in their endorsements of its legitimacy. One of these was Mansergh's party colleague, Labhrás Ó Murchú, who also endeavoured to settle some old scores with those who still adhered to the revisionist perspective on Irish history. 'To some extent', he pronounced, 'Members should feel pleased that 1916 is being discusses in these Houses. It is no longer a matter of "Who Fears to Speak of Easter Week"'. 'In many ways', he added, 'we must be careful that in endeavouring to build bridges, we do not become apologetic for what we are'. Ó Murchú then took a swipe at the output of revisionists, alleging that their works had 'denigrated and misrepresented the patriots of the past'. 'I am worried', he said, 'that we are not mature enough to recognise those great people [of 1916] who gave leadership at a time it was necessary'. Another Fianna Fáil senator, Jim Walsh, concurred with Ó Murchú's anti-revisionist views, arguing that members of the Seanad 'should not join the revisionist brigade by confusing issues' in debates generated by the 90th anniversary. In praising the Proclamation for being 'an outstandingly far-sighted document for its time', he was also eager to commend 'those men who paid the supreme sacrifice during the Easter Rising'. Fianna Fáil's Camillus Glynn also went on record to say that he strongly identified with the pro-1916 sentiments expressed by Ó Murchú and Walsh.[7]

At this point, the senators' debate moved on to the discussion on 1916 that was taking place in the pages of *The Irish Times*. In reference to 'a strong debate taking place via a particular columnist [namely Kevin Myers]', Fianna Fáil's Mary O'Rourke told the Seanad: 'I hope we recognise that the President has a much

---

6   Ibid., Vol. 182, 2 February 2006.
7   Ibid.

broader agenda than what was detailed in the column which suggested that she was entering into a narrow focused debate.'[8] The Fianna Fáil senators were by no means the only people capable of speaking out in public against those who held hostile views about what the President had said in Cork about 1916. On 4 February, a lengthy written response to some of Myers's harsh criticisms of the President, along with a defence of his questioning of the actual make-up of the proceedings at University College Cork, was published in the letters page of *The Irish Times* by the conference's organisers, historians Dermot Keogh and Gabriel Doherty. They protested 'in the strongest possible terms' against the language that he had used in his two columns, finding his usage 'of the term "imbecilic" ... particularly objectionable'. The two historians also took issue with the journalist's 'employment of the term "underlying agenda" in his assessment of the line-up of speakers', and felt that he had reduced academic discourse on 1916 to rather simplistic binary terms – namely 'misleading and mutually exclusive labels such as "pro" and "anti"'.[9]

Notwithstanding the conference organisers' misgivings over the reduction of the 1916 debate to clear-cut polarities, most reactions by the public and media to the President's speech in the days and weeks that followed typically varied from either categorical support to downright hostility. A plethora of opinions were published in the letters page of *The Irish Times*, debating not only the merits of the President's speech, but also offering feedback on the writings of Myers. As per usual, the 'paper of record' published a representative sample of divergent viewpoints. Support for what the President had said in Cork was by no means lacking, and for one letter-writer from Dún Laoghaire, her words had the potential to generate 'an approving response in the hearts of many people of different political views'.[10] Another Dubliner wrote in to offer support for a celebration of the Rising, highlighting that it had led directly to 'an awakening of national consciousness which ultimately led to the defeat of the British Empire by our small nation'.[11] Politicians also entered the debate. Labour's Dermot Lacey asked why Myers had not condemned 'the bloody, ugly and ultimately counter-productive premeditated murder of the leaders of the Rising'.[12] Fianna Fáil's Barry Andrews took Myers to task for his criticism of the fact that none of the signatories of the Proclamation had stood for parliament. In offering an answer, Andrews argued that the Act of Union had been 'imposed and maintained against the will of the Irish people' and that it was the

8   Ibid.
9   *The Irish Times*, 4 February 2006.
10   Ibid., 31 January 2006.
11   Ibid., 1 February 2006.
12   Ibid., 2 February 2006.

privilege of the minor ruling class that it could enforce 'constitutional change without bloodshed'. 'It is only right and fitting', he added, 'that the President of this country should declare herself proud of the achievements of the men and women of 1916'.[13]

On the other hand, the content of McAleese's speech was openly criticised by a range of letter-writers (sometimes in quite vitriolic tones) as being propagandist, narrow-minded and exclusive – contrary to her self-proclaimed Presidential mission of 'building bridges' between communities. A letter to *The Irish Times* from a Galwegian went so far as to ponder whether her speech was 'a case of the mask slipping and the real person coming through'.[14] Likewise, a letter-writer from Peterhouse in Cambridge charged that the President's 'cult worship of founding fathers' went contrary to her work in promoting reconciliation on the island of Ireland. A philosopher from the Milltown Institute added that the President's quoting of the Proclamation would not 'prove that the signatories were inclusive, anxious to embrace unionists', because leaders such as Patrick Pearse had 'refused to accept that the northern unionists could not be coerced into an independent united Ireland'.[15] A letter-writer from Belfast accused the President of 'playing with fire' and reinforcing 'distorted views of the past, making reconciliation harder', while a letter-writer from Killester remarked that it was 'sad that, not for the first time, the President has allowed her thoughts to wing her on the wind of emotional attachment to an Ireland of idealistic folklore'.[16] Fine Gael's Ronan Guckian likened the speech to 'the old formula of republican ideology, violence and martyrdom' and complained that its was 'like an adjunct' of the speech that Ahern had delivered at the Fianna Fáil Ard Fheis the previous year.[17] Many other negative reactions to the Cork speech were published in *The Irish Times*, variously criticising the President for 'listening to what she wants to hear', acting 'as corporate cheerleader for the nation' and peddling 'sub-de Valeran tribal fantasies', and 'disappointing' people.[18]

Writing in *History Ireland*, Paul Bew also critiqued McAleese's speech, maintaining that she had overexaggerated the influence of 'landlord elements in the Kildare Street Club ... over Dublin Castle' and that she had unfairly projected 'a politically correct ideal' onto the 1916 rebels, namely 'an egalitarian opposition to élitism'.[19] The President was also criticised in *The Sunday Independent* by Ruth

---

[13]   Ibid., 3 February 2006.

[14]   Ibid.

[15]   Ibid., 1 February 2006.

[16]   Ibid., 2 February 2006.

[17]   Ibid., 13 February 2006.

[18]   See ibid., 6 February 2006, 7 February 2006 and 27 February 2006.

[19]   P. Bew, '"Why Did Jimmie Die"', *History Ireland*, Vol. 14, No. 2 (2006), pp. 38–39.

Dudley Edwards, who argued that she had demonstrated 'no sign of having read any modern Irish history except that produced by a gaggle of counter-revisionists who repackage old myths in modern jargon'. In the same newspaper, Eoghan Harris suggested that the President had played a part in putting 'the pluralist clock back 30 years' and had also 'muddied the moral clarity which the Irish Republic had reached on Irish nationalism'.[20] A negative reaction to McAleese's sentiments was also aired in *The Irish Independent* by Bruce Arnold, who registered his unease at 'the absurdity of what she had to say about 1916' and labelled her speech's content as both 'quasi-political nonsense' and 'offensive at many levels'.[21] In *The Sunday Times*, John Burns surmised that the President's eulogy of the 1916 rebels had been designed to clear 'an ideological path' for the forthcoming Easter Sunday military parade.[22] Unionist politicians also expressed concern about the President's speech. Writing for *The Irish Times*, Lord Laird of Artigarvan registered his disbelief that the President was 'endeavouring to persuade us that the 1916 Rising was not sectarian and narrow'. Furthermore, he proclaimed that McAleese 'ought to be extremely wary of Pearse's almost mystical views on republicanism's potential'.[23]

On 11 February, a fortnight after the ending of 'The Long Revolution' conference, an Editorial appeared in *The Irish Times*, entitled 'Going Beyond the Myths of 1916'. This revisited how the President had ended her keynote speech in Cork with the comment: 'Enjoy the conference and the rows it will surely rise.' McAleese, according to the Editorial, could not 'have been disappointed by the subsequent rows', although it did add that 'whether she or anyone else should "enjoy" them is another matter'. 'Until recently', it warned, 'people were being killed on this island in its name ... It is still a live political issue, not just in terms of whether Fianna Fáil or Sinn Féin can best claim its mantle – a debate from which President McAleese should stand clear – but also in terms of this country's view of itself and its values'.[24] The following day, an Editorial in *The Sunday Independent*, entitled 'No Role for Mary in 1916 Debate', likewise conveyed a cautionary message. It criticised McAleese's speech as 'unwise ... and ill-advised', adding that it was her duty 'to stay above public controversy, and to stay out of political debate'.[25] Later that month in *The Sunday Independent*, Paul Bew added a further contribution to the debate, declaring that the row over how to commemorate 1916 had turned 'into a subjective gabfest'. McAleese, he

20   *The Sunday Independent*, 5 February 2006.
21   *The Irish Independent*, 4 February 2006.
22   *The Sunday Times*, 5 February 2006.
23   *The Irish Times*, 4 February 2006.
24   Ibid., 11 February 2006.
25   *The Sunday Independent*, 12 February 2006.

pointed out, was 'an upwardly mobile Northerner of the Civil Rights era'. By looking back at 1916, he felt that she had seen her own reflection in the mirror and had seen 'Catholics smashing through the glass ceiling', especially by her speech's reference to the Kildare Street Club. The President and others, he contended, had thus far displayed 'a common lack of concern with the men of 1916 and their world view'.[26]

As the controversy over the President's Cork speech gradually abated in the run-up to Easter Sunday, politicians in the Dáil turned their attention to more mundane matters such as the logistics behind the preparations that were being put in place for the upcoming commemoration. When asked to make a statement about the matter, the Minister for Defence, Wille O'Dea, noted on 23 February that an interdepartmental working group, chaired by the Department of the Taoiseach, had been established to oversee arrangements for route of the parade, road closures, health and safety issues, publicity, and the issuing of invitations to the review stands outside the GPO. Its membership, stated O'Dea, included officials in his own department, Defence Force officers, officials from Dublin City Council, members of An Garda Síochána, fire services personnel, and officials from the Office of Public Works. The group, he confirmed, was 'meeting regularly'. Rather significantly, O'Dea also noted that tentative preparations had already commenced for the Rising's centenary in 2016: 'On the invitation of the Taoiseach, all parties in the Oireachtas have nominated spokespersons to offer advice on the appropriate scope and content of a 1916 Centenary Commemoration Committee to be put in place in coming years.'[27]

Within the Dáil itself, TDs were not shy in bringing forth proposals as to how the 90th anniversary could be remembered. On 28 March, for example, Tom McEllistrim sought to demonstrate his patriotic credentials by calling upon the Minister for Education and Science, Mary Hanafin, to distribute a Tricolour and flagpole to every primary school in the country. 'It would help', he said, 'to add a local element to the national commemoration ceremony'. To his disappointment, however, no new flags or flagpoles were forthcoming from the government at such short notice. Instead, the Tánaiste, Mary Coughlan, stated: 'the focus will be on sending out educational materials'.[28] On a small number of occasions, some fresh controversies arose in the lead-up to Easter Sunday. The political squabbling that followed on from the Taoiseach's Ard Fheis announcement came to the surface once again in early March, when Fine Gael and Labour councillors in Dún Laoghaire-Rathdown County Council narrowly

---

[26] Ibid., 26 February 2006.
[27] *Dáil Debates*, Vol. 615, 23 February 2006.
[28] Ibid., Vol. 617, 28 March 2006.

defeated a Fianna Fáil proposal (by nine votes to eight) to display a copy of the 1916 Proclamation in their council chamber. Explaining his own objection to the idea, Labour councillor Denis O'Callaghan accused Fianna Fáil of seeking to gain political advantage: 'This is part of a Fianna Fáil attempt to take ownership of 1916, a race with Sinn Féin to see who can claim it.'[29]

Then, in the days before and after Easter Sunday, the Taoiseach was again subjected to a barrage of criticism from political opponents and certain sections of the media, following a speech at the National Museum Collins Barracks on 9 April. In the speech, entitled 'Remembrance, Reconciliation, Renewal', Ahern emphasised the need 'to honour the dead generations who have gone before us' and declared that citizens had 'a solemn duty to vindicate the living generations who will come after us, to leave to them, as was left to us, a country that has profited from the continuing dedication, generosity and commitment of its people'. More controversially, however, he referred to 'four cornerstones of independent Ireland in the twentieth century', which he said were 'the foundations of the future we are building today and tomorrow'. These 'cornerstones', he argued, consisted of the 1916 Proclamation and three momentous political initiatives that ensued under the tenure of Fianna Fáil Taoisigh – namely the passing of de Valera's 1937 Constitution, Jack Lynch's ratification of the Treaty of Rome in 1972 (thus enabling entry into the EEC the following year) and his own role in the brokering of the Good Friday Agreement of 1998.[30]

Some commentators interpreted Ahern's speech as an extremely narrow-minded view of modern Ireland's development, not least Alan Ruddock of *The Sunday Times*, who labelled it 'remarkably blinkered'.[31] In *The Irish Times*, a letter-writer from Castletroy contended that the Taoiseach's speech smacked 'of petty, partisan grudge-bearing, something to which we have become accustomed from Fianna Fáil these days'.[32] Fine Gael leader Enda Kenny was equally chagrined and complained that no political party had 'a monopoly on Irish history'.[33] Fine Gael reaction to Ahern's historical assessment also appeared in a letter from a member of the party's Taylor's Hill-Kingston branch in Galway city to *The Galway City Tribune*, which maintained that the Taoiseach had 'obliterated Fine Gael from Irish history' by ignoring political milestones such as the Anglo-Irish Treaty of 1922, the initial declaration of the Republic in 1948 by John A. Costello, the Sunningdale Agreement of 1973, and the

[29] *The Irish Times*, 6 March 2006.
[30] For the full text of the Taoiseach's speech at the National Museum, see ibid., 10 April 2006.
[31] *The Sunday Times*, 16 April 2006.
[32] *The Irish Times*, 20 April 2006.
[33] *The Sunday Independent*, 16 April 2006.

signing of the Anglo-Irish Agreement in 1985 by Garret Fitzgerald.[34] Senator Joanna Tuffy of Labour also weighed in on the debate, complaining in a letter to *The Irish Times* that the Taoiseach's 'most blatant revision of history' had ignored the historic role of former Labour leader Thomas Johnson in the drafting of the Democratic Programme adopted by the first Dáil in 1919.[35] Two days before the Easter Sunday commemoration, however, Ahern rejected complaints that Fianna Fáil had turned the 90th anniversary into a party political event by airbrushing Fine Gael and other political parties from history. 'The one thing Fianna Fáil never has to do', he said, 'is to prove its republicanism ... I certainly do not want the commemorations and the remembrance of the men and women of 1916 to be anything other than an inclusive event for all of the people of Ireland'. On other matters, the Taoiseach also defended his belief in the legitimacy of 1916, arguing that the rebels 'were driven to what they did to bring international recognition to their cause'.[36]

Ahern was by no means alone when it came to defending the Rising against attacks on its validity. In the lead-up to Easter Sunday, a number of distinguished writers went on record to offer a historiographical perspective that sought to reaffirm both the achievements of the 1916 rebels and the ideals enshrined in the Proclamation. As he had done on previous occasions such as 1966 and 1991, Garret Fitzgerald, revisited the legacy of 1916 as an elderly (but still intellectually agile) statesman four days before the military parade. He distanced himself from John Bruton's Mansion House speech and restated his belief that the 1916 Rising had been a historical necessity. Fitzgerald also wrote with nostalgic pride about the part that his father Desmond had played in the GPO during Easter Week. 'Ninety years later', wrote Fitzgerald, 'it seems to me absurd that people should be concerned to sit in judgment on men who, in circumstances unimaginably different from the world of today, in my father's words "with calm deliberation decided on a course ... that ... meant their own inevitable death"'. He also repeated his long-held and well-rehearsed retrospective approbation of the actions of the 1916 rebels, using econometric insights and wealth indices to proclaim that 'without the independence ... secured in the aftermath of the Rising we could never have become a prosperous and respected state and

---

[34] *The Galway City Tribune*, 5 May 2006.

[35] *The Irish Times*, 29 April 2006.

[36] *The Irish Independent*, 15 April 2006. In a letter to *The Irish Times*, 6 May 2006, Councillor Michael McGrath of Fianna Fáil registered his disappointment about Senator Tuff's remarks, noting that the Taoiseach had stated a number of times during the previous month 'that almost all of the parties ... had their roots in 1916 and that accordingly the commemoration could not be the preserve of any one political party'.

member of the EU'.[37] With such an overwhelming endorsement from the main Opposition party's former leader, Ahern and other Fianna Fáilers must have felt privately reassured as the big event on Easter Sunday finally approached.[38]

Naturally enough, the looming commemoration was a hot topic of discussion in newspaper Editorials – which registered either guarded or optimistic opinions about the military parade that was about to take place in the capital. On Easter Saturday, the two leading broadsheets were somewhat tetchy about the prospect of viewing such a display of militarism on O'Connell Street. An Editorial in *The Irish Times*, entitled 'Commemorating the 1916 Rising', drew attention to the lessons learnt from the Troubles in Northern Ireland. 'Nobody', it stated, 'would argue with the Taoiseach's desire for an inclusive, celebratory and forward-looking commemoration'. However, it wondered whether a military parade was 'the most appropriate symbol to celebrate our sovereignty', even though 'the barrel of a gun ... is no longer in our faces'. Despite its circumspection about the 'uneasy legacy' of 1916 and the restoration of 'an old-fashioned military parade', it did concede that 'the creation and maintenance of ... democracy ... is something to celebrate, something which can unite and reconcile old divisions ... [and] accommodate newcomers'.[39] *The Irish Independent* was both guarded and hopeful in an evenly-balanced Editorial, entitled 'Now is the Time to Build a Pedestal for All Our Heroes'. Many people, it acknowledged, believed that military parades were 'an anachronism, owing more to the Wagnerian tradition of triumphalism than to the realities of the Ireland of today' – one that underplayed the 'transcendent power' and historic role played by constitutional politicians such as Daniel O'Connell, Charles Stewart Parnell and John Redmond. Nevertheless, it commended Ahern for reclaiming 'some of the more emotive symbols of 1916' for the state and for highlighting the fact that the Republic had 'only one legitimate army'. 'But', it concluded, 'we should also embrace at this time of memory and reflection a more hard-headed resolution ... no more violence and bloodshed in the pursuit of political objectives'.[40]

37 *The Irish Times*, 12 April 2006. Fitzgerald's annoyance over those who expressed critical views about the Rising was later criticised in the letters page of ibid., 20 April 2006 by John A. Murphy, who registered his dissatisfaction that people 'who have expressed unease about 1916 have been excoriated by Rising enthusiasts'.

38 Writing in ibid., 15 April 2006, Senator Mansergh sought to put the military parade's impending revival in perspective, by arguing that it would set about securing 'the long overdue task of separating out what is valuable, noble and enduring in the Irish republican tradition, and what can by any international standard be regarded as legitimate'.

39 Ibid.

40 *The Irish Independent*, 15 April 2006.

A wide range of standpoints were again evident in the Editorials of the Easter Sunday newspapers. The Editorial in *The Sunday Tribune*, entitled 'Unease about Military Parade for Rising Anniversary', was rather wary about the militaristic trappings of the parade taking place that day. While admitting that 'it is only proper that the state today takes the lead role in organising the celebrations' and conceding that the Taoiseach 'was entirely correct in reconstituting a significant national event to mark the Easter Rising', it speculated that 'maybe he was wrong to simply pick up the pieces where the state left off with its last military parade over 30 years ago'.[41] *The Sunday Independent's* Editorial, entitled 'Ahern has 1916 Lessons to Learn', took a dim view of the Taoiseach's decision to revive the military parade, arguing that he was 'ill-advised'. By virtue of announcing the parade's comeback at an Ard Fheis and by focusing almost exclusively upon his own party's achievements during his speech at Collins Barracks, the newspaper declared that Ahern 'has given this celebration a somewhat partisan political flavour'.[42]

The Sunday Times, by contrast, ran a more sanguine Editorial, entitled 'Reason to Celebrate'. However, it offered only 'two cheers for today's commemorations' and observed that the people of Ireland 'no longer need the dubious myths and shibboleths of the past to bolster their identity'. Nevertheless, it did acknowledge that 'there is a sense in which 2006 is an auspicious moment for Irish people to remember and reexamine the events of Easter Sunday 1916, and all that followed'. Looking back on the 40 years that had passed since the Golden Jubilee, it reflected: 'The popular desire to turn the patriot dead into plaster saints ... has diminished, its decline hastened by a healthy disenchantment with official pieties'. Irish citizens, it concluded, 'are ... fortunate enough to be free to appreciate all that was liberating and outward-looking about the Easter Rising while simultaneously rejecting all that was destructive and narrow-minded'.[43] As the clock ticked down to the commencement of the military parade, it soon became apparent that *The Sunday Times* appeared to be on the mark with its prudent appraisal of Ahern's political gamble with the ghosts of 1916.

---

[41]   *The Sunday Tribune*, 16 April 2006.
[42]   *The Sunday Independent*, 16 April 2006.
[43]   *The Sunday Tribune*, 16 April 2006.

Dublin Remembers

Despite the reservations of unionists and certain high-profile journalists, none of the awarkwardness that marked the low-key 75th anniversary was evident during the more peaceful and prosperous times that existed when the 90th anniversary was commemorated in 2006. On Easter Sunday 16 April, the government's commemoration ceremonies got off to an early start at Kilmainham Gaol – days before the centenary celebrations of playwright Samuel Beckett's birth, a week before census night, around a month before the 80th anniversary of the foundation of Fianna Fáil, and a month and a half before the 100th anniversary of Michael Davitt's death. Given its intimate associations with the executions in 1916 and its prominence as one of the capital's main heritage tourism attractions, the Inchicore site seemed an obvious choice for the Taoiseach to commence the Easter Sunday ceremonies. Much had happened to Kilmainham since de Valera's visit in 1966 and a brief review of its fortunes is worthwhile.

Since coming under state control in 1986, the site had quickly gained renown as a must-see attraction for many a school tour to Dublin.[44] It also rose to prominence as one of the top tourist attractions in the capital city – which by Easter 2006 ranked as the sixth most popular city in Europe for overseas visitors. Much of the increase in Kilmainham's footfall can be attributed to wider factors, most especially the tourism boom that much of Western Europe experienced from the late twentieth century onwards. It was during these years that tourism emerged as one of the world's largest industries, blending 'exploration with an innate hunger for knowledge'.[45] While the heritage strategy for Kilmainham was emphasising wider aspects of its history by Easter 2006 (for example, its child prisoners during the nineteenth century), the story of 1916 still remained the central focus of tours of the site. These typically started with the viewing of a short film, followed by a guided tour of the prison cells in the 1916 leaders' corridor and the execution spot in the Stonebreakers' Yard. These tours usually concluded by giving visitors the opportunity of browsing through a panel exhibition on the history of the prison and the opportunity to purchase postcards, copies of Pearse's poems and the text of Pearse's oration at

---

44   Writing in ibid., Fiona Looney recalled: 'The first time I was in the prison [at Kilmainham] was when I was about 12 years old – the most impressionable age of them all – and on that occasion, when I stood in the Stonebreakers' Yard where the signatories of the Proclamation met their end, the tears came easily and unexpectedly ... This week, I was back in the yard ... and again, the emotion rose up in this stillest of terrible places.'

45   A. Wright and M. Linehan, *Ireland: Tourism and Marketing* (Blackhall Publishing, Dublin, 2004), p. xi.

O'Donovan Rossa's funeral. Although first published in 1992, an educational document pack for secondary schools, entitled *Kilmainham Gaol Document Pack: The 1916 Rising*, still remained a popular seller in 2006.[46]

After arriving there early in the morning of Easter Sunday, Ahern laid a wreath to hushed tones in the Stonebreakers' Yard, just like de Valera had done 40 years previously (albeit at a much later time in the day). In attendance were members of the Army, the leaders of the two main Opposition parties and 92-year-old Father Joe Mallin (the son of the Irish Citizen Army's Deputy Commander, Michael Mallin – who had flown all the way from Hong Kong to attend the ceremony). In a speech filled with optimism, Ahern put special emphasis on justifying 'the cause' of 1916 and said that the commemoration of its 90th anniversary was about discharging one generation's debt of honour to the bravery of another – especially to those who had been executed in the very space that he was standing and speaking in. 'By gathering here today', he said, 'our presence is testimony to the fact that our generation still cherishes the ideals of the courageous men and women who fought for Ireland in Easter Week ... that we honour and respect their selfless idealism and patriotism, and that we remember with gratitude the great sacrifices they made for us'. 'The potential for progress', he added, 'has never been greater. Independent Ireland is now in full stride and beginning to fulfil the hopes and expectations that all the patriots of the past knew we possessed'. Looking to the future and to the wider context of foreign affairs, Ahern said: 'we must be generous and inclusive so that all of the people of Ireland can live together with each other and with our neighbours in Great Britain on a basis of friendship, respect, equality, and partnership'.[47]

Whereas de Valera had antagonised unionists with his pronouncements about unification in 1966, the difference in emphasis in Ahern's diplomatic discourse at Kilmainham was marked. As one of the major architects of the Good Friday Agreement, Ahern was extra keen to emphasise the all-important principle of bridge-building that morning: 'we will continue to work for peace, for justice, for prosperity, and for reconciliation between all who share and who love this special island'.[48] Privately though, the Taoiseach's thoughts were also very much focused on the gruesome nature of what had transpired at Kilmainham 90 years beforehand. As he later recounted in his autobiography, his visit to the site of the executions had proved 'the most telling moment' of an emotional day of commemoration:

---

[46]   See Office of Public Works and Blackrock Teachers' Centre, *Kilmainham Gaol Document Pack: The 1916 Rising* (Office of Public Works and Blackrock Teachers' Centre, Dublin, 1992).

[47]   *The Irish Times*, 17 April 2006.

[48]   Ibid.

There was a short ceremony held in the yard at Kilmainham Gaol where Pearse and his compatriots had been brought out and shot. That was such an eerie moment. The walls of the yard are so high, you could imagine how the shots must have sounded as they reverberated around. Inevitably my thoughts were with the fifteen men who were executed over nine days, but for a moment I found myself wondering about the emotions of the young soldiers who shot them. When the wounded Connolly was brought in on a stretcher, strapped to his chair and shot, it must have been an image that stayed with his executioners forever.[49]

In his speech at Kilmainham, the Taoiseach also pointed to the links that he saw between the accomplishments of the Army of the present and the rebels of the past, indicating how proud he was that the generations who followed 1916 had 'used ... freedom to support peace across the world through the efforts of our Defence Forces, Óglaigh na hÉireann'.[50]

After the ceremony at Kilmainham, the focus of attention switched to Dublin city centre, where the sun shone intermittently and a festive atmosphere prevailed as around 2,500 Army personnel paraded through O'Connell Street (Plate 6.1). The military spectacle also featured a fly-past by the government's jets and various Air Corps helicopters and aircraft. Both *The Irish Times* and the *Irish Examiner* estimated that at least 100,000 spectators turned up to watch the parade, while *The Irish Independent* put the figure at around 120,000.[51] Participants in the procession, which was commanded by Brigadier General Gerry McNamara, included members of the Army (including the elite Army Ranger unit who wore masks, goggles and camouflage coats), the Air Corps, the Naval Service, An Garda Síochána, the Organisation of Ex-Servicemen, and the Irish UN Veterans Association. It also included a small number of pipe bands and the Garda Band. Six members from the equitation school also paraded on horseback, but unlike the parade in 1966, no community groups or cultural organisations were allowed to participate. A wide variety of Army wares were displayed in the parade – including four 25-pounder artillery pieces, a HOBO bomb disposal robot, DURO armoured vehicles, 105 mm artillery guns, MOWAG personnel carriers, Panhard AML armoured cars fitted with cannons, an Aardvark mine-clearing vehicle, Scorpion tanks, L70 anti-aircraft guns, and a Javelin anti-tank missile.

---

[49]   B. Ahern with R. Aldous, *Bertie Ahern: The Autobiography* (Hutchinson, London, 2009), p. 294.

[50]   *The Irish Independent*, 17 April 2006. For the full text of Ahern's speech, see the *Irish Examiner*, 17 April 2006.

[51]   *Irish Examiner*, 17 April 2006; *The Irish Independent*, 17 April 2006; *The Irish Times*, 17 April 2006.

Plate 6.1   Members of the Irish Army marching past the GPO for the
              Rising's 90th anniversary commemoration on 16 April 2006

Source: The Irish Times, 17 April 2006.

In addition to the Tricolour and flags bearing the Starry Plough and old republican colours, the military heritage of the more recent past was also prominent through the carrying of flags representing all the nations that the Defence Forces had served in with the UN.[52]

Geographically, the parade commenced at Dublin Castle and then proceeded along a route via Dame Street and College Green to Westmoreland Street, and then continued northwards across the River Liffey to O'Connell Street. In line with the custom and tradition of previous commemorations, the parade paused halfway up O'Connell Street for a special ceremony outside the GPO. The 'grand old lady of O'Connell Street', as the Assistant Secretary of An Post described it later that year, was nice-looking once again, having had its interior and exterior specially refurbished by the Office of Public Works in time for the 90th anniversary.[53] At 12.00 noon the National Flag above the portico of the GPO was lowered to half mast, to the sound of a piper's lament. Captain Tom Ryan of the Sixth Infantry Battalion, a former Roscommon footballer (originally from town of Boyle), then read the Proclamation, which was greeted with applause. Ahern then invited President McAleese to lay a wreath on behalf of the people of Ireland. To the airs of 'Mise Eire' she moved forward and laid the wreath, which was followed by a minute's silence for all those who had died in the fighting in 1916. After the sounding of the 'Last Post', the National Flag was then returned to full mast and the ceremony then concluded with the playing of 'Reveille' and the National Anthem, which was also greeted with applause and cheering. The parade then resumed and for around an hour it filed past the three reviewing stands on O'Connell Street and finished by the Garden of Remembrance at Parnell Square.

For those who could not turn up to witness the spectacle first hand, RTÉ offered two hours of live television coverage of the event from 11.30 am that morning (followed by highlights later that night), while Sky News also carried live reports of the event. As had been hoped for, the parade passed off without any major incidents. The only reported problems on the day were of a technical nature. The *Irish Examiner* reported that sound problems were experienced with a big video screen that had been installed on Westmoreland Street.[54] Although many of the public managed to obtain good vantage points from behind metal safety barriers on O'Connell Street, *The Irish Times* reported that the shutting of roadworks at its northern end had drawn complaints from others, as this had resulted in restricted views of the parade between The Spire and Cathal

[52]   *The Irish Times*, 17 April 2006.
[53]   S. Ferguson, *At the Heart of Events: Dublin's General Post Office* (An Post, Dublin, 2007), pp. 5, 35.
[54]   *Irish Examiner*, 17 April 2006.

Brugha Street. The sealing of rubbish bins by Dublin City Council also drew criticism, as it resulted in much littering. But this action was defended by the authorities, who pointed to how republican dissidents had used them to conceal weapons during riots two months previously at a Love Ulster parade. Despite these minor nuisances, the overall security operation on Easter Sunday proved to be an overwhelming success. More than 1,200 Gardaí participated, including armed detectives, members of the Public Order Unit and four horses from the Mounted Unit. Armed troops also were on standby throughout the day in 4x4 vehicles.[55]

The presence beside the barriers along the main thoroughfare of a large number of uniformed Gardaí clad in luminous yellow jackets also ensured that there was no repeat of the scenes of crushing or fainting that had been witnessed amongst the crowd that lined O'Connell Street in 1966. While everything flowed freely along the main route of the military parade, a small amount of people used the occasion to stage miscellaneous protests from the sidelines (thus reviving memories of how Misneach and other groups had used previous commemorations to air grievances over a range of contentious matters). As the Army paraded, one particular group of activists known as 'The Unmanageables' (who were dressed up as women), mounted a relatively unobtrusive demonstration from O'Connell Bridge. From there they handed out copies of the Proclamation to passers-by, so as to highlight concerns that they had with Ireland's sovereignty. So as to register their distaste at US military flights using Shannon Airport as a stopover point to and from the Middle East, the group also distributed badges emblazoned with black shamrock logos, which symbolised mourning for people who they claimed had perished 'as a result of Irish collaboration in the Iraq and Afghanistan wars'.[56] The protesters also sought to draw people's attention to areas in which they felt that Irish society needed to examine its conscience – including the controversial proposal to build a motorway near the sacred heritage site of the Hill of Tara.

Although the last of the 1916 veterans had passed away in the 1990s, raw sentiment was still evident amongst parts of the crowd gathered along O'Connell Street. *The Irish Independent* observed that some 'old soldiers' from other eras 'took off their spectacles and wiped away the tears' during the ceremony, while the Taoiseach's eyes also had tears.[57] Some of the current generation of Irish soldiers also seemed to be profoundly moved by the magnitude of the occasion. Speaking to *The Mayo News*, the 25-year-old Ballaghaderreen native

---

[55] *The Irish Times*, 17 April 2006.

[56] Ibid.

[57] *The Irish Independent*, 17 April 2006.

and County Mayo ladies footballer, Lieutenant Denise McDonagh, recounted how she had been given the honour of lowering and raising the Tricolour over the GPO. This role, she noted, had been her career highlight to date: 'I was absolutely delighted to be chosen as the person ... because it was one of the few parts of the parade that involved an individual.' 'To be that individual', she added, 'was a huge honour for me, and after I raised the flag over the GPO and Amhrán na BhFiann was playing, naturally I saluted, but there was a shiver going down my spine the whole time'.[58] Those watching the parade from the reviewing stands included the entire Cabinet, backbenchers, Opposition leaders, senators, councillors, diplomats, relatives of the 1916 rebels, and Army veterans.[59] While contemplative reflection was etched on many of their faces as the military spectacle ensued, there was no repeat of the emergencies that had unfolded in 1966, when some of those in the reviewing stands had been so overcome with emotion that they passed out and subsequently required medical attention.

Reactions from Politicians, Journalists and Historians

Many of Ireland's leading politicians, journalists and historians were quickly to the fore in registering their opinions about the military parade. At receptions afterwards in both the GPO and the Gresham Hotel, many of the political leaders who were present eagerly hailed the commemoration as a triumph – one that remembered and commemorated the events of Easter Week with demureness. The Minister for Defence, Willie O'Dea, called it 'an outstanding military display', one which 'I am sure has made the people of this country immensely proud'. The Taoiseach said that the day was 'about discharging one generation's debt of honour to another'.[60] 'The turn-out', he added, 'was smashing considering it was a long weekend and so many people were away from the city. The best way you can judge this is by talking to the relatives of the Rising leaders and they are all very happy'.[61] In the end, the Taoiseach spent hours speaking to relatives of the 1916 leaders at the Gresham, where he shared stories and stood in for photographs. 'In a way', he later recalled, the reception 'was like the meeting of a great extended family'. While the relatives had not 'been afforded anything like this in a long time', they were now 'together in a national day of commemoration

---

58   *The Mayo News*, 19 April 2006.
59   *The Irish Independent*, 17 April 2006.
60   *The Irish Times*, 17 April 2006.
61   *The Irish Independent*, 17 April 2006.

and celebration, with a whole new generation taking up the message of Easter 1916.[62]

In addition to celebrating the heritage of 1916, it was also clear that the commemoration had also been designed so as to evoke newer and more wider meanings – most especially as a mature platform for the proud track-record that the Defence Forces had built up in international peacekeeping missions with the United Nations since 1958 (in the Congo, the Lebanon, the Sinai, Cyprus, Costa Rica, Honduras, El Salvador, Guatemala, Nicaragua, Bosnia, Kosovo, Cambodia, Iran, Iraq, Afghanistan, Kuwait, Namibia, Western Sahara, Somalia, Tahiti, and East Timor).[63] For the watching public, all of the high-tech equipment on show proved to be a real eye-opener, and seemed to be well appreciated. As O'Dea confirmed later on in the day: 'The Army is obviously very popular with the public. The Irish people obviously appreciates what the Army does, particularly its peacekeeping role overseas.'[64]

Equally impressed with the military spectacle was another Cabinet member, Eamon Ó Cuív, who said that 1916 had been 'the seminal moment' when aspirations for a Republic had become a possibility and that the anniversary itself represented 'a great day for which the public came out in huge numbers'. Just as jovial was the Minister for Justice (and grandson of Eoin MacNeill), Michael McDowell, who described the commemoration as 'a spectacular success' and pointed out that the 'open, inclusive' debate that had preceded it had brought a new generation of younger people into contact with 1916. The day's events, he said, would encourage the Republic 'to be confident about celebrating 1916'.[65] In a thinly-veiled swipe at critics of the parade, the Progressive Democrats politician also stated: 'I know some people doubted the wisdom of all this but I think that, just as with the Rising itself, it was the results that showed. We have one Óglaigh na hÉireann.' 'The Defence Forces', he added, 'are our Defence Forces and they are the successors to the Volunteers. Their cap badge is the same. We have to remember that there is one State, one Army, one Constitution'.[66] Days later, McDowell further asserted his republican credentials by announcing the abolition of traditional British legalspeak in the High Court and the Supreme Court, and the replacement of forms of address such as 'My Lord' with 'Judge'.

---

[62]   Ahern with Aldous, *Bertie Ahern*, pp. 294–95.

[63]   For a short overview of the role of the Irish Defence Forces in international peacekeeping missions, see D. MacCarron, *The Irish Defence Forces Since 1922* (Osprey Publishing Ltd., Oxford, 2004), pp. 33–35.

[64]   *The Irish Independent*, 17 April 2006.

[65]   *The Irish Times*, 17 April 2006.

[66]   *The Irish Independent*, 17 April 2006.

Plate 6.2   Labour and SIPTU's 'Liberty 1916–2006' banner, unveiled on
               Dublin's Liberty Hall in the lead-up to Easter 2006
Source: Author.

'I think it is appropriate', he explained, 'that, under a republican constitution, the old-fashioned mode of address has been ended'.[67]

Unlike the 50th and 75th anniversary commemoration ceremonies, the leaders of the main Opposition parties were all present at the reviewing stand outside the GPO for the 90th anniversary. After the military parade had finished, Fine Gael leader Enda Kenny seemed less miffed than he had been a week beforehand, following Ahern's controversial speech at Collins Barracks. Kenny accepted that the parade had been 'a great showcase for the Irish Army, for the services at home and abroad, and great to see them all'.[68] However, he did go on record to suggest that the centenary in 2016 should entail a broader range of events.[69] Labour leader Pat Rabbitte – whose party had earlier joined forces with the SIPTU trade union to erect a gigantic, red-coloured 'Liberty 1916–2006'

---

67   Ibid., 21 April 2006.
68   *The Irish Times*, 17 April 2006.
69   *Irish Examiner*, 17 April 2006.

banner on Liberty Hall (Plate 6.2) – said that the parade 'was well-organised and provided an opportunity to the Defence Forces to put their capabilities on display'.[70] Equally pleased was fellow party member Dermot Lacey. He declared: 'The important task of re-establishing the legitimacy of Irish republicanism has been furthered.'[71] While registering his concern about the marginalisation of the Irish language, Green Party leader Trevor Sargent remarked that the parade was a symbol of a modern Ireland and that the state's service in UN peacekeeping operations 'was a modern example of the self-sacrifice espoused by the 1916 leaders'.[72] Lesser-ranking politicians from around the country also conveyed a positive impression about the parade in the capital. In an interview with the *Roscommon Champion*, the Mayor of Roscommon, Councillor Tom Crosby, who had watched the proceedings from behind the Taoiseach and President in the main reviewing stand, stated: 'I was delighted to be there on the day and it was a spectacular display and I wish to congratulate all the organisers, especially the Army and Gardaí.'[73]

Although Downing Street steered clear of directly commenting on the Easter Sunday commemorations in Dublin, the presence of British Ambassador Stewart Eldon amongst the dignitaries in the reviewing stand pointed to the enhanced levels of mutual friendliness that now existed in the diplomatic ties between Ireland and England. The Ambassador's presence also raised media speculation afterwards about the possibility of a follow-on royal visit to Ireland by Queen Elizabeth II. President McAleese remarked that the Ambassador's presence at the commemoration 'says it all'.[74] Also in attendance at the event were church representatives, Lord Mayor of Dublin Catherine Byrne, and Lord Mayor of Cork Deirdre Clune. Despite the absence of Charles Haughey (due to ill health), three former Taoisigh were present on the day – Liam Cosgrave, Garret Fitzgerald and Albert Reynolds. For Reynolds, the commemoration was 'a tremendous day and high time a parade like this was held'. 'Too much time', he added, 'had elapsed. It was fantastic for the Defence Forces and the public'.[75]

In addition to Fr. Mallin, other relatives of the 1916 rebels who attended the commemoration included Cathal Brugha's grandsons, brothers Austin and Rossa. Some of Northern Ireland's most prominent politicians also attended on

---

[70] *The Irish Times*, 17 April 2006. On Easter Monday, Liz McManus and other Labour TDs donned costumes to participate in a reenactment of the Irish Citizen Army's march from Liberty Hall to the GPO. This was organised by the City Pavement Pageants Collective.

[71] Ibid., 20 April 2006.

[72] *The Irish Independent*, 17 April 2006.

[73] *Roscommon Champion*, 18 April 2006.

[74] *The Irish Independent*, 17 April 2006.

[75] Ibid.

the day. One of these was the leader of the SDLP, Mark Durkan, who registered his satisfaction with the proceedings: 'It is good to see people reclaiming the memory of 1916 in a positive and responsible way.' 'The task of national reconciliation', he added, 'in many ways can be summed up in that challenge of how we do that, because carrying our divided history is very important and it is part of making sure we have a shared future'.[76] Also present was the SDLP's chief negotiator, Seán Farren, who remarked: 'The role for all political representatives as we remember the 90th anniversary of the Easter Rising must now be to maximise reconciliation between all the people of Ireland.' Besides Durkan and Farren, other Northern Ireland Assembly members who attended included Alex Attwood, Dominic Bradley, John Dallat, Tommy Gallagher, Dolores Kelly, Dr. Alasdair McDonnell, and Patsy McGlone.[77] On the whole, though, representation from other political parties in the North was poor. Those who did not show up included the leaders of the Ulster Unionist Party and the Democratic Unionist Party, who had earlier declined formal invitations from the Irish government to watch the 1916 commemorations.

In explaining the Ulster Unionist Party's decision to decline the Taoiseach's invitation the month beforehand, party member and parades spokesman Michael Copeland had reiterated its belief that the Rising had been an act of sedition. 'It is up to the people of the Republic', he said carefully, 'to celebrate their own past in whatever way they think appropriate'. 'We have no problem with that', he added, 'but their version of history would not be our version of history'.[78] Neither did any sympathy come from Ian Paisley Junior of the Democratic Unionist Party. Days before Easter Sunday, he had let it be known that unionists took 'a very different view of those involved in organising a rebellion against the United Kingdom in 1916', arguing that the rebels had been motivated by a 'hatred of all things British'. In an attack on McAleese's view that the 1916 rebels and the soldiers who died on the Somme had both helped a new generation to be raised in freedom, Paisley Junior pointed out that the former had collaborated with the Germans during a time when many other Irish had been 'fighting on the battlefields of Europe for the real cause of freedom'.[79] In the end, a number of SDLP politicians registered their disappointment at the absence of the unionists in Dublin. One of these was McDonnell, who remarked that he could not comprehend their 'failure to attend and obtain any insight

---

[76]   Ibid.
[77]   *The Irish Times*, 17 April 2006.
[78]   Ibid., 7 March 2006.
[79]   Ibid., 15 April 2006.

into what exactly was going on'.[80] Despite the no-show by the unionists, Fianna Fáil politicians refrained from making any pronouncements about Irish unity, which had so provoked a young Ian Paisley back in 1966.

For the most part, reaction from the print media to the deftly orchestrated and non-triumphalist parade in Dublin was predominantly positive. Writing for *The Irish Independent*, Miriam Lord noted that Dublin's twenty-first century response to 1916 'was warm and heartfelt' and that the parade had touched 'a chord among the most cynical among us'. For a while 'on this special Easter Sunday', she observed, 'eyes misted over and the tens of thousands of people who lined the streets of Dublin found a common bond, united in a sense of national pride and belonging'. Beyond 'all the pomp and political posturing', she noted that 'an extraordinary groundswell of public emotion' had been evident from people's reaction to the parade. The mood, Lord added, 'was celebratory, yet reverential, good-natured and good-humoured, patriotic but not triumphalist'. She also suggested that 'we have recovered our appetite for looking back', as the commemoration 'was about reinstating a pilfered past to all the people of Ireland, allowing the nation to slip back in step with history again'.[81] Equally cheery was Caroline O'Doherty of the *Irish Examiner*, who maintained that 'the spirit of 1916 prevailed' during the parade. 'The bands', she added, 'had feet tapping, the peacekeepers had hands clapping and various exotic species of specialist machinery with names like Flycatcher and The Beast had mouths gaping'.[82]

Kathy Sheridan of *The Irish Times* also reported on how warmly the Irish people had reacted to the parade, remarking that 'the crowd's collective heart was stirred' and that 'the outing was an undiluted success'. Following the conclusion of the rituals outside the GPO, she noted that 'the mood moved into a kind of friendly St. Patrick's Day parade' and that the sight of 'the very big guns' on display 'stirred the soul' for many.[83] In the same paper, John Waters also registered a positive verdict. Notwithstanding 'some good and understandable reasons why 1916 became discredited' in the past, he argued that its rehabilitation had demonstrated 'that we may be awakening from the sleep of unreason that rendered us unconscious for a generation'. As he had done for many years, Waters again rejected revisionist criticisms of the Rising and dismissed the notion that it had been an insular and narrow-minded enterprise. The 'emerging

---

[80]   *The Irish Independent*, 17 April 2006. Writing in the same paper, John Cooney cleverly bemoaned that 'the absence of Ulster Unionists offered the spectacle of a Celtic Hamlet without the Orange Prince'.

[81]   Ibid.

[82]   *Irish Examiner*, 17 April 2006.

[83]   *The Irish Times*, 17 April 2006.

Ireland' of 2006, he argued, with its diversity of people, cultures, creeds, and skin colours, 'is a precise flowering of the promises of the Proclamation'.[84] The veteran broadcaster and journalist, Vincent Browne, also endorsed the revival of the parade, stating: 'The city looked good, the GPO imposing, the Army drilled and disciplined, military music not too bad, fine weather, great crowds ... [and] no speeches.'[85]

The upbeat mood of the reporters was also reflected in a number of optimistic Editorials that appeared in the Easter Monday newspapers. One such commentary in the *Irish Examiner*, entitled 'Rising Can Inspire the New Ireland', articulated the view that the Rising had been directly responsible for 'the nascent Ireland, which the rest of the world has witnessed ascend to command its own sovereignty and sit with other nations to influence the course of world history'. Reflecting upon the state's political maturity and the overwhelming endorsement of 'constitutional and inclusive politics' by the Good Friday Agreement of eight years previously, it stated that the 90th anniversary 'was an occasion to publicly regain and acknowledge the memory of Ireland's inheritance from those patriots of 1916 for the modern day inheritors of their sacrifice'.[86] Another Easter Monday Editorial in *The Irish Independent*, entitled '1916: In Step with History', offered an equally buoyant assessment of the previous day's events. It remarked that the anniversary had been commemorated 'with dignity, solemnity and style' and surmised that the government had 'reclaimed it [1916] for everyone'. The mood surrounding the parade, it added, was 'a world removed from the sometimes bitter and petty arguments about the revival of the event after a lapse of more than three decades'. 'For now', it concluded, 'we can take pride in the spectacle of an army that threatens nobody, in a country committed to settling every political question by democratic means'.[87]

Reaction from a range of regional newspapers was also quite positive during the week that followed. In the *Limerick Leader*, for example, the commemoration was warmly reviewed in an Editorial entitled 'Proud Heritage'. This determined that 'the evidence of Easter 2006 is that in spite of everything the spirit of Pearse lives on'. The large crowd that attended the parade in Dublin, it added, was proof that 'the silent people have spoken'.[88] Equally upbeat assessments of the military parade were evident in the letters pages of various newspapers. In a letter to *The Irish Times*, a Cavanman remarked that the commemoration 'struck exactly the right note' and 'showcased our Defence Forces in an extremely positive

---

84   Ibid.
85   Ibid., 19 April 2006.
86   *Irish Examiner*, 17 April 2006.
87   *The Irish Independent*, 17 April 2006.
88   *Limerick Leader*, 22 April 2006.

and professional way'.[89] A week after Easter Sunday, another letter-writer in *The Sunday Independent* rejoiced that the commemoration had passed off as 'a dignified and touching occasion' and that Irish citizens no longer seemed to feel 'uncomfortable about their history'. In his *ad hoc* 'A Log' column in the same newspaper, Brendan O'Connor described the parade as 'a moving and proud event ... about Ireland, and our army, and not being ashamed of them, or the fact that they are an army and use real guns – something many people would have us ignore'. Elsewhere in the newspaper, Eoghan Harris complimented the Defence Forces 'on their fine parade' and concluded that the Taoiseach had 'got the best out of the 1916 business'. On the other hand, though, he sought to retain some degree of detachment from the alluring thrill of the spectacle, by declaring that 'the Irish people want to move on from 1916' and observing 'that military ceremony is not likely to sustain an annual interest in 1916'. 'No need to emote', he wrote rather drolly, 'about how we enjoyed the parade. Like children with a big cone of green ice cream. We loved the first lick but felt a bit sick by the last'.[90] Writing for *The Sunday Tribune* on the same day, the ever-redoubtable Nuala O'Faolain observed that the commemoration 'felt much more like a new beginning than a looking backward'. She also sensed 'that there was a definite element of relief in the sudden pride in our own armed forces, and the relief was at getting the national question off our backs – not just getting the Provos off our backs'.[91]

The reaction of the British media to the commemoration was mostly composed and reflective. These were changed times and while David Lister and David Charter of *The Times* of London cautiously remarked that 'it is no surprise that the parade's revival is controversial', they still highlighted the fact that 1916 was 'viewed as the pivotal moment in modern Irish history'. They also appeared to take some comfort from what they interpreted as Ahern's 'deliberate move ... to "out-republican" Sinn Féin, the IRA's political wing, and to halt its seemingly unstoppable advances across the Irish political landscape'.[92] Writing in *The Guardian*, Brian Whitaker (with the assistance of news agencies) alluded to how the revival of the 'politically sensitive' parade had 'stirred debate' throughout the Republic over the issue of 'whether the rebels were romantic heroes or a band of thugs'. Pointing to the Provisional IRA's pledge in the previous year 'to lay down its arms', he reflected positively upon how the changed

---

89   *The Irish Times*, 19 April 2006.

90   *The Sunday Independent*, 23 April 2006.

91   *The Sunday Tribune*, 23 April 2006.

92   *The Times*, 17 April 2006. Over two months beforehand, in an interview with *The Sunday Tribune*, 12 February 2006, Ahern steadfastly ruled out Sinn Féin as a potential coalition partner in the next government, stating: 'it's because practically everything we stand for they're opposed to'.

political climate had facilitated the attendance of the British Ambassador at the Dublin parade. Whitaker also observed how the parade's revival had enabled Ahern's government 'to reassert its nationalist credentials' on the issue of 1916. Additionally, he remarked that the Taoiseach was going 'some way towards recognising calls to view the Rising in a broader context by honouring the many Irish soldiers who died in the Battle of the Somme, in the same year'.[93]

The absence of the unionists from the reviewing stands at the GPO, however, must have been very disappointing to the Irish government, especially to those working behind the scenes in the Department of the Taoiseach and the Department of Defence. The Taoiseach himself had certainly been anxious to make the 90th anniversary commemoration as inclusive as possible. Two days before the parade, for example, he let it be known that he was hopeful that the unionists would respect the Rising as part of the island of Ireland's shared history. So as to further emphasise the inclusiveness of the commemoration in the lead-up to Easter Sunday, a special section of the website of the Department of the Taoiseach (which had been accessed 1,300 times by Good Friday) went so far as to offer translations of the Proclamation in the Polish and Chinese languages.[94] In a Dáil debate on the commemoration, which took place on 25 April, Ahern explained that whilst the invitation to the unionists had been made as 'a gesture of both friendship and respect', the government still 'fully respected' their decision not to attend. Overall, the general consensus within the Dáil was that Easter Sunday had been a resounding success, regardless of the unionists' absence. 'I am sure', added Ahern, that TDs 'will agree the parade was a wonderful spectacle'. Enda Kenny concurred, noting that it 'was a successful occasion' and conceding that it 'was entirely apolitical'. Kenny also complimented the Army on its appearance, noting that 'it still retains absolute neatness and gives a lesson to everybody on how to shine shoes'. Pat Rabbitte likewise commended the Army for its 'performance'. Trevor Sargent also registered a positive verdict, but asked the Taoiseach whether he would consider broadening future commemorations beyond the military theme, to include an 'emphasis on equal rights'. Having taken flak from some newspapers for not having gone in person to the Dublin parade, Caoimhghín Ó Caoláin explained to the Dáil that both himself and Gerry Adams had been absent from the review stand at the GPO due to 'prior commitments' elsewhere. However, he was anxious to stress that at least five prominent Sinn Féin politicians (including Arthur Morgan and

93　*The Guardian*, 17 April 2006.
94　*The Irish Independent*, 15 April 2006; http://www.taoiseach.gov.ie/eng/ (accessed on 6 July 2006).

Mary Lou MacDonald) had attended the Dublin proceedings, 'amply indicating our intent to be fully involved and support the event'.[95]

The next day in the Seanad, a number of senators registered their own positive verdicts on the parade. Whilst conceding that he had been 'mildly sceptical of the 1916 commemoration', Brian Hayes stated that he was 'very proud of the Irish Army on Easter Sunday', adding that it was 'great to see ... the goose stepping black beret brigade [of paramilitaries]' consigned to 'the second division' for the day. Elaborating upon the significance of this point, Joe O'Toole pointed out that the presence of the Sinn Féin politicians on the review stand outside the GPO had been a historic event insofar as it was the first occasion since 1926 that Sinn Féin members had apparently 'acknowledged Óglaigh na hÉireann as being the Irish Army'. 'I also agree', he added, 'that the second parade of balaclavas, black berets and so forth looked tired and very yesterday, as it were'. The Army itself received fulsome praise from Labour's Geraldine Feeney, who commended its 'professionalism'. Fianna Fáil senators also spoke glowingly of the commemoration. Mary O'Rourke was in no doubt that the Taoiseach's gamble to restore the military parade had ultimately paid off. 'While many doubted whether the commemoration should proceed', she said, 'they quickly limped off after we had our day of celebration'. The Taoiseach, she added, had 'stuck to his guns and persisted'. In joining with all those who had commented in favourable terms 'on the solemn occasion of the Easter 1916 commemoration', Camillus Glynn felt that it was appropriate to conclude that 'all fair-minded people would agree it was a great success'.[96]

For the most part, the general consensus seemed to be that the Irish Army had stolen the show during the 90th anniversary. Lieutenant General Jim Sreenan, the Chief of Staff of the Defence Forces, was obviously a pleased man after the Easter Sunday parade, given the warm reception that his troops had received from the general public. Three weeks before the parade took place, he appeared on RTÉ Radio 1's *This Week* show, and said that any army operating in a democracy needed to have the support of its citizens for it to be successful. 'It's a very significant parade', he stated, to commemorate 'those who have gone before us' and also to 'say thank you to the people of Ireland for the wonderful support they have given us, particularly in our peacekeeping role down the years'.[97] In a special commemorative issue of the magazine, *An Cosantóir*, Sreenan reflected further upon the significance of the 90th anniversary and also recalled with

95   *Dáil Debates*, Vol. 618, 25 April 2006.
96   *Seanad Debates*, Vol. 183, 26 April 2006.
97   *The Irish Times*, 3 April 2006.

pride how he had been part of the Colour Party that had led the military parade down O'Connell Street for the Golden Jubilee commemorations in 1966.[98]

A small minority of individuals, however, remained decidedly unimpressed with the politics behind the parade. Once the initial euphoria had faded, assortments of cynical views were also aired in the national newspapers about the Taoiseach's recasting of the Rising. In a sceptical letter published in *The Irish Times* three days after the parade, Reverend David Frazer of the Church of Ireland registered his distaste at the legacy of 1916, remarking that while the government may sense 'a smug satisfaction at pulling off a good show ... not all the people are fooled by the glamour of the event'.[99] Following his various criticisms of the President and 1916 in 'An Irishman's Diary' in *The Irish Times*, Kevin Myers moved to rival newspaper *The Irish Independent* in the second quarter of 2006, during which his denunciations of the story of 1916 continued unabated. In one particularly razor sharp criticism of the bloodshed that stemmed from the Rising, which appeared in print at the end of May, he cited the cases of a range of ambiguous killings of RIC officers in 1916 – which he deemed to have been 'thoroughly and intrinsically evil'. 'From where', he asked, 'did the 1916 insurgents derive the moral authority to kill innocent Irishmen and women in a cause for which they had never even attempted to seek the approval of the electorate?' The Rising's 'actual individual deeds', he argued, 'were morally-answerable actions by individual human beings who were not immune to the Ten Commandments merely because they were self-proclaimed "republicans"'.[100]

At the other end of the spectrum, an assortment of individuals from within the ranks of the Irish nationalist tradition also conveyed a certain amount of reticence about the politics behind the Taoiseach's attempt at reclaiming the 'spirit of 1916' on Easter Sunday. In a letter to *The Sunday Business Post* a week after the parade, for example, an individual who had attended 1916 commemorations organised by the National Graves Association on a yearly basis since the mid-1960s, was in no mood to congratulate Fianna Fáil.

---

[98] J. Sreenan, 'Message from the Chief of Staff', *An Cosantóir: The Defence Forces Magazine*, Vol. 66, No. 3 (2006), p. 5. The General also emphasised how the revival of the military parade offered 'an opportunity to deepen our [the Army's] connection with the people of this country'. 'Today', he added, 'the men and women of Óglaigh na hÉireann must strive to fulfil the lofty and noble ideals of the Volunteers of 1916 and of the peacekeepers who down the years have served in some of the world's most troubled areas with great distinction'. 'So as we march down O'Connell Street this Easter Sunday', he continued, 'we do so in a spirit of remembrance and respect to those who have gone before us ... and in the spirit of spreading and fulfilling the ideals of the Proclamation ... at home and far beyond these shores'.

[99] *The Irish Times*, 19 April 2006.

[100] *The Irish Independent*, 31 May 2006.

'In the past', he charged, '1916 appeared to be an embarrassment to those who have suddenly woken up to the realities of its legacy', so as to obtain 'party benefit'.[101] Writing in a similar vein in the *Cork Independent* in the following month, Fintan Cullen sardonically reflected that there had been a hint of irony to the Easter 2006 proceedings, when compared to the restrained commemoration of the 75th anniversary in 1991: 'Republicanism is hip, thanks to El Berto's burst of born-again nationalism and it's all a far cry from the 75th anniversary of the Easter Rising when the slightest manifestation of republicanism could lead to a stint in Portlaoise [Jail].'[102]

Like those in the world of politics and journalism, the occasion of the 90th anniversary saw Ireland's leading historians air a mixture of opinions about both the Rising's place in past and its commemoration in the present. In an article written for *The Sunday Times* on Easter Sunday, entitled 'A Battle for Truth, Not Historical Myths', Joe Lee observed that even though Ireland had been too small for 1916 'to rank as an event of global significance', its 'ripples' had 'spread far and wide' into various corners of the British Empire such as India. Then, in seeking to dispel what he saw as some harmful legends about 1916, he pronounced that the occasion of the 90th anniversary 'is really for Ireland to consider, and to commemorate'.[103] Writing for *The Sunday Independent* on the same day, Lee's former colleague at University College Cork, John A. Murphy, produced a characteristically clinical look at the meanings enshrined in the Proclamation – which shed some light on the absence of the unionists from the GPO that same day. The Proclamation, he declared, 'is an outdated text, reflecting a political situation now consigned to history, and putting forward historical interpretations that were never sustainable'. Whilst lauding its progressive views on gender equality, he steadfastly declared that its 'most misunderstood and misapplied phrase ... is "cherishing all the children of the nation equally"'. This, he claimed, was a 'reference ... to the nationalist-unionist divide'. Much to his distaste, there had been an 'underlying, and unwarranted, assumption ... that there is only one nation in Ireland'. Nationalist assumptions about the North, he reminded readers, had been 'clouded ... almost down to the present day'.[104]

But much had changed, of course, in North-South relations in the decade leading up to the 90th anniversary. In a book review written for the 'Weekend' supplement of *The Irish Times* over six months after the Dublin parade had taken place, Charles Townshend observed that most people had 'felt a real sense

---

[101] *The Sunday Business Post*, 23 April 2006.

[102] *Cork Independent*, 18 May 2006.

[103] *The Sunday Times*, 16 April 2006.

[104] *The Sunday Independent*, 16 April 2006.

of relief that the legacy of 1916, ambiguous though it may still be, no longer needed to be awkwardly suppressed'.[105] Not that this meant any end to habitual divisions between different camps of historians writing about the Irish past. Roy Foster, for example, took a cautious stance regarding what had transpired on the streets of Dublin. In his book *Luck & the Irish*, he worryingly reflected that the anniversary's celebration had been characterised by 'aggressive government backing', with the cult of 'blood sacrifice and Christian iconography' being 'proclaimed once more with gusto, particularly by President McAleese'. Furthermore, he expressed concern that it had been 'considered safe, and even desirable, to stress cultural difference from Britain and to glow with pride in the liberation struggle of nearly a century before – particularly as the IRA's traditional liberationist strategies were, for the moment, tactfully in abeyance in the North'.[106] Others felt rather differently. In *1916: The Long Revolution*, the collection of essays that emanated from the conference at University College Cork, Gabriel Doherty suggested that the 90th anniversary commemoration had offered proof, if ever needed, that the traditions of Irish republicanism were still 'widely ... cherished' by an 'overwhelming majority of the public'. 'It was the revisionists', he claimed, 'who were found to be guilty of living in the past', relying upon 'stale arguments' and 'incapable of moving with the times'.[107]

Regional Events and Anecdotes

In addition to the official ceremonies held in Dublin, the Rising's 90th anniversary was also marked by the staging of a wide range of events throughout the island of Ireland. While the protests of the 'The Unmanageables' served as a minor distraction to the main event in Dublin, anti-war activists were far more to the fore at the Galway city commemoration, where members of the

---

[105] *The Irish Times*, 2 December 2006.

[106] R. F. Foster, *Luck & the Irish: A Brief History of Change, 1970–2000* (Allen Lane, London, 2007), p. 144.

[107] G. Doherty, 'The Commemoration of the Ninetieth Anniversary of the Easter Rising', in Doherty and Keogh (eds), *1916: The Long Revolution*, p. 407. One group who clearly cherished these traditions was the Cork-based Aubane Historical Society. In June 2006, it published a book that accused revisionists of denigrating the memory of those who had devoted themselves to achieving Irish independence (an argument which evoked memories of Father Brendan Bradshaw's critique of Ruth Dudley Edwards's ideas in 1991). See, for example, J. Herlihy, 'Preface', in B. Clifford and J. Herlihy (eds), *Envoi: Taking Leave of Roy Foster: Reviews of His Made Up Irish Story* (Aubane Historical Society, Millstreet, 2006), p. 7, in which she accuses revisionists of 'subverting memory', thereby reducing 'the nation ... to a cypher which can easily be manipulated to an external agenda'.

Galway Alliance Against War mounted a special protest before a Fianna Fáil commemoration at the Liam Mellows statue in Eyre Square. So as to highlight the group's opposition to the government's policy of permitting US military aircraft to use Shannon Airport, a guest speech was delivered by Abubaker Deghayes from Britain (the brother of a Guantanamo Bay detainee from Libya), who proclaimed that the best way that Irish people could mark the will of 1916 was by helping 'other oppressed peoples'. They could do this, he suggested, by opposing the landing of US warplanes in Shannon. Later on, when a large group of Fianna Fáil supporters marched to the statue of Mellows, the group of activists remained with their banners.[108] In their presence was Rosa Plunkett O'Laoghaire, the great grandniece of the 1916 leader, Joseph Plunkett.[109] The *Galway Advertiser* reported that 'an element of farce crept in' at the day's proceedings, as the activists had draped the statue of Mellows in a large white T-shirt smeared with mock bloodstains and emblazoned with the words 'Mise Iraq'. Nobody appeared willing to remove it for the start of the Fianna Fáil commemoration, until a plucky woman rectified the situation by pulling the contentious garment off the statue of the 1916 rebel.[110]

As tempers abated, a wreath was finally laid at the base of the statue and an oration was delivered by the Minister of State for Justice, Frank Fahey. The 'Galway of today', he said, owed a lot to 'the courage and tenacity of the men of 1916'. The Proclamation was then read by Fianna Fáil's Connaught-Ulster MEP, Seán Ó Neachtáin.[111] According to the *Galway Independent*, however, the celebrations 'almost turned sour', with Councillor John Connolly of Fianna Fáil accusing Galway Alliance Against War members of booing during the ceremony. Speaking to the newspaper afterwards, he said: 'They let themselves down really. There were people there on the day whose parents and grandparents had fought in 1916 and have been attending the commemoration for many years.' In response, a spokesperson for the Galway Alliance Against War denied the accusations and blamed Fianna Fail members for being 'upset that we have expressed our disgust' about the Shannon Airport situation.[112] In County Galway, the Easter Sunday commemorations proved to be far less

---

[108] *The Irish Times*, 17 April 2006.

[109] *The Irish Independent*, 17 April 2006.

[110] *Galway Advertiser*, 20 April 2006.

[111] *The Irish Times*, 17 April 2006.

[112] *Galway Independent*, 19 April 2006. A few weeks later in the *Galway Advertiser*, 4 May 2006, a member of the Socialist Workers' Party also defended the demonstration, describing it as 'most sincere, inclusive ... and poignant', and one that was 'representative of Ireland today, standing together to honour the men and women of 1916 who rose up against occupation, colonisation, war, and empire'.

contentious. A commemorative walk was held in the parish of Castlegar, to
reenact the route travelled by 60 local men who had joined up with Mellows to
partake in the Rising. Afterwards, a lecture was delivered at Castlegar Hurling
Club by local historian William Henry, while the Proclamation was read by
Stephen Casserly, the grandson of an Irish Volunteer from Two Mile Ditch.[113]

Elsewhere in the west of Ireland, members of Fianna Fáil laid a wreath at the
Nally monument in Balla, County Mayo on Easter Sunday. In attendance were
local TD John Carty and the Chairman of the Organising Committee, Noel
Langan. Poetry was recited by Feila McEllin and music was rendered by the Balla
Pipe Band. The Proclamation was read by Padraic Cunnane.[114] A commemoration
was also held nearby in the county town of Castlebar by members of Sinn Féin,
including Councillor Noel Campbell. In Achill Island, Sinn Féin marked the
anniversary by marching behind the Pollagh Pipe Band from Dookinella church
to the Father Sweeney monument, where wreaths were laid and the Proclamation
was read. Speeches were delivered by Mayo Councillor Gerry Murray and MLA
(Member of the Legislative Assembly) for South Down, Catriona Ruane. The
presence of Micheal Ó Seighin, one of the so-called Rossport Five (who had
served time in prison for resisting Shell's plans to bring natural gas ashore in
Mayo), added an environmental slant to the 1916 commemorations. Councillor
Murray linked the Mayoman's battle against Shell to earlier struggles – most
especially the way in which Michael Davitt had struggled for land reform over
a century beforehand. 'The resources of Ireland', said Murray, belong 'to the
people of Ireland'.[115] Other Easter Sunday commemorations in the west and
northwest were held at Drumshanbo Cemetery in Leitrim, at Elphin in County
Roscommon, at the republican plot in Sligo Cemetery, and at Drumboe in
County Donegal.

Republicans in the province of Munster also held a variety of commemorations
on Easter Sunday. In County Limerick, some of these were held at locations linked
to the War of Independence, such as the Seán Wall monument in Bruff and the
republican plot at Killeagh Cemetery in Broadford. Amidst windy conditions,
an intriguing private ceremony watched by only seven others was also conducted
at 12.00 noon at the 1916/War of Independence memorial at Sarsfield Bridge
in Limerick city by Billy McGuire from Cappagh – a grandnephew of Tom

[113] *The Galway City Tribune*, 21 April 2006.
[114] *The Connaught Telegraph*, 19 April 2006.
[115] *The Mayo News*, 19 April 2006. Environmental issues were also raised a letter to *The Irish Times*, 29 April 2006, by Brian Murphy of Glenstal Abbey, who briefly outlined Patrick Pearse's teachings on the environment and made the eco-friendly suggestion that 'it would possibly be a good idea for various interest groups to start planting 16 trees in memory not only of those who died in 1916 but also to show their concern for the environment'.

McGuire (who was involved in declaring the War of Independence at Vaughan's Hotel in Dublin on 21 January 1919). As he had done each year for nearly four decades, McGuire raised a small harp (symbolising the sovereign seal of the first Dáil) and turned it from the location of the rising sun in the east to the sinking sun in the west, thus 'symbolising transition from pagan to Christian to sovereign'. His solemn gesture reenacted the way that his granduncle, during a private session of the first Dáil, had performed the same basic ceremony with a 12-string harp.[116] Elsewhere in Limerick city, crowds assembled at the Munster Tavern on Mulgrave Street for a parade to the republican plot at Mount St. Lawrence Cemetery. In County Kerry, Easter Sunday parades took place to Killavarogue Cemetery in Caherciveen, Rath Cemetery in Tralee and the republican plot in Listowel.

In County Cork, Easter Sunday commemorations were held in towns such as Clonakilty, Youghal, Bandon, and Bantry. In the deep waters of Cork Harbour, yet another quirky commemoration took place when members of the Sovereign Club dived 115 feet to the seabed in order to place a plaque at the site of where the scuttled bow of the Aud lay. This was inscribed in English, Irish and German and dedicated to the memory of 'Roger Casement, Captain Karl Spindler and the crew of the Aud'.[117] A spokesman for the club described the Aud's Captain Spindler as a forgotten hero of the Rising, who in the face of surrender had 'decided to scuttle the ship rather than give the British 20,000 guns'.[118] The 90th anniversary was also marked in Cork city by the holding of a 'Corcaigh 1916' exhibition at Cork Public Museum in Fitzgerald's Park. Among the memorabilia on display were an Irish Volunteer membership form, old photographs and guns.[119] As they had done for years, Sinn Féin supporters in Cork city held an Easter Sunday parade from the National Monument at the Grand Parade to the republican plot at St. Finbarr's Cemetery. In speaking at the commemoration, Martin McGuinness seemed rather annoyed with Fianna Fáil, alleging that 'establishment parties' had been 'attempting to rebrand themselves as republicans and the inheritors of the legacy of 1916'. 'It is wonderful', he complained, 'the effect that that the prospect of losing Dáil seats can have on election-year republicans'.[120]

In the province of Leinster, Easter Sunday commemorations were held at the republican plot at St. Patrick's Cemetery in Dundalk, at Ardbracken in Navan, at St. Ibar's Cemetery in Wexford town, and at the grave at Frank

---

[116] *Limerick Leader*, 22 April 2006.

[117] *The Irish Times*, 17 April 2006.

[118] *The Irish Independent*, 17 April 2006.

[119] *Cork Independent*, 13 April 2006.

[120] *The Irish Times*, 17 April 2006.

Drivers in Ballmore Eustace in County Kildare. The following day, a wreath-laying ceremony took place at the 1798 monument in the town square at Portarlington in County Laois. In Northern Ireland, Gerry Adams spoke at a Sinn Féin commemoration in Belfast the day beforehand. During the event, he put his own introspective spin on the imminent Dublin spectacle, calling for a national coalition – regardless of political attachment – to bring about Irish unity. 'The Taoiseach', he said, 'has called for a return to the core values of Irish republicanism. I welcome that call. That is what the Easter commemoration is about. So, I urge the Taoiseach to follow through on the logic of what he has said'.[121] Elsewhere in the North, a range of commemorations took place on Easter Sunday at the republican plot in Belfast's Milltown Cemetery, at Sandyhill Cemetery in Armagh, St. Colman's Cemetery in Lurgan, at the City Cemetery in Derry, St. Mary's Cemetery in Newry, Roslea in County Fermanagh, and Edendork Cemetery in east Tyrone.

In addition to outdoor ceremonies, the 90th anniversary of 1916 was also marked throughout the provinces by a range of indoor events before and after Easter Sunday. Besides the academic conference at University College Cork in January, another anniversary event was held at the Town Hall Theatre in Galway city on 20 February. This involved the screening of *Mise Éire*, followed afterwards by a scholarly appraisal by the NUI Galway historian, Gearóid Ó Tuathaigh. Three days later in Dublin, Trinity College Dublin's Philosophical Society hosted a debate on 1916, featuring Fianna Fáil's Noel Treacy and Sinn Féin's Michelle Gildernew speaking in favour of the Rising, and journalist Eoghan Harris and Chris Hudson making a case against it. The debate, according to Harris, 'was full of fire and brimstone before a packed house'.[122] The college's History Society also played host to a two-day conference on the theme of 'The 1916 Rising: Then and Now'. This took place on 21–22 April and was held in conjunction with the Ireland Institute.

A few weeks later, a public seminar on the theme of '1916: Local Dimensions' was held in NUI Galway from 19–20 May. This was held under the auspices of the university's History Department and the Centre for Human Settlement and Historical Change. Its chief organiser was the labour historian, John Cunningham, who explained that 'in order to understand 1916, it is necessary to take into account the various regional mobilisations, and the impact of the Easter Week events on communities throughout Ireland'.[123] The seminar was officially opened by Ó Tuathaigh, while a keynote speech on the opening night

---

[121] *The Irish Independent*, 17 April 2006. For an extended treatment of Adams's thinking at this time, see G. Adams, *An Irish Eye* (Brandon, Dingle, 2007), pp. 187–93.

[122] *The Sunday Independent*, 26 February 2006.

[123] *Galway Advertiser*, 18 May 2006.

was given by Charles Townshend. On the second night, Seosaimh Ó Cuaig introduced the documentary film, *Taibhsi na Staire*, which gave an overview of Patrick Pearse's links to Rosmuc in Connemara. To coincide with the seminar, a special 1916 exhibition was also held in the foyer of the university's James Hardiman Library. One of the main exhibits on display was the restored Sun motorcycle that Mellows had used to organise the Irish Volunteers in Galway in 1916. This was loaned to the university by Sergeant Brian Smyth, the Museum Curator at the Dún Uí Mhaoilíosa Barracks.[124] Display panels with texts and illustrations about Galway's role in 1916 also formed part of the exhibition, as did a range of old and rare books which were put on show in glass display cases.

Rebel Irish Heritages

Although most heritage, according to Brian Graham and Peter Howard, has little inherent worth, 'it is meaning that gives value, either cultural or financial, to heritage and explains why certain artefacts, traditions and memories have been selected from the near infinity of the past'.[125] Over the course of time, the story of 1916 has been invested with significant meaning. Notwithstanding the firing of occasional salvos at its cultural legacy and historic worthiness, memorabilia associated with the Rising remained in very high demand in the lead-up to the 90th anniversary. As Easter Sunday approached, the financial value attached to various heritage objects associated with Easter Week rocketed dramatically. Although this rampant price inflation was spurred on no doubt both by the enhanced nostalgia brought about by the widening passage of time and the onset of a post-revisionist paradigm in Irish historiography (which made the Rising more acceptable again), it was also symptomatic of the high levels of consumer spending and unbridled financial speculation that characterised the heyday of the Celtic Tiger era. Unsurprisingly, as had been the case with many previous commemorations of 1916, Irish auction houses offered a range of rare memorabilia from the revolutionary years for sale to the public in the days leading up to Easter Sunday. To the annoyance and disappointment of certain citizens, however, a range of old manuscripts fell into the hands of private collectors, who time and again outbid representatives of state institutions such as the National Library and the National Museum.

In Dublin city, Adam's Auctioneers ran a special 'Independence' auction on 12 April. The event, however, did not pass off without controversy as members

---

[124] *The Galway City Tribune*, 19 May 2006.

[125] B. Graham and P. Howard, 'Heritage and Identity', in B. Graham and P. Howard (eds), *The Ashgate Research Companion to Heritage and Identity* (Ashgate, Aldershot, 2008), p. 2.

of Ógra Sinn Féin – the party's youth wing – attempted to interrupt the proceedings by bursting into the auctioneer's premises during the sale of the lots. After having been eventually expelled, the demonstrators continued to picket the sale from outside on the street. According to a report in *The Sunday Times*, an attendee at the auction alleged that elderly republicans 'were insulted ... by people yelling ... over their shoulders, shouting about things they know nothing about'.[126] In spite of the hullabaloo, the Adam's auction still went ahead as scheduled and vast prices were obtained on several items of historic note. One of the prime lots, the first draft of the National Anthem, sold for €760,000. To the delight of many, some 300 letters exchanged between Thomas Clarke and his wife Kathleen from 1899–1915 were snapped up by the National Library of Ireland for €90,000 – an acquisition which its Director, Aongus Ó hAonghusa, described as 'an investment for scholars'. However, Clarke's handwritten farewell note to his wife prior to his execution was bought by an anonymous buyer for €75,000. Three days after Easter Sunday, it was reported in *The Connaught Telegraph* that the auction total for the sale of 460 out of 482 lots had fetched a staggering €3.5 million.[127]

Amongst the other items that sold at the Adam's auction were a handful of the 2,000 or so medals that were previously presented to those who had fought in the Rising. One posthumously-awarded 1916 medal fetched a price of €105,000, while sums between €3,200 and €14,000 were obtained for other 1916 and War of Independence medals. Although it came under pressure from certain quarters to intervene in the heritage market and acquire certain artefacts for the state, the government was unable to quell the inflationary bubble in prices being paid for the purchase of historical artefacts from the revolutionary era at the time of the 90th anniversary. However, in instances where families whose ancestors had fought in the revolutionary era came looking for replacements for medals that had gone astray, the government did operate an official policy. This was confirmed by the Minister for Defence in a Seanad debate on 17 May. For many years, according to O'Dea, it had been the policy of the Department of Defence not to issue replacement medals to those families of veterans who were looking for them, given both their 'intrinsic value' and 'their monetary value on the open market'. When asked if he had any intentions of altering this policy, the Minister said that he had not, but did confirm that he had initiated a departmental examination 'of the possibility of issuing some form of official certificate for such cases'.[128]

[126] *The Sunday Times*, 16 April 2006.
[127] *The Connaught Telegraph*, 19 April 2006.
[128] *Seanad Debates*, Vol. 183, 17 May 2006.

Not everyone seemed intent on profiteering from the auctioning off of 1916-related heritage objects. In a notable gesture of generosity, a rare original copy of the Proclamation (with fold marks from 90 years beforehand still evident) was presented to the National Museum on 7 March by Joseph McCrossan, whose grandmother had obtained it from O'Connell Street shortly after the Rising had started. Explaining his generosity, McCrossan stated: 'We [his family] believed that it was important that the Proclamation should stay in Ireland and that it be in public ownership.'[129] The same philanthropic spirit was evident amongst the family of the late Jackie Clarke from Ballina in County Mayo, who donated his entire personal library – containing 200 boxes and numerous books and documents (including another original copy of the Proclamation) – to Mayo County Council in April. When asked to describe his feelings about the collection, the Cathaoirleach of Mayo County Council, Fine Gael's Councillor Henry Kenny, remarked that his heart had burned 'within him with joy'.[130] The government itself made its own benevolent gesture when O'Dea announced in May that two surviving veterans of the War of Independence (104-year-old Seán Clancy from Dublin and 103-year-old Dan O'Donovan from Tralee) and 750 spouses of veterans were to obtain a 50% boost to their pensions, backdated to 1 April. Explaining the decision to award the rise, which worked out at an extra €120 per month, O'Dea stated that it 'was a timely opportunity to recognise the tremendous efforts and sacrifices they made for our country ... [and] to show our gratitude on this anniversary year'.[131]

Unlike the year of the Golden Jubilee, which saw the release of a plethora of official (and unofficial) commemorative souvenirs, only a very limited amount of such items were commissioned by the government for the 90th anniversary. One such souvenir was a special 48 cents postage stamp released by An Post. Designed by Ger Garland, it featured a contemporary photograph of the refurbished GPO by Donal Murphy, emblazoned with a white logo reading '1916–2006'. Upon launching the stamp, the Taoiseach said that the building's rejuvenation had represented 'a reflection of the renewal that has taken place in this country in the intervening years'.[132] The market for the retailing of 1916-related stories was also tapped into by commercial outlets such as the Trinity College Dublin-based company, Eneclann Ltd., which released a CD entitled *Personalities Files*. Costing €49.90, this contained copies of around 19,000 folio pages of Dublin Castle's Royal Irish Constabulary Special Branch intelligence reports

---

[129] *The Irish Times*, 8 March 2006.
[130] *The Mayo News*, 19 April 2006.
[131] *The Irish Times*, 15 May 2006.
[132] *Irish Examiner*, 17 April 2006.

on republican suspects from 1899–1921, including surveillance information on many of the leading figures involved in the Rising.[133]

Academic historians also offered up a range of newly-published works to neatly coincide with the 90th anniversary. One of these was Keith Jeffery's *The GPO and the Rising*, which furnished a history of 1916 from the perspective of those who were working in the GPO itself on Easter Monday. Drawing heavily from sources in the British Post Office Archives, the book illuminated the story of 'the bystanders, the uncommitted, the ordinary workers in the GPO, who were suddenly caught up in the dramatic events of Easter Week'. It was 'appropriate', wrote Jeffries, 'that stories of "soldiering on" or "derring do" during the Rising ... be told so that the full history of the Rising may be recorded'.[134] As with previous anniversaries, the lives of some of the 1916 leaders continued to preoccupy the minds of academics. In *Unstoppable Brilliance: Irish Geniuses and Asperger's Syndrome*, psychiatrists Antoinette Walker and Michael Fitzgerald produced new and intriguing insights into Pearse's life. They claimed that he suffered from Asperger's syndrome – a condition in which 'some areas of the brain are hyperdeveloped and others are underdeveloped', thus bringing 'both blessings and burdens'. Consequently, they noted that Pearse remained an iconic, elusive and eccentric figure, whose actions were full of contradictions. While recognising that he was a gifted educationalist and orator who expressed persuasive views on the Irish language, education and nationalism, they concluded that his historical legacy remained mixed: 'one man's revolutionary hero is another man's murderer'.[135]

The newspapers were also to the fore in marking the 90th anniversary. *The Irish Times* produced a special 16-page commemorative broadsheet supplement, in association with the Department of Education and Science. This was published on 28 March and despite adding an extra 10 cent to the price of that day's paper, sales went well. With the intention of informing existing debates about the Rising, the supplement featured a day-by-day recollection of the course of the events of Easter Week (based on its own substantial archive and Bureau of Military History Witness Statements). It commenced with an introductory article by Fintan O'Toole, entitled 'Witnesses to History'. Besides aiming to bring readers more closer to the human reality of the daily events of the

---

[133] *The Sunday Independent*, 2 April 2006; *The Irish Times*, 7 April 2006. For an insight into the content of the reports (held at the National Archives, Kew), see F. McGarry, 'Keeping an Eye on the Usual Suspects: Dublin Castle's "Personalities Files", 1899–1921', *History Ireland*, Vol. 14, No. 6 (2006), pp. 44–49.

[134] K. Jeffery, *The GPO and the Easter Rising* (Irish Academic Press, Dublin, 2006), p. 2.

[135] A. Walker and M. Fitzgerald, *Unstoppable Brilliance: Irish Geniuses and Asperger's Syndrome* (Liberties Press, Dublin, 2006), pp. 12, 56–57.

Rising and concurrently seeking to distance them from debates over the rights and wrongs of 1916, O'Toole noted that the supplement's aim was 'to return to the people who took part in or witnessed this event'. Other commentaries in the supplement were furnished by Stephen Collins (on 'How do Political Parties Lay Claim to the 1916 Legacy?') and Joe Carroll (on 'How *The Irish Times* Covered this "Desperate Episode"' and 'The Beginning of the End', the latter of which looked at the Rising's subsequent impact upon the fortunes of the British Empire). In another article in the supplement, entitled 'Bringing the Rising into the Classroom', secondary school teacher Peter Brennan offered practical suggestions as to how teachers could engage in active learning by organising student projects on 1916. A glossy A2-sized pull-out poster, designed by Francis Bradley, was also included. This featured a street map of Dublin in 1916, surrounded by an array of black and white images and text (in English on one side and Irish on the other).[136]

Months later, a much enlarged version of the supplement was published in hardback by Shane Hegarty and Fintan O'Toole, entitled *The Irish Times Book of the 1916 Rising*.[137] In the lead up to Easter Saturday, *The Irish Times* also ran a special six-part series, '1916: 90 Years Later', which presented the views of distinguished politicians for whom the Rising was part of family lore, along with the opinions of historians and journalists on the legacies and contemporary meanings of the story of 1916. Diarmuid Ferriter opened the series with an article entitled 'Aspirations of Rebels were Never Fulfilled'. This was followed by articles written by Michael McDowell (on 'Reconciliation: The Unfinished Business of the Rising'), Garret Fitzgerald (on 'Rising and Early Independence Brought Prosperity'), Fianna Fáil MEP Eoin Ryan (on 'The Rising was Phase in Ongoing Fight for Freedom'), Sinead McCoole (on 'Women were Among the Chief Sufferers in the Ensuing Conflict'), and Paul Bew (on 'Rising was a Catholic Revolt Against Redmondite Elite').[138]

The *Irish Examiner* also marked the 90th anniversary by running its own seven-part series of investigative articles about the Rising, which concluded on Easter Monday with a piece entitled 'Rebel Cork's Waiting War', written by Gerry White and Brendan O'Shea.[139] On Easter Saturday, *The Irish Independent*

---

[136] See 'The 1916 Rising: A Special Supplement to Mark the Rising's 90th Anniversary in Association with the Department of Education and Science', in *The Irish Times*, 28 March 2006.

[137] S. Hegarty and F. O'Toole, *The Irish Times Book of the 1916 Rising* (Gill and Macmillan, Dublin, 2006).

[138] See *The Irish Times*, 10 April 2006, 11 April 2006, 12 April 2006, 13 April 2006, 14 April 2006, and 15 April 2006.

[139] See *Irish Examiner*, 10 April 2006, 11 April 2006, 12 April 2006, 13 April 2006, 14 April 2006, 15 April 2006, and 17 April 2006. For an expanded treatment of some of the events covered

marked the anniversary by devoting over a third of its tabloid-sized 'Weekend Review' supplement to the theme of '90 Years On'. This featured contributions from the likes of Tim Pat Coogan, Conor Cruise O'Brien, Ruth Dudley Edwards, and Bruce Arnold.[140] The Easter Sunday issue of the Irish language newspaper, *Fionse*, also marked the anniversary with a number of articles devoted to the legacy of the Rising. Áine Ní Chiaráin, for example, looked back on the Misneach hunger strike of 1966, by drawing upon the recollections of Cian Ó hÉigeartaigh, Séamus Ó Tuathail and Eoin Ó Murchú.[141] Many local newspapers also revisited the events of 1916, sometimes bringing new evidence to light. In the *Clare Champion*, for example, Peadar McNamara offered a well-illustrated overview of the role played by people from County Clare in the Rising, and published a small selection of the 45 Witness Statements that Claremen had given to the Bureau of Military History.[142] In the *Western People*, James Laffey furnished an account of the newspaper's original reaction to the Rising and the impact that this subsequently had on public opinion in the west.[143]

New forms of commemoration also became more noticeable at the time of the 90th anniversary. As Roy Foster has noted, the shaping of history to instant goals in the 2000s took 'new and vivid forms, linked to the advances of media technology'.[144] Not surprisingly, there was no shortage of websites devoted to the Rising's history and legacy. Some of the state's principal cultural institutions also engaged with virtual forms of commemoration. The National Library, for example, launched an online exhibition, called 'The 1916 Rising: Personalities and Perspectives'. Featuring over 500 images connected to the fight for Irish independence, this study resource drew almost entirely upon its vast holdings of manuscripts, newspapers, photographs, drawings, and books.[145] More traditional forms of historical exhibition also featured in the programme of events staged by the main institutions. As already seen, the Taoiseach launched an exhibition at the National Museum Collins Barracks on 9 April (entitled 'Understanding 1916: The Easter Rising'). This featured uniforms of the Irish Volunteers and an original copy of the Proclamation.[146] Notwithstanding the

---

in the final article in the series, see G. White and B. O'Shea, *'Baptised in Blood': The Formation of the Cork Brigade of the Irish Volunteers 1913–1916* (Mercier Press, Cork, 2005).

[140] See the 'Weekend Review' supplement, in *The Irish Independent*, 15 April 2006.

[141] *Foinse*, 16 April 2006.

[142] *Clare Champion*, 14 April 2006.

[143] *Western People*, 25 April 2006.

[144] Foster, *Luck & the Irish*, p. 162.

[145] http://www.nli.ie/1916/ (accessed on 17 July 2008 and 25 October 2011).

[146] For further details on the 1916-related artefacts, documents and photographs held by the National Museum of Ireland, see M. Kenny, *The Road to Freedom: Photographs and Memorabilia*

political controversy generated by Ahern's reference to the 'four cornerstones' of modern Irish history, the speech was noteworthy in another regard, for he signalled the government's intention to facilitate more detailed scholarly research in future years on the nation's pathway towards independence. In this regard, he announced his intention to establish a working group, chaired by the Department of the Taoiseach, to look into the feasibility of making the Military Service Pensions Collection available to all researchers 'in good time' before the 2016 centenary.[147] This announcement went down very well with military historians in search of new data, as the collection included the records of around 17,000 successful applications for pensions by persons and/or their dependants, for service rendered between April 1916 and September 1923.

As there was no repeat during 2006 of an official event like *Aiséirí-réim na Cásca* (the pageant staged at Croke Park for the Golden Jubilee), it was left to others in the world of the creative and performing arts to tap into the upsurge in Irish nationalist nostalgia. Not least among them was the renowned Irish-American dancer, Michael Flatley. Having already achieved fame through his starring role in shows such as *Riverdance* and *Lord of the Dance*, he staged an emotional comeback with his *Celtic Tiger* dance show at Dublin's Point Depot, four nights after the Easter Sunday commemoration. Various scenes from Irish history – including evictions, burning cottages, redcoats, and the GPO in 1916 – featured significantly in the first half of the unrestrained production (which charged €85 for tickets). The extravaganza featured performances of rebel songs like 'The Wearing of the Green' and 'A Nation Once Again' – the former of which was played during a street scene in front of the GPO. The whole spectacle was described by *The Irish Independent's* Miriam Lord as 'a history of Ireland by way of dancing girls, pub republicanism and Las Vegas. Think the Wolfe Tones meet Liberace'.[148] Writing for *The Dublin Review*, Ann Marie Hourihan also seemed bewildered by all the razzmatazz of it all, most especially by the central role of Flatley himself, who was dressed up at the start as a Celtic warrior

---

from the 1916 Rising and Afterwards (Country House/The National Museum of Ireland, Dublin, 2001), especially pp. 5–6, 17–30, 32–40. A discussion of the challenges facing museums at the outset of the twenty-first century can be found in R. Vaughan, 'Images of a Museum', *Museum Management and Curatorship*, Vol. 19, No. 3 (2001), pp. 253–68.

[147] *The Irish Times*, 10 April 2006. A further boost to the work of historians came in the same year, when the government made a commitment to upscale research across the humanities, an official policy 'founded in a belief in the intrinsic value of scholarship, both to a democratic society and to the effective functioning of ... communities of knowledge and discourse'. See Government of Ireland, *Strategy for Science, Technology and Innovation 2006–2013* (The Stationery Office, Dublin, 2006), p. 30.

[148] *The Irish Independent*, 21 April 2006.

in leather breastplates and later on in high-waisted trousers and a tweed cap: 'To him, Irish history is a children's story, just like *The Wizard of Oz*. In a daring move, Flatley, who is pushing fifty, inserts himself within it, as Doherty.'[149]

Besides the official visit by the Taoiseach, Kilmainham Gaol played host to a number of artistic events to commemorate the 90th anniversary. One of these was an exhibition of digital photography by the visual artist, Elpida Hadzi-Vasileva, entitled 'Time Stands Still'. This ran from 9 February to 12 March and included contrasting imagery of war and memory from both Macedonia and Ireland. Employing a combination of 'digital photography and image manipulation', the exhibition partly explored the Rising's history by incorporating images of Kilmainham's architecture 'to investigate ... notions of identity, nationalism and cultural development'.[150] A week after Easter Monday, a dramatic production called *Operation Easter* opened for business in Kilmainham Gaol. Written by Donal O'Kelly and produced by the Calypso Theatre Company, the play probed the human intricacies of what theatre critic Sara Keating called a 'middle ground' approach to 'the political and ideological inheritance' of 1916. The central character of the play was Elizabeth O'Farrell (played by Mary Murray), the nurse who played a significant role in the Rising and had a public park along Dublin city's south quays named after her. The play also placed emphasis on four of the Rising's leaders, providing insights into why they had fought and eventually surrendered. Explaining the genesis of the production to Keating, writer O'Kelly stated: 'We're stuck in cloud-cuckoo-land pretending it [1916] didn't happen ... But three generations have passed and it's time we dealt with it ... It's taken 90 years.' By imaginatively comparing contemporary debate to a kind of therapy, O'Kelly further noted: 'It's like we've only just discovered a tumour and we're trying to excise it, and the only way that we can get over it as a nation is by talking it through.'[151]

The Rising's 90th anniversary also witnessed a flurry of creative activity from both the national broadcaster and independent filmmakers. Television viewers got a taste of some of the issues raised by the President's speech at University College Cork when RTÉ's *Primetime* programme, produced by Angela Daly and hosted by Mark Little, broadcast a live debate on the Rising on 9 February. This featured a vivacious discussion between the four guests, namely Diarmuid Ferriter, Martin Mansergh, Caoimhghín Ó Caoláin, and Eoghan Harris. Five days before Easter Sunday, RTÉ's Hidden History series revisited the issue of General Maxwell's draconian treatment of the 1916 leaders, in a programme

---

[149] A. M. Hourihan, 'A Series of Accidents', *The Dublin Review*, No. 23 (2006), p. 102.

[150] Typescript Entitled 'Time Stands Still: Elpida Hadzi-Vasileva Exhibition', 2006, Kilmainham Gaol Archives, 22MS-1G12-18.

[151] *The Irish Times*, 20 April 2006.

entitled *The Man Who Lost Ireland*. This combined original footage with contemporary reconstructions, interspersed with commentary from an array of talking heads, including Ferriter.[152] In a review for *History Ireland*, Eamon O'Flaherty commended the documentary for its 'quality', its focus on 'historical context' and its 'engagement with many different perspectives'. He also observed that by focusing on Maxwell, it had explained 'much about the larger context of the failure of Britain's Irish policy at a crucial juncture in the history of both countries'. 'The public appetite for analyses of the Easter Rising', he added, 'seems insatiable'.[153]

Other episodes from the revolutionary years also loomed large during 2006, as filmmakers tapped more and more into the prevailing nostalgia. The most notable success in this regard was *The Wind that Shakes the Barley*. This film, which was directed by Ken Loache from England, proved to be a hit at the Irish box office (having already won the prestigious Palme d'Or award at the Cannes Film Festival the year beforehand). Starring Cilian Murphy (of *Batman Begins* fame), it depicted the harrowing story of two brothers caught up in the War of Independence and Civil War in West Cork.[154] In addition to Loache's movie, a number of old films depicting the struggle for independence were also released on DVD for the first time ever in 2006. *Irish Destiny*, a silent film about the War of Independence which was first released in 1926, was reissued by the Irish Film Institute and RTÉ. Old films depicting the story of the Rising also resurfaced. The Irish language body, Gael Linn, rereleased *Mise Éire* from 1959 and *An Tine Bheo* from 1966 (followed a year later by the War of Independence drama, *Saoirse?*, from 1961). Although the time when Irish rebel songs topped the music charts had long since transpired, the 90th anniversary did result in an increase in market demand for samples of the genre. To cater for this, Dolphin Records released a special DVD of performances of rebel songs and poems, entitled *The 1916 Easter Rising: The 90th Anniversary Commemorative Collection* (featuring, for example, Ronnie Drew reciting Pearse's 'The Rebel'

---

[152] *The Man Who Lost Ireland* (RTÉ Hidden History, 2006).

[153] E. O'Flaherty, 'TV Eye: *The Man Who Lost Ireland*, RTÉ 1, 18 April 2006', *History Ireland*, Vol. 14, No. 3 (2006), p. 50.

[154] The film generated considerable historiographical debate about the complexities of the revolutionary period. See, for example, D. Carroll (ed.), *The Wind that Shakes the Barley* (Galley Head Press, Cork, 2006); R. Foster, 'The Red and the Green', *The Dublin Review*, No. 24 (2006), pp. 43–51; B. Hanley, 'Film Eye: *The Wind that Shakes the Barley*', *History Ireland*, Vol. 14, No. 5 (2006), pp. 50–51; and B. P. Murphy, 'The Wind that Shakes the Barley: Reflections on the Writing of Irish History in the Period of the Easter Rising and the Irish War of Independence', in R. O'Donnell (ed.), *The Impact of the 1916 Rising: Among the Nations* (Irish Academic Press, Dublin, 2008), pp. 200–220.

and Frances Black singing Canon O'Neill's 'The Foggy Dew').[155] Not all commercial ventures met with success. For example, rumours concerning the release of a new film about James Connolly's life never came to fruition.[156]

Despite the nationalist resurgence, new monuments commemorating the Rising were in short supply throughout 2006. The most significant of these was unveiled under pouring rain by the Minister for Defence at the Curragh Camp in County Kildare on 6 April, in the presence of General Sreenan. The monument consisted of a large free-standing stone plaque measuring around 13 feet in height, dedicated to the memory of the seven signatories of the Proclamation. Both Irish and English translations of the signatories' surnames were inscribed on the monument. The location of the memorial was chosen because it was where seven barracks were taken over by the Defence Forces from the British Army in 1922. These were subsequently renamed in honour of the seven signatories of the Proclamation in 1929.[157] Five months after the unveiling, another addition to the cultural landscape appeared in the grounds of the Rotunda Hospital in Dublin, when Lord Mayor of Dublin Vincent Jackson unveiled a plaque in memory of those captured at the hospital during the Rising. Measuring about one square metre, it bore the inscription: 'Seachtain na Cásca 1916: Easter Week 1916.' A small crowd gathered in the Autumn sunshine to watch the ceremony, which was held on 7 September. They included the hospital's Master, Dr. Michael Geary and Seán Bermingham of the National Graves Association.[158]

Although the newly-erected monuments did little to capture the public's imagination, its attention was certainly drawn to the continuous media coverage throughout 2006 (and the year beforehand) of the high-profile case of 'Save 16 Moore Street'. In the years leading up to Easter 2006, heritage campaigners (including An Taisce, the National Graves Association, the Academy for Heritage, Councillor Dermot Lacey, and Tim Pat Coogan) had fought strenuously to preserve and conserve Number 16 Moore Street –

---

[155] See *An Tine Bheo* (Gael Linn, 1966; Reissued 2006); *Irish Destiny* (Eppels Films Ltd., 1926; Reissued by The Irish Film Institute and RTÉ, 2006); *Mise Éire* (Gael Linn, 1959; Reissued 2006); *Saoirse?* (Gael Linn, 1961; Reissued 2007); *The 1916 Easter Rising. The 90th Anniversary Commemorative Collection* (Dolphin Records Ltd., 2006); and *The Wind That Shakes the Barley* (Sixteen Films, 2006).

[156] It was reported in *The Irish Times*, 13 May 2006, that Rascal Films were planning to commence shooting a film called *Connolly*, with financial backing from SIPTU. Previously, *The Sunday Times*, 7 November 2004, reported that plans were afoot to raise €15 million to finance a major biopic called *Nora's Journey*.

[157] *The Irish Times*, 7 April 2006.

[158] Ibid., 8 September 2006.

a Georgian era terraced townhouse (built in 1783) in Dublin city centre, where the 1916 leaders were believed to have signed their surrender (Plate 6.3).[159] Even though the building had fallen into a derelict condition, many Dubliners were still familiar with its historic significance at the time of the 90th anniversary, as denoted by a wall plaque bearing the words 'Éirí Amach na Cásca 1916'. As part of a statutory consultation process, outlined in Section 54 of the Planning and Development Act, 2000, Dublin City Council received many submissions in 2006 in favour of conserving the Moore Street structure (including one from the American branch of the Ancient Order of Hibernians).

But arising from background legal wrangling over who actually owned Number 16, a submission against the proposal to protect the site was submitted by the property developer, Joe O'Reilly (of Chartered Land) – one of the individuals claiming beneficial ownership of Number 16. In his submission, reference was made to research by a military historian, who determined that the evidence for Number 16 being the surrender headquarters was both ambiguous and incomplete. The developer's submission also argued that commemorating the surrender 'might be thought excessively morbid or "martyrological"'. Instead of protecting the townhouse, the submission put forward a suggestion that the local authority could retrospectively designate 1916 Rising ideals to The Spire on nearby O'Connell Street.[160] But in a letter to *The Irish Times* six days after Easter Sunday, James Connolly's descendant, James Connolly Heron, went on record to register his concern about Number 16's 'uncertain future'.

---

[159] What happened is worth briefly recounting. During the height of the property boom, a cloud was cast on the building's continued existence by regeneration plans for a large area of urban space between Moore Street and O'Connell Street. Upon hearing of the plans, the battle to save the surrender headquarters commenced. As reported by ibid., 10 August 2005, a future museum-like role for the building was first hinted at in Dublin City Council's *Development Plan*, which made it a future objective of the local authority to preserve Number 16 'as a commemorative centre marking the events of 1916'. However, uncertainty arose over whether Number 16 was in fact the surrender headquarters. According to *The Irish Times*, 5 December 2005, a report by consultants determined that the buildings beside Number 16 formed part of a 'distinctive streetscape'. Consequently, as reported by ibid., 6 December 2005, councillors on Dublin City Council then voted unanimously under their reserve function to set in motion procedures to put Number 16 on the Record of Protected Structures, along with Numbers 14, 15 and 17.

[160] *The Irish Times*, 19 April 2006. For details on the legislation concerning the record of protected structures, see Government of Ireland, *Planning and Development Act, 2000* (The Stationery Office, Dublin, 2000), p. 87. This stipulates that additions could be made if the planning authority deemed that 'the addition is necessary or desirable in order to protect a structure, or part of a structure, of special architectural, historical, archaeological, artistic, cultural, scientific, social, or technical interest'.

Plate 6.3   Numbers 14–17 Moore Street, Dublin
Source: Author.

Clearly incensed, he described the decrepit state of the building and its surroundings as 'a national disgrace'. He also went so far as to propose that all of Moore Street should be preserved, in light of the 'dreadful dilemma of historical uncertainty' raised by the military historian's findings.[161] In the months following the military parade, the Moore Street saga then took a new twist, as campaigners began lobbying the government to confer National Monument status on Number 16.

Remembrance, Citizenship and Electioneering

Notwithstanding some lingering ambiguities about the violence that was intrinsic to the story of Easter Week, the occasion of the 90th anniversary appeared to win overwhelming approval from the Irish public, falling as it did at a time in which Ireland was finally at peace. Many citizens appeared to be very

---

161   *The Irish Times*, 22 April 2006.

much at ease with the use of military heritage as a means of fostering collective patriotism and pride about the accomplishments of the revolutionary period. One of the attendees at the GPO on Easter Sunday was the economist and blogger, Gerard O'Neill. In a book exploring the relevance of the Proclamation for the twenty-first century, he recalled his experience of travelling to O'Connell Street for both 'the spectacle' and 'the historical significance of the occasion'. 'It isn't often these days', he wrote, 'that one gets to feel patriotic, outside of sporting events'. 'And yet it was patriotism', he added, 'that I, and no doubt others, felt that day – a sense of pride born out of seeing my country and my fellow citizens celebrating our past and looking forward optimistically to the future'.[162] Despite its militaristic trappings, the emphasis that the parade placed on the Army's peacekeeping accomplishments served to deflect much attention away from the heated public debates that had followed in the wake of the speeches delivered by the President at University College Cork and the Taoiseach at Collins Barracks. Indeed for many of the people who turned out in the capital city on Easter Sunday, the commemoration had a certain feelgood factor to it and seemed more akin to a family-friendly St. Patrick's Day parade.

One thing that was especially noticeable by Easter 2006 was the radically changed ethnic composition of Irish society. By this stage, Western society in general had become 'more self-consciously socially and culturally diverse' – a phenomenon that had implications for 'heritage selection, interpretation and management'.[163] This fragmentation was especially noticeable in the Republic of Ireland, whose increasingly heterogeneous population represented a radical departure from the situation that had prevailed during the Golden Jubilee 40 years beforehand. Results of the 2006 Census (the 15th since the state's foundation), taken exactly a week after the Easter Sunday parade, revealed that the Republic's population enumerated on 23 April was 4,239,848 persons (with around 61% living in urban areas). This compared with a total of just 2,884,002 persons in 1966, representing an increase of 47%. The 2006 figure also represented an impressive 20.25% increase on the 3,525,719 persons recorded in 1991 – caused by both natural increase and considerable net inward migration.[164] The large-scale immigration, according to Piaras Mac Éinrí and Allen White, was 'simultaneously new and exceptional', and reflected Ireland's 'modern reality

[162] G. O'Neill, *2016: A New Proclamation for a New Generation* (Mercier Press, Cork, 2010), pp. 7–8.

[163] G. J. Ashworth, B. Graham and J. E. Tunbridge, *Pluralising Pasts: Heritage, Identity and Place in Multicultural Societies* (Pluto Press, London, 2007), pp. 2–3.

[164] Central Statistics Office, *Census 2006: Principal Demographic Results* (The Stationery Office, Dublin, 2007), pp. 11–12, 14, 37.

... as a multi-ethnic state'.[165] In many respects, the strong performance of the Celtic Tiger economy was a vital ingredient of the buoyant nature of the 90th anniversary. The boom in the construction sector resulted in thousands of new job opportunities, many of which were filled by non-nationals. In total, it was estimated that the amount of new jobs created in the year leading up to May 2006 was a staggering 90,000, while the GDP growth rate for the year as a whole measured 4.8%.[166] The continued growth of the financial and hi-tech sectors also contributed to the strong economic performance, as did the climate of industrial harmony that was forged under various social partnership agreements.

Remarkably, it was revealed that around 420,000 of the population usually resident in the 2006 Census were non-Irish nationals and that EU accession state migrants were the fastest-growing within this group, with the numbers of Poles exceeding 63,000 and Lithuanians 24,000.[167] Between the years 2002–2006, the number of non-Irish nationals in the total population had risen sharply from 5.8% to 10%, which worked out at a significant net immigration flow of 47,800 on an average annual basis. A wider factor behind the dramatic increase was the accession of 10 new member states to the EU in May 2004.[168] In addition to those who had moved from mainland Europe to Ireland for jobs, the state's population at the time of the 90th anniversary included a fair amount of asylum seekers, especially from the African continent. United Nations figures showed that Ireland ranked tenth overall for asylum applications among developed countries during the decade 1995–2005.[169] After reviewing the ethnic make-up of the country in the 2006 Census figures, Marc Coleman highlighted the fact that within a very short space of time, the Republic had managed to create 'a new home for the Diaspora of other lands ... most of them with no ancestral ties to Ireland'.[170] Or as Des Geraghty observed, the arrival of those 'carried on the tide of necessity' and 'seeking a better life' had juxtaposed 'an impressive array of other colours' upon the so-called 'Green Isle'.[171]

[165] P. MacÉinrí and A. White, 'Immigration into the Republic of Ireland: A Bibliography of Recent Research', *Irish Geography*, Vol. 41, No. 2 (2008), pp. 161, 164.

[166] N. Whelan, *Showtime or Substance? A Voter's Guide to the 2007 Election* (New Island, Dublin, 2007), pp. 58–59.

[167] Central Statistics Office, *Statistical Yearbook of Ireland 2007* (The Stationery Office, Dublin, 2007), p. 1.

[168] Office of the Minister for Integration, *Migration Nation: Statement on Integration Strategy and Diversity Management* (The Stationery Office, Dublin, 2008), p. 69.

[169] Whelan, *Showtime of Substance?*, p. 111.

[170] M. Coleman, *The Best is Yet to Come* (Blackhall Publishing, Dublin, 2007), p. 2.

[171] D. Geraghty, *40 Shades of Green: A Wry Look at What it Means to be Irish* (Real Ireland Design, Kilcoole, 2007), p. 111.

Another sign of generational change was the fact that the number of visitors representing the Irish Diaspora on O'Connell Street on Easter Sunday was markedly down on the 1966 figures. Although the number of Irish-Americans who returned to Dublin to watch the military parade and rekindle patriotic aspirations was well down on the numbers that visited for the Golden Jubilee, their sentiments and affections towards Ireland still remained pretty strong. The Ben & Jerry ice-cream company found this out earlier on in the month, when it was on the receiving end of several complaints about its launch in the USA of a scrumptious-sounding (yet politically incorrect) 'black-and-tan' flavour. For the most part though, the age-old differences between nationalists and unionists on the island of Ireland must have seemed increasingly like old hat not only to the newer generation of immigrants that had arrived from Eastern Europe, Asia and Africa, but also to Irish teenagers and twenty-somethings, who had become more accustomed to peacetime prosperity (and who were more likely to measure the track records of successive governments through figures for job creation rather than through geopolitical/ideological aspirations). The warming of Anglo-Irish relations in the years leading up to the parade, coupled with the confidence instilled by the Celtic Tiger, resulted in a marked decline in Anglophobic sentiments. Such was the degree of wealth created by international investment in Ireland at the outset of the twenty-first century that one Belfast businessman-turned-historian, Francis Costello, remarked that 'one of the most conservative societies has become one of the most outward looking'. 'The sheer volume of ... investment', he contended, had 'spawned an Ireland that would be unrecognisable to the founders of the Irish independence movement and perhaps quite at variance with Arthur Griffith and the original Sinn Féin movement's adherence to its doctrine of "Ourselves Alone"'.[172]

Writing for a special issue of *History Ireland*, which was published to mark the 90th anniversary, Paul Bew contended that the passing of time had diluted the significance of the story of 1916:

> The truth is that modern Ireland today has a complex, distanced relationship to 1916. It does not represent the simple fulfilment of that dream ... we are now in a place defined only very partially by the men of 1916 ... The Good Friday Agreement's articulation of the consent principle on the North is obviously more Redmondite than classic Irish republican – why not admit it? Then we might be able finally to close the door on that cult of the gun, which ... so clearly ... disfigured modern Irish democratic life.[173]

---

[172] F. Costello, *The Irish Revolution and its Aftermath 1916–1923* (Irish Academic Press, Dublin, 2003), p. 337.

[173] P. Bew, "'Why Did Jimmie Die?'", p. 39.

In other respects, the very nature of the setting for the 90th anniversary parade offered further proof of change. In contrast to the abundance of bunting that had been on show during the commemorative spectacle in 1966, O'Connell Street displayed a far more minimalist appearance on Easter Sunday 2006. Trees bearing bullet holes from the Rising had already been removed a few years earlier, as part of a regeneration programme for the main thoroughfare. Although some members of the public waved miniature Tricolours on Easter Sunday, no papal flags were to be seen fluttering from masts. In the space previously occupied by the Anna Liva/'Floozie in the Jacuzzi' monument (and Nelson's Pillar beforehand), O'Connell Street now boasted a shiny and flamboyant piece of Celtic Tiger sculpture, namely the high-rise metal structure known as The Spire (nicknamed by some Dubliners as 'The Stiletto in the Ghetto'). Directly opposite the refurbished façade of the hallowed GPO was the Ann Summers shop – a British adult emporium whose very presence reflected not only the secular turn in Irish society, but also the squeezing out of so many domestic retailers from the fabric of O'Connell Street (as a consequence of growing international competition and homogenising global trends).

Materialism and consumerism reigned supreme amongst many elements of the populace in the years and months leading up to Easter 2006 – owing to the widespread availability of cheap credit and the rapid inflation of the bubble in the property market. The cultural impact of the surge in consumer spending was the subject of much commentary. In a book entitled *The Pope's Children: Ireland's New Elite* (which was also broadcast as a three-part documentary series on RTÉ and first released on DVD in 2006), David McWilliams noted that Ireland had become 'the full-on nation' – awash with money that was being spent in a consumer frenzy. 'We are', he declared, 'Europe's hedonists and the most decadent Irish generation ever'.[174] Fintan O'Toole later reflected that a 'substitute identity' had accompanied the Celtic Tiger years, in place of the old pillars of nationalism and Catholicism. At its best, the new Ireland was optimistic, confident and willing to embrace 'a new openness and ease'. At its worst, he felt that 'mad consumerism' had led to 'an arrogance towards the rest of the world' and even 'a wilful refusal of all ties of history and tradition'.[175] This latter theme was also picked up by Thomas Bartlett, who deduced that unbridled consumerism had 'swiftly elbowed an earlier generation's goal of "frugal comfort"

---

[174] D. McWilliams, *The Pope's Children: Ireland's New Elite* (Gill and Macmillan, Dublin, 2005), pp. 3–4. Also see the accompanying DVD, *The Pope's Children* (RTÉ, 2006).

[175] F. O'Toole, *Enough is Enough: How to Build a New Republic* (Faber and Faber, London, 2010), p. 3.

out of the way', so much so that shopping represented 'the new religion' and the malls 'its cathedrals'.[176]

Although the economy was booming in 2006, some politicians were anxious to draw attention to the gap between the 'haves' and the 'have nots'. In addition to a resurgence of interest in the writings of James Connolly, this focus on social inequalities was no doubt also inspired by the fact that the year also marked the centenary of Michael Davitt's death. In a Dáil debate on 23 February, the Green Party's John Gormley addressed the inequality issue head on by drawing attention to the Proclamation's concern with 'equal rights and equal opportunities' for all citizens. A particular source of ire to Gormley was the matter of 'blatant inequalities ... in our health system'. He also took a swipe at Fianna Fáil, by asking how the 1916 rebels would 'feel about the movers and shakers' gathering each year in the party's tent at the Galway Races, directly availing of 'access and opportunity'. During the same debate, the Independent TD, Tony Gregory, said that even though he welcomed the news of the military parade's revival, he felt that having 'a solely military approach to the 90th anniversary is wrong and does not commemorate in a fitting way what the men and women of 1916 fought for'.[177]

Even though the state was able to boast of the miracle of more or less full employment at the time of the 90th anniversary, some of those in power did seem genuinely concerned with sociological matters. On certain occasions, for example, a call for more active citizenship was bound up with rhetoric that embraced themes such as patriotism and the lessons of history. During her speech at University College Cork, the President sought to drive home the message that Ireland had come to a moment in its history where social inequalities could be actually be addressed head on. 'For many years', said McAleese, 'the social agenda of the Rising represented an unrealisable aspiration, until now that is, when our prosperity has created a real opportunity for ending poverty and promoting true equality of opportunity for our people'. The Proclamation's 'idealistic words', she claimed, 'have started to become a lived reality and a determined ambition'.[178] As Taoiseach, Ahern was also very much to the fore in espousing a sociological doctrine inspired by the words of the

---

[176] T. Bartlett, *Ireland: A History* (Cambridge University Press, Cambridge, 2010), p. 547. In his insider's account of the period of the property bubble, one of the big developers, Simon Kelly, claimed that the nation's preoccupation with owning homes could be partly explained by Irish history. See S. Kelly, *Breakfast with Anglo* (Penguin Books, London, 2011), p. 208, who writes that property 'was the rock on which Irish society was built', a phenomenon that could be linked, 'at least in part', to 'our previous existence as landless tenant farmers'.

[177] *Dáil Debates*, Vol. 615, 23 February 2006.

[178] *The Irish Times*, 28 January 2006.

Proclamation. The occasion of the 90th anniversary, he believed, represented a chance to renew the nation's social capital by underpinning the core values of active citizenship and social responsibility. In a Dáil debate on 1 March, he announced that he was appointing Mary Davis, the Chief Executive Officer of Special Olympics Ireland, as Chair of a Task Force on Active Citizenship. 'We are trying to foster', he explained, 'the spirit of activism such that people will volunteer some of their time, be it to the scouts, junior schoolboy football, the Irish Girl Guides, or any other activity'. Fondly recalling his own upbringing, the Taoiseach credited the Home Farm soccer club for educating him about the merits of volunteerism. 'I would be lifted out of it', he remembered, 'if I was two minutes late for voluntary training'.[179]

In his 'Remembrance, Reconciliation, Renewal' speech at the National Museum Collins Barracks on 9 April, Ahern again spoke passionately about the patriotic virtues of active citizenship. Just like Lemass and de Valera had done 40 years beforehand, he issued his own clarion call for the citizens of the Republic of Ireland to engage in patriotic duty, hinting that the past was the key to the present and future:

> Patriots today are people who are at least as fully aware of the needs of their community as they are of their own individual rights. Whereas the rights of the individual have been hard won and are rightly deserving of protection, Ireland now needs to develop a strong and corresponding sense of duty and of community ... At the beginning of the 21st century, Ireland needs to reimagine a new culture of active citizenship to build a vibrant civic society ... In this Republic we are citizens, not subjects. And, it is as citizens that we remember our past, reconcile our differences and renew our hope for the future. Our civic duty calls on us to look beyond our purely private roles and rights as consumers to our active roles and responsibilities as citizens ... We can sit back and allow this new Ireland to happen to us and hope for the best. Or like the patriots we commemorate today, we can commit ourselves to shaping a better future for our children and our grandchildren.[180]

The upbeat mood at the museum was subsequently commented upon by a number of attendees. Writing for *The Dublin Review*, Ann Marie Hourihan observed that Ahern had 'swept' into the museum 'on his usual wave of charm', and that the assembled crowd had then listened attentively as he issued his call 'on us to do voluntary work in our communities'.[181]

---

179   *Dáil Debates*, Vol. 615, 1 March 2006.
180   *The Irish Times*, 10 April 2006.
181   Hourihan, 'A Series of Accidents', p. 102.

When the 2006 Census was taken on 23 April, the government's interest in active citizenship was evident from the insertion of a question about voluntary activities for the first time ever. The response showed that over 553,000 persons, representing only 16.4% of the population aged 15 and over, were involved in at least one of the five categories of voluntary activity listed on the census form in the four weeks prior to census night. Such categories included helping or voluntary work with a social or charitable organisation, a religious group or church, a sporting organisation, a political or cultural organisation, or any other kind of voluntary pursuit.[182] Whilst swathed as it was in a twenty-first century Irish republican wrapping that extolled the need for more civic engagement at a time of unprecedented prosperity, the prominence accorded by Ahern to community needs cannot be explained by a sudden interest in sociological matters alone. Throughout the course of Irish history, it was common for generations of people who saw themselves as nationalists at heart to place great emphasis on the place of the community (especially the parish) within the overall fabric of society. As Richard English has noted, 'the necessity and appeal of community' illuminates 'what, at root, makes people into nationalists and makes nationalists of so many people'. The persistent popularity and durability of the nationalist tradition, he adds, stemmed from the way in which accounts of community resonated 'with key instincts, needs ... and dimensions of being human.'[183]

Over a month after the Collins Barracks event, many of the sentiments that Ahern had expressed were reaffirmed throughout the state's electoral divisions by a range of Fianna Fáil politicians commenting on the occasion of the 80th anniversary of the foundation of the party by de Valera. Writing for the *Cork Independent*, Cork North Central TD Billy Kelleher, whose grandfather had been involved in the party's formative years, reiterated the notion that 'the ideals of equality and social progress ... remain at the core of Fianna Fáil'. Obviously encouraged by the success of the 1916 commemoration, Kelleher also proclaimed: 'Fianna Fáil was – and still is – a radically idealistic political party aimed at achieving a united, republican Ireland within a constitution framework.'[184] The optimism that emanated from the Easter Sunday parade also seemed to feed into the negotiations between the social partners during the National Pay Talks later on in 2006. In the end, a 10-year deal was agreed upon by the government, employers and trade unions. This was called *Towards 2016* and was neatly timed to run for the decade leading up to the centenary of the

---

[182] Central Statistics Office, *Census 2006: Principal Socio-Economic Results* (The Stationery Office, Dublin, 2007), pp. 29, 101.

[183] R. English, *Irish Freedom: The History of Irish Nationalism* (Macmillan, London, 2006), p. 12.

[184] *Cork Independent*, 18 May 2006.

1916 Rising. As the Taoiseach himself acknowledged in his short 'Foreword' to the published version of the agreement, the 'date highlighted by the title' was 'historically significant'.[185]

Even though Ahern won many plaudits in the aftermath of Easter Sunday for his desire to showcase the Defence Forces as the *bona fide* descendants of the 1916 rebels (rather than the Provisional IRA), some of his most vocal critics remained persistent in charging that the military parade had smacked of political opportunism. The very fact that he had announced its revival at his party's Ard Fheis brought with it the lasting accusation that the Taoiseach was playing politics and seeking to construct a bulwark against the electoral threat posed by Sinn Fein in the next General Election. To all intents of purposes, however, the impassioned patriotism that Ahern engaged in by reviving the 'spirit of 1916' proved to be a stroke of political genius for both himself and his fellow 'Soldiers of Destiny' in Fianna Fáil – one that certainly bore resemblance to de Valera's dexterous outmanoeuvring of the IRA when the Rising's 19th anniversary was commemorated in 1935. Notwithstanding the tribulations that were simmering in the background as a consequence of the economy's overreliance on the construction industry, the political outlook after Easter Sunday looked good as far as progress with the Peace Process was concerned. While Ahern certainly took a gamble in reviving the 'spirit of 1916', the successful rehabilitation of its legacy was a political bet that passed off without any major incidents. It also reaped some short-term political dividends for Fianna Fáil.

In addition to winning the public's approval for reviving the military parade, the commemoration proved to be a very big hit with Fianna Fáilers, who saw the recasting of the story of 1916 as a chance not only to rekindle the party's deep-rooted patriotic fortitude, but also as an opportune moment to salvage the essence of the Rising's legacy from Sinn Féin and the Provisional IRA. In a General Election that was held 14 months after Easter Sunday 2006, Fianna Fáil returned to power in a Coalition government, winning 78 seats (including the automatic return of the Ceann Comhairle) or a total of 41.6% of the 2.06 million votes cast. In the Dublin Central constituency, Ahern topped the poll with 12,734 or 36.76% of the total first preference votes. In spite of its strong performances in various opinion polls, Sinn Féin's electoral hopes were severely squashed. After running a total of 41 candidates, it only managed to secure four seats or 6.9% of the total votes nationwide.[186] For both his most ardent political

---

[185] B. Ahern, 'Foreword', in Government of Ireland, *Towards 2016: Ten-Year Framework Social Partnership Agreement 2006–2015* (The Stationery Office, Dublin, 2006), p. 2.

[186] Statistical data derived from various tables in L. Weeks, 'Appendices', in M. Gallagher and M. Marsh (eds), *How Ireland Voted 2007: The Full Story of Ireland's General Election* (Palgrave Macmillan, Basingstoke, 2008), pp. 232–43.

supporters and opponents, Charles Haughey's earlier description of Ahern may have sprung to mind once again: 'He's the man. He's the best, the most skilful, the most devious, and the most cunning of them all.'[187]

---

[187] Haughey, cited in M. Clifford and S. Coleman, *Bertie Ahern and the Drumcondra Mafia* (Hachette Books Ireland, Dublin, 2009), p. 108.

Chapter 7

The Pathway to the Centenary, 2016

On many levels, the perennial renegotiation of the past in an ever-changing present may be equated to T. S. Eliot's poetic notion of 'exploring' and then arriving back at the starting point to 'know the place for the first time'.[1] Although first-hand participant recollection of the 1916 Rising faded away with the deaths of the last of the veteran rebels during the 1990s, memory of this determining episode of Irish history remained undiminished in the years leading up to the centenary anniversary. As 2016 drew closer, reflection on the story of 1916 still had the capacity to generate much debate and critical reflection amongst politicians, the media, historians, heritage campaigners, and the public in general. In trying to make some preliminary observations about the pathway towards 2016, it is worth posing four pertinent questions in this final chapter. Firstly, what impact did the passage of time have for the functioning of memory, heritage conservation/restoration, battlefield archaeology, and tourism? Secondly, to what degree did the enhanced remembrance of pluralist pasts serve not only as an agent for peace and reconciliation between Irish nationalists and unionists, but as a means for enhancing Anglo-Irish relations? Thirdly, what prospects unfolded for Irish heritage, culture and identity as the centenary of the Rising drew closer? And fourthly, how did Queen Elizabeth II's historic visit to the Garden of Remembrance, on 17 May 2011, contribute to the dawning of a new era of unprecedented positivity in Anglo-Irish relations?

The Passage of Time and its Implications

Going on the evidence furnished in the previous chapters, it seems reasonable to suggest that the trajectory of different episodes in the making and remembrance of the Rising was rather uneven to say the least. On lots of occasions, memories of the event were subjected to continuous renegotiation by different peoples with particular agendas. As the evidence presented suggests, an unmistakable blueprint to the Irish state's official attitude to 1916 materialised over shifting

---

[1]   Eliot, cited in P. Haggett, *Geography: A Global Synthesis* (Pearson Education, Harlow, 2001), p. 4.

temporal frameworks. Some of the Rising's key anniversaries were celebrated in a striking fashion on occasions of comparative peace (as happened in both 1966 and 2006) or used to reaffirm Irish sovereignty during periods of international turmoil (as transpired during the Emergency in 1941, when Ireland had declared a policy of neutrality in World War II). During periods of domestic tumult, however, a more contradictory pattern can be discerned in the nature and practice of historical commemoration. When the state faced mounting internal pressures from the IRA during the mid-1930s, Eamon de Valera responded by reasserting Fianna Fáil's own republican credentials through the staging of major commemorations for the Rising's 19th and 20th anniversaries in 1935 and 1936 respectively. But when it came to the 75th anniversary in 1991, Charles Haughey scaled down the commemorations. By then, the Troubles in Northern Ireland had been waging for over two decades and as a result of the Provisional IRA's much-trumpeted claim to the legacy of 1916, the Fianna Fáil-led government did not wish to be accused of being in the same boat as paramilitaries.

By the end of the 90th anniversary commemorations in 2006, however, it had become ever clearer that the shifting sands of memory had deposited more psychological sediment and critical distance between Ireland's hard-won path to independence and the matter of its more contemporary peacetime memorialisation. This phenomenon managed to push remembrance of 1916 a bit further into the realm of the less contentious commemorative pageantry that has typified older and more festive international celebrations, such as Bastille Day in France and the Fourth of July in the USA. This cultural phenomenon was commented on by the Minister for Justice, Michael McDowell, when he offered an insightful perspective on the Rising's 90th anniversary in an article that appeared in *The Sunday Independent* on Easter Sunday 2006, entitled 'Free from the Shackles of Nationalist Rhetoric'. 'But this Easter weekend', he reflected, '21st-century Ireland seems to be far more at ease with the 1916 Rising and far more accepting of it as history – as distinct from some form of liturgical sacrifice in nationalist ideology'. 'Irish culture', he concluded, 'reflects ... complexity and diversity. Like a lot of other things, 1916 is "part of what we are". It does not define us or enslave us. Realising that much liberates us'.[2] Without doubt, the impact of trends such as peace, secularisation and multiculturalism had huge repercussions for the way in which the Irish began to view the 1916 metanarrative from the 2000s onwards. The old adage of time being a healer certainly gained some ground, as the gap in time between 1916 and each successive Easter became more pronounced. In the process, traditional 'pro' and 'anti' 1916 stances were

---

[2]   *The Sunday Independent*, 16 April 2006.

increasingly supplemented by a more neutral constituency that possessed a more passive attitude, especially when it came to registering verdicts about whether (or how) 1916 should be remembered as the centenary approached.

Not that this served in any way to dilute the importance accorded to 1916 in public debates. The passage of time was important in other respects too. It meant that the buildings and landscapes linked to the Rising were getting older – a natural phenomenon that generated significant discussions about heritage conservation/restoration, battlefield archaeology and tourism. What follows is a synopsis of some of these issues. In the years that followed 2006, the campaign to conserve the 1916 surrender headquarters at Moore Street gathered pace. Early in 2007, the Minister for the Environment, Heritage and Local Government, Dick Roche, signed an official preservation order on Numbers 14–17 Moore Street, thus giving the four buildings National Monument status and full protection under the National Monuments Act. Roche's action meant that it then became a statutory offence, punishable by fine or imprisonment, for any person to 'damage, injure, remove or to carry out, or to cause or permit, work affecting the monument, without the written consent of the Minister'.[3] In explaining his decision to sign the preservation order for the surrender headquarters, Roche said that the government needed to resort to 'every protective weapon in its statutory arsenal to protect a building of such immense historical importance'. Furthermore, he maintained that it would have been 'unconscionable for the government not to close any potential legal loophole which might result in ... loss or destruction'.[4] As the accolade of National Monument status had usually been reserved for much older structures, typically archaeological sites predating 1700, some eyebrows were raised following Roche's pronouncement. But when viewed from another angle, it was clear that the passage of time was again proving significant in the forging of new ways of thinking about the Rising's built heritages. Ultimately, the magnitude of what had transpired did not escape the notice of the Academy. In a letter to *The Irish Times*, archaeologist Joe Fenwick described Roche's preservation order as 'an interesting and welcome interpretation of the National Monuments Act'. This, he noted, had raised 'an equally interesting legal precedent', for the 'otherwise unremarkable building' had been granted protection 'on the basis of an associated historical event alone' – namely its association 'with a pivotal event in Irish history ... the 1916 Rising'.[5]

---

3 *The Irish Times*, 20 January 2007.

4 Ibid.

5 Ibid., 29 January 2007.

Roche's concern for the fate of the Moore Street buildings was also echoed in public statements made by some equally-patriotic voices in Fianna Fáil. In a Seanad debate on 4 November 2008, for example, Labhrás Ó Murchú spoke affectionately about the 'major historical significance' of the surrender headquarters and went so far as to suggest that it had the potential to 'be the equivalent of the Alamo'.[6] On the face of it, the ongoing campaign to conserve the surrender headquarters served as a powerful reminder of the intensity of certain citizens' emotional attachments to the storylines surrounding the Rising's heritages. Within the broader context of the twin processes of development control and forward planning, the Moore Street saga also demonstrated the great prominence that social activists were giving to heritage matters in the capital city during the early years of the twenty-first century. Although some Dubliners were well acquainted with prolonged conservation campaigns, having witnessed the fight to preserve the extensive Viking site at Wood Quay during the late 1970s and early 1980s, the fight to save the Moore Street buildings did much to raise awareness about heritage conservation amongst a younger generation growing up in the early twenty-first century (as did conservation campaigns at treasured archaeological sites such as Carrickmines Castle and the Hill of Tara). Most notably perhaps, the Moore Street saga highlighted the degree to which legislators were prepared to go to enshrine memories of 1916 into the canon of Irish law. If anything, the whole episode seemed to offer further proof of Barack Obama's general dictum, in *Dreams from My Father*, that 'law is ... memory', recording 'a long-running conversation, a nation arguing with its conscience'.[7]

With the date of 2016 looming, heritage enthusiasts also began to point towards the potential of transforming sites such as Numbers 14–17 Moore Street and the GPO into heritage tourism attractions – contending that the buildings had the capability of rivalling existing sites like Kilmainham Gaol and the Pearse Museum. Given the extent to which heritage tourism had boosted economic activity in Ireland in the late twentieth and early twenty-first centuries, the calls to add to the existing range of 1916 attractions came as no surprise. When it reported to the government in 2003, a special 14-member task force known as the Tourism Policy Review Group drew attention to two of the country's most distinctive assets, namely 'the intrinsic attractiveness of the landscape, culture and people of Ireland', along with the enterprise and enhanced professionalisation of the people working 'in a largely Irish-owned service industry'. It also highlighted the type of action needed to cater for the requirements and expectations of potential tourists, by identifying a number of

---

[6]   *Seanad Debates*, Vol. 191, 4 November 2008.

[7]   B. Obama, *Dreams from My Father* (Canongate Ltd., Edinburgh, 2007), p. 437.

pertinent factors. In addition to usual considerations such as affordability and travel expediency, the task force also suggested that tourism 'will, in future, be increasingly demand driven with greater emphasis being placed on ... personal fulfilment, unique experiences, authenticity, ... [and] emotional involvement'.[8] By the time of the Rising's 90th anniversary, the tourism industry had become the biggest 'Irish-owned' and 'internationally-traded' part of Ireland's economy, accounting for nearly 4% of GNP and supporting over 150,000 direct jobs (nearly one tenth of all employment in the economy). Between the years 1990–2006 alone, the numbers of international visitors coming to Ireland increased from 3.6 million to over 6 million, while total foreign revenue income swelled from €1.4 billion to €4.6 billion.[9]

In its *National Development Plan, 2007–2013*, which was published in January 2007, the government gave a commitment to advance all aspects of tourism enterprise during the life of the plan. It further stated that the tourism industry possessed 'the capacity and the capital stock to achieve further growth in the future' so as 'to help promote regional development at a time when many indigenous sectors face structural and trading difficulties'.[10] Most ambitiously, the government plan also endorsed the Tourism Policy Review Group's aim to double overseas tourism revenue to €6 billion over the period 2003–2012. After the onset of the global financial crisis in September 2008, however, this target became completely unrealistic. The domino effect sparked by the collapse of Lehman Brothers in the USA produced, as John Lanchester has written, 'a colossal wreck' where losses had to be socialised after 'an unprecedented self-generated financial implosion'. This was especially the case with the Irish economy, which endured a 'horrific crash' after a 'grotesque bubble'.[11] Property prices went into freefall, the main banks had to be bailed out at the expense of

---

[8]   Tourism Policy Review Group, *New Horizons for Irish Tourism: An Agenda for Action* (Department of Arts, Sport and Tourism, Dublin, 2003), pp. viii–ix.

[9]   Tourism Policy Review Group, *New Horizons for Irish Tourism*, p. viii; R. McManus, 'Identity Crisis? Heritage Construction, Tourism and Place Marketing in Ireland', in M. McCarthy (ed.), *Ireland's Heritages: Critical Perspectives on Memory and Identity* (Ashgate, Aldershot, 2005), p. 236; Government of Ireland, *National Development Plan 2007–2013: Transforming Ireland, A Better Quality of Life for All* (The Stationery Office, Dublin, 2007), p. 171. For an international perspective on the growth of tourism, based on tourism satellite accounting systems, see C. M. Hall and S. J. Page, *The Geography of Tourism and Recreation: Environment, Place and Space*, 3rd Edition (Routledge, London, 2006), p. 1, who note that the travel and tourism industry directly and indirectly constituted 11% of global GDP by the mid-2000s and supported around 200 million jobs worldwide.

[10]   Government of Ireland, *National Development Plan 2007–2013*, p. 171.

[11]   J. Lanchester, *Whoops! Why Everyone Owes Everyone and No One Can Pay* (Allen Lane, London, 2010), pp. xiv, 76, 199–200.

taxpayers and unemployment levels soared to levels not seen since the recession of the 1980s. Whilst the impact of the economic downturn on the overall tourism sector was profound, it by no means led to total despair about what the future held for Irish heritage tourism. Although specific policy statements on plans to nurture 1916-themed tourism by 2016 were absent from the *National Development Plan*, due to its finish date of 2013, the creation of the all-party 1916 Centenary Commemoration Committee served as evidence that the political will was there to forward plan well in advance of 2016. Unsubstantiated reports began to surface in the media during the summer of 2007 that a large site behind the GPO was set to be transformed into a major tourist attraction in time for 2016, boasting its own special underground 1916-themed museum. By the end of the 2000s, however, the government was severely financially challenged and its actual intentions in relation to tapping into the tourism potential of the Rising's story remained clouded in uncertainty.

In an article published in the Autumn 2009 edition of *The Property Valuer*, Senator David Norris suggested the idea of moving the Abbey Theatre into the GPO in time for the Rising's 100th anniversary. An edited version of this article subsequently appeared in *The Irish Times* on 15 October, in which the rationale behind Norris's vision was rearticulated. 'The advantages of the GPO ... as the new home for the Abbey Theatre', he suggested, 'are manifest and inarguable'. 'Culture', he noted, had 'provided the imaginative spark' for the Rising, which 'was, after all, an insurrection of poets'. 'In these grim times', he concluded, the country needed 'a boost, a bit of visionary oomph to raise our spirits. Redeveloping the Abbey Theatre in the GPO would demonstrate our national pride at a very reasonable cost ... and if we start now it will be ready by 2016'.[12] When the matter was debated in political and media circles in the weeks afterwards, some endorsed the idea but others felt that it was best to leave the GPO as it was – namely as a functioning post office (incorporating a small exhibition on its history). In the Programme for Government that was renegotiated between Fianna Fáil and the Green Party in October 2009, a commitment was given to look into the possibility of moving the Abbey into the GPO. By the end of the year, the Minister for Arts, Sports and Tourism, Martin Cullen, seemed to endorse the idea of the move, when he suggested that it would cost much less than the original plan to relocate the Abbey to George's Dock, and would also prove good for the rejuvenation of the capital's main thoroughfare.[13]

---

[12] *The Irish Times*, 15 October 2009.
[13] Ibid., 12 and 13 December 2009.

The government's plans became a bit clearer on 28 July 2010, during the opening of a new An Post Museum, located beside the Philatelic Shop in the GPO complex. Based on the theme of 'Letters, Lives and Liberty' and curated by Stephen Ferguson, the exhibition used a combination of digital technology (including an 11-minute film featuring scenes from the Rising) and traditional displays to trace the postal system's evolution as part of the state's history. In addition to retelling stories gleaned from the eyewitness accounts of postal staff on Easter Monday 1916, the exhibition also featured an original copy of the Proclamation (bearing the following inscription in one corner: 'Found in Dublin Easter 1916 by John Phillips').[14] As he presided over the opening of the museum, the Minister for Communications and Natural Resources, Eamon Ryan, hinted that the space occupied by the GPO could be used for a variety of purposes in the future. Besides declaring that An Post 'can and will thrive as a publicly-owned institution', Ryan also suggested that the idea of moving the Abbey into the GPO would represent a 'very good development' if the right funding was available. Describing the GPO as 'our best building' and one of 'huge importance to us', he also recommended that 'we should also keep it as a post office'. With 2016 in mind, An Post Chairman John Fitzgerald commented that the launch of the new museum (which was installed with the intention of being a permanent fixture) was especially timely, given his company's ongoing commitment to being 'custodians of an important heritage, not just for the city but for the nation'.[15] In the end, the much-publicised plan to move the Abbey to the GPO was discarded in July 2011, after it was ascertained that 'it would cost €293 million and would be unlikely to be completed in time for 2016'.[16] There was little public outcry following this decision. The historian, Gillian O'Brien, felt that the abandonment of the relocation idea was the wisest course of action, given the many question marks that remained over the GPO's suitability as a functioning theatre space. Moreover, she also questioned the wisdom of linking Patrick Pearse and William Butler Yeats 'in perpetuity', when it was a known fact that the two 'had little time for each other' when they were alive.[17]

Just around the corner from the GPO, the fate of Numbers 14–17 Moore Street continued to make headlines in the early 2010s, as more precise details emerged about Chartered Land's plans for the redevelopment of parts of Moore Street and adjoining areas (including Upper O'Connell Street,

---

[14]　S. Ferguson, *Letters, Lives & Liberty at the An Post Museum* (An Post, Dublin, 2011), pp. 6, 9, 36.

[15]　*The Irish Times*, 29 July 2010.

[16]　Ibid., 22 July 2011.

[17]　G. O'Brien, 'The Future of the Past', in L. Sirr (ed.), *Dublin's Future: New Visions for Ireland's Capital City* (The Liffey Press, Dublin, 2011), pp. 213–14.

Henry Street, O'Rahilly Parade, and Parnell Street). In March 2010, An Bord
Pleanála (the planning appeals board) granted permission for the construction
of an 800,000 square foot development – comprising retail units, restaurants,
cafes, apartments, and car parking spaces. Although it was stipulated that the
façade of Numbers 14–17 would have to be preserved in the redevelopment,
James Connolly Heron complained in a letter to *The Irish Times* on 8 July that
up to 60% of the buildings' internal structures were going to be appropriated
'for private commercial use'.[18] Having been taken on a tour of Moore Street
two weeks later by Connolly Heron and other campaigners, the Labour leader,
Eamon Gilmore, added his support for the development of a commemorative
museum at Moore Street:

> There is a lot of talk now about the 100th anniversary of the 1916 Rising coming up, and
> the Labour Party supports the campaign which has been initiated by the descendants
> of the leaders of the 1916 Rising to have these buildings properly preserved, properly
> commemorated, and an appropriate museum and commemorative centre developed
> there in conjunction with the GPO. Our commemoration of 1916 should not just
> be a token flag-waving commemoration, but it should be real, and I think there is an
> obligation on the state to respond positively to the relatives of the 1916 leaders to go
> with this project.[19]

When asked about his own hopes for the future, Connolly Heron suggested
that the 'district or area' of Moore Street should be developed separately 'from
the proposed retail development' and transformed into 'an historic or cultural
quarter as a fitting memorial to the 1916 Rising'.[20] Although question marks
persisted over the viability of the 800,000 square foot development due to the
crash in the property market, it was noted in a Dáil debate on 22 June 2011,
that plans had been prepared by Chartered Land to conserve Numbers 14–17
'as a commemorative centre to facilitate interpretation of the significant cultural
history relating to the events of Easter 1916'. In the following month, the
relatives of the 1916 leaders added a new angle to their campaign, by suggesting
that the centenary could also be marked by erecting a bronze sculpture of the
Proclamation in Moore Street.[21]

In the decade leading up to Easter 2016, one of the most absorbing
contributions to public debates about the cultural landscapes of 1916 came
from archaeologist Damien Shiels. Writing for *Archaeology Ireland*, he argued

---

18   *The Irish Times*, 8 July 2010.
19   Ibid., 23 July 2010.
20   Ibid.
21   *Dáil Debates*, Vol. 735, 22 June 2011; *The Irish Times*, 20 July 2011.

that Ireland 'has a great deal of potential when it comes to conflict archaeology, and much can be learned from exploring our twentieth-century sites'. Still bearing remnants of the fierce warfare that occurred at Easter 1916, he noted that various sites throughout Dublin city centre still contained archaeological footprints of a conflict landscape, although with rebuilding and repair works to damage, 'only tantalising glimpses' remained. These included the bullet holes in the O'Connell monument on O'Connell Street and the pock-marked Royal Dublin Fusiliers Memorial Arch at the entrance to St. Stephen's Green facing Grafton Street. He also pointed to the likelihood that numerous lead-based metal bullets still existed in the soil, along with the remains of trenches dug with pickaxes by the rebels inside the gates. The latter, he speculated, may have survived under the modern-day tarmacadam. Shiels also went on to suggest possible strategies that could be utilised if St. Stephen's Green 'was to be archaeologically investigated for traces of the 1916 fighting'. His suggestions included the commissioning of a licensed metal-detector survey to recover bullets and a geophysical prospection or a ground-penetrating radar survey to locate trenches used during Easter Week.[22]

While Shiels did not go as far as to recommend a full-scale archaeological excavation of the sites of the trenches dug by the rebels at St. Stephen's Green, the idea certainly aroused the curiosity of readers of *Archaeology Ireland* at the time. Indeed by the mid-2000s, the practice of excavating battlefield sites dating from the early part of the twentieth century had gained much ground in different parts of continental Europe – especially so in relation sites associated with the Great War. At Thiepval Wood in France, for example, a series of reconstructed trenches that had been used by the 36th Ulster Division in the Battle of the Somme were reopened to the public on 1 July 2006 for both tourism and educational purposes, having been excavated by the Somme Association and the No Man's Land European Group for Great War Archaeology. The excavations proved particularly informative in relation to the methods that had been used in their construction and added a new dimension to nearby heritage tourism attractions such as the Ulster Tower. Could a similar restorative approach be taken for some of the 1916 Rising's battlefields?

'Restoration', as David Lowenthal has avowed, 'implies going back to an earlier condition', frequently 'the pristine original' in which the preceding 'is held better – healthier, safer, purer, truer, more enduring, beautiful, or authentic – than what now exists'. 'Envisaged restoration', he adds, 'deploys humanity's

---

[22] D. Shiels, 'The Archaeology of Insurrection: St. Stephen's Green, 1916', *Archaeology Ireland*, Vol. 20, No. 1 (2006), pp. 8–11.

continuing creative along with its conserving instincts'.[23] Assuming that they have survived the passage of time in the first place, the excavation, restoration and conservation of rebel trenches at St. Stephen's Green could, if conducted, add a new military heritage attraction to Dublin in the future (albeit in more miniscule proportions to overseas attractions such as the World War I trenches at Thiepval Wood, the World War II remains at Berlin's Topography of Terror or the Vietnam War tunnels at Cu Chi). Alternatively, the identification of the locations of the trenches at St. Stephen's Green through non-evasive methods might at least facilitate their ongoing preservation *in situ*, while the erection of signposts of various kinds (such as commemorative plaques, ground markings and/or information panels) could potentially enhance the public's knowledge of what could be described as the Rising's 'heritagescape'.[24] In addition to St. Stephen's Green and other key battlefield sites around Dublin city centre, heritage tourism possibilities may also arise in years to come at other locations that witnessed rebel activity in 1916, such as Ashbourne, Castlelyons, Enniscorthy, and the east of County Galway.

Going forward, an appreciation of the broader comparative perspective may offer some clues as to the challenges that may arise from any future endeavor to create tourist attractions out of the archaeologies of Irish conflict. In many locations around the world, heritage spaces in the form of battlefield sites have often assumed iconographic status, vested with ideological meanings. Sometimes, as Petri Raivo has observed, they have become part of nationalist metanarratives, assuming the status of 'mystical places where it is still possible to experience imagined visions and sounds of the past'. But more often than not, the 'spirit' of a battlefield has depended upon the 'ghosts' that people 'are willing to see, hear ... and feel'.[25] Although many Irish people seemed more positively disposed towards the 1916 Rising's place in history following the success of the 90th anniversary commemorations, some politicians did not shy away from writing about the importance of coming to terms with the contentious legacies

---

[23]   D. Lowenthal, 'Restoration: Synoptic Reflections', in S. Daniels, D. DeLyser, J. Nicholas Entrikin, and D. Richardson (eds), *Envisioning Landscapes, Making Worlds: Geography and the Humanities* (Routledge, London, 2011), pp. 209, 220.

[24]   The notion of a 'heritagescape', as put forward by M.-C. E. Garden, 'The Heritagescape: Looking at Landscapes of the Past', *International Journal of Heritage Studies*, Vol. 12, No. 5 (2006), pp. 396, 407–408, is a means for thinking about heritage sites as landscapes. Heritage sites, she notes, can represent 'a complex social space constructed by the interaction and perceptions of individuals who visit the site', while the 'heritagescape' can advance recognition of the sites' characteristics 'as dynamic, changing spaces', thus offering the prospect of locating them 'in the context of their larger environment' and acknowledging their 'relationship to other sites'.

[25]   P. J. Raivo, 'Landscaping the Patriotic Past: Finnish War Landscapes as a National Heritage', *Fennia*, Vol. 178, No. 1 (2000), p. 145.

emanating from its spirit and ghosts. In a book published in 2010, for example, Eamon Gilmore drew attention to the theme of Seán O'Casey's 1926 play, *The Plough and the Stars*, and called for a rediscovery in the lead-up to 2016 of the playwright's 'interpretation of armed conflict and the suffering it brings'. Whilst not doubting that the Rising 'was central to the birth of the independent Ireland of which we are rightly proud', Gilmore added a *caveat*, arguing that 'we will also have to confront our tendency to glorify the violence and conflict that is such a significant part of our history'.[26]

Perennial reminders about the physical destruction and killings that occurred in 1916 may explain why much of the tourism potential of buildings and spaces linked to the Rising remained relatively underdeveloped by the outset of the 2010s. The fact that 100 years had not yet passed since Easter Week may also elucidate why the Rising's built heritage was not always deemed to be on a par with the likes of ancient monuments. Ultimately, Ireland was by no means alone at the time in lacking a coherent heritage tourism strategy for battlefield sites less than 100 years old. Indeed a perennial problem with the utilisation and commodification of twentieth-century war as a twenty-first century tourist attraction, as Joan Henderson has pointed out in an analysis of military heritage attractions in Southeast Asian countries such as Vietnam, Cambodia and Laos, has been the way in which 'dark tourism or thana-tourism' – that is, 'where the motivation is encountering death and the macabre at sites such as prisons' – has sometimes allowed entertainment, overcommercialisation or political propaganda to take 'precedence over education'. From time to time, she argues, this has caused 'the obscuring of historical truths' and the trivialisation 'of the subject matter'.[27] Whilst dark tourism can be a tricky and risky business, there have of course been instances around the world where present-centred heritage interpretations of past conflicts (and efforts to conserve/restore battle sites) have reaped socio-economic dividends. Some attractions, for example, have been successful in supporting local economies, fostering community pride

---

[26]   E. Gilmore with Y. Thornley, *Leading Lights: People Who've Inspired Me* (Liberties Press, Dublin, 2010), pp. 80–92.

[27]   J. C. Henderson, 'The Meanings, Marketing and Management of Heritage Tourism in Southeast Asia', in D. J. Timothy and G. P. Nyaupane (eds), *Cultural Heritage and Tourism in the Developing World: A Regional Perspective* (Routledge, London, 2009), p. 80. Within an Irish context, Roy Foster has expressed disquiet about some intersections between military history and heritage – citing examples from the 1798 Rebellion's bicentennial commemoration in 1998. See R. F. Foster, *The Irish Story: Telling Tales and Making it up in Ireland* (Allen Lane, London, 2001), p. 33, in which he registers concern about 'the idea of historical memory as a feelgood happy-clappy therapeutic refuge' or as 'a fantastical theme-park', and conveys bewilderment with 'a boom in pop history, with a distinctly make-believe feel to it'.

and creativity (without automatically glorifying), educating about the horrors of warfare (and its impact upon peoples and landscapes), promoting racial tolerance, encouraging cultural diversity, and championing the merits of peace-making and reconciliation.

Reconciliation by Way of Remembrance

In an overview of some of the key issues facing the international heritage sector in the early twenty-first century, George Corsane arrived at the illuminating conclusion that heritage and museum outputs could no longer be 'viewed to hold the same authority that they once had'. Because of the fact that they were unable to 'present and transmit absolute knowledge', he determined that the most that they could do was to 'only provide representations and interpretations of the world'. 'Rather than being monovocal', he also argued that 'outputs should allow for a polyvocality with representation being focused on all, no matter what their age, gender, class, ethnic background, religious persuasion or political allegiance'. Corsane also pointed to the inherent difficulties in trying to accommodate a version of the past that would prove unproblematic. Heritage and museum outputs, added Corsane, 'need to be prepared to engage with topical and sometimes difficult issues'. 'Where risks are taken in order to produce outputs and public programmes', he noted that 'there is always the potential for controversy'. 'Yet when carefully negotiated in responsible and sensitive ways', he added, 'this potential for outputs to attract controversy may be reduced, whilst at the same time the possibilities of encouraging challenging engagement and opportunities for learning can be enhanced'.[28]

In the case of Ireland, calls for a polyvocal approach to commemorationist policy gathered considerable pace in the years following the Good Friday Agreement. Martin Mansergh, who played a key advisory role in the government's approach to the Peace Process, proved to be a strong advocate of the merits of remembering pluralist pasts. In a book that pondered the legacies bequeathed by history, he argued that acts of historical commemorations were best conducted 'in a way that is open to different traditions, experiences and points of view'.[29] This way of thinking resurfaced again in the work of geographers Catherine Kelly and Caitríona Ní Laoire. In an article on Irish cultural policy-making,

---

[28] G. Corsane, 'Issues in Heritage, Museums and Galleries: A Brief Introduction', in G. Corsane (ed.), *Heritage, Museums and Galleries: An Introductory Reader* (Routledge, London, 2005), pp. 9–10.

[29] M. Mansergh, *The Legacy of History for Making Peace in Ireland: Lectures and Commemorative Addresses* (Mercier Press, Cork, 2003), p. 16.

they illuminated how representations of the past in the public realm impinged upon constructs of heritage, culture and identity. Overviewing the results of an educational fieldtrip to a popular tourist attraction located in one of the Border counties, they noted that different groups of third-level students from the North and South had been able to 'engage with heritage space on different levels yet see the perspectives of other possible interpretations'. With 'new, inclusive concepts of culture' gaining much ground throughout the island at the time, Kelly and Ní Laoire conscientiously sketched out their own roadmap for the future of Irish cultural policy-making, maintaining that 'heritage professionals will need to pay greater attention to how politically-sensitive constructs of the past might best be represented', so as to play a fuller role in supporting 'cultures of tolerance'.[30]

With the rapid warming of both North-South and Anglo-Irish relations, themes such as cooperation, inclusiveness and reconciliation became commonplace in Irish commemorationist policy in the early twenty-first century. Nonetheless, those in government circles remained insistent that the part played in history by those who forcefully resisted British rule during the revolutionary era of 1916–1921 should not be denied either. After the decade-long countdown to 2016 began in earnest, newer opportunities for revisiting, reinterpreting or reconfiguring the age-old story of 1916 (and the storylines of Ireland's involvement in the Great War) presented themselves. In the process, the matter of historical commemoration retained a prime position within an evolving military heritage management strategy that sought to nurture the prospects and possibilities for stronger levels of understanding and compromise between different political traditions across the island of Ireland. While commemorations of 1916 had long been rather one-sided affairs that celebrated the actions of the rebels, certain politicians started pointing to the need, especially from 2006 onwards, for the staging of bilateral ceremonies that would give due recognition to the fact that there were also opponents and victims of the past struggle for independence. Typically, there was a common strand to their arguments. Many civilians, they noted, had lost their lives because of the violence in 1916, while the casualties amongst the British Army and the police had included many Irishmen, whose lives were lost (and later forgotten) in the course of duty, while suppressing the Rising.

The possibility of commemorating the British Army casualties during 1916 commemorations was mooted twice in early 2006 by the Deputy Leader of the Labour Party, Liz McManus. On 18 February, she stated: 'we should not be

---

[30] C. Kelly and C. Ní Laoire, 'Representing Multiple Irish Heritage(s): A Case Study of the Ulster-American Folk Park', *Irish Geography*, Vol. 38, No. 1 (2005), pp. 72, 81–82.

afraid to remember all who died at Easter 1916 ... The Royal Dublin Fusiliers were among the regiments called upon to put down the Rising and Irishmen served in many of the other units involved'.[31] Over a week later, she again called for 'all future ceremonies' of 1916 to be reconciliatory by honouring 'all combatants involved in the Rising no matter what uniform they wore', and also highlighted the need to remember the innocent civilians who died.[32] In a Seanad debate on 26 April, Brendan Ryan also called for greater recognition of all those who had died in the fighting. Whilst stating that he did not wish 'to cause the slightest bit of rancour about the 1916 commemorations', he did register a bit of unease about the theme of an upcoming religious service that the Army was organising for the 90th anniversary:

> Like others, I have received an invitation to attend a Solemn Requiem Mass for the souls of those who died for Ireland in the 1916 Rising ... Some 90 years on we could offer the Mass for everybody who died, not just those who took a particular view. It is long enough behind us now to recognise that civilians were also killed, who did not think they were dying for Ireland at the time, in addition to policemen and British soldiers. A Mass for those who died in 1916 would be a small gesture in the direction of that word 'inclusiveness'.[33]

McManus and Ryan were by no means the only Irish politicians to invoke the merits of pluralist pasts when making pronouncements in 2006 about a roadmap for future commemorations of the Rising.

Writing in the April 2006 issue of the political magazine, *Magill*, a political perspective from nationalists in Northern Ireland was offered by the SDLP's Seán Farren, who outlined his wishes for future commemorations. 'As we approach the hundredth anniversary', he wrote, 'there is a compelling case for dedicating the next decade to a process of national reconciliation'. The Ireland of the twenty-first century, he advised, would do best to focus 'on the lessons to be drawn' from the past, not least the underestimation by the leaders of the Rising 'of the sharp division' that existed in 1916 between nationalists and unionists, and 'the scale of its implications for Ireland's future'. Calling to mind the spirit of consensus enshrined within the Good Friday Agreement, Farren argued that 'nationalism must engage with unionism much more than it does at present' in order to 'progress to ever-closer relationships'. 'Commemorating the Rising', he concluded, 'will entail a great deal of looking back'. 'But to do so', he

---

[31]   *The Irish Times*, 18 February 2006.
[32]   *The Sunday Independent*, 26 February 2006.
[33]   *Seanad Debates*, Vol. 183, 26 April 2006.

warned, 'without an eye to the lessons to be learnt, and without a commitment to building a future in which all will be truly reconciled and cherished, would be tragic.'[34]

Ideas like these continued to gain increasing levels of popularity and acceptability from 2006 onwards, especially in relation to how memories of both the Rising and the Great War could be reinterpreted so as to conjure up fresh relationships between war remembrance and peace/reconciliation initiatives across the island of Ireland. Once again, calls emerged for a more inclusive agenda that could address the traditional unionist boycott of 1916 commemorations. As official preparations continued behind the scenes, it was clear to certain people that the scope of the 100th anniversary could be broadened by involving unionists in some sort of official remembrance ceremony in one of Dublin's military cemeteries. Examples that sprung to mind included Grange Gorman and Dean's Grange, where many British soldiers and Royal Irish Constabulary officers who perished in the line of duty in 1916 were laid to rest. Curiosity in the stories behind these long-forgotten heritage spaces was aroused by a government-backed fieldwork survey of British war graves located across the island of Ireland. In *Remembering the War Dead: British Commonwealth and International War Graves in Ireland since 1914*, which was published in 2007, Fergus D'Arcy showed that for the years 1914–1918 as a whole, there were approximately 2,882 British Army graves to be found at a variety of locations throughout Ireland, with 2,007 of these located in the Republic and 875 in the North. The all-island figure for the years 1914–1921, he demonstrated, amounted to 3,829 (inclusive of 36 unidentified graves) – with 2,602 in the Republic and 1,227 in Northern Ireland. Through the efforts of the Imperial War Graves Commission, which was renamed the Commonwealth War Graves Commission in 1960, these graves were initially marked with simple wooden crosses. After the foundation of the Free State, the crosses in the South were eventually replaced on a permanent basis with standard headstones by the Office of Public Works, which from the late 1920s onwards 'was brought formally and fully into the humanely delicate, politically sensitive and administratively demanding business of identifying ... commemorating and caring for these war graves.'[35]

The fate of those who suppressed the Rising also featured more prominently in the pages of the official magazine of the Defence Forces, *An Cosantóir*. In an article entitled 'Discovering the Past', which was published to mark the

---

[34] S. Farren, '1916 Rising: The View from the North', *Magill: Ireland's Political and Cultural Monthly* (April, 2006), pp. 54–55.

[35] F. A. D'Arcy, *Remembering the War Dead: British Commonwealth and International War Graves in Ireland since 1914* (The Stationery Office, Dublin, 2007), pp. 2, 14.

90th anniversary, Mike Whelan estimated that around 106 of the occupants of British war graves in Ireland had died during the course of the Rising (and estimated that a third of these were from Irish regiments). One of the British soldiers buried at Dean's Grange, he showed, was 19-year-old Private Alfred Ellis from Leeds. He was transferred in 1916 from the Royal Field Artillery to the Royal Dublin Fusiliers and died during the fighting in Easter Week. The remains of Private Charles Sanders of the 2/6 South Staffords, who was also killed during the Rising, were also interred in the same cemetery. Graves such as these, he wrote, 'represent a minute cross section of the belligerents of 1916' and 'help to remind us of the events of the time, and the human cost paid by those who fought on both sides'.[36] By the late 2000s, local historians in Dublin were also showing more interest in these forgotten aspects of the capital's built and cultural heritage.[37]

Noteworthy too is the fact that the 90th anniversary commemoration was swiftly followed by a series of other cultural initiatives, aimed at both enhancing Anglo-Irish relations and healing old wounds and divisions between nationalists and unionists. A key facilitator in this regard was the Taoiseach, Bertie Ahern. Although he was to later admit in a major TV documentary about his life (screened on RTÉ in 2008) that he had harboured animosity in his younger years against the British government after the Bloody Sunday killings, and had been part of a large crowd that witnessed the burning of the British Embassy in Dublin in February 1972, this had long since abated.[38] By the time he was a middle-aged man, Ahern, by virtue of his role in the Peace Process, had established a very cordial working relationship with the British government. His diplomatic willingness to reach out to the British was clear to see within a week and a half of the military parade on Easter Sunday 2006, when he welcomed the Duke of Edinburgh, Prince Philip, on a surprise courtesy visit to Government Buildings in Dublin on 26 April (Plate 7.1). Later that night, Prince Philip visited the National Concert Hall, where he and President McAleese partook in a joint ceremony to honour the young recipients of Gaisce: The President's Awards and the Duke of Edinburgh's Awards.

---

[36]   M. Whelan, 'Discovering the Past', *An Cosantóir: The Defence Forces Magazine*, Vol. 66, No. 3 (2006), p. 21.

[37]   See, for example, P. O'Brien, *Blood on the Streets: 1916 and the Battle for Mount Street Bridge* (Mercier Press, Cork, 2008), p. 9, who outlines how some British Army casualties ended up in the grounds of the Royal Hospital Kilmainham and how two soldiers of the Nottingham and Derbyshire regiments (who fell during fighting at Mount Street Bridge) were eventually interred in a public area within the grounds, known as Bully's Acre.

[38]   See *Bertie* (RTÉ, 2008).

Plate 7.1   The Duke of Edinburgh, Prince Philip, and the Taoiseach, Bertie
Ahern, meeting at Government Buildings on 26 April 2006
Source: *The Irish Times*, 27 April 2006.

Commenting on the significance of the royal presence, the President remarked
that it 'added a unique and inspirational dimension' to the proceedings, thus
strengthening an important building-block 'of the exceptional relations that
now exist between Ireland and the United Kingdom'.[39]

The warming of Anglo-Irish relations was also demonstrated in 2006 by the
attendance of members of the Irish Navy at a Royal Navy event commemorating
the 200th anniversary of Admiral Nelson's victory over French and Spanish
fleets at Trafalgar and by the presence of Irish Air Corps personnel at a Royal Air
Force commemoration of the 65th anniversary of the Battle of Britain. Then, on
21 June, the added importance attached to remembering and commemorating
pluralist pasts was again evident when Ahern unveiled a commemorative stamp
to mark the 90th anniversary of the Battle of the Somme (featuring J. P. Beadle's
painting of the first day of the fighting). 'The commemorative stamp,' explained
Ahern, 'is part of an overall programme that reflects the shared history and
shared experience of the people of this island, from all traditions, in the year

---

[39]   *The Irish Times*, 27 April 2006; M. McAleese, *Building Bridges: Selected Speeches and
Statements* (The History Press Ireland, Dublin, 2011), pp. 186–87, 189.

of 1916'. 'It is important', he added, 'that the history of all of the people, and all of the traditions, of this island is acknowledged and commemorated in an appropriate and respectful manner'.[40] On 1 July, yet another significant pluralist and reconciliatory gesture was made by the Irish government when it staged a high-profile public commemoration of the 90th anniversary of the Battle of the Somme at the National War Memorial Gardens in Islandbridge (which included a recital of Lieutenant-Colonel John McCrae's evocative poem, 'In Flanders Fields'). 'The time had come', as Ahern later wrote, 'to acknowledge the shared history and shared experience of the people of this island, from all traditions, in the year of 1916 ... in an appropriate and respectful manner'.[41]

Over three months later on 8 October, on a rainy day in north County Cork, history was made again when Ahern joined with party colleague Ned O'Keeffe, Michael Flatley, British Ambassador Stewart Eldon, and members of both the Royal British Legion and Royal Irish Guards, to unveil a €20,000 monument in memory of 131 soldiers from the Fermoy area who fell in the Great War. At the event, Ahern said that it was fitting that the heroism of Irishmen from 1914–1918 should be properly recognised: 'As a country, we owe it to the many ... who fought and died in that war to remember the part they played.'[42] He later recalled that it 'was deeply moving to witness such a noble spirit of remembrance and commemoration' at a location situated within 'the heart' of a 'traditional' republican area.[43] The reconciliatory genuineness in all of Ahern's actions indicated how far Fianna Fáil leaders had moved on since the difficult days of the early 1980s – when Charles Haughey as Taoiseach instructed the then President, Patrick Hillery, not to attend a Remembrance Day service for the Great War dead in Dublin's Pro-Cathedral. As 2006 drew to a close, the British government also moved on from historically-entrenched positions by granting an official pardon to 28 Irish soldiers 'Shot at Dawn' during the Great War – for misdemeanours such as disobeying orders or going absent without leave.[44] The granting of this pardon followed the waging of a lengthy campaign by activist Peter Mulvan, who was supported in his efforts by the Department of Foreign Affairs and the Irish Embassy in London – the latter of which had

40   *The Irish Times*, 22 June 2006.
41   B. Ahern with R. Aldous, *Bertie Ahern: The Autobiography* (Hutchinson, London, 2009), pp. 295–96.
42   *The Irish Independent*, 9 October 2006.
43   Ahern with Aldous, *Bertie Ahern*, p. 296.
44   For a full list of the 'Irish-Born Soldiers and Members of Irish Regiments Executed', from 28 January 1915 to 12 September 1918, see S. Walker, *Forgotten Soldiers: The Irishmen Shot at Dawn* (Gill and Macmillan, Dublin, 2007), p. 214.

presented a report on the matter to the British Foreign and Commonwealth Office over two years earlier.

Throughout the first half of 2007, there were many more signs of a greater willingness to create a better future in North-South and Anglo-Irish relationships. Most noteworthy were the following occurrences: the convivial reception accorded to the English rugby team's appearance at the hallowed turf of the GAA's Croke Park headquarters in February (which featured the playing of 'God Save the Queen'), the awarding of an honorary Knighthood to U2's Bono by Ambassador Eldon in March, Ahern's historic handshake with Ian Paisley at Farmleigh House in April, the restoration of power-sharing in the executive of the Stormont Assembly in May, a momentous address by Ahern to a joint sitting of the Houses of Parliament at Westminster in the same month, and the sight of Paisley and Martin McGuinness (who became affectionately known as 'The Chuckle Brothers') smiling together as they posed with cricket bats at a reception in June to mark the unanticipated success of the Irish cricket team's performance at the World Cup in Barbados.

One of the most significant expressions of reconciliation transpired on 6 May 2008, when Ahern and Paisley bowed out as leaders of their respective parties by opening a new visitor centre at the site of the Battle of the Boyne at Oldbridge Estate near Drogheda. By then, memory not only evoked echoes of past divisions and old struggles within the island of Ireland, but had acquired the additional capacity to act as a social mechanism for putting old divisions aside in the vastly improved political climate. This yielded a more shared and polyvocal heritage experience that reflected the unprecedented peacetime transformations that had occurred in just a decade. The site of Paisley smiling as he greeted Ahern with a warm handshake and pat on the shoulder at Oldbridge was the ultimate sign of a more happier relationship between North and South – and offered a stark contrast to the unsavoury scenes 43 years earlier when a hostile Paisley threw snowballs at Sean Lemass when he travelled to the North to meet Terence O'Neill. These were changed times indeed. When he published his autobiography in 2009, Ahern was forthright in explaining the reasoning behind his belief in pluralist politics, thus offering an insight into the nature of his new-found friendship with Paisley: 'That ability to reflect on our history in an open and tolerant way was a central priority of my period as Taoiseach.' The island of Ireland's 32 counties, he continued, 'had been a divided society in so many different ways, but we had a shared past'. 'If that history could be commemorated respectfully', he added, 'I believed that would make an impact on our shared future'.[45]

---

[45]   Ahern with Aldous, *Bertie Ahern*, pp. 292–93.

Another significant gesture to mark the warming of Anglo-Irish relations was evident on 7 May 2009, when both President McAleese and Queen Elizabeth II met members of Ireland's Grand Slam-winning rugby team at a reception in Hillsborough Castle in County Down, which was also held to coincide with the 30th anniversary celebrations of the Cooperation Ireland organisation. Half a year later, another bridge-building milestone was accomplished on Armistice Day, 11 November, when the President of the Royal British Legion travelled to Glasnevin Cemetery in Dublin to witness the unveiling, by the Commonwealth War Graves Commission, of the first of a series of headstones that it planned to erect on the graves of Irish soldiers who served in the Great War and who were subsequently buried in paupers' graves. The first headstone to be unveiled was that of Martin Carr of the Connaught Rangers, who died in the trenches on 4 July 1916. Watching the ceremony on the day was Martin Mansergh, in his new position as Minister of State with responsibility for the Office of Public Words. Referring to the sacrifices of those Irishmen who died while fighting for the British Army, Mansergh stated that they 'should not be left out of the nation's consciousness'. 'In their final resting place', he added, 'they are entitled to our respect'.[46]

A more nuanced tale of the complexities behind the history of Anglo-Irish relations during the 1910s also loomed large in the literary world. Yet another indication of the escalation of public interest in Ireland's place in the Great War, and indeed the overwhelming complications that the outbreak of the Rising had caused for Irish soldiers serving in the trenches, was the success in the mid-2000s of Sebastian Barry's novel *A Long Long Way*, which ended up being shortlisted for the Man Booker Prize. This novel portrayed the complex personal struggles that those fighting with the Royal Dublin Fusiliers against the Germans had felt as news of the Rising filtered through to the continent.[47] Alan Monaghan's *The Soldier's Song*, published in 2010, also echoed many of the themes that featured in Barry's novel.[48] Taken together, even though *A Long Long Way*

[46] *The Irish Times*, 12 November 2009.

[47] S. Barry, *A Long Long Way* (Faber and Faber, London, 2005). An illuminating passage (on p. 138) lucidly captures the ironies of the time. The novel's central character, Willie Dunne, conveys his confusion about the events in Dublin to his fellow officer, Pete: 'So that means, like it or lump it, we're the fucking enemy. I mean, we're the fucking enemy of the fucking rebels!'

[48] See A. Monaghan, *The Soldier's Song* (Macmillan, London, 2010), which focuses on the fictional story of Stephen Ryan of the Royal Dublin Fusiliers, who returns to Ireland from duty in Turkey and wonders where his allegiances lie as he finds Dublin city in the grip of the Rising. Monaghan (on p. 129) describes how a dumbfounded Stephen, with 'his heart beating harshly against his chest' and 'his throat closing with fear', eventually encounters his wounded brother, Joe, shot in the back whilst fighting for the rebels.

and *The Soldier's Song* were works of historical fiction, their wider significance within the realm of memory was that they both succeeded in demonstrating the personal quandry that many ordinary Irishmen had faced whilst fighting for the British Army, both during and after the Rising. In many senses, the novels' wider significance was that they did much to alert Irish citizens to the different sides of the complex human dilemmas associated with the story of 1916. Moreover, the pluralist threads woven into the thematic backbone of each of the novels played a part in advancing a wider appreciation of some of the collaborative bonds that had existed between Ireland and Britain in the past, thereby opening up further possibilities of reconciliation by way of shared remembrance. However, as shown at the end of this chapter, the best was yet to come in this regard, during four historic days in May 2011.

Towards 2016: Prospects for Heritage, Culture and Identity

While the soldiers and police who were killed during the Rising (and subsequently buried in Irish cemeteries) were still being ignored during state commemorations of 1916 at the outset of the 2010s, there was certainly a growing realisation that these men were part of the story of 1916 too and were deserving of greater recognition. Outside the realm of memories relating to those who took up arms during the revolutionary era, there was also a dawning realisation in the lead-up to 2016 that large elements of the constitutional tradition of parliamentary politics – including figures such as Daniel O'Connell, Charles Stewart Parnell and John Redmond – had been unjustly ignored when it came to commemorating the history of Irish nationalism. Calls also emerged for a greater recognition of the democratic institutions ushered in by the legal enactment of the Free State on 6 December 1922 (following the Dáil's passing of the Irish Free State Constitution Bill) and the anniversary of Ireland becoming a Republic under John A. Costello's leadership on 18 April 1949. Some amends were made in this regard early in 2009 when the Fianna Fáil and Green Party Coalition government, led by Ahern's replacement as Taoiseach, Brian Cowen, sanctioned a more formal celebration to mark the 90th anniversary of the enactment of the First Dáil on 21 January 1919.[49]

Altogether though, as the evidence presented throughout this book has shown, commemorations during the twentieth and early twenty-first centuries of the achievement of Irish independence were very heavily skewed in favour of the

---

[49] The all-party commemoration of nine decades of Irish parliamentary democracy was staged in the Dáil a day earlier on 20 January 2009 – with the events in Dublin neatly timed to coincide with Barack Obama's inauguration as the first black President of the USA.

events of the revolutionary era. Anniversaries of the Rising typically featured a lot of military pomp and ceremony – aspects that proved increasingly problematic when the Troubles were waging. Conversely, military parades proved to have their advantages on certain occasions, as was the case during the Emergency in 1941, when de Valera was intent on sending a signal to the outside world. With the reinstation of the military parade in 2006, attention also focused on additional themes such as active citizenship. Ireland was finally at peace, and the government was widely commended for making the peacekeeping achievements of Óglaigh na hÉireann on United Nations missions a central feature of the rejuvenated parade. Less than two years afterwards, however, priorities changed sharply as Ireland faced its biggest challenge since independence, after becoming engulfed in the global economic and banking crisis. But as the going got tough and politicians focused on the challenges posed by the exchequer deficit and the rising cost of borrowing in the international bond markets, the legacy of 1916 was by no means cast aside by Cowen's government.

This was especially clear in the early months of 2010, when the upcoming centenary of the Rising began to take on additional significance for the Fianna Fáil-Green Party government, not long after it had passed through a series of extremely harsh budgetary measures, in the hope of restoring some order to the public finances. In an unscripted speech to the AGM dinner of the Dublin Chamber of Commerce early in February 2010, Cowen outlined his hopes for the country's future by making several references to the Rising's forthcoming centenary. In a rallying call that urged people to make short-term sacrifices in order to permit a return to progress by 2016, the Taoiseach argued that it was imperative to do this so as not to let down those whose efforts had led to independence: 'Yes, we were a generation that lived at a time and place of prosperity, but when challenged we looked to the future and looked beyond our own self-interest.' He then called to mind the sacrifices made during the 1916 Rising. 'Yes', he stated, 'we can say in 2016 when we get to O'Connell Street and look up at those men and women of idealism that gave us the chance to be the country we are that: "Yes, we did not fail our children, but we did not fail our country either".'[50]

Like so many of his predecessors, Cowen's impassioned words were indicative of a recurrent trend within Irish politics in modern times, namely the invoking of the story of 1916 so as to link the sacrifices of the past to the sometimes difficult choices facing people in the present. This trend was again evident in Dublin on Easter Sunday 4 April 2010, when the Rising's 94th anniversary was commemorated at a humble 20-minute ceremony at the GPO, watched by around 2,000 spectators, 300 members of the Defence Forces, politicians

50   *The Irish Times*, 5 February 2010.

(including a new Minister for Defence, Tony Killeen), and relatives of the rebels. In spite of the 'bright spring sunshine that warmed the occasion', the *Irish Examiner* despondently noted that the commemoration 'had a solemn air to it' as the Army's chaplain, Monsignor Eoin Thynne, led the gathering in prayers that concentrated 'more on present day upheavals than those recalled from 94 years ago' and appealed 'for efforts to build a more ethical society'.[51] According to the *Irish Daily Mail*, there 'was little in the way of clapping or cheering' outside the GPO following the lowering of the Tricolour. As the chaplain spoke his 'harsh words', it noted that some of the gathered politicians appeared to be 'shifting uncomfortably' as they 'kept their heads firmly bent low'.[52] Compared to the upbeat commemoration of the 90th anniversary at the GPO only four years beforehand, the contrast in moods between 2006 and 2010 could not have been greater. Referring directly to the shocking announcement that had been made, only six days before the 94th anniversary commemoration, regarding the transfer of Anglo-Irish Bank's toxic loans to the state's newly-established bad bank, NAMA (the National Asset Management Agency), *The Irish Independent* lamented that 'now, a new, bullet-free battle is waging in Ireland, with thousands falling victim to the economic fallout'.[53]

Later on in the year, a couple of well-known writers hinted that an Irish rebirth was possible, in spite of the fateful wounds that had been inflicted on the public finances by the bursting of the housing and credit bubble, and the subsequent bailout of the banks. In *Enough is Enough: How to Build a New Republic*, Fintan O'Toole attempted to offer up some critical reflections on the state's difficulties, especially regarding the profound sense of 'shock and disorientation' caused by the economy's exposure to 'a blinding snowstorm of woes'. In setting out a roadmap for political transformation towards 'a new idea of a republic', he argued that the 1916 Proclamation's ambition of treasuring the nation's citizens could still become a reality:

> There has to be a belief that enough is enough – both in the sense of no longer being willing to put up with a system that has done so much harm and in the sense of moving towards a society that can give everyone enough to enjoy the dignity of citizenship ...
> There has to be a new sense of national pride – not the weepy sentimentality of pub patriotism ... but a belief in our collective capacity to create a country to be proud of
> ... In the second decade of the twentieth century, many Irish people decided that a

---

51 *Irish Examiner*, 5 April 2010.
52 *Irish Daily Mail*, 5 April 2010.
53 *The Irish Independent*, 5 April 2010.

republic was worth dying for. In the second decade of the twenty-first century, many more can decide that a republic is worth living in.[54]

The veteran American reporter, David J. Lynch, also hinted that a national resurrection was achievable in the years leading up to 2016. 'Independent Ireland', he wrote, 'is now roughly as old as was the United States at the time of its great national trial, the Civil War'. 'As it prepares to celebrate the centennial anniversary of the 1916 Easter Rising', he counselled that 'Ireland needs finally to shed the anachronistic habits of mind – including the ethic of the "cute hoor" – that have kept it from fully maturing'.[55]

Notwithstanding the scholarly perils and potential hazards of seeking to gaze into the future with some sort of crystal ball, it is tempting to draw this book to a conclusion by speculating tentatively upon what the future might hold, as far as memory-making is concerned in Ireland. Although it has been 'part of life ... to question ... where the future is leading', Eric Hobsbawm once advised that 'predicting the future must necessarily be based on knowledge of the past', whilst also acknowledging that 'much of the future is entirely unpredictable'.[56] In the specific case of memory itself, Geoffrey Cubitt has written that it 'has never been changeless', for its history has been 'one of shifting definitions and evolving uses, as human beings, individually and collectively, have adapted themselves to new social and technological conditions'. 'That our sense ... of ... the past ... is ... evolving', he adds, 'seems certain'.[57] As far as the Republic of Ireland is concerned, both continuity and change in the evolution of historical memory seems plausible enough, with transformation being steered by trends such as economic challenges, globalisation, multiculturalism, pluralism, social networking, and secularism. Within the wider context of the island of Ireland as a whole, some movement away from traditional binary (or tribal) distinctions between nationalists and unionists may continue to unfold in the not-too-distant future. The potential for greater cross-border economic cooperation and cultural common ground may prove a significant catalyst in this regard.[58]

[54] F. O'Toole, *Enough is Enough: How to Build a New Republic* (Faber and Faber, London, 2010), pp. 6, 236–37.

[55] D. J. Lynch, *When the Luck of the Irish Ran Out: The World's Most Resilient Country and its Struggle to Rise Again* (Palgrave Macmillan, New York, 2010), p. 215.

[56] E. Hobsbawm with A. Polito, *On the Edge of the New Century* (The New Press, New York, 2000), p. 1.

[57] G. Cubitt, *History and Memory* (Manchester University Press, Manchester, 2007), p. 249.

[58] At the commencement of the 2010s, D. Tapscott and A. D. Williams, *Macrowikinomics: Rebooting Business and the World* (Atlantic Books, London, 2010), p. 20, optimistically predicted that the emergence of 'macrowikinomics' would signal more mass collaboration and social

Then again, as others have predicted, the continual shrinking of space within the 'global village' may actually serve to reinforce some of the most cherished metanarratives of Westernised nation states in the years that lie ahead. As Rodney Harrison and others have argued, 'authenticity gains a new value' if items 'become increasingly easily reproducible', while the concept of place 'becomes more important' for people who 'experience many landscapes in a single lifetime'.[59] In pondering the question of how the world might look after the mid-point of the twenty-first century, the futurist thinker, Richard Watson, has produced some intriguing reflections on the theme of 'history influencing the future', including the intriguing prediction that 'nationalism will certainly be a feature' in the 2050s. In contrast to the thinking of others, he hints at the possibility of deglobalisation taking place, noting that 'global provincialism is taking over from global cooperation as a dominant theme of modern politics'. What many ordinary voters want to preserve, he argues, 'are ... the old ways', such as the lifestyles and values that make countries 'special' and endow them with some degree of 'international prestige'.[60]

How the story of 1916 is commemorated and remembered in years to come will no doubt reflect a complex intermingling of political, social, cultural, economic, and technological forces intersecting at a variety of spatial scales – from the household to the parish and from the region to the nation (and beyond). But these 'what mights' are part of another story and best left to the historians of future years. After all, the art of prediction, as Tim Footman has written, can be 'a highly unpredictable science'.[61] What is clearer, however, is what the Czech novelist, Milan Kundera, once wrote in *Ignorance*: 'for memory to function well, it needs constant practice – if recollections are not evoked again and again ... they go'.[62] The logic behind such a way of thinking seemed to be well ingrained in the mind of Brian Cowen on 20 May 2010, when he visited his *alma mater*, University College Dublin, to speak at the annual conference of the Institute of British-Irish Studies. In delivering a major policy speech that outlined a future

---

networking in the global financial system, thus leading to a future in which governments and society could 'harness the explosion of knowledge, collaboration ... and business innovation to lead richer, fuller lives and spur prosperity and social development for all'.

[59] R. Harrison, G. Fairclough, J. H. Jameson Jnr., and J. Schofield, 'Introduction: Heritage, Memory and Modernity', in G. Fairclough, R. Harrison, J. H. Jameson Jnr., and J. Schofield (eds), *The Heritage Reader* (Routledge, London, 2008), p. 2.

[60] R. Watson, *Future Files: The 5 Trends that Will Shape the Next 50 Years* (Nicholas Brealey Publishing, London, 2008), p. 70.

[61] T. Footman, *The Noughties: A Decade that Changed the World 2000–2009* (Crimson Publishing, Richmond, 2009), p. 158.

[62] M. Kundera, *Ignorance. Translated from the French by Linda Asher* (Faber and Faber, London, 2003), p. 33.

vision of how the centenary anniversaries of events during the influential decade of 1912–1922 could be commemorated, the Taoiseach seemed particularly keen on stressing the need to remember and not to forget during the years that lay ahead. Building upon the groundwork that had been done by Ahern in 2006, Cowen adopted a pluralist positionality by arguing that the forthcoming centenaries of both the 1916 Rising and the Battle of the Somme presented chances to 'reflect on' and 'better understand our shared identities'. Stressing that it was important to recognise the authenticity of all traditions on the island of Ireland that drew their identity and collective memory from the past, he added that the practice of commemoration warranted 'mutual respect' and 'historical accuracy', and had 'to recognise the totality of the history of the period, and all of the diversity that this encompasses'. The Taoiseach also expressed hope that the Rising's centenary would be remembered 'with respect and dignity in every part of this island', an aspiration which represented 'a challenge that must be considered by the leaders of unionism'. Then, in a stark warning about the potential of republican splinter groups to 'hijack history' and 'fight again the old battles', he forthrightly expressed his concerns about anyone looking 'to use the memory of the dead to bring suffering to the living'. 'To them', he said, 'I say ... Count me out'.[63]

Alas, the security threat that Cowen alluded to reared its ugly head in Northern Ireland on 2 April 2011, when Constable Ronan Kerr, a 25-year-old Catholic member of the Police Service of Northern Ireland (PSNI), was killed in a terrorist attack in the outskirts of Omagh, County Tyrone. This act of aggression was widely condemned across the political spectrum – both North and South. As Don Anderson has noted, the incident 'brought together all the political parties ... in a show of solidarity against the violence'.[64] Kerr's death also weighed heavily on the minds of many of the *c.* 3,000 people who attended the Republic's commemoration of the Rising's 95th anniversary at the GPO on Easter Sunday 24 April 2011, and the 300 or so guests who turned up for another official ceremony at Arbour Hill a week and a half later. Besides the killing of Kerr, the Republic's loss of economic sovereignty at the end of the previous year also contributed to the despondency that permeated much of this anniversary. So too did the high level of unemployment throughout the state. Results of the 2011 Census, taken on 10 April, later revealed that the number of people unemployed had reached 424,843, a staggering increase of 136.7%

---

[63] *The Irish Times*, 21 May 2010. Cowen partook in another 1916-related event in September 2010, namely the launch of *1916 Seachtar na Cásca* – a major television series on TG4, profiling the lives of each of the signatories of the Proclamation. The series was produced by Abú Media and directed by Daithí Keane.

[64] D. Anderson, *Special and Different: The History of the Police Federation for Northern Ireland* (The Police Federation for Northern Ireland, Belfast, 2012), p. 18.

since 23 April 2006 – amounting to an extra 245,387 people.[65] In a most telling sign of the adversity of the times, a copy of the Proclamation failed to sell at an auction by Adam's and Mealy's five days before Easter Sunday 2011, having only attracted bids that stalled at 10% short of the €100,000 estimate (which was itself far short of the exorbitant prices paid for other copies during the consumer frenzy that coincided with the occasion of the 90th anniversary).[66]

The modest commemoration ceremony on Easter Sunday 2011 was presided over by a newly-elected Fine Gael-Labour Coalition government, which had swept to power in a landslide electoral victory weeks earlier (winning a combined total of 113 seats).[67] As per usual, the Tricolour was lowered to half mast during the ceremony, the Proclamation was read aloud, the 'Last Post' and 'Reveille' were sounded, the President laid a wreath, the National Anthem was played, and there was a fly-past by the Air Corps. No speeches were delivered during the course of the ceremony, with the exception of a few words of solace from the Army chaplain, Monsignor Eoin Thynne. He echoed the sentiments of an Easter message from Archbishop Diarmuid Martin, by praying that the Irish people would find 'strength and courage' to 'stand up and speak out' against 'acts of terrorism'. Furthermore he prayed that the people would 'always show our love for our neighbour'.[68] Although the solemnity that characterised much of the 95th anniversary was a sobering reflection of the broken spirit of the populace, the official Easter Sunday proceedings were bathed in some welcome rays of spring sunshine. Moreover, there was also a thin veil of optimism in the air, with certain attendees hoping that the state would be in a better place by the time of the Rising's centenary. There was cause for hopefulness in other respects too. Following years of speculation and hearsay, reports about an official state visit to

---

[65] Central Statistics Office, *This is Ireland: Highlights from Census 2011, Part 2* (The Stationery Office, Dublin, 2012), p. 15.

[66] *The Irish Times*, 23 April 2011; *The Irish Independent*, 5 May 2011. One of the few pieces of memorabilia released for the anniversary was a double-DVD box set, *Michael Collins and the Easter Rising* (Go Entertainment Group Ltd., 2011). Of significance too was the unveiling in Dublin's Railway Street on Easter Monday, by the North Inner City Folklore Project, of a plaque in memory of the architects of the National Anthem – including its lyricist, Peadar Kearney (who fought in 1916).

[67] For details on the General Election, which was held on 9 March 2011, see S. Donnelly, 'How the Nation Voted', in D. McCarthy (ed.), *RTÉ The Week in Politics: Election 2011 & the 31st Dáil* (RTÉ Publishing, Dublin, 2011), pp. 14–19. Fine Gael secured 36.1% of first preference votes, while Labour obtained 19.45%. Fianna Fáil acquired only 17.45% of first preferences from a disillusioned electorate. Between 2007 and 2011, its number of seats fell dramatically from 78 to 20. Sinn Féin, by contrast, saw its number of seats grow from four to 14.

[68] *Irish Examiner*, 25 April 2011; *The Irish Independent*, 25 April 2011; *The Irish Times*, 25 April 2011.

the Republic of Ireland by Britain's Queen Elizabeth II were confirmed by the new government – which was still enjoying its honeymoon, having been formed just 47 days beforehand.

After the conclusion of the Easter Sunday commemoration, which lasted for 45 minutes, a short reception was held inside the GPO, where the new Taoiseach, Enda Kenny, met relatives of the 1916 rebels. Afterwards, he called for a rekindled sense of national purpose by telling the assembled media that it was his hope that the Republic could look 'to a new future'. Kenny also touched upon the theme of reconciliation by way of remembrance, by stating that it was his hope that the Queen's forthcoming visit would bring some sort of 'conclusion to centuries of division and dissent and difficulties'.[69] As a precursor to the royal visit, the Rising's 95th anniversary was of diplomatic significance in its own right, in that the sword and hat of Sir Roger Casement was finally put on display at the National Museum Collins Barracks four days before Easter Sunday, having been returned by the London Metropolitan Police after nine and a half decades. This thoughtful diplomatic gesture by the British rekindled memories of similar initiatives concerning the Casement question in the lead-up to the Golden Jubilee. In a similar vein, a report appeared in the media about a rumoured proposition that the British Prime Minister 'might be invited to attend the centenary of the 1916 Rising'.[70] However, the magnitude of this bridge-building possibility seemed to pale in significance when compared to the sequence of events that were soon to transpire in the capital city.

Queen Elizabeth II's Visit to the Garden of Remembrance

In spite of the immense security precautions that were necessitated by the threat from a small minority of dissident republicans, the Queen's historic four-day visit to the Republic of Ireland, which lasted from 17–20 May 2011, proved to be a resounding political and cultural success – a giant leap in the bridge-building journey of peace, reconciliation, mutual aid, and bilateral friendship. On the first day of the visit, anticipation was high after the Queen's plane finally touched down on Irish soil at Casement Aerodrome – a military airfield named after Sir Roger Casement. The magnitude of the occasion, falling as it did only

---

[69] *The Irish Independent*, 25 April 2011. In an article written for ibid., 17 May 2011, Kenny elaborated further on the merits of the royal visit, describing it as 'a new chapter in Irish-British relations', one already 'based on mutual respect, on partnership and on friendship'. The parts of the itinerary relating to the heritage of the 1916 Rising and the Great War, he predicted, 'will enrich different understandings of identity'.

[70] *The Sunday Independent*, 24 April 2011.

two and a half weeks after the high-profile wedding of Prince William to Kate Middleton at Westminster Abbey, was not lost on the royal biographer, Sarah Bradford. The Queen, she wrote, emerged from her aircraft sporting 'emerald green and a radiant smile' and stepped 'into history as the first monarch to visit the Republic of Ireland'. Despite some security alerts (including bomb scares and minor street disturbances), the Queen remained unaffected 'by any possible threat to her security' and proceeded to carry out her duties with 'courage and dignity', making 'no attempt to disguise the past'.[71]

In an illuminating article that was published in *The Irish Times* to coincide with the start of the visit, the British Prime Minister, David Cameron, elaborated upon its diplomatic aims, explaining 'that part of the intention ... is to pay respect to those who suffered through the course of our shared history'. Additionally, he wrote that 'it is right and appropriate that the Queen sees those places that still resonate with a difficult past'.[72] Much forward planning and thinking went into the geographical make-up of the Queen's highly-orchestrated itinerary, which necessitated performances of rituals in key symbolic spaces and the delivery of a defining speech. The whole spectacle, as Nuala Johnson has noted, was in essence 'an exercise in political and cultural diplomacy, choreographed around a set of spaces and translated through a series of ritualised events that would mark out new relational territory between Britain and Ireland'. Various acts of remembrance, she adds, sought to relocate the past 'in ways which recognised conflict but also aspired towards reconciled understandings of how that past could be more peacefully calibrated'.[73]

One of the undoubted highlights of the action-packed visit – which was the first visit to Dublin by a British monarch in 100 years and included trips to Islandbridge, Croke Park, two universities (Trinity College Dublin and University College Cork), and tourist attractions such as the Rock of Cashel and Cork's English Market – was a wreath-laying ceremony at the Garden of Remembrance in the afternoon of 17 May.[74] When she arrived at the Garden of Remembrance, the 85-year-old Queen, who had changed into a white and green dress for the occasion, displayed great humility by laying a sizeable laurel wreath (decorated with a blue and red ribbon) at the foot of the Children of Lir monument (Plate 7.2).

---

[71] S. Bradford, *Queen Elizabeth II: Her Life in Our Times* (Viking, London, 2012), pp. 253–54.

[72] *The Irish Times*, 18 May 2011.

[73] N. C. Johnson, 'A Royal Encounter: Space, Spectacle and the Queen's Visit to Ireland 2011', *The Geographical Journal*, Vol. 178, No. 3 (2012), p. 194.

[74] A televised recording of the proceedings, with commentary by John Bowman, can be found in a special commemorative DVD, entitled *A Week of Welcomes: The Visits to Ireland of Queen Elizabeth II and President Barack Obama 17th–23rd May 2011* (RTÉ, 2011).

Plate 7.2   Queen Elizabeth II and President Mary McAleese after laying
                wreaths at the Garden of Remembrance on 17 May 2011
Source: Maxwell Photography.

Then, in a hugely symbolic gesture that won overwhelming public approbation
and praise, the Queen bowed her head whilst observing a minute's silence for those
who had died fighting for Irish freedom – including the rebels of 1916.

     In a momentous speech delivered at a state dinner in Dublin Castle the
following night, which opened with five words in Irish, the unassuming
Queen won further plaudits from across the political spectrum with her use
of therapeutic rhetoric. Dressed in a silk outfit adorned with over 2,000 hand-
embroidered shamrocks, she acknowledged 'the complexity of our history' and
articulated a memorable soundbite – highlighting 'the importance of forbearance
and conciliation' and of 'being able to bow to the past but not being bound by
it'. Additionally, she welcomed 'the strength of the bonds that are now in place
between the governments and the people of our two nations'. So too did President
McAleese, who was in the final year of her second seven-year term as President. In
describing the Queen's visit as 'a culmination of the success of the Peace Process'
and welcoming the dawning of 'a new chapter' in the 'contemporary British-Irish
relationship', McAleese also insinuated that the royal presence was firm proof 'that
Ireland and Britain are neighbours, equals ... and friends', in spite of the difficult
and complicated geopolitical legacies of the past. Within this context, she added

a further touch of common sense by suggesting that the Queen's stopover was 'an acknowledgement that while we cannot change the past, we have chosen to change the future'.[75]

A few days later, an opinion poll was published in *The Sunday Independent*, which indicated the seismic levels of goodwill and glee that had been generated by the friendly and matriarchal nature of the visitation. This illustrated that 95% of those polled agreed that the Queen had 'won the hearts of the Irish people', while 89% agreed that the visit had 'improved our national self-esteem as a people'.[76] The leading piece in the politics magazine, *Village*, also registered a positive verdict: 'The Queen's visit was a triumph and, like most reconciliations, worth the effort.'[77] Equally impressed was the British Ambassador, Julian King. In an address to mark the official opening of the 2011 MacGill Summer School, he spoke of how the Queen's 'successful visit' had signified a 'remarkable transformation' in Anglo-Irish relations since 'those dark, difficult days' of the 1980s.[78] Historians were also quick off the mark in delivering their own judgment on the historic connotations of the visit. Writing for *History Ireland*, Edward Madigan concluded that the Queen's 'gracious bearing' had truly impressed the Irish people. Additionally, he pronounced that her visit was 'a resounding diplomatic success', as 'so many of her official engagements involved a sensitive but direct and unflinching acknowledgement of the very mixed history of Anglo-Irish relations'. The Queen's 'reverential demeanour' at the Garden of Remembrance, he added, 'was particularly unexpected and moving', while the bow of her head proved to be 'truly historic'.[79]

Much praise was also heaped upon the President, both for the good-natured manner in which she had received the Queen and for the tireless dedication that she had given to 'building bridges' during her 14-year spell in Áras an Uachtaráin. Despite 'the human tragedy of ... post-boom bust' that saw 'economic gloom

---

[75] For the full text of both speeches, see *The Irish Times*, 19 May 2011. A transcript of the President's remarks can also be found in McAleese, *Building Bridges*, pp. 282–86.

[76] *The Sunday Independent*, 22 May 2011.

[77] Anon., 'Leader: Courage, Enda!', *Village: Ireland's Political Magazine*, Issue 14 (2011), p. 4.

[78] J. King, 'The Ties Linking Ireland and Britain Are Significant', in J. Mulholland (ed.), *Transforming Ireland, 2011–2016: Essays from the 2011 MacGill Summer School* (The Liffey Press Ltd., Dublin, 2011), p. 2.

[79] E. Madigan, 'Commemoration and Conciliation During the Royal Visit', *History Ireland*, Vol. 19, No. 4 (2011), pp. 10–11. Another positive verdict on the visit can be found in B. M. Walker, *A Political History of the Two Irelands: From Partition to Peace* (Palgrave Macmillan, Basingstoke, 2012), pp. 197–98, who notes that it impacted 'on many and in different ways' and was 'extremely successful' for the way in which history was repeatedly invoked 'in a way that included regret for past conflict, an acknowledgement of each other's traditions and history ... and a determination to move together to the future'.

overshadowing Ireland in 2011', Patrick Kiely and Dermot Keogh observed that McAleese had acted as 'a strong symbol of probity at a very difficult time in the history of the country'.[80] The renowned poet, Séamus Heaney, also portrayed the royal visit in a positive light. He described it as 'the greatest moment' in McAleese's Presidency and imaginatively reckoned that it 'set a crown – almost literally – upon a lifetime's effort'. The building 'of this particular arch', he added, meant that 'the British-Irish bridge was in place as never before'.[81] To mark her last day in office as President, McAleese contributed an article to *The Irish Times* on 10 November 2011, in which she too contemplated the positive legacy of the Queen's 'happy and healing visit'. This, she reflected, had 'showcased the extent of the mutual desire for a new and healthier relationship between our countries and how far we have travelled in creating it'.[82] In her Christmas Day 2011 broadcast from Buckingham Palace, which started and ended with performances by the Band of the Irish Guards and featured 21 seconds of footage from her trip to the Garden of Remembrance, the Queen also spoke fondly of her 'memorable and historic' visit to the Republic. 'Relationships that years ago were once so strained', she observed, 'have through sorrow and forgiveness blossomed into long term friendship', thus giving 'hope for tomorrow'.[83] In a documentary broadcast on RTÉ a couple of nights later, entitled *The Queen's Speech*, the British Prime Minister also spoke about the visit in complimentary terms, calling it 'a game-changer'. Additionally, Cameron pronounced that it had put the 'strong' Anglo-Irish friendship 'into a massive new perspective'. In the same documentary, the former President of Ireland, Mary Robinson, spoke emotively about how 'compelling' it was 'to see the way in which the Queen, at the Garden of Remembrance, which symbolises so much of the history between our countries', bow 'in a way that conveyed ... her own sense of healing'.[84]

When all was said and done, it was quite clear that moderate opinion had prevailed in the public debate over whether or not Queen Elizabeth II should visit the Republic of Ireland in 2011. Within the wider context of the pathway to 2016 and the peace framework presented by the Good Friday Agreement (and subsequently refined by the St. Andrew's and Hillsborough Castle Agreements), other realities also seemed to hit home. The widespread denunciation of the murder of Constable Kerr in the weeks before the Rising's 95th anniversary,

---

[80]   P. Kiely with D. Keogh, 'Turning Corners: Ireland 2002–11', in T. W. Moody and F. X. Martin (eds), *The Course of Irish History*, New Edition (Mercier Press, Cork, 2011), pp. 358, 377.

[81]   S. Heaney, 'Foreword', in McAleese, *Building Bridges*, p. 15.

[82]   *The Irish Times*, 10 November 2011.

[83]   The full text of the Queen's Christmas Day broadcast can be found at http://www.royal. gov.uk/imagesandbroadcasts/ (accessed on 25 December 2011).

[84]   *The Queen's Speech* (RTÉ, 2011).

coupled with the bridge-building momentum that was subsequently provided by the Queen's visit to the Garden of Remembrance, suggested that the times were indeed changing and that a new diplomatic seed had been sown for the times that lay ahead. As the old wounds inflicted to Anglo-Irish relations by the outbreak and ensuing suppression of the 1916 Rising were consigned even further to the realm of historical memory, the building-blocks for a dignified, inclusive and openminded commemorative heritage appeared to be securely in place with just a few years remaining until Easter 2016.

Bibliography

Manuscript Sources

Allen Library, Dublin
Monteith/Casement Papers.
P. H. Pearse Letters, Newscuttings, Etc.

Army Museum of Western Australia, Freemantle
Pte. Martin O'Meara Papers.
Thomas Carberry Papers.

*Bodleian Library, Oxford University, Department of
Special Collections and Western Manuscripts*
MS. Asquith.

Boole Library, University College Cork, Special Collections
Letter from Rev. James Campbell to his Parents.
Press Censorship Reports (Ireland).

British Library, London
Letter from Ivor Churchill Guest to General Maxwell.
Viscount George Cave Papers.

Cobh Heritage Centre, Cobh
Letter from Winifred Hull to Mrs. Swanton.

Cork City and County Archives, Cork
Catalogue of an Exhibition of Objects Associated with the Young Ireland
     Movement.
Minute Book of the County Cork Old IRA Benevolent Association.
Printed Republican Poster.

Dublin City Library and Archive, Dublin
Birth of the Republic: A Collection of Irish Political Ephemera and Photographs.

Dublin Diocesan Archives, Archbishop's House, Drumcondra
Archbishop William J. Walsh Papers (Laity).
Monsignor Michael Curran Papers (Political).

Imperial War Museum, London, Sound and Archive Section
The Papers of A. C. Hannat.
The Papers of A. E. Slack.
The Papers of Captain C. A. Brett.
The Papers of Captain Edward Frederick Chapman.
The Papers of Captain J. L. Horridge.
The Papers of Captain John Lowe.
The Papers of Captain O. L. Beater.
The Papers of H. L. Franklin.
The Papers of Lieutenant-Colonel James Melville Galloway.
The Papers of Richard Albert Pedler.
The Papers of W. Jones.
The Papers of W. L. Lynas.

James Hardiman Library, NUI Galway, Special Collections Reading Room
Pádraig Ó Mathúna Papers.

J. S. Battye Library of West Australian History, State Library of Western Australia, Perth
Sir John Joseph Talbot Hobbs Papers.

Kilmainham Gaol Archives, Dublin
'Kilmainham News: Bulletin of Work in Progress'.
Lorcan C. G. Leonard Papers.
Typescript Entitled 'Time Stands Still: Elpida Hadzi-Vasileva Exhibition'.

Kilmainham Gaol Museum, Dublin
Joseph McGill's Internment Order.

Library of Congress, Washington, DC, Manuscript Division
Charles Edward Russell Papers.
Robert Lansing Papers.

Military Archives, Cathal Brugha Barracks, Dublin
Department of Defence Papers.

National Archives (formerly the Public Record Office), Kew
Colonial Office Papers.
Home Office Papers.
War Office Papers.

National Archives of Ireland, Dublin
Bureau of Military History, 1913–1921, Witness Statements.
Department of External Affairs Papers.
Department of the Taoiseach Papers.

National Army Museum, Chelsea
Appeal for Funds to Support Pte. Thomas Hughes, VC, Connaught Rangers.
Documents Relating to the Royal Munster Fusiliers.
Last Address to the Connaught Rangers by King George V.

National Library of Ireland, Dublin, Department of Manuscripts
Colonel Maurice Moore Papers.
Diarmuid Lynch Papers.
Letter from Patrick O'Connor to Frank W. Poulter.
Minute Book of the Roger Casement Committee, London.
Pearse Papers.
War Bulletin by P. H. Pearse.

National Library of Scotland, Edinburgh, Manuscript and Map Collections
Correspondence of Major Arthur C. Murray.
War Diary of Lance-Corporal George Ramage.

National Maritime Museum, Greenwich
Arnold White Papers.
Journal of Walter S. Burt.

New York Public Library, Manuscripts and Archives Division
Clara Barrus Papers.
John Quinn Papers.

Parliamentary Archives, Houses of Parliament, London
The Bonar Law Papers.
The Lloyd George Papers.
Papers of John St. Loe Strachey.

Pearse Museum, Dublin
Letter from Margaret Pearse to Mr. and Mrs. McGrath.
P. H. Pearse's Oration at the Graveside of O'Donovan Rossa.
Typescript Entitled 'Brief Chronology of St. Enda's Park'.

Pontifical Irish College Archives, Rome
The Papers of John Hagan.

Princeton University Library, Department of Rare Books and Special Collections
Sir John Grenfell Maxwell Papers.

Public Record Office of Northern Ireland, Belfast
Cabinet Minutes.

Royal Irish Academy, Dublin
1916 Police Pass of Wm. Houston.
Scrapbook of Newscuttings.

Trinity College Dublin, Manuscripts and Archives Research Library
Commemorative Card for Thomas J. Clarke, P. H. Pearse and Thomas MacDonagh.
Diary of Easter Week Dictated by Douglas Hyde.
Dublin University Officer Training Corps Correspondence.
E. L. (Nelly) O'Brien's Account of Her Experiences during Easter Week.
James A. Glen's Account of Trinity College in Easter Week.
Last Letter of Thomas MacDonagh.
Letter from Gerard Fitzgibbon to William Hume Blake.
Memorial Card for Thomas J. Clarke, John Daly and John Edward Daly.
Poems by E. E. Speight and Séamus O'Sullivan.
Recollection of Easter Monday 1916 by A. A. Luce.
Samuel Bolingbrooke Reede's Recollection of the Easter Rising.

United States National Archives and Records Administration,
College Park, Maryland
General Records of the Department of State.
Records of the Foreign Service Posts of the Department of State.

University College Dublin Archives
Archives of the Fianna Fáil Party.
Archives of the Fine Gael Party.
Bernard O'Rourke Papers.
Captain Jack O'Carroll Papers.
Eoin MacNeill Papers.
Seán MacEntee Papers.

Newspapers and Periodicals

An Claideamh Soluis.
An Talam.
Belfast Telegraph.
Clare Champion.
Cork Independent.
Éire-Ireland.
Fianna.
Financial Times.
Foinse.
Gaelic Weekly.
Galway Advertiser.
Galway Express.
Galway Independent.
Honesty: An Outspoken Scrap of Paper.
Irish Catholic.
Irish Daily Mail.
Irish Examiner.
Irish Freedom.
Irish Opinion.
Irish War News.
Limerick Leader.
New Ireland: An Irish Weekly Review.
Notes from Ireland.
Roscommon Champion.
RTV Guide.
Scissors and Paste.
Sinn Féin.
The Connaught Telegraph.
The Connacht Tribune.

The Cork Examiner.
The Cork Weekly News.
The Daily Telegraph.
The Derry Journal.
The Eye-Opener.
The Factionist.
The Freeman's Journal.
The Gael.
The Gaelic American.
The Galway City Tribune.
The Guardian.
The Harp.
The Hibernian.
The Irish Citizen.
The Irish Independent.
The Irish Nation.
The Irish Press.
The Irish Times.
The Irish Volunteer.
The Irishman.
The Leader.
The Mayo News.
The Northern Standard.
The Phoenix.
The Republic.
The Sligo Champion.
The Spark.
The Sunday Business Post.
The Sunday Independent.
The Sunday Times.
The Sunday Tribune.
The Sunday World.
The Times.
The Tuam Herald.
The United Irishman.
The Workers' Republic.
Western People.

Parliamentary Debates

Houses of the Oireachtas (Dáil Éireann and Seanad Éireann),
Leinster House, Dublin
Dáil Debates.
Seanad Debates.

Websites

http://debates.oireachtas.ie (accessed on 20 July 2011 and 30 July 2012).
http://historical-debates.oireachtas.ie (accessed on 20 June 2009).
http://www.irishstatutebook.ie/1946/ (accessed on 17 January 2012).
http://www.nli.ie/1916/ (accessed on 17 July 2008 and 25 October 2011).
http://www.oireachtas-debates.gov.ie (accessed on 18 June 2009, 20 June 2009 and 21 June 2009).
http://www.president.ie/speeches/ (accessed on 6 August 2012).
http://royal.gov.uk/imagesandbroadcasts/ (accessed on 25 December 2011).
http://www.taoiseach.gov.ie/eng/ (accessed on 6 July 2006).

Filmography

1916 Seachtar na Cásca (TG4, 2010).
An Tine Bheo (Gael Linn, 1966; Reissued 2006).
A Week of Welcomes: The Visits to Ireland of Queen Elizabeth II and President Barack Obama 17th–23rd May 2011 (RTÉ, 2011).
Bertie (RTÉ, 2008).
Irish Destiny (Eppels Films Ltd., 1926; Reissued by The Irish Film Institute and RTÉ, 2006).
Making History: The Irish Historian (RTÉ Hidden History, 2007).
Michael Collins and the Easter Rising (Go Entertainment Group Ltd., 2011).
Mise Éire (Gael Linn, 1959; Reissued 2006).
Pearse: Fanatic Heart (RTÉ True Lives, 2001).
Rattle and Hum (Paramount, 1988).
Saoirse? (Gael Linn, 1961; Reissued 2007).
Somme Journey (BBC, 2002).
The 1916 Easter Rising: The 90th Anniversary Commemorative Collection (Dolphin Records Ltd., 2006).
The Man Who Lost Ireland (RTÉ Hidden History, 2006).

The Pope's Children (RTÉ, 2006).
The Queen's Speech (RTÉ, 2011).
The Wind that Shakes the Barley (Sixteen Films, 2006).

Printed Primary Sources (Official Records, Contemporary Accounts, etc.)

Ahern, B., 'Foreword', in Government of Ireland, *Towards 2016: Ten-Year Framework Social Partnership Agreement 2006–2015* (The Stationery Office, Dublin, 2006), p. 2.

Anon., *The Sinn Féin Leaders of 1916. With Fourteen Illustrations and Complete Lists of Deportees, Casualties, Etc.* (Cahill & Co., Dublin, 1917).

Anon., *Souvenir of the Official Opening of Roger Casement Park, Anderstown, Belfast, From 14th till 21st June 1953* (Conor Publications, Belfast, 1953).

Anon., *Cuimhneachán Mhuineacháin, 1916–66. Souvenir Programme* (Clogher Historical Society, Monaghan, 1966).

Anon., *A Journey of Reconciliation: The Island of Ireland Peace Park, Messines, Flanders, Belgium* (DBA Publications, Dublin, 1998).

Berresford Ellis, P. (ed.), *James Connolly: Selected Writings*, New Edition (Pluto Press, London, 1997).

Carty, J., *Bibliography of Irish History 1912–1921* (The Stationery Office, Dublin, 1936).

Central Statistics Office, *Census 2006: Principal Demographic Results* (The Stationery Office, Dublin, 2007).

Central Statistics Office, *Census 2006: Principal Socio-Economic Results* (The Stationery Office, Dublin, 2007).

Central Statistics Office, *Statistical Yearbook of Ireland 2007* (The Stationery Office, Dublin, 2007).

Central Statistics Office, *This is Ireland: Highlights from Census 2011, Part 2* (The Stationery Office, Dublin, 2012).

Department of External Affairs, *Cuimhneachán 1916–1966: A Record of Ireland's Commemoration of the 1916 Rising* (An Roinn Gnóthaí, Dublin, 1966).

Government of Ireland, *Heritage Act, 1995* (The Stationery Office, Dublin, 1995).

Government of Ireland, *National Development Plan 2007–2013: Transforming Ireland, A Better Quality of Life for All* (The Stationery Office, Dublin, 2007).

Government of Ireland, *Nelson Pillar Act, 1969* (The Stationery Office, Dublin, 1969).

Government of Ireland, *Planning and Development Act, 2000* (The Stationery Office, Dublin, 2000).

Government of Ireland, *Strategy for Science, Technology and Innovation 2006–2013* (The Stationery Office, Dublin, 2006).

Hamilton Norway, M. L., *The Sinn Féin Rebellion As I Saw It* (Smith, Elder & Co., London, 1916).

Hopkinson, M. (ed.), *Frank Henderson's Easter Rising: Recollections of a Dublin Volunteer* (Cork University Press, Cork, 1998).

MacAonghusa, P. (ed.), *What Connolly Said: James Connolly's Writings* (New Island Books, Dublin, 1995).

MacLochlainn, P. F. (ed.), *Last Words: Letters and Statements of the Leaders Executed After the Rising at Easter 1916* (Office of Public Works, Dublin, 2005).

Martin, F. X. (ed.), 'Select Documents XX. Eoin MacNeill on the 1916 Rising', *Irish Historical Studies*, Vol. 12, No. 47 (1961), pp. 226–71.

Martin, F. X. (ed.), 'Easter 1916: An Inside Report on Ulster', *Clogher Record*, Vol. 12, No. 2 (1986), pp. 192–208.

Municipal Council of the City of Dublin, *Minutes of the Municipal Council of the City of Dublin, from the 1st January to 31st December 1916* (Sealy, Bryers & Walker, Dublin, 1918).

National Gallery of Ireland, *Cuimhneachán 1916: A Commemorative Exhibition of the Irish Rebellion 1916* (Dolmen Press Ltd., Dublin, 1966).

Office of Public Works and Blackrock Teachers' Centre, *Kilmainham Gaol Document Pack: The 1916 Rising* (Office of Public Works and Blackrock Teachers' Centre, Dublin, 1992).

Office of the Minister for Integration, *Migration Nation: Statement on Integration Strategy and Diversity Management* (The Stationery Office, Dublin, 2008).

O'Hea O'Keeffe, J. and O'Keeffe, M. (eds), *Recollections of 1916 and its Aftermath: Echoes from History* (Privately Published, Kerry, 2005).

Oifig an tSoláthair, *Oidhreacht 1916–1966* (Oifig an tSoláthair, Baile Átha Cliath, 1966).

Royal Commission on the Rebellion in Ireland, *Report of the Commission: Presented to both Houses of Parliament by Command of His Majesty* (His Majesty's Stationery Office, London, 1916).

Sawyer, R. (ed.), *Roger Casement's Diaries 1910: The Black and the White* (Pimlico, London, 1997).

Stephens, J., *The Insurrection in Dublin* (Colin Smythe Ltd., Gerrard's Cross, 2000).

Tourism Policy Review Group, *New Horizons for Irish Tourism: An Agenda for Action* (Department of Arts, Sport and Tourism, Dublin, 2003).

Warwick-Haller, A. and Warwick-Haller, S. (eds), *Letters from Dublin, Easter 1916: Alfred Fannin's Diary of the Rising* (Irish Academic Press, Dublin, 1995).

Weekly Irish Times, *Sinn Féin Rebellion Handbook, Easter 1916* (Fred Hanna Ltd., Dublin, 1917).

Wells, W. B. and Marlowe, N., *A History of the Irish Rebellion of 1916* (Maunsel & Company Ltd., Dublin, 1916).

Secondary Sources

Adams, G., *Who Fears to Speak …? The Story of Belfast and the 1916 Rising*, Revised Edition (Beyond the Pale Publications, Belfast, 2001).

Adams, G., *An Irish Eye* (Brandon, Dingle, 2007).

Ahern, B. with Aldous, R., *Bertie Ahern: The Autobiography* (Hutchinson, London, 2009).

Alison Phillips, W., *The Revolution in Ireland 1906–1923*, 2nd Edition (Longmans, Green & Co. Ltd., London, 1926).

Anderson, D., *Special and Different: The History of the Police Federation for Northern Ireland* (The Police Federation for Northern Ireland, Belfast, 2012).

Anon., 'Easter Week: A Distinguished Visitor', *Irish Republican Bulletin*, Vol. 5, No. 1 (April 1947), p. 1.

Anon., *Prelude to Freedom 1916–1966: 50th Anniversary Review* (Irish Art Publications Ltd., Dublin, 1966).

Anon., 'Wilson, Harold (Baron Wilson of Rievaulx) (1916–95)', in Gardiner, J. . (ed.), *The History Today Who's Who in British History* (Collins & Brown Ltd., London, 2000), pp. 848–49.

Anon., 'Leader: Courage, Enda!', *Village: Ireland's Political Magazine*, Issue 14 (2011), p. 4.

Ashworth, G. J., Graham, B. and Tunbridge, J. E., *Pluralising Pasts: Heritage, Identity and Place in Multicultural Societies* (Pluto Press, London, 2007).

Augusteijn, J., 'Accounting for the Emergence of Violent Activism Among Irish Revolutionaries, 1916–21', *Irish Historical Studies*, Vol. 35, No. 139 (2007), pp. 327–44.

Augusteijn, J., *Patrick Pearse: The Making of a Revolutionary* (Palgrave Macmillan, Basingstoke, 2010).

Ballagh, B., '1916: Goodbye to All That?', *Irish Reporter*, No. 2 (1991), pp. 6–8.

Bannon, M. J., 'Development Planning and the Neglect of the Critical Regional Dimension', in Bannon, M. J. (ed.), *Planning: The Irish Experience 1920–1988* (Wolfhound Press, Dublin, 1989), pp. 122–57.

Bardon, J., 'Hunger Strike', in Lalor, B. (ed.), *The Encyclopaedia of Ireland* (Gill and Macmillan, Dublin, 2003), p. 507.

Barry, S., *A Long Long Way* (Faber and Faber, London, 2005).

Bartlett, T., *Ireland: A History* (Cambridge University Press, Cambridge, 2010).

Barton, B., *From Behind a Closed Door: Secret Court Martial Records of the 1916 Easter Rising* (The Blackstaff Press, Belfast, 2002).

Bateson, R., *They Died by Pearse's Side* (Irish Grave Publications, Dublin, 2010).

Beckett, I. F. W., 'Review Article: War, Identity and Memory in Ireland', *Irish Economic and Social History*, Vol. 36 (2009), pp. 63–84.

Beck, P. J., *Presenting History: Past and Present* (Palgrave Macmillan, Basingstoke, 2012).

Beiner, G., *Remembering the Year of the French: Irish Folk History and Social Memory* (The University of Wisconsin Press, Madison, 2007).

Bew, P., *Ideology and the Irish Question: Ulster Unionism and Irish Nationalism 1912–1916* (Clarendon Press, Oxford, 2002).

Bew, P., '"Why Did Jimmie Die?" A Critique of Official 1916 Commemorations', *History Ireland*, Vol. 14, No. 2 (2006), pp. 37–39.

Bew, P., *Ireland and the Politics of Enmity 1789–2006* (Oxford University Press, Oxford, 2007).

Bhreathnach-Lynch, S., 'Commemorating the Hero in Newly Independent Ireland: Expressions of Nationhood in Bronze and Stone', in McBride, L. W. (ed.), *Images, Icons and the Irish Nationalist Imagination* (Four Courts Press, Dublin, 1999), pp. 148–65.

Blair, T., *A Journey* (Hutchinson, London, 2010).

Bolger, D., 'Introduction', in Bolger, D. (ed.), *Letters from the New Island: 16 on 16. Irish Writers on the Easter Rising* (Raven Arts Press, Dublin, 1988), pp. 7–8.

Bolger, D., 'Milestone to Monument: A Personal Journey in Honour of Francis Ledwidge', in Bolger, D. (ed.), *The Ledwidge Treasury: Selected Poems* (New Island, Dublin, 2007), pp. 72–128.

Bolger, M., *Statues and Stories: Dublin's Monuments Unveiled* (Ashfield Press, Dublin, 2006).

Bonar, H., 'National Commemoration: 75th Anniversary of the Easter Rising, 1916', *An Cosantóir: The Irish Defence Journal*, Vol. 51, No. 4 (1991), pp. 4–6.

Borgonovo, J., "'Thoughtless Young People" and "The Battle of Patrick Street": The Cork City Riots of June 1917', *Journal of the Cork Historical and Archaeological Society*, Vol. 114 (2009), pp. 10–20.

Bort, E. (ed.), *Commemorating Ireland: History, Politics, Culture* (Irish Academic Press, Dublin, 2004).

Bourke, M., *The Story of Irish Museums 1790–2000: Culture, Identity and Education* (Cork University Press, Cork, 2011).

Bower, P., 'Appendix. Paper History and Analysis as a Research Procedure' in Daly, M. (ed.), *Roger Casement in Irish and World History* (Royal Irish Academy, Dublin, 2005), pp. 243–52.

Bowman, J., *De Valera and the Ulster Question 1917–1973* (Clarendon Press, Oxford, 1982).

Bowman, J., *Window and Mirror. RTÉ Television: 1961–2011* (The Collins Press, Cork, 2011).

Bowman, T., 'The Ulster Volunteer Force and the Formation of the 36th (Ulster) Division', *Irish Historical Studies*, Vol. 32, No. 128 (2001), pp. 498–515.

Bradford, S., *Queen Elizabeth II: Her Life in Our Times* (Viking, London, 2012).

Bradshaw, B., 'Nationalism and Historical Scholarship in Ireland', *Irish Historical Studies*, Vol. 26, No. 104 (1989), pp. 329–51.

Brady, C., "'Constructive and Instrumental": The Dilemma of Ireland's First "New Historians"', in Brady, C. (ed.), *Interpreting Irish History: The Debate on Historical Revisionism, 1938–1994* (Irish Academic Press, Dublin, 1994), pp. 3–31.

Breen, D., *My Fight for Irish Freedom* (Anvil Books, Dublin, 1989).

Brendon, P., *The Decline and Fall of the British Empire 1781–1997* (Alfred A. Knopf, New York, 2008).

Brown, M., *The Imperial War Museum Book of the First World War* (Pan Books, London, 2002).

Bruce, S., *Paisley: Religion and Politics in Northern Ireland* (Oxford University Press, Oxford, 2009).

Burke, T., "'Poppy Day" in the Irish Free State', *Studies: An Irish Quarterly Review*, Vol. 92 (2003), pp. 349–58.

Busteed, M. A. and Mason, H., 'The 1973 General Election in the Republic of Ireland', *Irish Geography*, Vol. 7 (1974), pp. 97–106.

Byatt, A. S., 'Introduction', in Harvey Wood, H. and Byatt, A. S. (eds), *Memory: An Anthology* (Vintage Books, London, 2009), pp. xii–xx.

Campbell, F., 'The Easter Rising in Galway', *History Ireland*, Vol. 14, No. 2 (2006), pp. 22–25.

Campbell, F., "'Reign of Terror at Craughwell": Tom Kenny and the McGoldrick Murder of 1909', *History Ireland*, Vol. 18, No. 1 (2010), pp. 26–29.

Campbell, M., 'Emigrant Responses to War and Revolution, 1914–21: Irish Opinion in the United States and Australia', *Irish Historical Studies*, Vol. 32, No. 125 (2000), pp. 75–92.

Campbell-Smith, D., *Masters of the Post: The Authorised History of the Royal Mail* (Allen Lane, London, 2011).

Carman, J. and Sørensen, M. L. S., 'Heritage Studies: An Outline', in Carman, J. and Sørensen, M. L. S. (eds), *Heritage Studies: Methods and Approaches* (Routledge, London, 2009), pp. 11–28.

Carroll, D. (ed.), *The Wind that Shakes the Barley* (Galley Head Press, Cork, 2006).

Caulfield, M, *The Easter Rebellion*, New Edition (Gill and Macmillan, Dublin, 1995).

Clifford, M. and Coleman, S., *Bertie Ahern and the Drumcondra Mafia* (Hachette Books Ireland, Dublin, 2009).

Coleman, M., *The Best is Yet to Come* (Blackhall Publishing, Dublin, 2007).

Collingwood, R. G., *The Idea of History* (Oxford University Press, Oxford, 1970).

Collins, L., *16 Lives: James Connolly* (The O'Brien Press, Dublin, 2012).

Collins, S., 'The Election Background', in Collins, S. (ed.), *Nealon's Guide to the 30th Dáil & 23rd Seanad* (Gill and Macmillan, Dublin, 2007), pp. 6–7.

Connell Jnr., J. E. A., *Where's Where in Dublin: A Directory of Historic Locations, 1913–1923. The Great Lockout, The Easter Rising, The War of Independence, The Irish Civil War* (Dublin City Council, Dublin, 2006).

Coogan, T. P., *The Troubles: Ireland's Ordeal 1966–1995 and the Search for Peace* (Hutchinson, London, 1995).

Coogan, T. P., *1916: The Easter Rising* (Cassell & Co., London, 2001).

Coogan, T. P., *A Memoir* (Weidenfeld and Nicolson, London, 2008).

Cooke, P., *Scéal Scoil Éanna: The Story of an Educational Adventure* (Office of Public Works, Dublin, 1986).

Cooke, P., *A History of Kilmainham Gaol 1796–1924* (Dúchas The Heritage Service, Dublin, 2001).

Cooney, J., *'Battleship Bertie': Politics in Ahern's Ireland* (Blantyremoy Publications, Dublin, 2008).

Corsane, G., 'Issues in Heritage, Museums and Galleries: A Brief Introduction', in Corsane, G. (ed.), *Heritage, Museums and Galleries: An Introductory Reader* (Routledge, London, 2005), pp. 1–12.

Costello, F., *The Irish Revolution and its Aftermath 1916–1923* (Irish Academic Press, Dublin, 2003).

Cottrell, P., *The Anglo-Irish War: The Troubles of 1913–1922* (Osprey Publishing Ltd., Oxford, 2006).

Crawford, H. C., *Outside the Glow: Protestants and Irishness in Independent Ireland* (University College Dublin Press, Dublin, 2010).

Cronin, S., *Our Own Red Blood: The Story of the 1916 Rising* (Irish Freedom Press, Dublin, 2006).

Crooke, E., *Politics, Archaeology and the Creation of a National Museum of Ireland: An Expression of National Life* (Irish Academic Press, Dublin, 2000).

Cruise O'Brien, C., *States of Ireland* (Hutchinson & Co. Ltd., London, 1972).

Cruise O'Brien, C., *Neighbours: The Ewart-Biggs Memorial Lectures 1978–1979* (Faber and Faber, London, 1980).

Cruise O'Brien, C., *Memoir: My Life and Times* (Poolbeg, Dublin, 1998).

Cubitt, G., *History and Memory* (Manchester University Press, Manchester, 2007).

Curtin, N. J., '"Varieties of Irishness": Historical Revisionism, Irish style', *Journal of British Studies*, Vol. 35 (1994), pp. 195–219.

Curtis, L. P., 'The Greening of Irish History', *Éire-Ireland*, Vol. 29 (1994), pp. 7–28.

Daly, M., 'Irish Nationality and Citizenship since 1922', *Irish Historical Studies*, Vol. 32, No. 127 (2001), pp. 377–407.

Daly, M. (ed.), *Roger Casement in Irish and World History* (Royal Irish Academy, Dublin, 2005).

Daly, M., 'Forty Shades of Grey?: Irish Historiography and the Challenges of Multidisciplinarity', in Harte, L. and Whelan, Y. (eds), *Ireland Beyond Boundaries: Mapping Irish Studies in the Twenty-First Century* (Pluto Press, London, 2007), pp. 92–110.

Daly, M. and O'Callaghan, M. (eds), *1916 in 1966: Commemorating the Easter Rising* (Royal Irish Academy, Dublin, 2007).

D'Arcy, F. A., *Remembering the War Dead: British Commonwealth and International War Graves in Ireland since 1914* (Office of Public Works, Dublin, 2007).

Davies, N., *Europe at War 1939–1945: No Simple Victory* (Macmillan, London, 2006).

Davies, N., *Vanished Kingdoms: The History of Half-Forgotten Europe* (Allen Lane, London, 2011).

De Rosa, P., *Rebels: The Irish Rising of 1916* (Bantham Press, London, 1990).

Desmond Greaves, C., *The Easter Rising in Song and Ballad* (Kahn and Averill, London, 1980).

Desmond Greaves, C., *The Life and Times of James Connolly*, New Edition (Lawrence and Wishart Ltd., London, 1986).

Desmond Greaves, C., *1916 as History: The Myth of Blood Sacrifice* (The Fulcrum Press, Dublin, 1991).

Desmond Greaves, C., *Liam Mellows and the Irish Revolution* (An Ghlór Gafa, Belfast, 2004).

Desmond Williams, T., 'Ireland and the War', in Nowlan, K. B. and Desmond Williams, T. (eds), *Ireland in the War Years and After, 1939–51* (Gill and Macmillan, Dublin, 1969), pp. 14–27.

Dillon, A., *Casement* (Haus Publishing, London, 2003).

Doherty, G., 'Modern Ireland', in Duffy, S. (ed.), *Atlas of Irish History*, 2nd Edition (Gill and Macmillan, Dublin, 2000), pp. 96–131.

Doherty, G., 'The Commemoration of the Ninetieth Anniversary of the Easter Rising', in Doherty, G. and Keogh, D. (eds), *1916: The Long Revolution* (Mercier Press, Cork, 2007), pp. 376–407.

Doherty, G. and Keogh, D. (eds), *1916: The Long Revolution* (Mercier Press, Cork, 2007).

Dolan, A., 'An Army of Our Fenian Dead: Republicanism, Monuments and Commemoration', in McGarry, F. (ed.), *Republicanism in Modern Ireland* (University College Dublin Press, Dublin, 2003), pp. 132–44.

Dolan, A., *Commemorating the Irish Civil War: History and Memory, 1923–2000* (Cambridge University Press, Cambridge, 2003).

Dolan, J. P., *The Irish Americans: A History* (Bloomsbury Press, New York, 2008).

Doolan, L., 'Foreword', in O'Donohue, J., *Echoes of Memory*, New Edition (Transworld Ireland, London, 2009), pp. 1–11.

Donnelly, M. and Norton, C., *Doing History* (Routledge, London, 2011).

Donnelly, S., 'How the Nation Voted', in McCarthy, D. (ed.), *RTÉ The Week in Politics: Election 2011 & the 31st Dáil* (RTÉ Publishing, Dublin, 2011), pp. 14–19.

Doorley, M., 'The Friends of Irish Freedom: A Case Study in Irish-American Nationalism, 1916–21', *History Ireland*, Vol. 16, No. 2 (2008), pp. 22–27.

Downing, J., *'Most Skilful, Most Devious, Most Cunning': A Political Biography of Bertie Ahern* (Blackwater Press, Dublin, 2004).

Dudley Edwards, O. and Pyle, F. (eds), *1916: The Easter Rising* (MacGibbon & Kee, London, 1968).

Dudley Edwards, R. with Hourican, B., *An Atlas of Irish History*, 3rd Edition (Routledge, London, 2005).

Dudley Edwards, R., *Patrick Pearse: The Triumph of Failure* (Irish Academic Press, Dublin, 2006).

Dunne, T., 'New Histories: Beyond Revisionism', *The Irish Review*, No. 12 (1992), pp. 1–12.

Dunne, T. and Geary, L. M., 'Introduction', in Dunne, T. and Geary, L. M. (eds), *History and the Public Sphere: Essays in Honour of John A. Murphy* (Cork University Press, Cork, 2005), pp. 3–5.

Edwards, N., *The Archaeology of Early Medieval Ireland* (B. T. Batsford Ltd., London, 1990).

Elliott, S. and Flackes, W. D., *Northern Ireland: A Political Directory 1968–1999*, 5th Edition (The Blackstaff Press, Belfast, 1999).

Ellis, S. G., 'Writing Irish History: Revisionism, Colonialism, and the British Isles', *The Irish Review*, No. 19 (1996), pp. 1–21.

English, R., *Armed Struggle: A History of the IRA* (Macmillan, London, 2003).

English, R., *Irish Freedom: The History of Irish Nationalism* (Macmillan, London, 2006).

English, R., 'Directions in Historiography: History and Irish Nationalism', *Irish Historical Studies*, Vol. 37, No. 147 (2011), pp. 447–60.

Evans, E., *Prehistoric and Early Christian Ireland: A Guide* (B. T. Batsford, London, 1966).

Fact Pack Travellers' Guides, *Glasnevin Cemetery* (Morrigan Books, Killala, 1997).

Farrell Moran, S., *Patrick Pearse and the Politics of Redemption: The Mind of the Easter Rising, 1916* (The Catholic University of America Press, Washington, 1997).

Farren, S., '1916 Rising: The View from the North', *Magill: Ireland's Political and Cultural Monthly* (April, 2006), pp. 54–55.

Father Henry, 'Reamhrá', *The Capuchin Annual 1966*, No. 33 (1966), p. 152.

Fedorowich, K., 'The Problems of Disbandment: The Royal Irish Constabulary and Imperial Migration, 1919–29', *Irish Historical Studies*, Vol. 30, No. 117 (1996), pp. 88–110.

Feeney, B., *Pocket History of the Troubles* (The O'Brien Press, Dublin, 2004).

Ferguson, N., *The Pity of War* (Penguin Books, London, 1999).

Ferguson, N., *Empire: How Britain Made the Modern World* (Allen Lane, London, 2003).

Ferguson, N., 'Introduction. Virtual History: Towards a "Chaotic" Theory of the Past', in Ferguson, N. (ed.), *Virtual History: Alternatives and Counterfactuals* (Pan Books, London, 2003), pp. 1–90.

Ferguson, N., *The War of the World: History's Age of Hatred* (Allen Lane, London, 2006).

Ferguson, S., *'Self Respect and a Little Extra Leave': GPO Staff in 1916* (An Post, Dublin, 2005).

Ferguson, S., *At the Heart of Events: Dublin's General Post Office* (An Post, Dublin, 2007).

Ferguson, S., *Letters, Lives & Liberty at the An Post Museum* (An Post, Dublin, 2011).

Ferriter, D., *The Transformation of Modern Ireland 1900–2000* (Profile Books Ltd., London, 2004).

Ferriter, D., *What If? Alternative Views of Twentieth-Century Ireland* (Gill and Macmillan, Dublin, 2006).

Ferriter, D., 'Commemorating the Rising, 1922–65: "A Figurative Scramble for the Bones of the Patriot Dead"?', in Daly, M. and O'Callaghan, M. (eds), *1916 in 1966: Commemorating the Easter Rising* (Royal Irish Academy, Dublin, 2007), pp. 198–218.

Ferriter, D., *Judging Dev: A Reassessment of the Life and Legacy of Eamon de Valera* (Royal Irish Academy, Dublin, 2007).

Ferriter, D., *Occasions of Sin: Sex and Society in Modern Ireland* (Profile Books, London, 2009).

Finnan, J. P., '*Punch's* Portrayal of Redmond, Carson and the Irish Question, 1910-18', *Irish Historical Studies*, Vol. 33, No. 132 (2003), pp. 424–51.

Fitzgerald, D., *Desmond's Rising: Memoirs, 1913 to Easter 1916* (Liberties Press, Dublin, 2006).

Fitzgerald, G., 'The Significance of 1916', *Studies: An Irish Quarterly Review*, Vol. 55, No. 217 (1966), pp. 29–37.

Fitzgerald, G., *Towards a New Ireland* (Charles Knight & Co. Ltd., London, 1972).

Fitzgerald, G., 'Eamon de Valera: The Price of His Achievement', in Doherty, G. and Keogh, D. (eds), *De Valera's Irelands* (Mercier Press, Cork, 2003), pp. 185–204.

Fitzgerald, G., *Reflections on the Irish State* (Irish Academic Press, Dublin, 2003).

Fitzhenry, E. C., *Nineteen-Sixteen: An Anthology* (Browne and Nolan Ltd., Dublin, 1935).

Fitzpatrick, D., '"Decidedly a Personality": De Valera's Performance as a Convict, 1916–1917', *History Ireland* Vol. 10, No. 2 (2002), pp. 40–46.

Footman, T., *The Noughties: A Decade that Changed the World 2000–2009* (Crimson Publishing, Richmond, 2009).

Forest, B. and Johnson, J., 'Unravelling the Threads of History: Soviet-Era Monuments and Post-Soviet National Identity in Moscow', *Annals of the Association of American Geographers*, Vol. 92, No. 3 (2002), pp. 524–47.

Foster, J. K., *Memory: A Short Introduction* (Oxford University Press, Oxford, 2009).

Foster, R. F., 'We Are All Revisionists Now', *The Irish Review*, No. 1 (1986), pp. 1–5.

Foster, R. F., *Modern Ireland 1600–1972* (Allen Lane, London, 1988).

Foster, R. F., *The Irish Story: Telling Tales and Making it up in Ireland* (Allen Lane, London, 2001).

Foster, R. F., *W. B. Yeats: A Life. II. The Arch-Poet 1915–1939* (Oxford University Press, Oxford, 2005).

Foster, R., 'The Red and the Green', *The Dublin Review*, No. 24 (2006), pp. 43–51.

Foster, R. F., *Luck & the Irish: A Brief History of Change, 1970–2000* (Allen Lane, London, 2007).

Foy, M. and Barton, B., *The Easter Rising* (Sutton Publishing Ltd., Stroud, 1999).

Fraser, T. G., *Ireland in Conflict 1922–1998* (Routledge, London, 2000).

Fussell, P., *The Great War and Modern Memory*, 25th Anniversary Edition (Oxford University Press, Oxford, 2000).

Gailey, A., 'King Carson: An Essay on the Invention of Leadership', *Irish Historical Studies*, Vol. 30, No. 117 (1996), pp. 66–87.

Garden, M.-C. E., 'The Heritagescape: Looking at Landscapes of the Past', *International Journal of Heritage Studies*, Vol. 12, No. 5 (2006), pp. 394–411.

Garvin, T., 'Introduction', in McDonald, W., *Some Ethical Questions of Peace and War: With Special Reference to Ireland* (University College Dublin Press, Dublin, 1998), pp. xi–xvii.

Garvin, T., *Preventing the Future: Why was Ireland So Poor for So Long?* (Gill and Macmillan, Dublin, 2005).

Garvin, T., *Judging Lemass: The Measure of the Man* (Royal Irish Academy, Dublin, 2009).

Garvin, T., *News from a New Republic: Ireland in the 1950s* (Gill and Macmillan, Dublin, 2010).

Geary, L. M. (ed.), *Rebellion and Remembrance in Modern Ireland* (Four Courts Press, Dublin, 2001).

George Boyce, D., '1916, Interpreting the Rising', in George Boyce, D. and O'Day, A. (eds), *The Making of Modern Irish History: Revisionism and the Revisionist Controversy* (Routledge, London, 1996), pp. 163–87.

Geraghty, D., *40 Shades of Green: A Wry Look at What it Means to be Irish* (Real Ireland Design, Kilcoole, 2007).

Gibbons, L., *Transformations in Irish Culture* (Cork University Press, Cork, 1996).

Gilmore, E. with Thornley, Y., *Leading Lights: People Who've Inspired Me* (Liberties Press, Dublin, 2010).

Girvin, B., *The Emergency: Neutral Ireland 1939–45* (Macmillan, London, 2006).

Githens-Mazer, J., *Myths and Memories of the Easter Rising: Cultural and Political Nationalism in Ireland* (Irish Academic Press, Dublin, 2006).

Gough, P., 'Corporations and Commemoration: First World War Remembrance, Lloyds TSB and the National Memorial Arboretum' *International Journal of Heritage Studies*, Vol. 10, No. 5 (2004), pp. 435–55.

Gove, P. B. (ed.), *Webster's Third New International Dictionary of the English Language Unabridged* (G. and C. Merriam Company, Springfield, 1976).

Grace, D., 'Soldiers from Nenagh and District in World War 1', *Tipperary Historical Journal*, No. 12 (1999), pp. 44–51.

Graff-McRae, R., *Remembering and Forgetting 1916: Commemoration and Conflict in Post-Peace Process Ireland* (Irish Academic Press, Dublin, 2010).

Graham, B. J., 'Preface', in Graham, B. J. (ed.), *In Search of Ireland: A Cultural Geography* (Routledge, London, 1997), pp. xi–xii.

Graham, B., Ashworth, G. J. and Tunbridge, J. E., *A Geography of Heritage* (Arnold, London, 2002).

Graham, B. and Howard, P., 'Heritage and Identity', in Graham, B. and Howard, P. (eds), *The Ashgate Research Companion to Heritage and Identity* (Ashgate, Aldershot, 2008), pp. 1–15.

Graham, T., 'From the Editor', *History Ireland*, Vol. 14, No. 2 (2006), p. 4.

Graham, T., 'Counting Down to 2016', *History Ireland*, Vol. 19, No. 1 (2011), p. 3.

Haggett, P., *Geography: A Global Synthesis* (Pearson Education, Harlow, 2001).

Hall, C. M. and Page, S. J., *The Geography of Tourism and Recreation: Environment, Place and Space*, 3rd Edition (Routledge, London, 2006).

Hamilton, H., *The Speckled People* (Harper Perennial, London, 2004).

Hanley, B., 'Film Eye: *The Wind that Shakes the Barley*', *History Ireland*, Vol. 14, No. 5 (2006), pp. 50–51.

Hanley, B. and Millar, S., *The Lost Revolution: The Story of the Official IRA and the Workers' Party* (Penguin Ireland, Dublin, 2009).

Harrison, R., Fairclough, G., Jameson Jnr., J. H. and Schofield, J., 'Introduction: Heritage, Memory and Modernity', in Fairclough, G., Harrison, R., Jameson Jnr., J. H. and Schofield, J. (eds), *The Heritage Reader* (Routledge, London, 2008), pp. 1–12.

Hart, P., 'What Did the Easter Rising Really Change?', in T. E. Hachey (ed.), *Turning Points in Twentieth-Century Irish History* (Irish Academic Press, Dublin, 2011), pp. 7–20.

Harvey, D., 'The History of Heritage', in Graham, B. and Howard, P. (eds), *The Ashgate Research Companion to Heritage and Identity* (Ashgate, Aldershot, 2008), pp. 19–36.

Hayes-McCoy, G. A., 'A Military History of the 1916 Rising', in Nowlan, K. B. (ed.), *The Making of 1916: Studies in the History of the Rising* (The Stationery Office, Dublin, 1969), pp. 255–338.

Hay, M., 'The Foundation and Development of Na Fianna Éireann, 1909–16', *Irish Historical Studies*, Vol. 36, No. 141 (2008), pp. 53–71.

Healy, A., *Athenry: A Brief History and Guide* (The Connacht Tribune Ltd., Galway, 1989).

Heaney, S., 'Foreword', in McAleese, M., *Building Bridges: Selected Speeches and Statements* (The History Press Ireland, Dublin, 2011), pp. 11–15.

Hegarty, S. and O'Toole, F., *The Irish Times Book of the 1916 Rising* (Gill and Macmillan, Dublin, 2006).

Henderson, J. C., 'The Meanings, Marketing and Management of Heritage Tourism in Southeast Asia', in Timothy, D. J. and Nyuapane, G. P. (eds), *Cultural Heritage and Tourism in the Developing World* (Routledge, London, 2009), pp. 73–92.

Henry, W., *Supreme Sacrifice: The Story of Éamonn Ceannt 1881–1916* (Mercier Press, Cork, 2005).

Herlihy, J., 'Preface', in Clifford, B. and Herlihy, J. (eds), *Envoi: Taking Leave of Roy Foster: Reviews of His Made Up Irish Story* (Aubane Historical Society, Cork, 2006), pp. 6–8.

Higgins, M. D., *Renewing the Republic* (Liberties Press, Dublin, 2011).

Higgins, R., '"The Constant Reality Running through Our Lives": Commemorating Easter 1916', in Harte, L., Whelan, Y. and Crotty, P. (eds), *Ireland: Space, Text, Time* (The Liffey Press, Dublin, 2005), pp. 45–56.

Higgins, R., Holohan, C. and O'Donnell, C., '1966 and All That: The 50th Anniversary Commemorations', *History Ireland*, Vol. 14, No. 2 (2006), pp. 31–36.

Higgins, R. and Uí Chollatáin, R. (eds), *The Life and After-Life of P. H. Pearse* (Irish Academic Press, Dublin, 2009).

Higgins, R. and Uí Chollatáin, R., 'Introduction', in Higgins, R. and Uí Chollatáin, R. (eds), *The Life and After-Life of P. H. Pearse* (Irish Academic Press, Dublin, 2009), pp. xvii–xxiii.

Hill, J., *Irish Public Sculpture: A History* (Four Courts Press, Dublin, 1998).

Hobsbawm, E. with Polito, A., *On the Edge of the New Century* (The New Press, New York, 2000).

Hopkinson, M., 'Civil War', in Lalor, B. (ed.), *The Encyclopaedia of Ireland* (Gill and Macmillan, Dublin, 2003), p. 202.

Hopkinson, M., *Green Against Green: The Irish Civil War* (Gill and Macmillan, Dublin, 2004).

Hopkinson, M., *The Irish War of Independence* (Gill and Macmillan, Dublin, 2004).

Horgan, J., *Seán Lemass: The Enigmatic Patriot* (Gill and Macmillan, Dublin, 1999).

Horgan, T., *Christy Ring: Hurling's Greatest* (The Collins Press, Cork, 2008).

Hourihan, A. M., 'A Series of Accidents', *The Dublin Review*, No. 23 (2006), pp. 96–110.

Howard, P., *Heritage: Management, Interpretation, Identity* (Continuum, London, 2003).

Howe, S., *Ireland and Empire: Colonial Legacies in Irish History and Culture* (Oxford University Press, Oxford, 2000).

Hussey, G., *Ireland Today: Anatomy of a Changing State* (Town House, Dublin, 1993).

Ivory, G., 'The Meanings of Republicanism in Contemporary Ireland', in Honohan, I. (ed.), *Republicanism in Ireland: Confronting Theories and Traditions* (Manchester University Press, Manchester, 2008), pp. 85–106.

Jeffery, K., *Ireland and the Great War* (Cambridge University Press, Cambridge, 2000).

Jeffery, K., *The GPO and the Easter Rising* (Irish Academic Press, Dublin, 2006).

Jenkins, R., *Churchill* (Pan Books, London, 2002).

Johnson, N. C., *Ireland, the Great War and the Geography of Remembrance* (Cambridge University Press, Cambridge, 2003).

Johnson, N. C., 'A Royal Encounter: Space, Spectacle and the Queen's Visit to Ireland 2011', *The Geographical Journal*, Vol. 178, No. 3 (2012), pp. 194–200.

Johnston-Liik, E. M., *MPs in Dublin: Companion to the History of the Irish Parliament 1692–1800* (Ulster Historical Foundation, Belfast, 2006).

Jordan, A. J., *To Laugh or To Weep: A Biography of Conor Cruise O'Brien* (Blackwater Press, Dublin, 1994).

Joyce, J. and Murtagh, P., *The Boss* (Poolbeg Press Ltd., Dublin, 1997).

Judt, T., *Postwar: A History of Europe since 1945* (Pimlico, London, 2007).

Kautt, W. H., 'Studying the Irish Revolution as Military History: Ambushes and Armour', *The Irish Sword: The Journal of the Military History Society of Ireland*, Vol. 27 (2010), pp. 253–72.

Kay, S., *Celtic Revival? The Rise, Fall and Renewal of Global Ireland* (Rowman & Littlefield Publishers, Lanham, 2011).

Kearney, H. F., 'Visions and Revisions: Views of Irish History', *The Irish Review*, No. 27 (2001), pp. 113–20.

Kearns, G., 'Bare Life, Political Violence and the Territorial Structure of Britain and Ireland', in Gregory, D. and Pred, A. (eds), *Violent Geographies: Fear, Terror and Political Violence* (Routledge, New York, 2007), pp. 7–35.

Kearns, K. C., *The Bombing of Dublin's North Strand, 1941: The Untold Story* (Gill and Macmillan, Dublin, 2009).

Keena, C., *Bertie: Money and Power* (Gill and Macmillan, Dublin, 2011).

Kelly, C. and Ní Laoire, C., 'Representing Multiple Irish Heritage(s): A Case Study of the Ulster-American Folk Park', *Irish Geography*, Vol. 38, No. 1 (2005), pp. 72–83.

Kelly, S., *Rule 42 and All That* (Gill and Macmillan, Dublin, 2008).

Kelly, S., *Breakfast with Anglo* (Penguin Books, London, 2011).

Kelly, S. P., *Pictorial Review of 1916: A Complete and Historically Accurate Account of the Events which Occurred in Dublin in Easter Week, Fully Illustrated* (The Parkside Press Ltd., Dublin, 1946).

Kenneally, I., *The Paper Wall: Newspapers and Propaganda in Ireland 1919– 1921* (The Collins Press, Cork, 2008).

Kennedy, M., *Guarding Neutral Ireland: The Coast Watching Service and Military Intelligence, 1939–1945* (Four Courts Press, Dublin, 2009).

Kenny, M., *The Road to Freedom: Photographs and Memorabilia from the 1916 Rising and Afterwards* (Country House/The National Museum of Ireland, Dublin, 2001).

Kenny, M., *Crown and Shamrock: Love and Hate Between Ireland and the British Monarchy* (New Island, Dublin, 2009).

Kenny, T., *Galway: Politics and Society, 1910–23* (Four Courts Press, Dublin, 2011).

Keogh, D., *Jack Lynch: A Biography* (Gill and Macmillan, Dublin, 2008).

Keogh, D., 'Jack Lynch and the Defence of Democracy in Ireland, August 1969– June 1970', *History Ireland*, Vol. 17, No. 4 (2009), pp. 28–31.

Kiberd, D., 'The Elephant of Revolutionary Forgetfulness', in Ní Dhonnchadha, M. and Dorgan, T. (eds), *Revising the Rising* (Field Day, Derry, 1991), pp. 1–20.

Kiberd, D., '1916: The Idea and the Action', in Devine, K. (ed.), *Modern Irish Writers and the Wars* (Colin Smythe Ltd., Gerrard's Cross, 1999), pp. 18–35.

Kiberd, D., *Inventing Ireland: The Literature of the Modern Nation* (Vintage Books, London, 2006).

Kiely, P. and Keogh, D., 'Turning Corners: Ireland 2002–11', in Moody, T. W. and Martin, F. X. (eds), *The Course of Irish History*, New Edition (Mercier Press, Cork, 2011), pp. 358–97.

Killeen, R., *A Short History of the Irish Revolution 1912–1927* (Gill and Macmillan, Dublin, 2007).

Killeen, R., *A Short History of the 1916 Rising* (Gill and Macmillan, Dublin, 2009).

Kilmainham Jail Restoration Society, *Ghosts of Kilmainham* (Kilmainham Jail Restoration Society, Dublin, 1963).

King, J., 'The Ties Linking Ireland and Britain Are Significant', in Mulholland, J. (ed.), *Transforming Ireland, 2011–2016: Essays from the 2011 MacGill Summer School* (The Liffey Press Ltd., Dublin, 2011), pp. 1–5.

King, L., 'Text as Image: The Proclamation of the Irish Republic', in Sisson, E. (ed.), *History/Technology/Criticism: A Collection of Essays* (Dun Laoghaire Institute of Art Design and Technology, Dublin, 2001), pp. 4–7.

Kissane, B., 'Defending Democracy? The Legislative Response to Political Extremism in the Irish Free State, 1922–39', *Irish Historical Studies*, Vol. 34, No. 134 (2004), pp. 156–74.

Küchler, S., 'Landscape as Memory', in Bender, B. (ed.), *Landscape Politics and Perspectives* (Berg Publishers Ltd., Oxford, 1993), pp. 85–106.

Kundera, M., *The Book of Laughter and Forgetting. Translated from the Czech by Michael Henry Heim* (Penguin Books Ltd., Harmondsworth, 1984).

Kundera, M., *Ignorance. Translated from the French by Linda Asher* (Faber and Faber, London, 2003).

Laffan, M., *The Resurrection of Ireland: The Sinn Féin Party 1916–1923* (Cambridge University Press, Cambridge, 2005).

Lanchester, J., *Whoops! Why Everyone Owes Everyone and No One Can Pay* (Allen Lane, London, 2010).

Leahy, P., *Showtime: The Inside Story of Fianna Fáil in Power* (Penguin Ireland, Dublin, 2009).

Lee, J. J., *Ireland 1912–1985: Politics and Society* (Cambridge University Press, Cambridge, 1989).

Lee, J. J., 'Irish History', in Buttimer, N., Rynne, C. and Guerin, H. (eds), *The Heritage of Ireland* (The Collins Press, Cork, 2000), pp. 117–36.

Lee, J. J., '1916 as Virtual History', *History Ireland*, Vol. 14, No. 2 (2006), pp. 5–6.

Lemass, S., 'I Remember 1916', *Studies: An Irish Quarterly Review*, Vol. 55, No. 217 (1966), pp. 7–9.

Lemass, S., 'The Meaning of the Commemoration', *Comorú na Casca Digest: The Easter Commemoration Digest*, Vol. 8 (1966), pp. 12–19.

Leonard, A. (ed.), *The Junior Dean R. B. McDowell: Encounters with a Legend* (The Lilliput Press, Dublin, 2003).

Le Roux, L. N., *La Vie de Patrice Pearse* (Imprimerie Commerciale de Bretagne, Rennes, 1932).

Limond, D., '[Re]moving Statues', *History Ireland*, Vol. 18, No. 2 (2010), pp. 10–11.

Litton, H. (ed.), *Revolutionary Woman: Kathleen Clarke 1878–1972. An Autobiography* (The O'Brien Press, Dublin, 1991).

Lloyd, D. *Ireland after History* (Cork University Press, Cork, 1999).

Lowe, W. J., 'Irish Constabulary Officers, 1837–1922: Profile of a Professional Elite', *Irish Economic and Social History*, Vol. 32 (2005), pp. 19–46.

Lowenthal, D., *The Past is a Foreign Country* (Cambridge University Press, Cambridge, 1985).

Lowenthal, D., *The Heritage Crusade and the Spoils of History* (Cambridge University Press, Cambridge, 1998).

Lowenthal, D., 'Restoration: Synoptic Reflections', in Daniels, S., DeLyser, D., Nicholas Entrikin, J. and Richardson, D. (eds), *Envisioning Landscapes, Making Worlds: Geography and the Humanities* (Routledge, London, 2011), pp. 209–26.

Lowry, D., 'The Captive Dominion: Imperial Realities behind Irish Diplomacy, 1922-49', *Irish Historical Studies*, Vol. 36, No. 142 (2008), pp. 202–26.

Lynch, B., 'TV Eye: "Through the Eyes of 1916", RTÉ, 10–17 April 1966, Insurrection', *History Ireland*, Vol. 14, No. 2 (2006), pp. 54–57.

Lynch, D. J., *When the Luck of the Irish Ran Out: The World's Most Resilient Country and its Struggle to Rise Again* (Palgrave Macmillan, New York, 2010).

Lyons, F. S. L., *Ireland since the Famine* (Fontana Press, London, 1986).

MacCarron, D., *The Irish Defence Forces since 1922* (Osprey Publishing Ltd., Oxford, 2004).

MacDonald, I., *The People's Revolution* (Pimlico, London, 2003).

MacÉinrí, P. and White, A., 'Immigration into the Republic of Ireland: A Bibliography of Recent Research', *Irish Geography*, Vol. 41, No. 2 (2008), pp. 151–79.

MacGiollarnáth, S., 'Patrick H. Pearse: A Sketch of His Life', *Journal of the Galway Archaeological and Historical Society*, Vol. 57 (2005), pp. 139–50.

MacLaughlin, J., *Reimagining the Nation-State: The Contested Terrains of Nation-Building* (Pluto Press, London, 2001).

McAleese, M., '1916: A View from 2006', in Doherty, G. and Keogh, D. (eds), *1916: The Long Revolution* (Mercier Press, Cork, 2007), pp. 24–29.

McAleese, M., *Building Bridges: Selected Speeches and Statements* (The History Press Ireland, Dublin, 2011).

McBride, I. (ed.), *History and Memory in Modern Ireland* (Cambridge University Press, Cambridge, 2001).

McBride, L. W. (ed.), *Images, Icons and the Irish Nationalist Imagination* (Four Courts Press, Dublin, 1999).

McCabe, C., *Sins of the Father: Tracing the Decisions that Shaped the Irish Economy* (The History Press Ireland, Dublin, 2011).

McCaffrey, B., *Alex Maskey: Man and Mayor* (The Brehon Press, Belfast, 2003).

McCarthy, C., *Cumann na mBan and the Irish Revolution* (The Collins Press, Cork, 2007).

McCoole, S., 'Philanthropy, History and Heritage', *History Ireland*, Vol. 17, No. 3 (2009), pp. 10–11.

McCormack, W. J., *Roger Casement in Death or Haunting the Free State* (University College Dublin Press, Dublin, 2002).

McCracken, J. J., *The Irish Parliament in the Eighteenth Century*, Irish History Series No. 9 (Dundalgan Press, Dundalk, 1971).

McDonald, W., *Some Ethical Questions of Peace and War: With Special Reference to Ireland* (University College Dublin Press, Dublin, 1998).

McDowell, R. B., *McDowell on McDowell: A Memoir* (The Lilliput Press, Dublin, 2008).

McGahern, J., *Memoir* (Faber and Faber, London, 2005).

McGarry, F., 'Keeping an Eye on the Usual Suspects: Dublin Castle's "Personalities Files", 1899–1921', *History Ireland*, Vol. 14, No. 6 (2006), pp. 44–49.

McGarry, F., *The Rising. Ireland: Easter 1916* (Oxford University Press, Oxford, 2010).

McGarry, F., *Rebels: Voices from the Easter Rising* (Penguin Ireland, Dublin, 2011).

McGarry, P., *First Citizen: Mary McAleese and the Irish Presidency* (The O'Brien Press, Dublin, 2008).

McGee, O., *The IRB: The Irish Republican Brotherhood from the Land League to Sinn Féin* (Four Courts Press, Dublin, 2005).

McIntosh, G., 'Symbolic Mirrors: Commemorations of Edward Carson in the 1930s', *Irish Historical Studies*, Vol. 32, No. 125 (2000), p. 93–112.

McKearney, T., 'Internment, August 1971: Seven Days that Changed the North', *History Ireland*, Vol. 19, No. 6 (2011), pp. 32–35.

McKittrick, D., Kelters, S., Feeney, B., Thornton, C. and McVea, D., *Lost Lives: The Stories of the Men, Women and Children who Died as a Result of the Northern Ireland Troubles*, 3rd Edition (Mainstream Publishing, Edinburgh, 2008).

McMahon, S., *Bombs over Dublin* (Currach Press, Dublin, 2009).

McManus, R., 'Heritage and Tourism in Ireland: An Unholy Alliance?', *Irish Geography*, Vol. 30, No. 2 (1997), pp. 90–98.

McManus, R., 'Identity Crisis? Heritage Construction, Tourism and Place Marketing in Ireland', in McCarthy, M. (ed.), *Ireland's Heritages: Critical Perspectives on Memory and Identity* (Ashgate, Aldershot, 2005), pp. 235–50.

McNally, M., *Easter Rising 1916: Birth of the Irish Republic* (Osprey Publishing Ltd., Oxford, 2007).

McWilliams, D., *The Pope's Children: Ireland's New Elite* (Gill and Macmillan, Dublin, 2005).

McWilliams, D., *The Generation Game* (Gill and Macmillan, Dublin, 2007).

Madigan, E., 'Commemoration and Conciliation During the Royal Visit', *History Ireland*, Vol. 19, No. 4 (2011), pp. 10–11.

Maguire, D., 'Introduction', in Maguire, D. (ed.), *Short Stories of Padraic Pearse: A Dual-Language Book* (Mercier Press, Cork, 1968), p. 7.

Maillot, A., *New Sinn Féin: Irish Republicanism in the Twenty-First Century* (Routledge, Oxon, 2005).

Mansergh, M., *The Legacy of History for Making Peace in Ireland: Lectures and Commemorative Addresses* (Mercier Press, Cork, 2003).

Mansergh, M., 'The Easter Proclamation of 1916 and the Democratic Programme', in Jones, M. (ed.), *The Republic: Essays from RTÉ Radio's The Thomas Davis Lecture Series* (Mercier Press, Cork, 2005), pp. 58–74.

Mansergh, M., 'Fianna Fáil and Republicanism in Twentieth-Century Ireland', in Honohan, I. (ed.), *Republicanism in Ireland: Confronting Theories and Traditions* (Manchester University Press, Manchester, 2008), pp. 107–14.

Martin, F. X. (ed.), *The Easter Rising, 1916 and University College, Dublin* (Browne and Nolan Ltd., Dublin, 1966).

Martin, F. X. (ed.), *Leaders and Men of the Easter Rising: Dublin 1916* (Methuen & Co. Ltd., London, 1967).

Martin, F. X., 'Foreword', in Martin, F. X. (ed.), *Leaders and Men of the Easter Rising: Dublin 1916* (Methuen & Co. Ltd., London, 1967), pp. ix–xii.

Martin, M., *Freedom to Choose: Cork & Party Politics in Ireland 1918–1932* (The Collins Press, Cork. 2009).

Matthews, A., 'Vanguard of the Revolution? The Irish Citizen Army, 1916', in O'Donnell, R. (ed.), *The Impact of the 1916 Rising: Among the Nations* (Irish Academic Press, Dublin, 2008), pp. 24–36.

Mays, M., *Nation States: The Cultures of Irish Nationalism* (Lexington Books, Plymouth, 2007).

Mesev, V., Shirlow, P. and Downs, J., 'The Geography of Conflict and Death in Belfast, Northern Ireland', *Annals of the Association of American Geographers*, Vol. 99, No. 5 (2009), pp. 893–903.

Mills, M., 'Arms Crisis (1970)', in Lalor, B. (ed.), *The Encyclopaedia of Ireland* (Gill and Macmillan, Dublin, 2003), p. 46.

Mitchell, A., *Casement* (Haus Publishing Ltd., London, 2003).

Moloney, E., *A Secret History of the IRA*, 2nd Edition (Penguin Books, London, 2007).

Moloney, E., *Paisley: From Demagogue to Democrat?* (Poolbeg, London, 2008).

Monaghan, A., *The Soldier's Song* (Macmillan, London, 2010).

Montague, J., 'Living for Ireland', in Bolger, D. (ed.), *Letters from the New Island: 16 on 16. Irish Writers on the Easter Rising* (Raven Arts Press, Dublin, 1988), pp. 17–19.

Moran, J., *Staging the Easter Rising: 1916 as Theatre* (Cork University Press, Cork, 2005).

Morrissey, J., 'A Lost Heritage: The Connaught Rangers and Multivocal Irishness', in McCarthy, M. (ed.), *Ireland's Heritages: Critical Perspectives on Memory and Identity* (Ashgate, Aldershot, 2005), pp. 71–87.

Morrissey, T. J., *William J. Walsh, Archbishop of Dublin, 1841–1921: No Uncertain Voice* (Four Courts Press, Dublin, 2000).

Murphy, B. P., '*The Wind that Shakes the Barley*: Reflections on the Writing of Irish History in the Period of the Easter Rising and the Irish War of Independence', in O'Donnell, R. (ed.), *The Impact of the 1916 Rising: Among the Nations* (Irish Academic Press, Dublin, 2008), pp. 200–220.

Murphy, D., *Irish Regiments in the World Wars* (Osprey Publishing Ltd., Oxford, 2007).

Murphy, J. A., 'A Look at the Past', in McLoone, J. (ed.), *The British-Irish Connection* (The Social Study Conference, Galway, 1986), pp. 11–16.

Murphy, J. A., *Ireland in the Twentieth Century* (Gill and Macmillan, Dublin, 1989).

Murphy, J. A., *The College: A History of Queen's/University College Cork* (Cork University Press, Cork, 1995).

Murray, P., 'Obsessive Historian: Eamon de Valera and the Policing of His Reputation', *Proceedings of the Royal Irish Academy*, Vol. 101C, No. 2 (2001), pp. 37–65.

Myers, K., *From the Irish Times Column 'An Irishman's Diary'* (Four Courts Press, Dublin, 2000).

Myers, K., *Watching the Door: A Memoir 1971–1978* (The Lilliput Press, Dublin, 2006).

Myers, K., *More Myers: An Irishman's Diary 1997–2006* (The Lilliput Press, Dublin, 2007).

Nash, C., '"Embodying the Nation": The West of Ireland Landscape and Irish Identity', in O'Connor, B. and Cronin, M. (eds), *Tourism in Ireland: A Critical Analysis* (Cork University Press, Cork, 1993), pp. 86–112.

Neeson, E., *Myths from Easter 1916* (Aubane Historical Society, Cork, 2007).

Neilan, M., 'The Rising in Galway', *The Capuchin Annual 1966*, No. 33 (1966), pp. 324–26.

Nevin, D., *James Connolly: 'A Full Life'* (Gill & Macmillan, Dublin, 2005).

Newell, Ú., 'The Rising of the Moon in Galway', *Journal of the Galway Archaeological and Historical Society*, Vol. 58 (2006), pp. 113–35.

Ní Dhomhnaill, N., 'The Black Box', in Bolger, D. (ed.), *Letters from the New Island: 16 on 16. Irish Writers on the Easter Rising* (Raven Arts Press, Dublin, 1988), pp. 30–33.

Ní Dhonnchadha, M. and Dorgan, T. (eds), *Revising the Rising* (Field Day, Derry, 1991).

Ní Dhonnchadha, M. and Dorgan, T., 'Preface', in Ní Dhonnchadha, M. and Dorgan, T. (eds), *Revising the Rising* (Field Day, Derry, 1991), pp. ix–x.

Nora, P., 'From *Lieux de Mémoire* to *Realms of Memory*', in Kritzman, L. D. (ed.), *Realms of Memory: The Construction of the French Past. Volume I: Conflicts and Divisions* (Columbia University Press, New York, 1996), pp. xv–xxiv.

Novick, B., 'Postal Censorship in Ireland, 1914–16', *Irish Historical Studies*, Vol. 31, No. 123 (1999), pp. 343–56.

Nowlan, K. B. (ed.), *The Making of 1916: Studies in the History of the Rising* (The Stationery Office, Dublin, 1969).

Nowlan, K. B., 'Introduction', in Nowlan, K. B. (ed.), *The Making of 1916: Studies in the History of the Rising* (The Stationery Office, Dublin, 1969), pp. ix–xiii.

Ó Cuív, 'Foreword', in Henry, W., *Supreme Sacrifice: The Story of Éamonn Ceannt 1881–1916* (Mercier Press, Cork, 2005), pp. ix–xi.

Ó Drisceoil, D., 'Conflict and War, 1914–1923', in Crowley, J. S., Devoy, R. J. N., Linehan, D. and O'Flanagan, T. P. (eds), *Atlas of Cork City* (Cork University Press, Cork, 2005), pp. 256–64.

Ó Gadhra, N., *Civil War in Connaught 1922–1923* (Mercier Press, Cork, 1999).

Ó hAodha, M., *Siobhán: A Memoir of an Actress* (Brandon, Dingle, 1994).

Ó Laoi, *History of Castlegar Parish* (The Connacht Tribune Ltd., Galway, Undated).

O'Brien, G., *Irish Governments and the Guardianship of Historical Records 1922–72* (Four Courts Press, Dublin, 2004).

O'Brien, G., 'The Future of the Past', in Sirr, L. (ed.), *Dublin's Future: New Visions for Ireland's Capital City* (The Liffey Press, Dublin, 2011), pp. 209–25.

O'Brien, J. A., 'The Irish Enigma', in O'Brien, J. A. (ed.), *The Vanishing Irish: The Enigma of the Modern World* (W. H. Allen, London, 1954), pp. 7–14.

O'Brien, P., *Blood on the Streets: 1916 and the Battle for Mount Street Bridge* (Mercier Press, Cork, 2008).

O'Brien, P., 'Honouring the Dead', *An Cosantóir: The Defence Forces Magazine*, Vol. 71, No. 5 (2011), pp. 26–27.

O'Carroll, D., *The Wind that Shakes the Barley* (Galley Head Press, Cork, 2006).

O'Connor, J., *The 1916 Proclamation*, Revised Edition (Anvil Books, Dublin, 1999).

O'Dea, W., 'Message from the Minister for Defence', *An Cosantóir: The Defence Forces Magazine*, Vol. 66, No. 3 (2006), p. 4.

O'Donnell, C., *Fianna Fáil, Irish Republicanism and the Northern Ireland Troubles 1968–2005* (Irish Academic Press, Dublin, 2007).

O'Donnell, C., 'Pragmatism Versus Unity: The Stormont Government and the 1966 Easter Commemoration', in Daly, M. and O'Callaghan, M. (eds), *1916 in 1966: Commemorating the Easter Rising* (Royal Irish Academy, Dublin, 2007), pp. 239–71.

O'Donohue, J., *Anam Cara: Spiritual Wisdom from the Celtic World* (Bantam, London, 1997).

O'Dwyer, R., 'On Show to the World: The Eucharistic Congress, 1932', *History Ireland*, Vol. 15, No. 6 (2007), pp. 42–47.

O'Dwyer, R., 'The Golden Jubilee of the 1916 Easter Rising', in Doherty, G. and Keogh, D. (eds), *1916: The Long Revolution* (Mercier Press, Cork, 2007), pp. 352–75.

O'Dwyer, R., *The Bastille of Ireland. Kilmainham Gaol: From Ruin to Restoration* (The History Press Ireland, Dublin, 2010).

O'Faolain, N., *Are You Somebody?* (New Island, Dublin, 2007).

O'Farrell, M., *A Walk through Rebel Dublin 1916* (Mercier Press, Cork, 1999).

O'Farrell, M., *50 Things You Didn't Know About 1916* (Mercier Press, Cork, 2009).

O'Flaherty, E., 'TV Eye: *The Man Who Lost Ireland*, RTÉ 1, 18 April 2006', *History Ireland*, Vol. 14, No. 3 (2006), pp. 50–51.

O'Flanagan, P., 'Colonisation and County Cork's Changing Cultural Landscape: The Evidence from Placenames', *Journal of the Cork Historical and Archaeological Society*, Second Series, Vol. 84, No. 239 (1979), pp. 1–14.

O'Halpin, E., 'Parliamentary Party Discipline and Tactics: The Fianna Fáil Archives', *Irish Historical Studies*, Vol. 30, No. 120 (1997), pp. 581–90.

O'Hanrahan, M., 'The Heritage Council', *Group for the Study of Irish Historic Settlement Newsletter*, No. 6 (1995), p. 19.

O'Hara, B., *Killasser: Heritage of a Mayo Parish* (Killasser/Callow Heritage Society, Swinford, 2011).

O'Leary, E., 'Reflecting on the "Celtic Tiger": Before, During and After', *Irish Economic and Social History*, Vol. 38 (2011), pp. 73–88.

O'Neill, E., 'The Battle of Dublin 1916: A Military Evaluation of Easter Week', *An Cosantóir: The Irish Defence Journal*, Vol. 26, No. 5 (1966), pp. 211–22.

O'Neill, G., *2016: A New Proclamation for a New Generation* (Mercier Press, Cork, 2010).

O'Neill, J., *Blood-Dark Track: A Family History* (Granta Books, London, 2000).

O'Shea, B. and White, G., 'The Road to Rebellion', *An Cosantóir: The Defence Forces Magazine*, Vol. 66, No. 3 (2006), pp. 6–11.

O'Toole, E., 'The 1916 Medal', *An Cosantóir: The Irish Defence Journal*, Vol. 26, No. 2 (1966), p. 66.

O'Toole, F., '1916: The Failure of Failure', in Bolger, D. (ed.), *Letters from the New Island: 16 on 16. Irish Writers on the Easter Rising* (Raven Arts Press, Dublin, 1988), pp. 41–42.

O'Toole, F., *Enough is Enough: How to Build a New Republic* (Faber and Faber, London, 2010).

Obama, B., *Dreams from My Father* (Canongate Ltd., Edinburgh, 2007).

Orr, P., '"Across the Hawthorn Hedge the Noise of Bugles"', *History Ireland*, Vol. 17, No. 1 (2009), pp. 34–38.

Pearse, M. M., 'Foreword', in The Brothers Pearse Commemoration Committee, *Cuimní na bPiarrac: Memories of the Brothers Pearse* (Mount Salus Press Ltd., Dublin, 1958), p. 3.

Pender, S., 'How to Study Local History', *Journal of the Cork Historical and Archaeological Society*, Second Series, Vol. 46, No. 164 (1941), pp. 110–22.

Pointing, C., *World History: A New Perspective* (Pimlico, London, 2001).

Powell, J., *Great Hatred, Little Room: Making Peace in Northern Ireland* (Vintage, London, 2009).

Power, P. J., 'The Kents and their Fight for Freedom', in Ó Conchubhair, B. (ed.), *Rebel Cork's Fighting Story 1916–21: Told by the Men Who Made It* (Mercier Press, Cork, 2009), pp. 106–12.

Power, V., *Voices of Cork* (Blackwater Press, Dublin, 1997).

Privilege, J., *Michael Logue and the Catholic Church in Ireland, 1879–1925* (Manchester University Press, Manchester, 2011).

Quinn, R., *Straight Left: A Journey in Politics* (Hodder Headline Ireland, Dublin, 2005).

Raivo, P. J., 'Landscaping the Patriotic Past: Finnish War Landscapes as a National Heritage', *Fennia*, Vol. 178, No. 1 (2000), pp. 139–50.

Rankin, K., 'The Search for "Statutory Ulster"', *History Ireland*, Vol. 17, No. 3 (2009), pp. 28–32.

Rees, R., *Ireland 1905–25: Volume 1, Text & Historiography* (Colourpoint Books, Newtownards, 1998).

Reid, B., 'Labouring Towards the Space to Belong: Place and Identity in Northern Ireland', *Irish Geography*, Vol. 37, No. 1 (2004), pp. 103–13.

Reilly, T., *Joe Stanley: Printer to the Rising* (Brandon, Dingle, 2005).

Ruane, J., 'Colonialism and the Interpretation of Irish Historical Development', in Gulliver, P. and Silverman, M. (eds), *Approaching the Past: Historical Anthropology through Irish Case Studies* (Columbia University Press, New York, 1992), pp. 293–323.

Ryan, A., *Witnesses: Inside the Easter Rising* (Liberties Press, Dublin, 2005).

Ryan, D., *The Rising: The Complete Story of Easter Week* (Golden Eagle Books Ltd., Dublin, 1949).

Ryle Dwyer, T., *Nice Fellow: A Biography of Jack Lynch* (Mercier Press, Cork, 2001).

Ryle Dwyer, T., *Behind the Green Curtain: Ireland's Phoney Neutrality During World War II* (Gill and Macmillan, Dublin, 2009).

Schama, S., *Landscape and Memory* (Fontana Press, London, 1996).

Shaw, F., 'The Canon of Irish History: A Challenge', *Studies: An Irish Quarterly Review*, Vol. 61, No. 242 (1972), pp. 113–53.

Sheeran, P. F., 'The Absence of Galway City from the Literature of the Revival', in Ó Cearbhaill, D. (ed.), *Galway: Town & Gown 1484–1984* (Gill and Macmillan, Dublin, 1984), pp. 223–44.

Sheerin, N., *Renmore and its Environs: An Historical Perspective* (Renmore Residents Association, Galway, 2000).

Shiels, D., 'The Archaeology of Insurrection: St. Stephen's Green, 1916', *Archaeology Ireland*, Vol. 20, No. 1 (2006), pp. 8–11.

Silinonte, J. M., 'Vivian de Valera: The Search Continues', *Irish Roots*, Issue No. 49, No. 1 (2004), pp. 17–19.

Sisson, E., *Pearse's Patriots: St. Enda's and the Cult of Boyhood* (Cork University Press, Cork, 2004).

Smith, M. K., *Issues in Cultural Tourism Studies* (Routledge, London, 2003).

Smyth, W. J., 'The Making of Ireland: Agendas and Perspectives in Cultural Geography', in Graham, B. J. and Proudfoot, L. J. (eds), *An Historical Geography of Ireland* (Academic Press, London, 1993), pp. 399–438.

Smyth, W. J., *Map-Making, Landscapes and Memory: A Geography of Colonial and Early Modern Ireland c. 1530–1750* (Cork University Press, Cork, 2006).

Sreenan, J., 'Message from the Chief of Staff', *An Cosantóir: The Defence Forces Magazine*, Vol. 66, No. 3 (2006), p. 4.

Sweeney, E., *Down Down Deeper and Down: Ireland in the 70s and 80s* (Gill and Macmillan, Dublin, 2010).

Sweetman, R., '"Waving the Green Flag" in the Southern Hemisphere: The Kellys and the Irish College, Rome', in Keogh, D. and McDonnell, A. (eds), *The Irish College, Rome and its World* (Four Courts Press, Dublin, 2008), pp. 205–224.

Tapscott, D. and Williams, A. D., *Macrowikinomics: Rebooting Business and the World* (Atlantic Books, London, 2010).

Temple, M., *Blair* (Haus Publishing, London, 2006).

The Editors, 'Preface', *Irish Historical Studies*, Vol. 1, No. 1 (1938), pp. 1–3.

Thornley, D., 'Patrick Pearse', *Studies: An Irish Quarterly Review*, Vol. 55, No. 217 (1966), pp. 10–20.

Titley, A., 'The Brass Tacks of the Situation', in Bolger, D. (ed.), *Letters from the New Island: 16 on 16. Irish Writers on the Easter Rising* (Raven Arts Press, Dublin, 1988), pp. 26–27.

Tobin, F., *The Best of Decades: Ireland in the 1960s* (Gill and Macmillan, Dublin, 1984).

Tosh, J. and Lang, S., *The Pursuit of History*, 4th Edition (Pearson Education Ltd., Harlow, 2006).

Tosh, J., *Why History Matters* (Palgrave Macmillan, Basingstoke, 2008).

Tosh, J., 'Introduction', in Tosh, J. (ed.), *Historians on History* (Pearson Education Ltd., Harlow, 2009), pp. 1–16.

Townshend, C., 'Historiography: Telling the Irish Revolution', in Augusteijn, J. (ed.), *The Irish Revolution, 1913–1923* (Palgrave, Basingstoke, 2002), pp. 1–16.

Townshend, C., *Easter 1916: The Irish Rebellion* (Allen Lane, London, 2005).

Townshend, C., 'Making Sense of Easter 1916', *History Ireland*, Vol. 14, No. 2 (2006), pp. 40–45.

Toye, R., *Lloyd George & Churchill: Rivals for Greatness* (Pan Books, London, 2008).

Travers, P., *Eamon de Valera* (Dundalgan Press Ltd., Dundalk, 1994).

Travis, C., '"Rotting Townlands": Peadar O'Donnell, the West of Ireland and the Politics of Representation in Saorstát na hÉireann (Irish Free State) 1929–1933', *Historical Geography: An Annual Journal of Research, Commentary and Reviews*, Vol. 36 (2008), pp. 208–24.

Treacy, M., 'Rethinking the Republic: The Republican Movement and 1966', in O'Donnell, R. (ed.), *The Impact of the 1916 Rising: Among the Nations* (Irish Academic Press, Dublin, 2008), pp. 221–40.

Vaughan, M. A., 'Finding Constable Whelan: An Incident from 1916', *Ossary, Laois and Leinster*, Vol. 4 (2010), pp. 231–36.

Vaughan, R., 'Images of a Museum', *Museum Management and Curatorship*, Vol. 19, No. 3 (2001), pp. 253–68.

Walker, A. and Fitzgerald, M., *Unstoppable Brilliance: Irish Geniuses and Asperger's Syndrome* (Liberties Press, Dublin, 2006).

Walker, B. M., *Past and Present: History, Identity and Politics in Ireland* (The Institute of Irish Studies, Belfast, 2000).

Walker, B. M., *A Political History of the Two Irelands: From Partition to Peace* (Palgrave Macmillan, Basingstoke, 2012).

Walker, S., *Forgotten Soldiers: The Irishmen Shot at Dawn* (Gill and Macmillan, Dublin, 2007).

Walsh, J., *Patrick Hillery: The Official Biography* (New Island, Dublin, 2008).

Waters, J., *Was It For This? Why Ireland Lost the Plot* (Transworld Ireland, London, 2012).

Watson, R., *Future Files: The 5 Trends that Will Shape the Next 50 Years* (Nicholas Brealey Publishing, London, 2008).

Weeks, L., 'Appendices', in Gallagher, M. and Marsh, M. (eds), *How Ireland Voted 2007: The Full Story of Ireland's General Election* (Palgrave Macmillan, Basingstoke, 2008), pp. 232–49.

Weeks, L. and Quinlivan, A., *All Politics is Local: A Guide to Local Elections in Ireland* (The Collins Press, Cork, 2009).

Whelan, D., *Conor Cruise O'Brien: Violent Notions* (Irish Academic Press, Dublin, 2009).

Whelan, K., 'Come all you Staunch Revisionists: Towards a Post-Revisionist Agenda for Irish History', *Irish Reporter* (1991), pp. 23–26.

Whelan, K., 'The Power of Place', *The Irish Review*, No. 12 (1992), pp. 13–20.

Whelan, K. and Masterson, E., *Bertie Ahern: Taoiseach and Peacemaker* (Blackwater Press, Dublin, 1998).

Whelan, M., 'Discovering the Past', *An Cosantóir: The Defence Forces Magazine*, Vol. 66, No. 3 (2006), pp. 20–21.

Whelan, N., *Showtime or Substance? A Voter's Guide to the 2007 Election* (New Island, Dublin, 2007).

Whelan, N., *Fianna Fáil: A Biography of the Party* (Gill and Macmillan, Dublin, 2011).

Whelan, Y., 'Monuments, Power and Contested Space: The Iconography of Sackville Street (O'Connell Street) before Independence (1922)', *Irish Geography*, Vol. 34, No. 1 (2001), pp. 11–33.

Whelan, Y., 'Symbolising the State: The Iconography of O'Connell Street after Independence (1922)', *Irish Geography*, Vol. 34, No. 2 (2001), pp. 135–56.

Whelan, Y., *Reinventing Modern Dublin: Streetscape, Iconography and the Politics of Identity* (University College Dublin Press, Dublin, 2003).

White, G. and O'Shea, B., *Irish Volunteer Soldier 1913–23* (Osprey Publishing Ltd., Oxford, 2003).

White, G. and O'Shea, B., *'Baptised in Blood': The Formation of the Cork Brigade of the Irish Volunteers 1913–1916* (Mercier Press, Cork, 2005).

White, J., 'Index of Articles Relating to 1916 Previously Published in *An Cosantóir*', *An Cosantóir: The Irish Defence Journal*, Vol. 51, No. 4 (1991), p. 30.

Whitmarsh, V., *Shadows on Glass. Galway 1895–1960: A Pictorial Record* (Privately Published, Galway, 2003).

Wills, C., *That Neutral Island: A Cultural History of Ireland During the Second World War* (Faber and Faber, London, 2007).

Wills, C., *Dublin 1916: The Siege of the GPO* (Profile Books, London, 2009).

Wilson Foster, J., *Colonial Consequences: Essays in Irish Literature and Culture* (The Lilliput Press, Dublin, 1991).

Withers, C., 'Monuments', in Johnson, R. J., Gregory, D., Pratt, G. and Watts, M. (eds), *The Dictionary of Human Geography*, 4th Edition (Blackwell Publishers, Oxford, 2000), pp. 521–52.

Withers, C., 'Memory and the History of Geographical Knowledge: The Commemoration of Mungo Park, African Explorer', *Journal of Historical Geography*, Vol. 30, Issue 2 (2004), pp. 316–39.

Woods, C. J., 'Clark, Sir George Anthony', in McGuire, J. and Quinn, J. (eds), *Dictionary of Irish Biography: From the Earliest Times to Year 2002. Volume 2: Burdy-Czira* (Cambridge University Press, Cambridge, 2009), pp. 537–38.

Wright, A. and Linehan, M., *Ireland: Tourism and Marketing* (Blackhall Publishing, Dublin, 2004).

Zimmermann, G. D., *Songs of Irish Rebellion: Irish Political Street Ballads and Rebel Songs, 1780–1900* (Four Courts Press, Dublin, 2002).

Zuelow, E., *Making Ireland Irish: Tourism and National Identity since the Irish Civil War* (Syracuse University Press, Syracuse, 2009).

Index

and maps 226

and medals 98, 155–8, 163–4, 166, 171, 198, 230, 247, 398

and pageants 155–6, 203–4, 207, 252–3, 328, 383, 403

and paintings 159, 219, 235, 244, 328

and Pearse cottage 18, 45, 323

and Pearse Museum 301–5, 316, 422

and photographs/pictures 124, 172, 207, 219, 226, 228–9, 322, 235, 345, 395, 399, 402, 404

and placenames 218

and plaques 178, 217–18, 221, 224, 248–9, 251, 322–3, 395, 406–7, 428, 445

and plays: *Operation Easter* (2006) 404; *The Plough and the Stars* (1926) 227, 429; *The Voice of the Rising* (1966) 227

and poems 44, 103, 109, 122, 148, 159, 163, 179, 220, 227–9, 234, 239, 244, 247, 374, 405

and postcards 74, 235, 374

and posters 132, 234, 401

Proclamation, symbolic value of 347–8

and public parks 177, 247, 251–2, 261–2, 301, 404

and radio programmes 225, 227

'Save Number 16 Moore Street', campaign for 406–8, 421–2, 425–6

and scholarships 205, 272, 304

and sculpture 146–7, 159, 182–3, 214, 216, 219, 244, 246–7, 249, 272, 285–7, 300, 302, 334–6, 343–4, 426

and songs 59, 67, 107, 126, 215, 228–30, 328, 403, 405–6

and stamps 21, 155–6, 166, 231, 304, 399

and stickers 230–31

story of 1916, resilience of

and Sword of Light badges 21, 231

and uniforms 51, 159, 219–20, 239, 322, 402

and walking tours 328, 348

and wall murals 281

and websites 22, 388, 402

Worbys Company Ltd., sale of commemorative medallions 230

Easter Rising, history of 25–99, 101–9

arrests, numbers of 104

Britain, reactions in 57, 80–81, 83, 86–94

and British rule: administration of 26–7; reorganisation of 89–91

businesses, impact on 77, 80–82

and censorship 19, 43, 46–7, 87

and civilian deaths and injuries 20, 28, 63, 77–80, 85–7, 92, 101–2, 364, 431–2

civilians, recollections of 56, 77–9, 83–4

and court martial proceedings 20, 84, 94, 101, 243, 347

and Defence of the Realm Act 85

disloyalty and sedition, growth of 42–50

drunkenness, incidences of 79

execution of leaders, reactions to and implications of 84–98

foreign nationals, impact on 84

GPO: damage to 74–5; morale inside 58–9, 77; occupation of 51, 53; shelling of 58, 74; symbolic value of 54

'Holy War', comparisons to 103–9

Irish troops fighting against the Germans, reactions from 84

local newspapers, reports from 59, 72, 82–3, 114

looting, incidences of 77, 81

loss of life, official figures 101

military strategy of rebels: aims and plans of 52, 62–4, 69–70, 72;

For Product Safety Concerns and Information please contact our
EU representative GPSR@taylorandfrancis.com Taylor & Francis
Verlag GmbH, Kaufingerstraße 24, 80331 München, Germany